SHEET METAL SHOP DRAWING

SHEET METAL SHOP DRAWING

HOWARD BRETZ

Chief Draftsman, World Trade Center (New York City)

Formerly, Head, Sheet Metal Department, New York Trade School (Now Voorhees Technical Institute)

Trustee of the Joint Apprenticeship Committee; Member of the Examining Committee. Sheet Metalworker's International Assn., Local 28

INDUSTRIAL PRESS INC., 200 Madison Avenue, New York, N.Y. 10016

SHEET METAL SHOP DRAWING

Copyright © 1971, by Industrial Press Inc., New York, N.Y. Printed in the United States of America. All rights reserved. This book, or parts thereof, may not be reproduced in any form without permission of the publishers.

Library of Congress Catalog Card Number: 72-112965
ISBN 0-8311-3022-9

Contents

PREFACE	xi
1. ELEMENTARY PRINCIPLES OF DRAFTING	1
Materials, Tools and Methods	1
Lettering	5
Dimensioning	6
Geometrical Construction	7
Explanation of Geometrical Constructions in Fig. 1-3	7
Explanation of Geometrical Constructions in Fig. 1-4	10
Shop Drawings	12
Lists of Abbreviations and Symbols	12
Checking Practices for General Drafting	15
2. DUCT CONSTRUCTION	16
Construction of Seams and Connections (Fig. 2-1)	17
Duct End-Stiffeners, Cap and Collar Connections (Fig. 2-2)	18
Types of Duct Connections (Fig. 2-3)	18
Other Types of Duct Connections (Fig. 2-4)	18
Flanged Duct Connections	22
Rectangular-Duct Straight Joints (Figs. 2-5 and 2-7)	24
Elbows (Figs. 2-8 and 2-9)	26
Bevel Construction (Fig. 2-9)	29
Offset Construction (Figs. 2-10 and 2-11)	29
Construction of Nesting Fittings	32
Fabrication of Y-Branch Fittings	32
Collars	36
Collar Fabrication	36
Round Duct Fabrication	36
Examples for Round Duct Fittings	37
Plan 1	37
Plan 2	37
Plan 3	37
Plan 4	37
Rolling 90° Elbows	37
Special Fittings	37

CONTENTS

Transitions	37
Acoustic Lining	37
Terminal Boxes	44
Sound Absorbers	44
Fire Dampers and Access Doors	44
Flexible Duct Connections	44
Reheat Coils	44
3. PRACTICAL MATHEMATICS FOR SHEET METAL DRAFTSMEN	**45**
GENERAL ARITHMETIC REVIEW	45
Addition	45
Subtraction	45
Multiplication	45
Division	46
Fractions	46
Addition and Subtraction of Fractions	46
Multiplication and Division of Fractions	47
Fractions and Decimals	47
Powers, Roots and Exponents	48
INSTRUCTIONS FOR OPERATING THE MANNHEIM TYPE SLIDE RULE	49
Description of the Slide Rule	49
Location of Numbers on the Scales	50
Description of the Scales	51
Slide Rule Operations	51
Extracting a Square Root	53
To Find a Square Root from Tables (see Table 3-2)	54
LOGARITHMS	54
Interpolation	65
MATHEMATICAL ABBREVIATIONS AND SYMBOLS	85
Right Triangle	85
Circle	86
Circular Sector	86
Flat Oval	86
Square	87
Rectangle	87
Square Prism	87
Cylinder	88
PROPORTION	88
TRIGONOMETRY OF THE RIGHT TRIANGLE	89
Computing Parts of a Right Triangle	106
USE OF TRIGONOMETRY IN COMPUTATIONS FOR OFFSETS	107
Legend Symbols	107
The "Ogee Set"	107

CONTENTS

ROLLING BEVELS OR LATERALS	113
Legend Symbols	113
Formulas	113
TRIGONOMETRY FUNCTIONS FOR STANDARD LATERALS	113
ANGLE OF A HIP	121
By Geometry	121
By Trigonometry	121

4. DUCT DESIGN   124

- Air Quantity for Heating — 124
- Air Quantity for Cooling — 124
- Air Quantity Required for Ventilation — 127
- Air Pressures in Buildings — 127
- Industrial Exhaust Systems — 127
- Pressure — 128
- Purpose of the Fan in the Duct System — 130
- Duct Fitting, Friction Losses — 133

FRICTION LOSSES IN FITTINGS — 133

ELBOWS — 134
- Equal-Friction Method of Duct Design — 136
- Resistance in Low-Pressure Duct Systems — 142
- Static-Regain Method — 142

DUAL-DUCT SYSTEMS — 143

CASINGS — 143

AIR FILTERS — 144

HEATING COILS — 145

VIBRATION CONTROL — 145

LINING — 145

ROOF AIR INTAKES AND DISCHARGES — 145

FANS — 146

FAN SCROLL LAYOUT — 146

ROOF EXHAUST FANS — 146

CABINET FANS — 148

AXIAL FANS — 148

5. DESIGN DRAWINGS AND SPECIFICATIONS   151

- Contract and Mechanical Drawings — 151
- HVAC Drawings — 151
- Floor Plans — 151
- Mechanical Equipment Rooms — 159

PRODUCT DEFINITIONS	189
Air Moving Device (AMD)	189
Central-Station Units	189
Industrial, Axial, and Propeller Fans	190
Power Roof Ventilators	190
Steam and Hot-Water Unit Heaters	190
Classification by Arrangement	190
Miscellaneous Equipment	191
HVAC AND PLUMBING PIPING	197
SPRINKLER PIPING	198
Pneumatic Tube Systems	200
Electrical Systems	200
ELECTRICAL ABBREVIATIONS AND SYMBOLS	200
Special Process Systems and Owner's Equipment	204
STRUCTURAL DRAWINGS	204
THE ARCHITECTURAL DRAWINGS	247
GENERAL NOTES FOR ARCHITECTURAL DRAWINGS	252
ARCHITECTURAL ABBREVIATIONS	253
SAMPLE SPECIFICATIONS	255
Ductwork — Low Pressure	255
Flush Seam — Duct Construction	255
Ductwork — Medium-Pressure and Airtight	255
Duct Hangers and Support	255
Turning Vanes in Elbows	257
Duct Access Doors	257
Belt Guards	259
Double-Wall Apparatus Casings	259
Panel Joints and Finishes	259
Coordination of Work	259
BALANCING AIR QUANTITIES	259
General	259
Items to Check before Balancing	260
Air-Balancing Procedure	260
6. PREPARATION OF THE SHOP DRAWING	**262**
To Change Design Drawing to Shop Drawing Size	262
Drawing Layout and Specifications	263
GENERAL NOTES FOR DRAFTING PROJECTS NOS. 1–7	263
Drafting Project No. 1	264
Drafting Project No. 2	264
TYPICAL COMPUTATION PROCEDURE	264

GENERAL AND PROCEDURAL NOTES	265
Drafting Project No. 3	270
Drafting Project No. 4	270
Drafting Project No. 5	273
Drafting Project No. 6	277
Drafting Project No. 7	282
FIELD MEASURING	288
Field Measuring a Wall True-Angle	295
APPENDIX	**296**
TERMINOLOGY FOR SHEET METAL DRAFTSMEN	296
HEAT AND POWER	300
TEMPERATURE	300
HEATING VALUE OF VARIOUS FUELS	300
WEIGHT AND VOLUME	300
PRESSURE	300
CONVERSION FACTORS	300
SPECIFIC HEAT	300
INDEX OF TABLES AND FIGURES	**302**

Preface

Sheet metal ducts for heating, ventilating, air-conditioning, and exhaust systems are almost always "custom built." Consequently, as they must be fabricated to fit particular job-site conditions, it becomes imperative that accurate shop drawings be prepared. To prepare these drawings, draftsmen should be familiar with shop standards and fabrication methods; be aware of materials handling and erection procedures at the job site; and be able to analyze and interpret the architectural, structural, and mechanical-design drawings.

Fitting-layout knowledge and practical sheet metal experience acquired during their apprenticeship, and as journeymen, are invaluable assets to student draftsmen. Various books on pattern layout and for the design of air-handling systems are available.

The purpose of this book is to provide an instructional text specifically for sheet metal draftsmen and to provide shop drawing reference material and standards to sheet metal apprentices, journeymen, contractors, architects, engineers, and student designers. The student draftsman should endeavor to fully understand explanations and examples, which are kept as brief as possible. They should become familiar with tables and charts before attempting actual work sessions on drawing plates and projects. Various drafting styles are shown on drawings and "cuts" and should be so noted by the reader.

Always remember that well-executed drawings must convey concise information for shop fabrication and field erection, coordination of work to be done by other contractors, and for system approval by the architect and engineer.

While most of the tables and charts in the book have been acquired or developed during my many years of teaching, engineering, and drafting, others are printed with grateful acknowledgment to their respective sources.

Also, my thanks to the manufacturers of air-handling components for their kind cooperation in lending their "cuts" for greater authenticity in the text. And finally, for hours, days, and weeks of patience and all the manuscript typing by my wife, Dorothy, I must give sincere thanks.

CHAPTER 1

Elementary Principles of Drafting

MATERIALS, TOOLS, AND METHODS

The sheet metal draftsman must be able to quickly and accurately prepare neatly executed scale drawings of air-handling systems and components. To this end he should attain proficiency in mechanical drawing techniques through the proper use of drafting instruments and linework principles, constantly practice lettering and numbering, and perfect the ability to develop geometrical figures.

Drawings should be made on a good grade of *tracing paper* so that reproduction prints, as usually required, can be made. Tracing paper is available in stock rolls, 24 to 36 inches wide, or in precut sheets. Sizes of 8½" x 11", 18" x 24", 24" x 36", and 36" x 48" are convenient for filing and handling as they can be folded in equal increments. However, sheets or rolls in various other sizes and qualities to satisfy any requirement, may also be used. Precut sheets with printed borders, border coordinates, and title boxes are also available.

Although the student draftsman may not desire prints, it is advisable for him to use tracing paper as soon as possible so that experience may be acquired to simulate professional conditions. A medium-heavyweight roll, 30" wide by 20 yards long, of good transparency, is recommended for the student.

A 31" x 42" *drawing board* or table and a 36" *T square* are convenient for the beginner. The board should be covered with base paper which has a smooth, non-glare surface. This will give the drawing paper a good foundation so that sharp, clear linework is more readily produced.

The professional drawing table may have a *parallel ruling unit,* which is a refinement of the T square, or a *drafting machine.* This machine has movable arms attached to two scales set at 90° to each other. The scales are used as straightedges and may be rotated to any baseline position. Therefore, the *hand-scale rule, triangle,* and *protractor* are used only occasionally.

The drawing paper is fastened to the board with *drafting adhesive tape.* Place the sheet with the long end parallel to the T square and secure one corner — then recheck alignment prior to final placement. Pieces of the tape, approximately 1" long, are practical; use one at each corner, and also at about 12" intervals along top and bottom of the drawing paper. A 10-yard roll of ¾"-width tape is an adequate supply for many drawings.

A small amount of *drafting cleaning powder* should be sprinkled on the paper and spread out evenly with a *dusting brush.* The powder prevents smudges and helps to keep a drawing clean.

A *single-pointed lead holder,* to contain a *drafting lead* from H to 4H hardness, is used for linework and printing. Insert a 4H lead in the holder for construction lines and preliminary layout lines. This lead is the hardest of the group, and, therefore, will make sharp and clear lines.

While drawing a line, pull and rotate the "pencil," with a light pressure. The lead thus wears to an even point and will not develop flat spots. When the point becomes too blunt, resharpen it in a rotary type *pencil-lead pointer.* Keep a small clean rag handy for wiping excess graphite from the lead after sharpening.

If any portion of linework is to be removed, use an *erasing shield* and a good quality *eraser.* Brush away erasure debris with the *dusting brush.* Scale dimensions are marked off with a 6" or 12" *architect's scale rule.* A flat four-bevel rule has eight scales at reduced ratios to the full scale: ⅛", ¼",

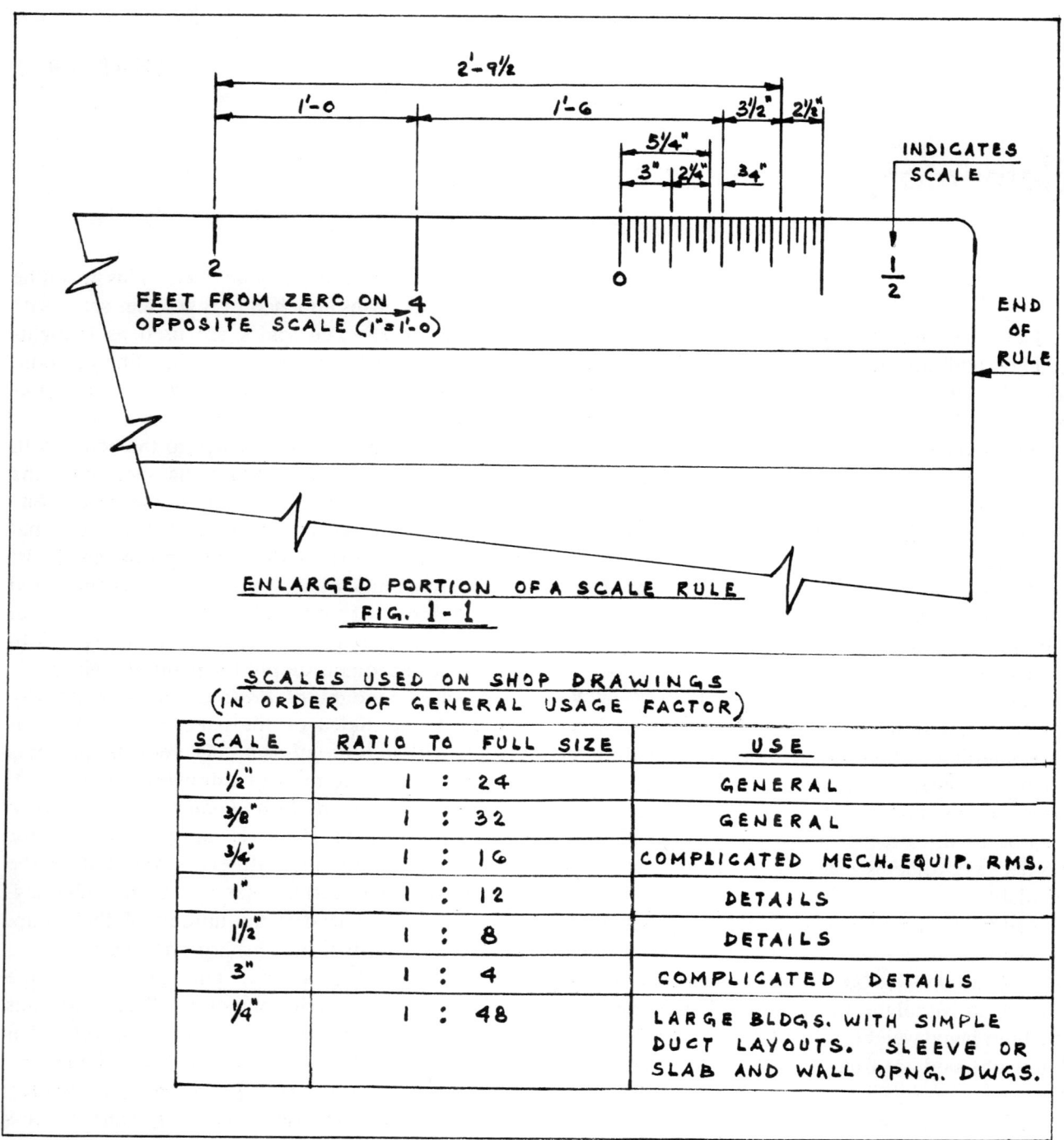

Fig. 1-1. Six-inch scale rule, enlarged detail.

½″, 1″, ⅜″, ¾″, 1½″, and 3″ = 1′-0. A 12″ triangular rule has ten scales: ³⁄₃₂″, ³⁄₁₆″, ⅜″, ¾″, 1½″, 3″, ⅛″, ¼″, ½″, and 1″ = 1′-0; and one edge fully divided, in inches to the 16ths.

Figure 1-1 is an enlarged detail of a portion of a 6″-scale rule. The scale is indicated near the end, adjacent to the divided equivalent, with each full size ½″ now equal to a scale-size 1′-0″. Measurements are read from zero to the right, for less than 12″; to the left, for feet and inches. Although scale dimensions in increments of ½″ can be read precisely, others, such as the 5¼″ shown, must be estimated. Various scales are read in the same manner, the larger ones in increments as small as ⅛″ (on the 3″ scale) and the smaller ones in divisions of 1″ or 2″ (on the ¼″ and ⅛″ scales), depending on the type of rule.

While the T square is used to draw baselines parallel to the long side of the drawing board, *plastic triangles* are used to draw vertical lines and sloping lines at increments of 15 degrees. A draftsman should have a 30°–60° and a 45° triangle of 10″ or 12″ size. Both triangles have 90° angles, therefore, either may be used to draw vertical lines in conjunction with the T square.

To draw a slope line of 30°, the long side of a 30°–60° triangle is placed on the upper edge of the T square. By placing the short side to the T square, the sloping side of the triangle is at 60° to the horizontal. Obviously, the 45° triangle may be used in either position for a 45° line, depending on the direction of slope required. As 45° is 15° more than 30°, and 15° less than 60°, it follows that the draftsman is able to draw lines of 15° or 75° by placing one side of a triangle against the T square and manipulating the 30°–60° and 45° triangles adjacent to each other, in proper positions.

When lines of other angles are required, a *protractor*, or *adjustable triangle* may be used. The former enables the draftsman to mark a rise-point in relation to a fixed base-point. The connecting line is then the slope required. To draw lines parallel to this, the edge of a triangle is placed along the line and held securely, while a *straightedge* (either another triangle or the T square) is placed adjoining any of the other sides of the triangle. Then, while holding the straightedge in position, the first triangle can be moved as required.

The adjustable triangle is a hinged combination protractor-triangle, with a scale and thumb setscrew so that it may be positioned to the desired angle. It is then used as a conventional triangle.

A standard 6″ *compass* and a 6″ combination *bow-beam-compass* enable the draftsman to draw arcs of from ⅛″ up to, approximately, a 14″ radius. For large radius requirements, *beam compasses* and *long bars* are available. *Circle templates* and *French curves* are useful to intensify linework initially drawn with a compass. As no portion of a French curve is a true arc, it is rotated to a position along the arc which most nearly conforms to the curve of the arc. Each end of this part of the French curve is then lightly marked with a pencil so that the same alignment can be repeated to intensify the entire arc. Care should be used during this procedure to avoid double-thickness linework. Various *templates* are used by the draftsman to attain rapidity as well as more elegance in drawing. These are available in the form of circles, ellipses, squares, hexagons, rectangles, and structural-steel shapes, large letters, numbers, and arrows, to mention a few. If a sheet-metal contractor has a developed system of drafting and shop-fabrication standards, duct symbols may also be drawn with templates to indicate different types of materials and duct construction.

Drafting instruments and equipment, other than those mentioned, may be acquired by the draftsman as further experience is gained in various drawing techniques.

Figure 1-2 is a partial plan and section of a shop drawing to indicate an *application of line standards*. (Circled numbers are for cross reference.)

1. *Construction lines* are light, thin, and solid — light for easy erasures, thin to provide accuracy, and solid for drafting rapidity. They should be drawn with a 3H or 4H lead and used for roughing-in duct layouts, lettering guides, and for initial construction details.

2. When the draftsman is assured that the layout is correct, the lightly-drawn construction linework should be intensified into sharp, clean, and dense *solid lines* for the visible outline. Turn the pencil while drawing, for good uniformity of line width.

3. Hidden objects are indicated by sharp, clear, *dashed lines* about ⅛″ long, evenly spaced, approximately ¹⁄₁₆″ apart. The

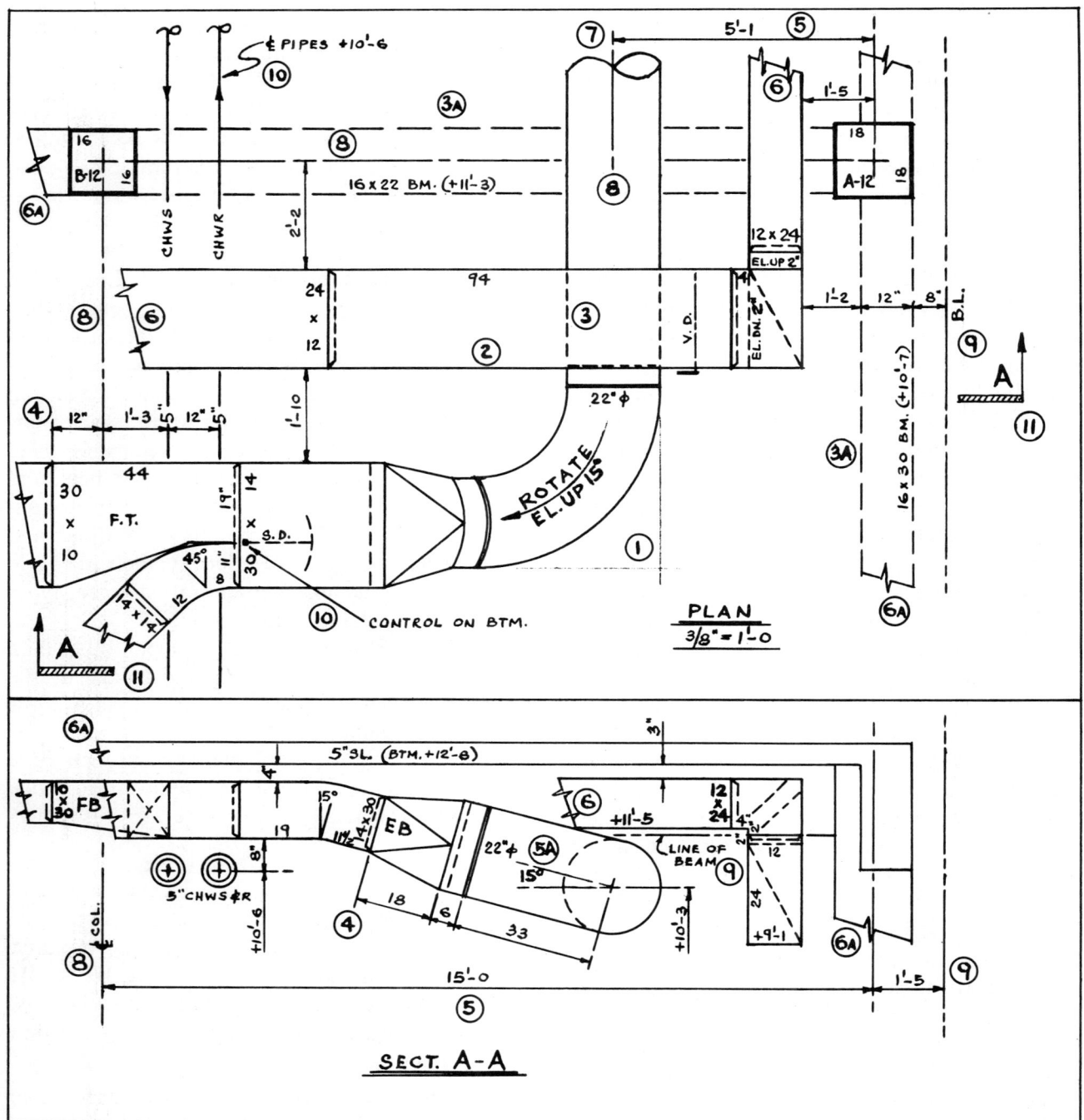

Fig. 1-2. Application of line standards and dimensions.

dashes should be slightly less dense than the solid lines.

3a. Outlines of beams at the slab above are also shown by *dashed lines* of an intensity similar to 3, but with longer dashes, about ¼" to ⅜", and evenly spaced as in 3, above.

4. The *witness,* or *extension line* is a fine, unbroken line of sharp intensity, to indicate the extent of dimensions. These should start about ¹⁄₁₆" from the object and ex-

tend to approximately $\frac{1}{16}''$ past the dimension line. Avoid crossing witness lines, but if it is necessary to do so, do not break lines at points of intersection.

5. *Dimension lines* are similar to extension lines, but are terminated at each end with *arrowheads.* Arrows should be sharp, dense, and end with a sharp point at the extension line. They are neatly drawn at a length-to-width ratio of 3 to 1. Draw dimension lines in true views at approximately $\frac{1}{4}''$ from the object. If more than one dimension is required from the same object, the smallest dimension is indicated closest to the object and lines are about $\frac{1}{4}''$ apart. (See Fig. 1-1.) When an object is only drawn in part, or out-of-scale for space reasons, the dimension line remains solid; but the abbreviation NTS (not to scale), is placed just below the dimension figure and line.

5a. A degree dimension for an angle is treated in the same manner as in 5, except that the dimension line is an arc, drawn from the angle apex.

6. A *break line* for a rectangular duct is indicated by an angular line drawn between the duct lines, broken by *two* free-hand zig-zags. All other qualities are the same as in the solid line.

6a. A *break line* for a building construction material is similar to 6, but is drawn with *one* zig-zag.

7. A *break line* for round ducts, pipes, columns, and other round objects is indicated by a symmetrical, high and low curve that intersects and joins at the centerline. All qualities are the same as the solid line.

8. The *centerline* is a sharp, clean, broken line composed of alternate long and short dashes, spaced evenly, about $\frac{1}{16}''$ apart. The long dashes are approximately $\frac{3}{4}''$ to $1''$ in length, depending on the scale-size of the drawing, while the short dashes are about $\frac{1}{8}''$ in length. Typical applications are to indicate centers for: round ducts, fans, casings, dampers, beams, columns, openings, lights, and rooms.

9. The *phantom line* is a sharp, fine, broken line of one long and two short dashes spaced evenly, about $\frac{1}{16}''$ apart. Typical uses, as shown in Fig. 1-2, are the building-line and line-of-beam in the section.

10. A *leader line* is a sharp, fine line which indicates the part of an object referred to by a note. A straight line is used when the line may be easily identified, although a snakelike, curved line is preferred in a complicated drawing or one having many straight lines. These leader lines end with arrowheads at the designated object.

11. When specific objects on a shop drawing cannot impart complete information in the plan view, it becomes necessary to draw auxiliary views or *sections.* A section is a partial, vertical view (elevation) of a particular portion of the plan, and is shown by *cutting-plane lines,* arrows, and letters. Cutting-planes are selected and located so that the desired information may be indicated clearly and precisely. Arrows show the direction of view; and letters, the same for plan and section, are used for identification. Two arrows and letters are needed to indicate the extent of the cut view.

In Fig. 1-2, the 22" diameter, 90° elbow rotates 15° upward. Dimensions are shown in Section A-A, where a true view of the sloping duct is indicated.

Sect. 'A-A' (Fig. 1-2), indicates advantages of a secondary view which clarifies the elbow for shop-fabrication requirements and field-installation information. Dampers, splitters, or other duct components need only be shown in one view, as plan and section are read simultaneously.

LETTERING

A drawing with sharp, clean linework is enhanced by careful, uniform, and legible lettering. Freehand *Gothic Capital letters* with equal line-density, neatly and evenly spaced, are used for all dimensions and notes. Styles may be either vertical, or slanted at approximately 20°. Recommended sizes are as follows:

1. 3/16" for drawing titles, prominent notes, and sections
2. 1/8" for dimensions and general notes
3. 3/32" for dimensions and general notes on small or crowded drawings
4. Single downward or horizontal strokes and well-rounded, counter-clockwise curves are used to draw round letters and numbers. B, D, J, P, R, S, 2, 3, and 5 are made with downward strokes and clockwise, or with combination curves.

Good lettering is not a simple accomplishment; only with practice, patience, and hard work will the beginning draftsman attain this goal. During a practice session, the pencil should not be held too tightly, as tenseness very quickly causes hand fatigue. Fasten a small sheet of drawing paper — approximately 8½" x 11" — to the drawing board. Using 2H lead in the lead holder, draw a series of light guide-lines for rows of 3/32" letters with 1/4" spaces between the rows. (The 1/8" scale is convenient for this; use the 9" marking for letter heights and 2'-0 for spaces.) With firm, single-stroke pencil action, practice making two or three rows of slant-type letters. Make sure that all letters are slanted parallel and have adequate spacing for clarity.

Do the same for a vertical series of letters and then objectively examine the two styles to determine which is most advantageous. Writing habits developed during childhood will have an important influence on whether a slant style or a vertical style will be better for the beginning draftsman. While a certain style is easy for one it may be difficult for another. Choose a lettering style which can be performed with ease and rapidity, and will be uniform and neat in appearance. Continue the practice session with the style of your choice. Use similar practice sheets for 1/8" and 3/16" letters and numbers. When a neat lettering style is acquired change to a 3H lead — this will cause less graphite smear and help to keep the paper cleaner — the slightly harder lead, however, will require firmer stroking.

Occasionally, a small or very crowded drawing requires 1/16" lettering. It is recommended that the student draftsman first become proficient in the larger sizes before attempting anything small.

DIMENSIONING

Duct size dimensions should be indicated in inches and located at connections, with an inch symbol (") for a round duct, but without the inch symbol for a rectangular duct. For rectangular ducts, the first dimension is that of the view shown. Duct-size numbers are indicated vertically on the drawing, except in crowded areas where greater clarity may be achieved by placing the numbers along the connection line. Use a diameter symbol (ϕ) with round-duct sizes. A duct run that does not change size need not be dimensioned at each connection. Duct sizes in end views are indicated along the direction of the sides.

Dimension lines are located outside of objects and in a continuous series. *Tie-ins* should be to the nearest permanent object such as a building structural member. Do not terminate these dimensions at partitions, walls, or phantom building-lines which may be nonexistent during the field erection. Dimensions from 0" to 12" are shown in inches, using the inch symbol. Dimensions over 12" are shown in feet and inches, separated by a dash, using the foot symbol (') only. *Out-of-scale dimensions* should be noted with the abbreviation *NTS* (not to scale).

Lineal duct dimensions are generally located in the duct, with no dimension lines or arrows, and the numbers are drawn along the side of the running length. *Rectangular-duct lengths* are in inches, with no inch marks unless required for clarity. Round-duct lengths are indicated by inches for shop-fabricated duct, and by feet and inches for spiral conduit. This method distinguishes a man-made piece from a machine-made item.

Elbows and bevel-leg lengths to connections are from the throat vertex for rectangular duct, and from the centerline vertex for round duct. The degree dimension for an elbow or bevel is indicated at the vertex. This is the amount of degrees past the straight-path direction. 90° elbows are obvious and need not be noted. Avoid duplication of dimensions. All dimensions for shop fabrication and field erection should be indicated on the shop drawing so that *scaling* the drawing will not be required.

GEOMETRICAL CONSTRUCTION

A series of examples of geometrical constructions most often used in mechanical drawings is shown in Figs. 1-3 and 1-4. Each of the problems and the explanatory procedures are identified with a prefix for the problem number, and with the suffix G, D, or P, to indicate **G**eometrical standard construction, **D**rafting-method usage, and **P**ractical application, respectively. The student draftsman should prepare a drawing and practice step-by-step procedures until all constructions are thoroughly understood. A 36″ x 30″ sheet of drawing paper with a *Title Box* and perimeter border shall be divided into 24 boxes as in Fig. 1-5. Construction linework shall be light and drawn to approximately twice the size of objects shown on the plates. Following a check for accuracy, object linework can now be intensified.

Explanation of Geometrical Constructions in Fig. 1-3

1-G. OBJECT: To bisect a straight line.
GIVEN: Line *AB*.
PROCEDURE:

1. From points *A* and *B* as centers, with a radius greater than $\frac{AB}{2}$, describe two arcs intersecting as at *C* and *D*.
2. Draw line *CD* which intersects *AB* at *E*.
3. *AE* is therefore equal to *BE*.

1-D. Using ¼″ = 1′-0 scale, draw line *AB* equal to 16′-10.
Measure equal distances *AE* and *BE* as $\frac{16'-10}{2} = 8'-5$ each.

1-P. Using 1″ = 1′-0 scale, draw a 30″ x 10″ duct with a 20″ ϕ collar on duct centerline.

2-G. OBJECT: To draw a perpendicular from a point on a line.
GIVEN: Line *AB*, and point *C* on the line
PROCEDURE:

1. From point *C*, set off equal distances *BC* and *DC*.
2. With *B* and *D* as centers, set compass at a distance greater than *BC* or *DC* and describe arcs intersecting at *E*.
3. Line *EC* is the perpendicular to *AB*.

2-D. Use both a 45° triangle and a 30°–60° triangle. Place one side of either triangle along line *AB* and hold firmly. Place the other triangle along this edge, and move the 90° corner to point *C*. Draw *CE* as the perpendicular to *AB*.

2-P. *AB* is shown as centerline of an offset. Set off $BC = \frac{AB}{4}$. From point *C*, draw a perpendicular which will intersect the end line extended to point *F*. Mark off *AG* = *BF*. Points *F* and *G* are then centers for offset curve.

3-G. OBJECT: To draw a perpendicular to a line from a point outside of the line.
GIVEN: Line *AB*, and point *C* outside of the line.
PROCEDURE: From point *C* as center, draw any radius extending below the line and cutting the line at points *B* and *D*. With points *B* and *D* as centers, draw arcs opposite point *C*, intersecting at point *E*. Connect *EC* as the perpendicular to *AB*.

3-D. (Same procedure as 2-D.)

3-P. To connect angular straight lines *AB* and *DE* with a radius, draw perpendiculars from *B* and *D* which will intersect at *C*, center of the desired radius.

4-G. OBJECT: To draw a straight line at a given distance parallel to another straight line.
GIVEN: Line *AB*, and distance *CE*.
PROCEDURE: From any two points *C* and *D* on line *AB*, describe arcs equal to distance *CE*. Draw the line *EF* tangent to arcs. *EF* is parallel to *AB*; *CE* and *DF* are perpendicular to both *AB* and *EF*.

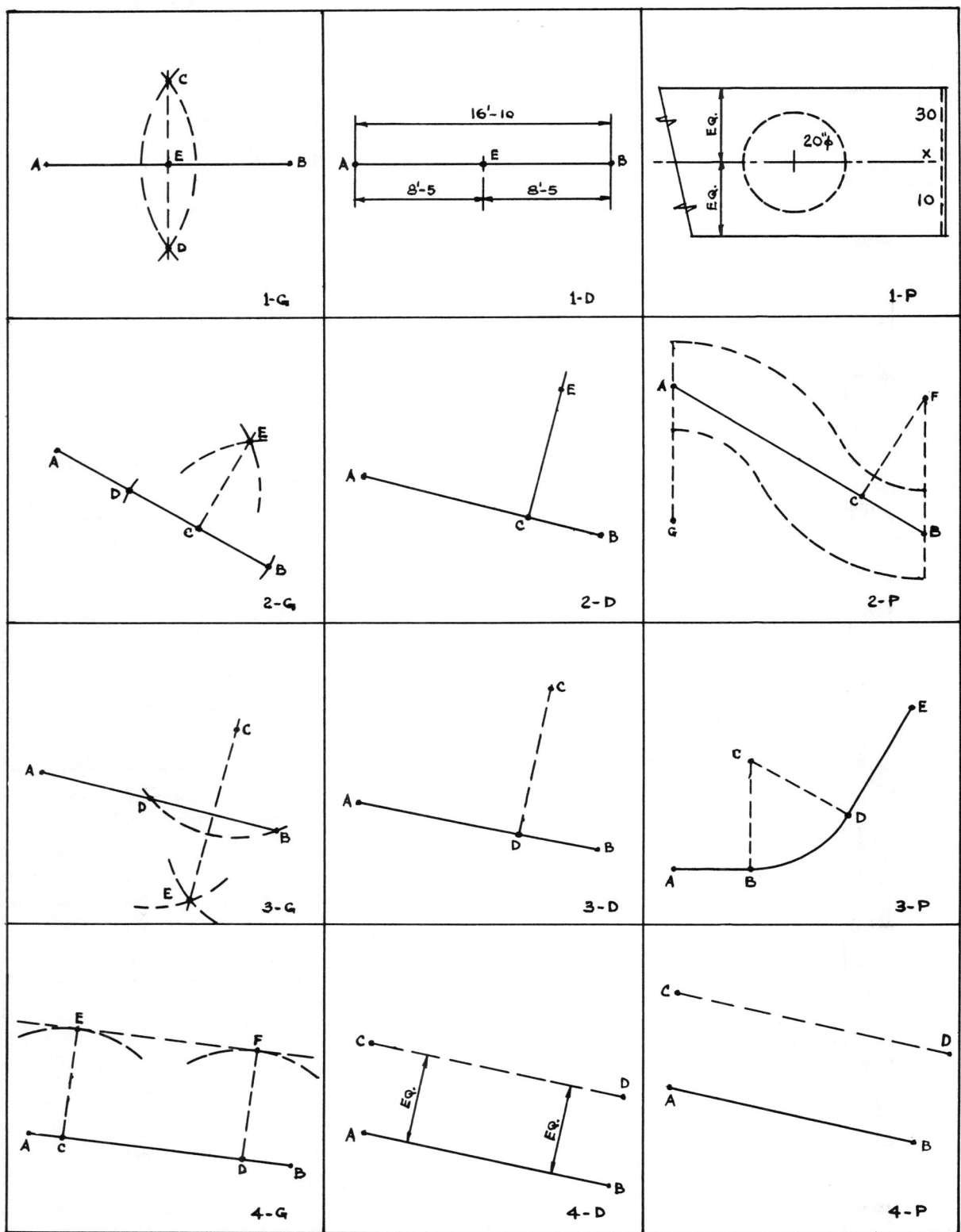

Fig. 1-3. Examples of geometrical construction.

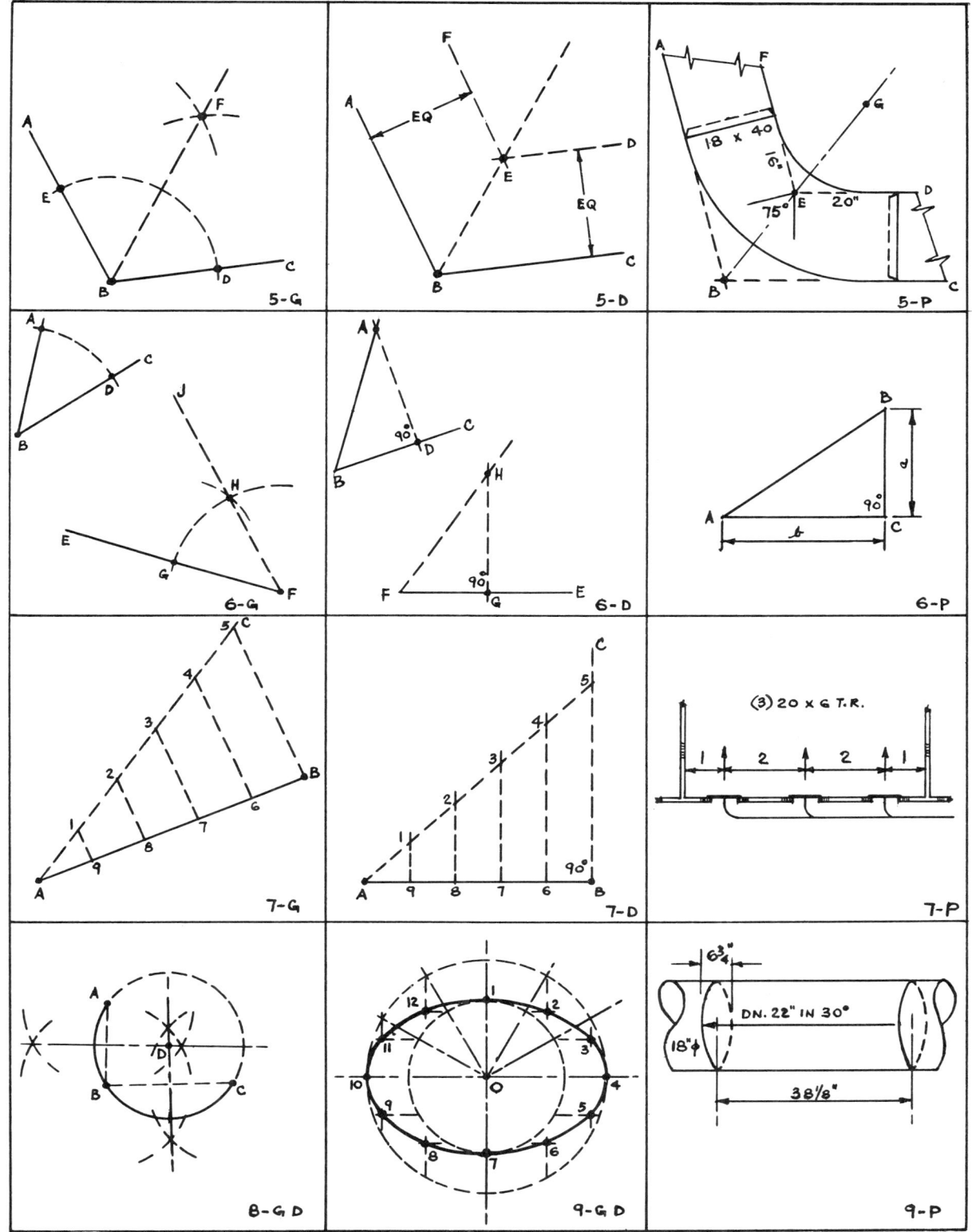

FIG. 1-4. Examples of geometrical construction (*continued*).

4-D. At right angles to line *AB,* measure two equal distances from line *AB.* Draw *CD* as parallel.

4-P. Draw *CD* parallel to *AB,* using one triangle in a stationary position, moving the other, as desired, for linework and guided by stationary triangle.

Explanation of Geometrical Constructions in Fig. 1-4.

5-G. OBJECT: To bisect a given angle.
GIVEN: Angle *ABC.*
PROCEDURE: With any convenient radius from point *B* as center, draw an arc intersecting *AB* and *BC* at *E* and *D,* respectively. From *E* and *D* as centers, again draw arcs, cutting each other at *F.* Connect points *B, F* as bisecting line.

5-D. Draw parallel lines to sides *AB* and *BC* of angle *ABC* at an equal distance and intersecting at *E,* as *FE* and *DE.* *BE,* therefore, bisects angle *ABC.*

5-P. Construct angle *ABC* equal to 105° with triangles. With 1″ = 1′-0 scale, draw lines *FE* and *DE,* 18″ from, and parallel to, *AB* and *BC.* Point *E* is throat vertex. The bisecting line is drawn indefinitely from point *B* through *E.* The center for the throat and heel radius is 18″ from *FE* and *DE,* and will be on line *BE* at point *G.* Complete elbow leg-length dimensions as shown.

6-G. OBJECT: To construct an angle equal to a given angle. (Method 1: Using arcs.)
GIVEN: Angle *ABC,* line *EF,* angle to be drawn at point *F.*
PROCEDURE: With any convenient radius from points *B* and *F* as centers, draw arcs intersecting *AB, BC,* and *EF* at points *A, D,* and *G,* respectively. From *D* as center, set compass equal to chord *AD.* With this setting, from point *G* as center, draw an arc intersecting previous arc at point *H.* Draw a line from *F* through *H* to *J.* Angle *EFJ* is equal to angle *ABC.*

6-D. (Method 2: Using perpendiculars.) Draw angle *ABC* and line *FE.* Construct *AD* perpendicular to *BC.* Set off *FG = BD.* Construct a perpendicular from *G.* From *G,* measure *GH = DA,* therefore, angle *EFH* = angle *CBA.*

6-P. When any two sides of a right-angle triangle are known, trigonometry may be used for determining angles. For example:

$$\tan \angle BAC = \frac{a}{b}$$

$$\tan \angle ABC = \frac{b}{c}$$

Angle *BAC* + angle *ABC* = 90° (see Chapter 3).

7-G. OBJECT: To divide a straight line into any desired number of equal parts.
GIVEN: Line *AB,* which is to be divided into five equal parts.
PROCEDURE: From either end of line, draw a straight line of indefinite length such as *AC.* Along line *AC,* with a convenient compass setting, step off an amount of spaces equal to the number of parts required, in this case, five, and connect points *B* and *5.* From points *1, 2, 3,* and *4,* draw lines parallel to *B5,* intersecting *AB* at *9, 8, 7,* and *6,* which will result in the required number of parts.

7-D. Draw a perpendicular from either end of given line *AB.*
Any convenient scale may be used (in this case, ¼″ = 1′-0). Pivot the scale, with zero at point *A,* to a multiple of the number of parts required, that will intersect on line *BC* at point *5* (*20* on scale). While holding the scale in a fixed position, mark off points *1, 2, 3,* and *4* from divisions *4, 8, 12,* and *16* on scale. Lines from *1, 2, 3,* and *4* parallel to *B5* will divide *AB,* as required.

7-P. Three supply registers are to be equally spaced in a room 15′ long by 10′ wide; use ¼″ =1′-0 scale and Method 7-D, for six spaces (*1-2-2-1*).

8-GD. OBJECT: To draw an arc through three

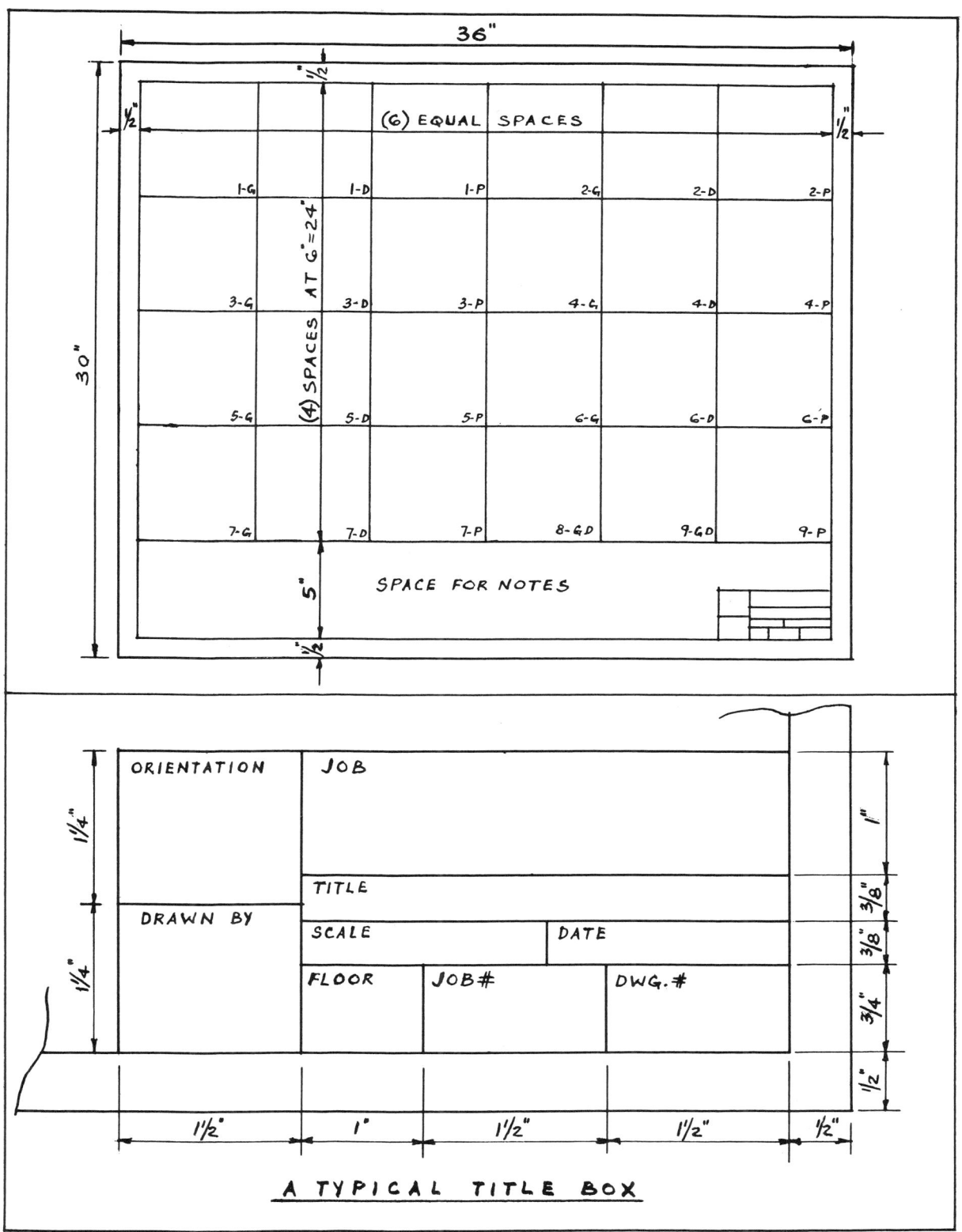

Fig. 1-5. Geometrical construction sheet layout and a typical title box.

given points which are not in a straight line.

GIVEN: Points *A*, *B*, and *C*

PROCEDURE: Bisect line *AB* and *BC*, extend lines to obtain point *D*. With *D* as a center, draw required arc through *A*, *B*, and *C*. Therefore, if the center of a circle is unknown, the intersection of bisecting lines of any two chords, as *AB* and *BC*, will locate the center, *D*.

9-GD. OBJECT: To draw an ellipse of given axis lengths

GIVEN: Major axis *10-4*

Minor axis *1-7*

Common center (point *0*)

PROCEDURE: Draw concentric circles with diameters *10-4* and *1-7*. Divide larger circumference into a convenient number of equal spaces (in this case, 12), and connect to point *0*; the two circles will now be divided equally. From outer circle divisions, draw lines parallel to *1-7*. From inner circle divisions, draw lines parallel to *10-4*. The intersections of these lines will be points of the ellipse. Connect three points at a time, with a curved template, to complete the ellipse.

9-P. Draw the 18″-diameter duct in 1″ = 1′-0 scale. Use 6¾″ for minor axis at miters. Elliptical drafting templates are convenient for close approximations of round-duct miters for fittings of this type.

SHOP DRAWINGS

Shop drawings should have a *dense*, ½″ border-line around the perimeter, a title box in the lower right-hand corner, schedules, and necessary notes. The title box contains a concise, specific title; drawing number; job number; floor number; drawing date; scale used; name, or initials, of draftsman; orientation of drawing; and a key plan.

Schedules are for revisions, drawing distribution records, reference drawings used in preparation of the drawing, fabrication and erection-symbol nomenclature, coordination dates, and names and addresses of the customer, architect, and design engineer. Additional schedules may be developed that are advantageous for fabrication or erection efficiency during the drawing process. *Notes* for specific data and information, preferably, are located on the right portion of the drawing at about mid-point, or slightly lower (see Fig. 1-5). Although accuracy cannot be overemphasized for scale drawings, it is occasionally desirable to show an object not-to-scale, and this should be so noted.

Details are often drawn to a larger scale than that of the main drawing, and these also should be so noted.

Sharp linework and lettering are important for clear print reproductions. Erase excessive construction lines and graphite smudges. Use standard abbreviations and symbols, details, and explanatory sections so that sufficient information is provided for fabrication and erection, for the coordination of other mechanical trades, for the architect and engineer, and for the general contractor.

Lists of Abbreviations and Symbols

The following list of abbreviations is useful for shop drawings. Periods are used only when the abbreviation spells a word such as: arch., auto., flex., etc.

Abbreviation	Definition
AA	All around
ABV	Above
AC	Air conditioning
ACN	Asbestos-cloth neck
AD	Access door
ADJ	Adjustable
AF	Angle frame, Air filter
AFF	Above finished floor
AHU	Air-handling unit
AL	Aluminum
ALD	Automatic-louver damper
AMBD	Automatic multiple-blade damper
AMT	Amount
APPROX	Approximately
ARCH.	Architect
ASB	Asbestos
ATC	Acoustical-tile ceiling
AUTO.	Automatic
B	Bottom
BD	Bottom down
BEV	Bevel
BG	Bottom grille
BI	Black iron

Abbreviation	Definition
BJ	Bar joist
BL	Building line
BLDG	Building
BLK	Black
BM	Beam
BO	By others
BOD	Bottom of duct
BOS	Back of slip
BR	Bottom register
BS	Below slab
BTM	Bottom
BU	Bottom up
CANV	Canvas
CCWBAD	Counterclockwise bottom angular down
CCWBAU	Counterclockwise bottom angular up
CCWBH	Counterclockwise bottom horizontal
CCWDB	Counterclockwise down blast
CCWTAD	Counterclockwise top angular down
CCWTAU	Counterclockwise top angular up
CCWTH	Counterclockwise top horizontal
CCWUB	Counterclockwise up blast
CD	Ceiling diffuser
CEIL, CLG	Ceiling
CFM	Cubic feet per minute
CG	Ceiling grille
CH	Ceiling height
CIRC	Circumference
CL COLL	Clinch collar
COD	Clean-out door
COL	Column
CONC	Concrete
CONN	Connection
CONT'D	Continued
COP	Copper
CR	Ceiling register
CS	Concrete slab
CT	Cooling tower
CWBAD	Clockwise bottom angular down
CWBAU	Clockwise bottom angular up
CWBH	Clockwise bottom horizontal
CWDB	Clockwise down blast
CWTAD	Clockwise top angular down
CWTAU	Clockwise top angular up
CWTH	Clockwise top horizontal
CWUB	Clockwise up blast
DEFL	Deflector
DEG	Degree
DET	Detail
DIA	Diameter
DMPR	Damper
DN	Down
DT	Drop top, Duct turn
DWDI	Double width, double inlet
DWG	Drawing

Abbreviation	Definition
EA	Each, Exhaust air
EB	Equal brake
EL	Elbow, Elevation
ELEV	Elevation
ENG, ENGR	Engineer
EQ	Equal
EQUIP	Equipment
EQUIV	Equivalent
ESH	Electric strip heater
ET	Equal taper
EXH	Exhaust
FA	Free area, Fresh air
FAI	Fresh air intake
FB	Flat bottom
FC	Flexible connection, Full corners
FD	Fire damper
F Fl, FIN FL	Finished floor
FIN.	Finished
FLBR	Fusible link bottom register
FLEX.	Flexible
FLG	Flange
FLTR	Fusible link top register
FP	Fire proofing
FPM	Feet per minute
FS	Flat slip
FSAA	Flat slips all around
FSB	Flat slip on bottom
FST	Flat slip on top
FT	Flat top
FTG	Fitting, footing
FVD	Friction volume damper
GA	Gage
GAL	Gallon
GALV	Galvanized
GC	General contractor
GI	Galvanized iron
GPM	Gallons per minute
HC	Hanging ceiling, Heating coil
HD	Head
HGR	Hanger
HP	High Pressure
HU	Humidifier
HV	Heating and ventilating
HVAC	Heating, ventilating and air conditioning
ID	Inside dimension
INCL	Inclusive
INT	Interior, Internal
J	Joggle
JT	Joint
KD	Knocked down
KE	Kitchen exhaust
KIT.	Kitchen

Abbreviation	Definition
LCC	Lead-coated copper
LDR	Leader
LE	Large end
LG	Long
LH	Left hand
LNG	Lining
LP	Low pressure
LVG	Leaving
MA	Matched angles
MAX	Maximum
MB	Mixing box
MD	Manual damper
MFG	Manufacturing
MIN	Minimum, Minute
MISC	Miscellaneous
MO	Masonry opening
NC	Normally closed
NEOP	Neoprene
NIC	Not in contract
NK	Neck
NO	Number, Normally open
NOM	Nominal
NR	Near
NTS	Not to scale
OA	Overall, Outside air
OAI	Outside air intake
OAL	Overall length
OBD	Opposed-blade damper
OC	On center
OD	Outside dimension
OH	Opposite hand
OPNG	Opening
OPP	Opposite
OS	Outside
PBD	Parallel-blade damper
R	Raw, Radius
RA	Return air
RE	Raw end
REQ	Required
RF	Roof fan
RH	Reheat, Right hand
RHC	Reheat coil
ROVD	Remotely-operated volume damper
RPM	Revolutions per minute
RT	Raise top
SA	Supply air
SD	Splitter damper
SE	Small end, Slip end
SECT.	Section
SK	Sketch

Abbreviation	Definition
SL	Slab
SLD	Slim-line diffuser
SPL	Splitter
SQ	Square
SS	Stainless steel
ST	Sound trap
STD	Standard
STL	Steel
STR	Straight
SWSI	Single width single inlet
T	Top
TC	Telescoping collar
TD	Top down
TG	Top grille
THK	Thick
THR	Throat
TOD	Top of duct
TOIL.	Toilet
TOS	Top of steel
TR	Top register
TU	Top up
TX	Toilet exhaust
TYP	Typical
UON	Unless otherwise noted
US.	Underside of slab
VD	Volume damper
VE	Vibration eliminator
VERT	Vertical
VOL	Volume
VOLFD	Volume-adjustable fire damper
WMS	Wire-mesh screen
WT	Watertight, Weight

The following list of symbols is useful for shop drawings:

Symbol	Represents
&	and
∡	angle
x	by
×	times
c/c	center to center
₵	centerline
₣	cubic feet per minute
[	channel
φ	diameter
↳	direction of view
=	equals
I	I beam
±	plus or minus
℞	plate
W⌐	wide flange beam

Checking Practices for General Drafting

1. Are lines and letters sharp, clean, and of the proper density for good reproduction quality?
2. Are all dimensions indicated? Do they add up correctly?
3. Is sufficient information provided for positive interpretation of the drawing?
4. Has any duplication of dimensions, notes, and information been avoided?
5. Have excessive construction lines and smudges been erased?
6. Are abbreviations and symbols correctly and advantageously used?
7. Are title boxes and schedules complete?

CHAPTER 2

Duct Construction

Job specifications generally designate materials, gages, types of connections, and other duct-construction details. However, the draftsman should have a thorough knowledge of these and other shop standards in order that economies in fabrication and erection can be achieved.

Construction of Seams and Connections (Fig. 2-1)

Various seams and connections are shown in Figs. 2-1 to 2-4. Prefix letters C, E, H, L, and T identify respectively: **C**ollars, **E**nd edges, **H**eavy-gage welded, **L**ongitudinal, and **T**ransverse seam connections. Airflow direction for transverse seams (T-) is from left to right. Details with more than one usage are designated accordingly, such as: *plain lap* L-1, T-1, H-4, which means that this type is used for **L**ongitudinal and **T**ransverse, and for **H**eavy gage.

The plain lap (L-1, T-1, H-4) and *flush lap* (L-2, T-2, H-5) are both used for various materials such as galvanized or black iron, copper, stainless steel, aluminum, or other metals and may be soldered, and/or riveted, as well as spot-, tack-, or solid-welded. Lap dimensions vary with the particular application, and since it is the duty of the draftsman to specify straight joints in lengths that use full-sheet sizes, transverse lap dimensions (T-1, T-2) must be known.

The *raw and flange corner* (L-3, H-6) is generally spot-welded, but may be riveted or soldered. For heavy gages it is tack-welded or solid-welded.

The *flange and flange corner* (L-4, H-7) is a refinement of the raw and flange corner. It is particularly useful for heavy-gage duct sections which require flush outside corners and must be field-erected.

The *standing seam* (L-5, T-3) is often used for large plenums, or casings. Before the draftsman is able to lay out a casing drawing, one of the items of information needed is seam allowance measurements, so that panel sizes can be detailed for economical use of standard sheets.

The *groove seam* (L-6, T-4) is often used for rectangular or round-duct straight joints, or to join some sheets for fittings which are too large to be cut out from standard sheets.

The *corner standing seam* (L-7) has similar usage to the standing seam, and also can be used for straight-duct sections.

The *double seam* (L-8) at one time was the most commonly used method for duct fitting fabrication. However, although it is seldom used because of the hand operations required for assembly, the double seam can be used advantageously for duct fittings with compound curves.

The *slide-corner* (L-8) is a large version of the double seam. It is often used for field assembly of straight joints, such as in an existing ceiling space, or other restricted working area where ducts must be built in place. To assemble the duct segments, opposite ends of each seam are merely "entered" and then pushed into position. Ducts are sent to job sites "knocked-down" for more efficient use of shipping space.

The *corner pocket* (L-9) is a flush-type seam which may be soldered or caulked. This seam can be modified slightly for use as a "snap lock."

The *Pittsburg* (L-10) is the most commonly used seam for standard-gage duct construction.

The *flange* (E-1, H-8) is an end edge stiffener. The draftsman must indicate size of the flange, direction of bend, degree of bend (if other than 90°) or when full corners are desired. Full corners are generally advisable for collar connections to con-

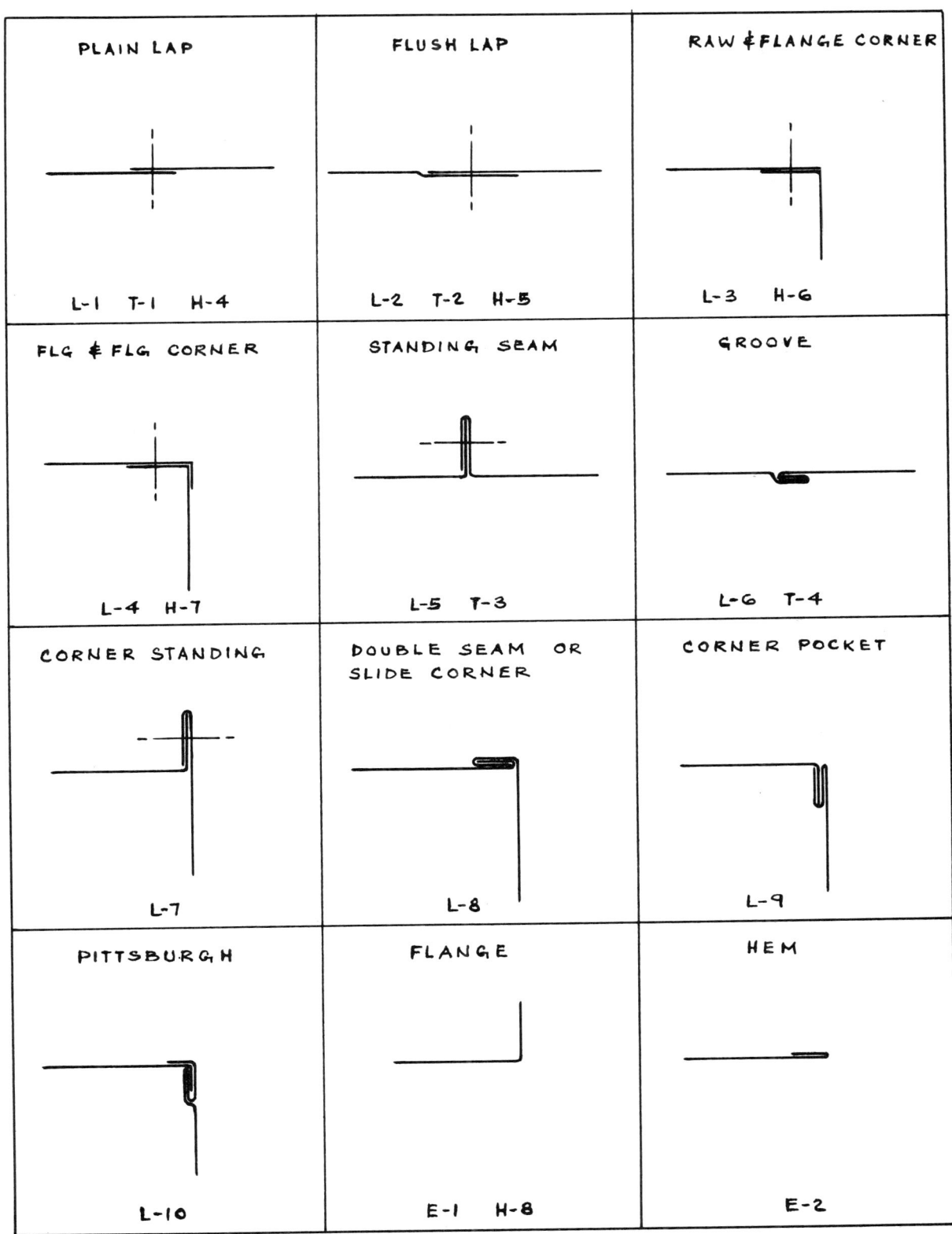

FIG. 2-1. Duct seams and edges.

crete or masonry wall-openings at louvers. For example, indicate on the drawing: 2″ FLG out AA with FC.

The *hem edge* (E-2) is a flat, finished edge. As with the flange, this must also be designated by the draftsman. For example, drawing should show: ¾″ hem out AA.

Duct End-Stiffeners, Cap and Collar Connections (Fig. 2-2)

The *hem and flange* (E-3), *double flange* (E-4), and *double hem* (E-5) are finished-end edges. These are used whenever additional sheet stiffening is needed. Sizes should be noted on the drawing, such as: ½″ hem and 1½″ FLG out T & B.

The *raw edge and angle* (E-6, H-9) and *flange and angle* (E-7) require that the sizes, materials, and fastening methods be indicated. For example, the note may read: 1½″ x 1½″ x ⅛″ GALV bolted outside (4) sides.

The *covered angle* (E-8) may be used as an end-stiffener or as a slip-fit connection to a building structure. Leg 'W' is fastened to the building. The entering edge of the sheet is inserted at arrow, is adjusted to suit, and then bolted through the angle. The draftsman should note on the drawing: 2″ x 2″ x ⅛″ BI COV ⊥ 118″ LG (4) REQ.

The *covered flat bar* (E-9) may be used to reinforce the end of large duct openings, or a connection such as a clinch-bar slip (see Fig. 2-4, T-12).

The *pan head* (E-10) is used for capping ends of rectangular duct. The draftsman should know the shop standard for the amount of head recess (D), as tap-ins may be required in the head.

A rectangular-duct *clinch collar* (Fig. 2-2, C-1) is used for fastening register collars, or branch duct-taps to trunk ducts. When the collar is flush to the duct corner, the flat clinch or "S" pocket is used. Allowances for clinch and S pocket are generally the same.

The rectangular-duct *inside clinch* collar (Fig. 2-2, C-2) may be used for installing registers on exposed duct to obtain a flush finish to the register flanges. This clinch is convenient for field installation of collars through existing wall openings.

Round clinch collars (in Fig. 2-2, C-3, C-4; and C-5 in Fig. 2-3) are for connections to rectangular ducts. The draftsman should indicate end-finish and diameters (*ID* or *OD*).

Types of Duct Connections (Fig. 2-3)

The rectangular *branch tap* with a positive internal *boot* (C-6) connects to the main with a three-sided angle frame as at *E*, and an 'S' pocket on the throat side as at *C*.

Methods for *wire-mesh-screen* connections on duct openings are shown in C-7, C-8, and C-9. C-7 and C-8 can be used for round or rectangular duct, while C-9 is for rectangular duct only. Screens in C-7 and C-9 are inserted into 'U' frames and are removable.

The *bead lap* connection (T-5) provides a shoulder and stiffener for round duct.

Normally, the 'S' slip (T-6) is used for small ducts. However, if the connection of a large duct is tight to a beam, column, or other object, and an 'S' slip is substituted for the shop standard slip, the draftsman should indicate FST, FSB, or whichever is required.

When a duct is specified as "flush seam construction," the 'S' pocket (T-7) and a joggle are used. A joggle is a projection or offset made by closely-formed kinks in opposite directions on the sheet. Notching and bending operations for an 'S' pocket on joints can be cumbersome and costly, especially for large sizes. To eliminate this problem, the double 'S' slip (T-8) is useful. For flush-type seams, all duct ends are joggled (not shown in T-8).

The *drive cleat* (T-9) is a slide-type connection generally used on small ducts in combination with 'S' slips (T-6) or bar slips (T-10, T-11) on the larger sides.

Other Types of Duct Connections (Fig. 2-4)

The *clinch-bar slip and flange* (T-12), uses the principle of the standing seam, but with a duct lap in the direction of airflow. These slips are generally assembled as a framed unit with full corners either riveted or spot-welded, which adds to the duct cross-section rigidity.

Reinforcement may be accomplished by spot welding the flat-bar to the flange of the large end. Accessibility to all four sides of the duct is required because the flange of the slip must be folded over

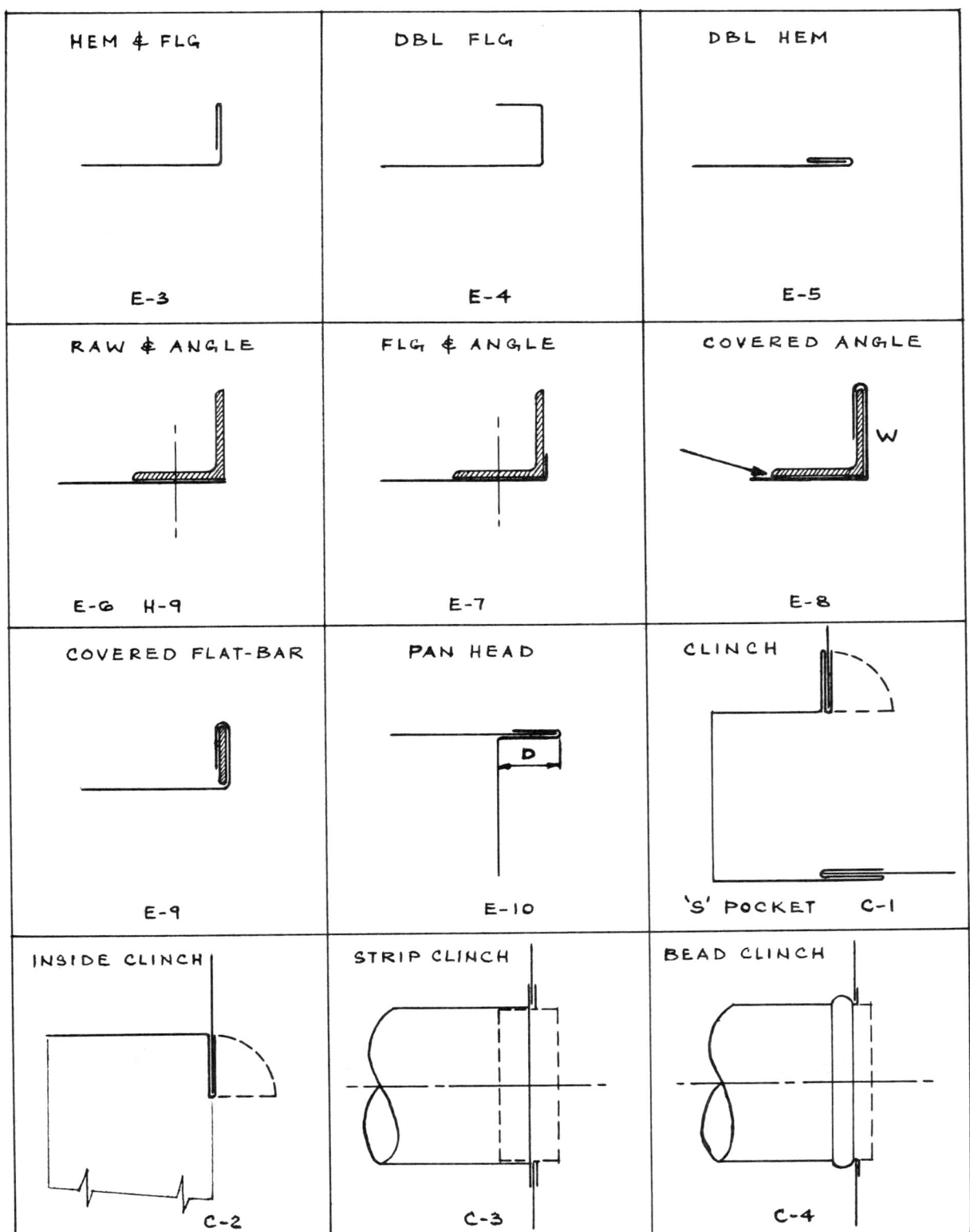

FIG. 2-2. Duct end-stiffeners, cap and collar connections.

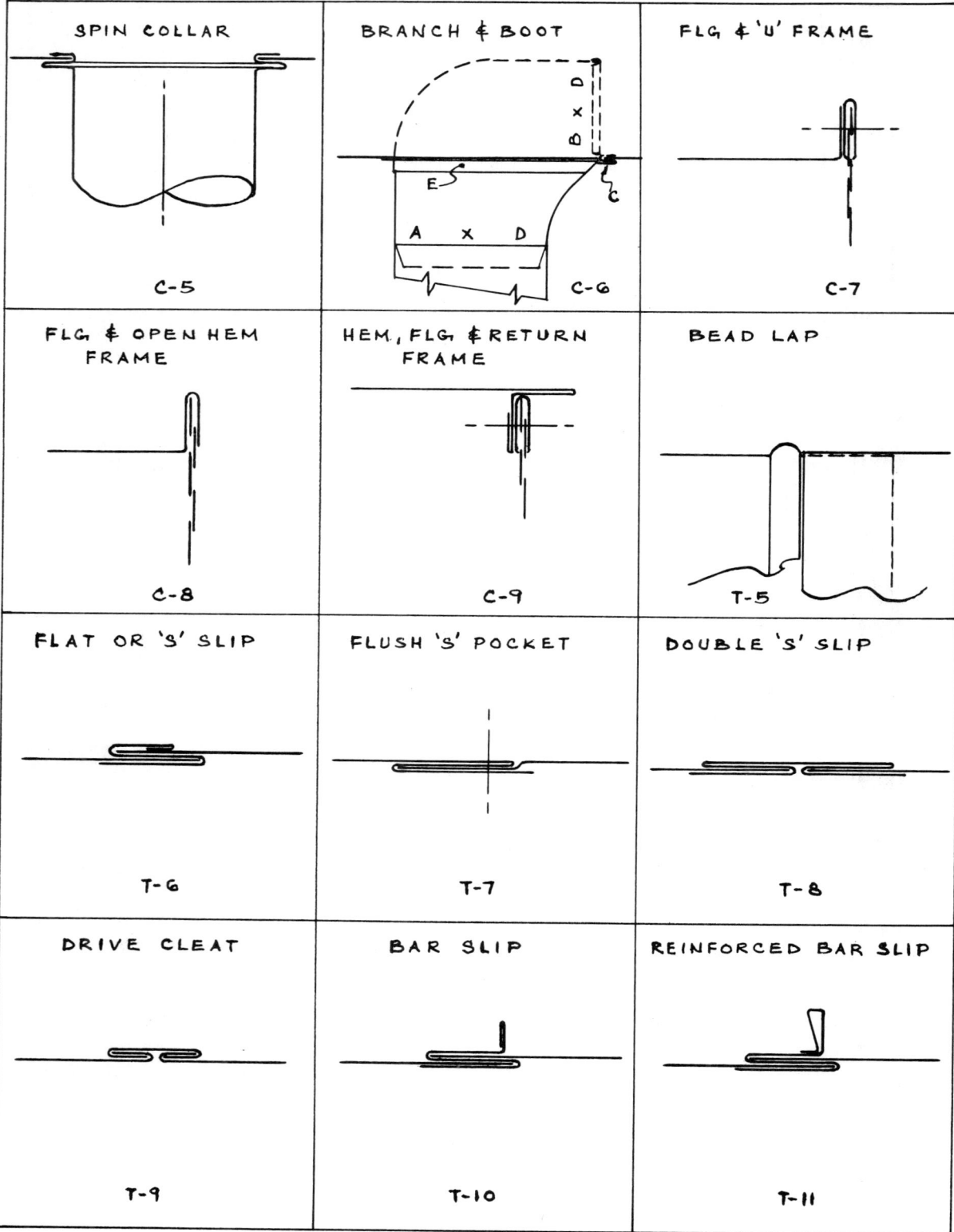

Fig. 2-3. Types of duct connections.

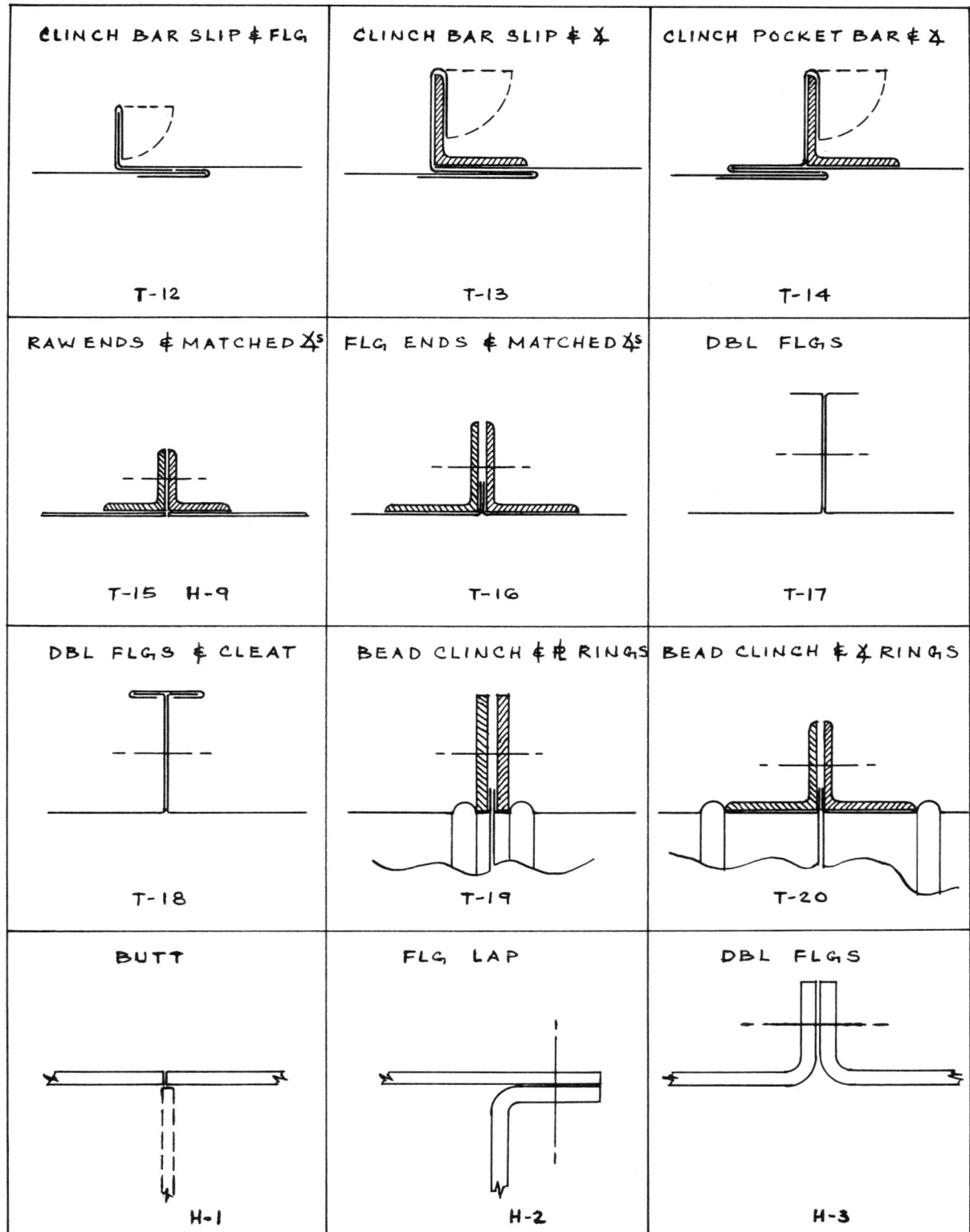

Fig. 2-4. Types of duct connections.

the flange on the large end after the ducts are connected.

The *clinch bar slip and angle* (T-13), is similar to clinch bar slip (T-12), but it has a riveted or spot-welded angle on the large end. This connection can also have a raw large end which is inserted into the space between the angle and the shop-fabricated slip. Matched angles (minimum of 16 ga) are riveted or spot welded to the smaller sides of the ducts, to pull the connection "home."

The *clinch pocket-bar slip and angle* (T-14), is a combination of bar slip (T-10) and clinch bar (T-13).

Flanged Duct Connections

Any of the following flanged connections may have gaskets. The draftsman should *not* allow for gasket thicknesses in calculations for running-length dimensions, nor should he indicate angle sizes, bolt centers, etc., as these items are established in job specifications and approved shop standards. Generally, angles are fastened to the duct sections in the shop. If conditions at the job site require consideration for length contingencies, the draftsman should specify "loose angles" such as at a connection to equipment which may be located later. The most common matched-angle-connection is the *angle frame, or ring* (T-15). The angles are fastened flush to the end of the duct.

The *flanged end and angle* (T-16), is often used for ducts 16 ga or lighter, as the flange provides a metal-to-metal gasket and holds the angle frame or ring on the duct without additional fastening. The draftsman may indicate in a field note that a round-duct fitting is to be "rotated as required." This type of angle-ring-connection is convenient for such a condition.

Double flanges (T-17), are similar to E-4, Fig. 2-2, except that the connecting flange has a series of matched bolt holes. This connection, caulked airtight, is ideal for single-wall apparatus casings or plenums. (T-18) is identical to (T-17), but has an airseal cleat.

Clinch-type flanged connections for round ducts, 16 ga or lighter, are shown in (T-19, T-20). The angles or rings can be loose, as explained in flanged end and angle (T-16). Seams for *heavy-gage ducts* are shown in H-1, H-2, H-3. The draftsman should indicate flange sizes, bend direction, and type of assembly. An example such as the flange

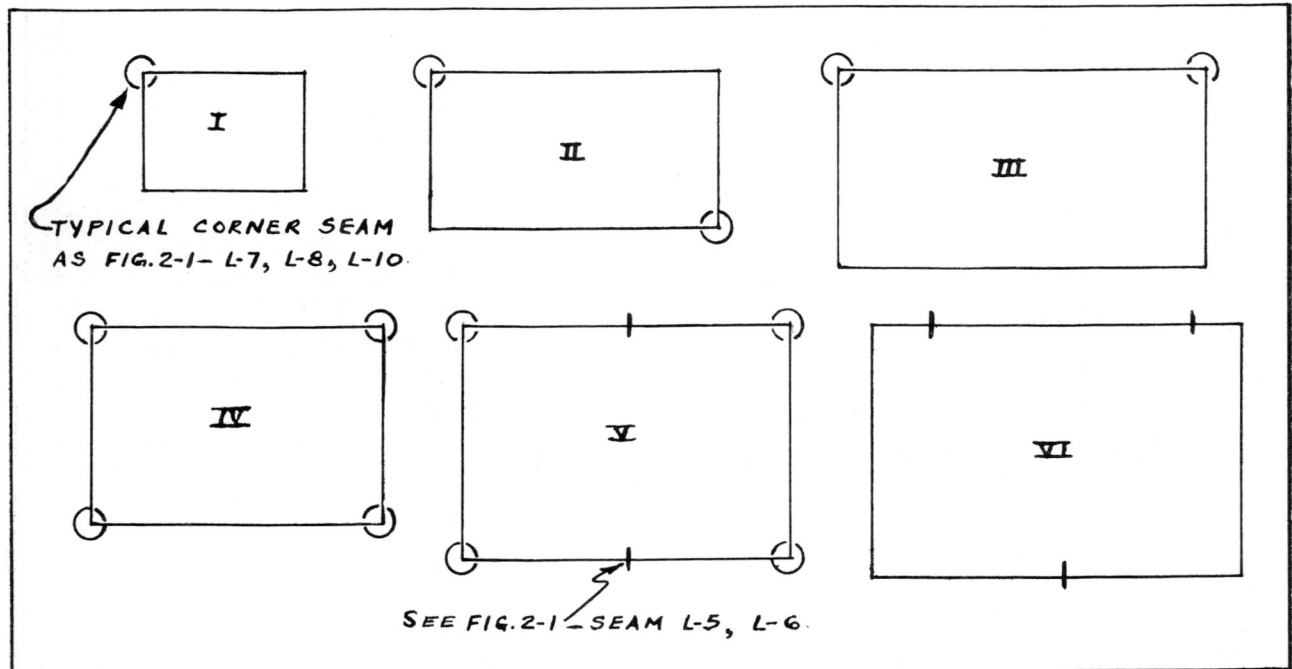

Fig. 2-5. Types of construction for rectangular duct straight joints.

Table 2-1 Running Lengths of Slip Connection Joints

No. of Joints	34"	46"	94"	118"
	Running Lengths, Ft, In.			
1	2' - 10"	3' - 10"	7' - 10"	9' - 10"
2	5 - 8	7 - 8	15 - 8	19 - 8
3	8 - 6	11 - 6	23 - 6	29 - 6
4	11 - 4	15 - 4	31 - 4	39 - 4
5	14 - 2	19 - 2	39 - 2	49 - 2
6	17 - 0	23 - 0	47 - 0	59 - 0
7	19 - 10	26 - 10	54 - 10	68 - 10
8	22 - 8	30 - 8	62 - 8	78 - 8
9	25 - 6	34 - 6	70 - 6	88 - 6
10	28 - 4	38 - 4	78 - 4	98 - 4
11	31 - 2	42 - 2	86 - 2	108 - 2
12	34 - 0	46 - 0	94 - 0	118 - 0
13	36 - 10	49 - 10	101 - 10	127 - 10
14	39 - 8	53 - 8	109 - 8	137 - 8
15	42 - 6	57 - 6	117 - 6	147 - 6
16	45 - 4	61 - 4	125 - 4	157 - 4
17	48 - 2	65 - 2	133 - 2	167 - 2
18	51 - 0	69 - 0	141 - 0	177 - 0
19	53 - 10	72 - 10	148 - 10	186 - 10
20	56 - 8	76 - 8	156 - 8	196 - 8
21	59 - 6	80 - 6	164 - 6	206 - 6
22	62 - 4	84 - 4	172 - 4	216 - 4
23	65 - 2	88 - 2	180 - 2	226 - 2
24	68 - 0	92 - 0	188 - 0	236 - 0
25	70 - 10	95 - 10	195 - 10	245 - 10

lap (H-2) for a field assembly of a 10-ga casing corner would be written: 1½" FLG out SQ on side with $9/32"$ ϕ bolt holes 12" CC.

The draftsman's aim in the preparation of a shop drawing is to indicate a duct system and related equipment in a manner which shall conform to the intent of the design engineer and architect and shall be fabricated and erected as economically as possible. To this end, careful selection of duct sizes and thorough coordination with other trades must be accomplished.

Ducts should be drawn with a minimum of compound fittings, simple fittings, and number of connections. Although various shop practices differ in fabrication methods, the student draftsman should attain a thorough knowledge of *standard duct construction*. Standard sheet sizes should be used to advantage in the selection of straight joints and fittings so that a minimum of seams is required for fabrication. See Fig. 2-5 for end views of six typical straight-joint fabrication methods.

Table 2-1 is for installed lengths of rectangular duct straight joints with a 1½" slip end allowance. Sheets are available in 30", 36", 42", and 48" widths by 96" or 120" lengths. Basically, the "layout piece" should fit on a standard sheet and seams in related stretchout pieces, such as the heel and throat of an elbow, should be kept to a minimum.

A series of *rectangular duct fittings* is shown in Figs. 2-7 to 2-14 with prefix letters J, E, B, S, N, Y, and C which represent **J**oints, **E**lbows, **B**evels, off**S**ets, **N**ested fittings, '**Y**' type branches, and **C**ollars, respectively. For proficiency in the art of drafting, the following exercises should be drawn to scale: 3/8" = 1'-0 on a sheet of paper 36" x 25½" (see Fig. 2-6):

1. Draw a ½" perimeter border
2. Mark off a 5½" x 2½" title box in lower right corner
3. Divide the 35" width into (8) equal spaces
4. Divide the 22" height into (6) equal spaces

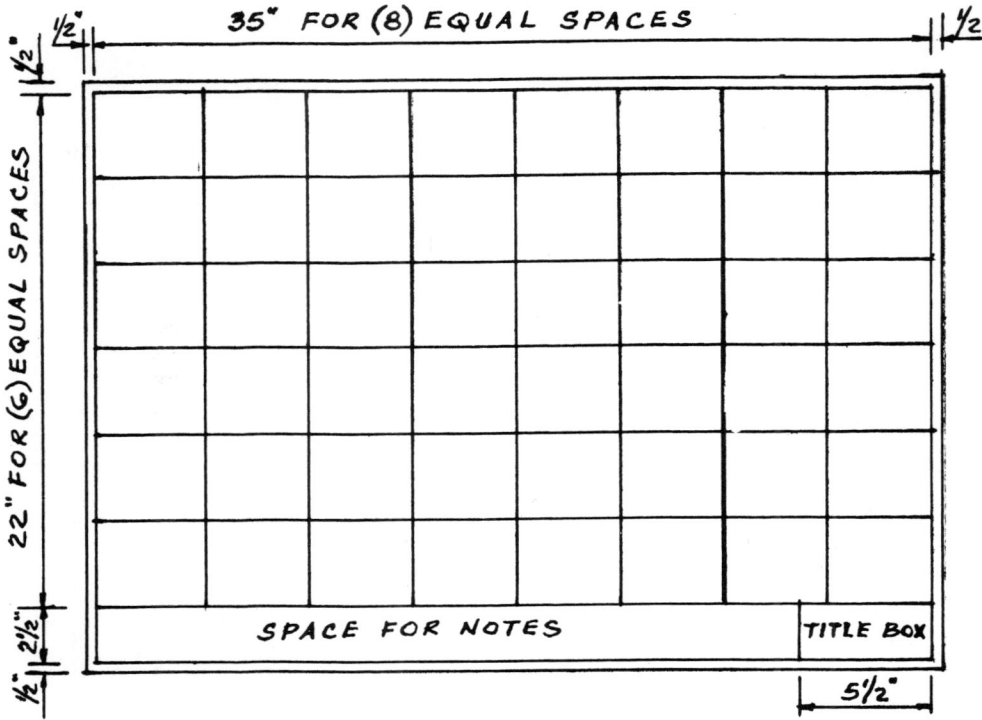

Fig. 2-6. Drawing sheet layout for rectangular duct fittings.

5. Paper will now have (48) rectangular boxes for the fittings
6. Start J-1 at the top left box and continue the series of rectangular duct fittings shown in Figs. 2-7 to 2-14, from left to right.

Rectangular-Duct Straight Joints (Figs. 2-5 and 2-7)

For joint selection, the draftsman must consider specification requirements, fabrication efficiency, shipping methods, materials handling at the job site, and duct-erection conditions. Any one of these items affects and determines the straight-joint lengths that will be used on the shop drawings.

When a job specification indicates duct transverse bracing to be on 4'-0 centers, maximum, the longest joints will use 96" sheets. However, if shop machinery capacities are adequate and bracing on 5'-0 centers is specified, 120" sheets can be used.

Generally, straight-joint lengths can be the full length of the sheet if the duct size segments — including seams and connection allowances — do not exceed the sheet width. A Type I joint for a 12" x 10" duct, 118" long will use a 48" x 120" sheet (see Fig. 2-5). For larger ducts (Type I), the joint length will be the *width* of the sheet if the duct-size perimeter can fit on the length of the sheet; similar to J-3 in Fig. 2-7, which can be cut from a 48" x 96" sheet.

J-1 in Fig. 2-7 can be fabricated from (2) 48" x 120" sheets (Type II, Fig. 2-5) and similarly, J-2 will use a 48" x 96" sheet. Note that J-5 uses (2) 48" x 96" sheets and is made from (2) L-shape pieces.

Joint Types I and III (Fig. 2-5) are usually selected for horizontal runs of watertight ducts as they have no longitudinal bottom seams which would need to be soldered, welded, or caulked.

Types IV and V (Fig. 2-5) are for large ducts which can be shop assembled or shipped KD (Knocked Down). (See J-6 in Fig. 2-7.)

Type VI (Fig. 2-5) may be KD if standing seams (L-5, Fig. 2-1) are used, but is shop-assembled when "grooved" as in L-6, Fig. 2-1.

It can be readily understood that, while preparing a drawing which will require duct systems of stainless steel, Monel, copper, galbestos, or other

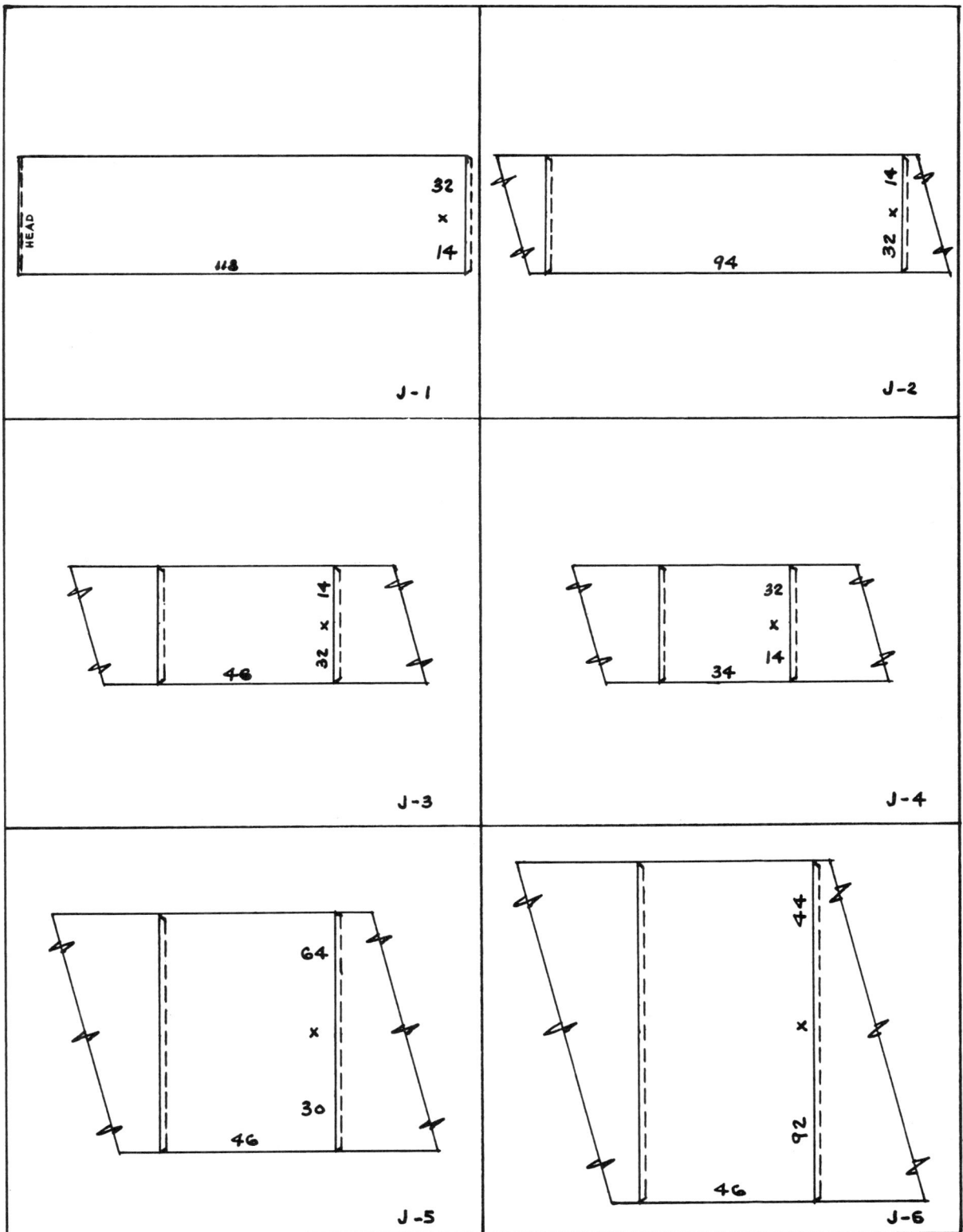

Fig. 2-7. Rectangular duct straight joints.

costly materials, the draftsman should select the type of straight joint most economical to fabricate and with the lowest waste factor, before a joint length is determined. This is most important if long, straight duct runs are probable.

Large ducts are often more efficiently shipped and handled at the job when left KD. Alteration or addition jobs often require special study for materials handling, such as for door or elevator sizes, and for ducts which must be assembled in place. (KD joints should be noted on the drawing by the draftsman.)

Elbows (Figs. 2-8 and 2-9)

Careful selection and design of *elbows* is of primary importance for the economical operation and function of a duct system. To minimize airflow friction, the engineering design drawings and specifications generally indicate that radius elbows shall have a minimum throat radius equal to the cheek width, and *square-throat square-heel elbows* must have turning vanes. Depending on the specification requirement, vanes may be single thickness or double thickness. Both types are indicated by the draftsman with parallel, double dashed lines (Fig. 2-8, E-1 and E-2).

Radius elbows are more economically fabricated than those with vanes and should be selected by the draftsman wherever possible. For cheek widths up to 24″, one leg-length should be limited so that the cheek pattern can be laid-out along the edge of a standard sheet by the shop layout man (E-4). For cheek widths of 25″ to 34″, both leg-lengths should be limited so that the full cheek-pattern may be positioned on a standard sheet (E-5). For cheek widths of 35″ to 45″, multiple elbows should be used with leg-lengths for each 45° cheek pattern to fit on a standard sheet (E-6). Elbows of cheek widths 46″, and greater, should be drawn as square throat and heel with vanes. Try to avoid transition elbows and radius elbows which require splitter-type vanes.

The throat radius of a *transition elbow* should be equal to one-half the sum of the cheek widths at both ends of the elbow.

The heel radius is equal to the sum of the throat radius and narrow cheek width (E-8, in Fig. 2-9).

For a square-throat, square-heel elbow with turning vanes and equal cheek-widths (Fig. 2-8, E-1) duct sizes need be indicated at one end of the elbow only.

For a square-throat, square-heel elbow with turning vanes and unequal cheek widths, E-2, duct sizes are required at both ends of the elbow. E-3 is a plan and sectional view of a square-throat, radius-heel elbow. The air-inlet opening penetrates one side of a masonry cavity wall, with the heel of the elbow at 8″ AFF (Above Finished Floor). This elbow has unequal cheek widths and the heel radius is equal to narrow cheek width, in this case, 12″. To achieve satisfactory flow characteristics, elbows of this type are often specified to have splitter-type vanes. (Refer to Fig. 4-3 in Chapter 4.) When standard shop practice uses flanges for field installation of registers, the draftsman should indicate: ¾″ FLGS out 4S.

A radius-throat, radius-heel elbow with equal cheek width, and with a throat radius equal to cheek width is commonly called a *full radius elbow* (E-4). To draw it, extend duct-run construction lines until they intersect. From the square corner of the throat or heel, draw a 45° line. Elbow leg-lengths are measured from the throat vertex which is shown by lines, approximately ¼″ to ⅜″ long, drawn towards the heel. With a slip end leg-length equal to the cheek width, the slip end line extended, would intersect the 45° line and locate the center for heel and throat radii.

Although some elbows may appear too large for a one-piece cheek pattern because overall dimensions are greater than standard sheet sizes (E-5), it is possible for full radius elbows up to 34″ cheek width to be laid out on a 48″ x 96″ sheet by turning the layout on an angle. When the leg lengths vary, the draftsman should check the elbow to accomplish a one-piece layout, if possible. Large elbows often must have intermediate bracing angles and will not fit on a sheet in any position. It is advantageous here to bisect it (E-6) with a connection so that the half-cheeks may be laid out in one piece and the duct connection will substitute for the required bracing. Double elbows of this type may be used, up to approximately 45″ cheek width, while square-throat elbows with turning

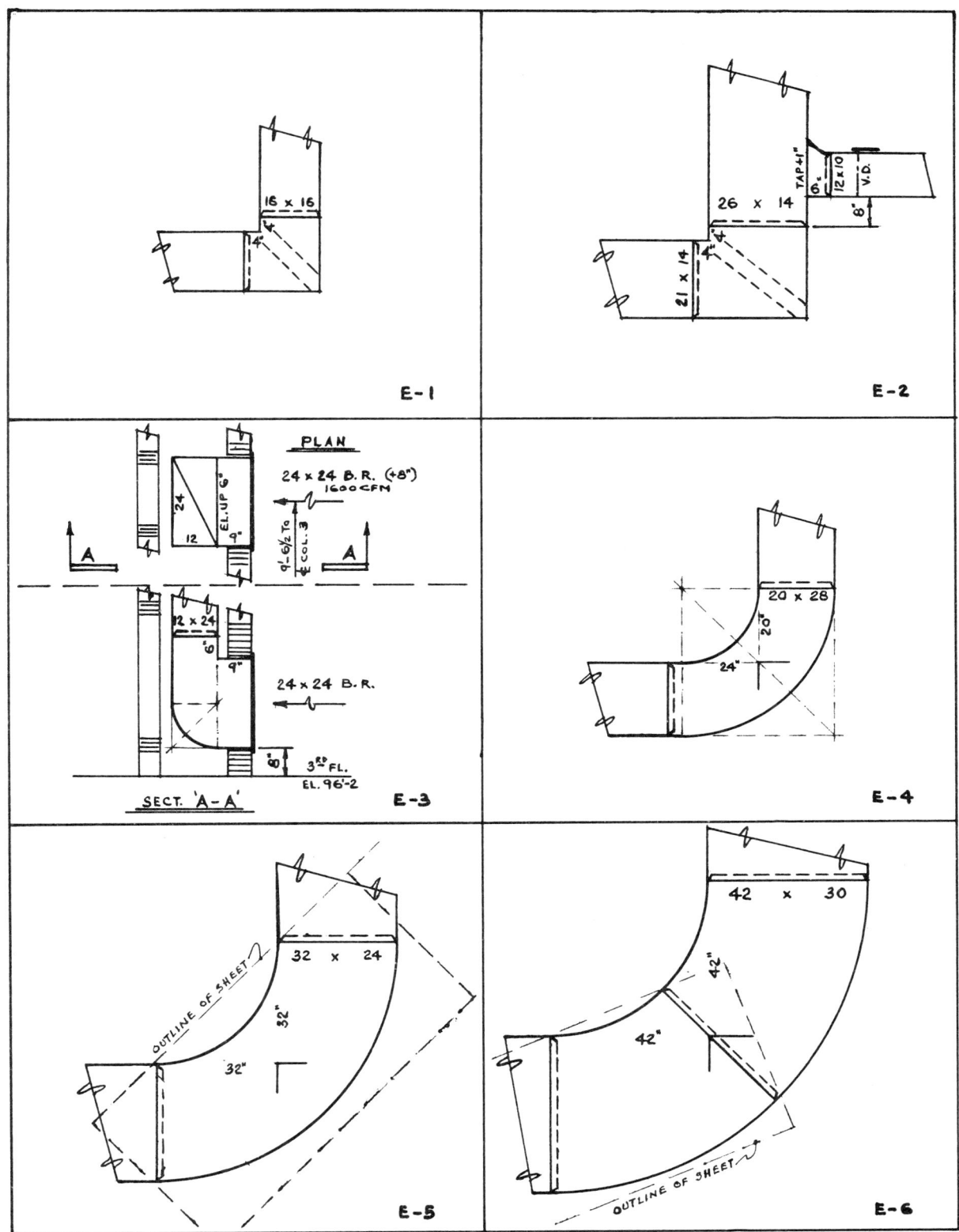

Fig. 2-8. Rectangular duct elbows.

28 DUCT CONSTRUCTION Ch. 2

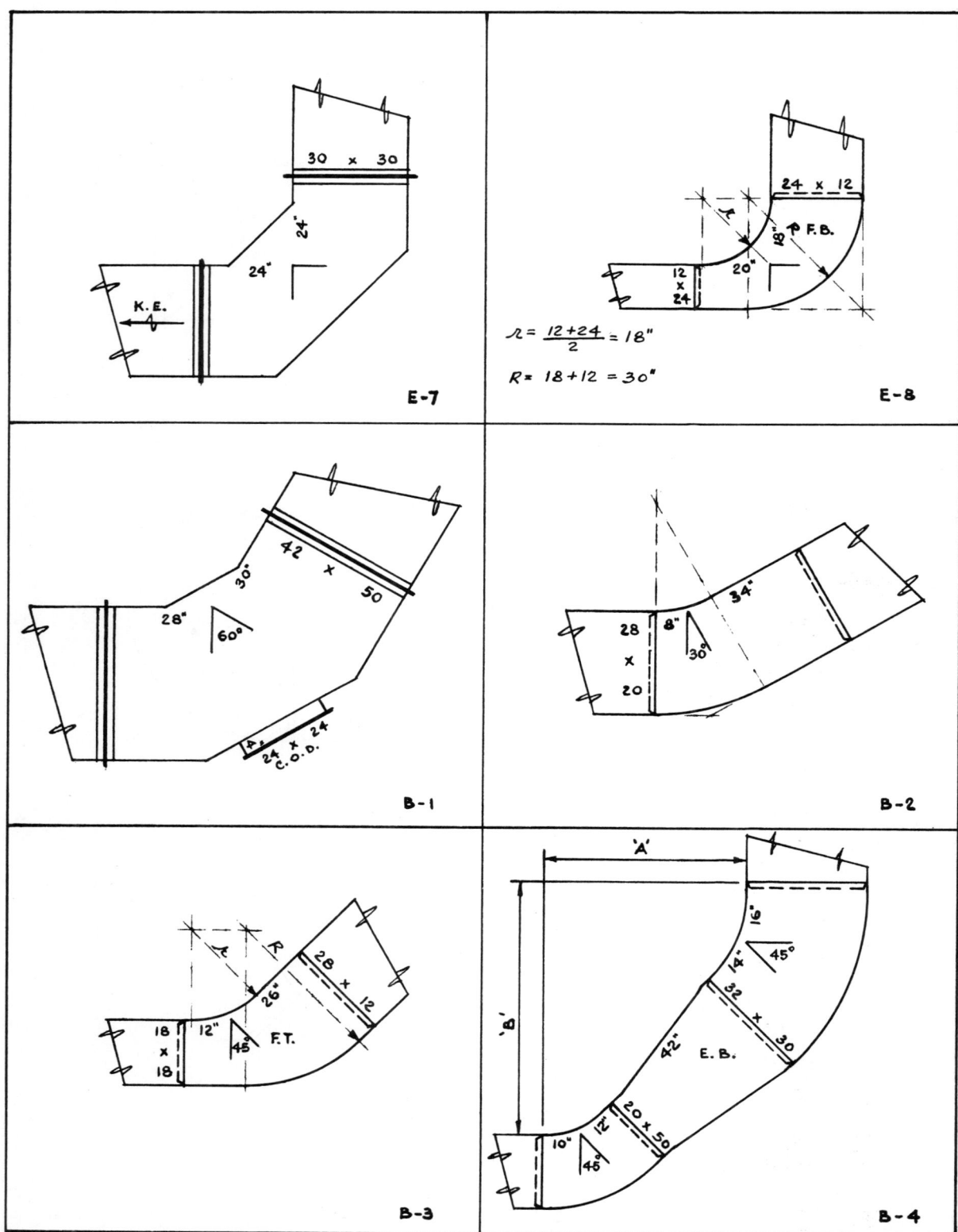

Fig. 2-9. Rectangular duct elbows and bevels.

vanes should be used for ducts 46″ wide, and larger.

Brake heel, and throat elbows are often used for heavy-gage systems such as kitchen-exhaust ducts or boiler breechings. To draw elbow (Fig. 2-9, E-7) scribe 30″ throat and 60″ heel radii and connect 45° tangents to each arc. Mark off 24″ elbow leg-lengths (this does not affect layout).

A transition elbow with the bottom cheek in a level plane (*FB*) is shown in Fig. 2-9, E-8. Throat and heel radii are 18″ and 30″. When specific job conditions require a similar fitting to satisfy an elevation change in the plane of a turn, duct heights should be adjusted before the junction if possible, so that either the bottom cheek, or the top cheek can be flat and thereby simplify shop layout. Second choice would be identical patterns for top and bottom cheeks (*EB*).

Bevel Construction (Fig. 2-9)

Similar in construction principle to 90° elbow, E-7, is a 60° bevel, smoke-breeching fitting, B-1. It is drawn similarly to a three-piece 60° elbow with a throat radius of 42″. Throat and heel lines are tangential to their respective radii. The leg lengths here prevent the cheek pattern from being laid out in one piece on a standard 48″-wide sheet. On a piece of scrap paper, draw a full-radius 60° elbow, 42″ x 50″, of the same type, in such a way that the full cheek can fit on a 48″ x 96″ sheet. What had to be changed on the fitting to gain this advantage? A 30°, full-radius bevel elbow with equal cheek widths is shown in B-2. A 45°-transition bevel elbow with a 24″ throat radius is shown in B-3. Typical of unequal cheek elbows, the heel radius is the sum of the throat radius and smaller cheek-width (see general comments for transition elbow E-8).

An arrangement of two 45° elbows with an intermediate equal-break transition as in B-4, is a convenient method to join ducts of unequal sizes requiring a 90° turn. Dimensions *A* and *B* will be computed in the trigonometry section of Chapter 3.

Offset Construction (Figs. 2-10 and 2-11)

An offset for a heavy-gage duct (Fig. 2-10, S-1), is a fixed combination of 30° elbows, with one having 6″ straight. The amount of straight on the other end of the set (Dimension *X*), is actually determined by the various dimensions indicated and may be computed by the draftsman, if desired.

A full throat-radius 'set' (S-2) has a fixed amount of offset and running length, and the angle of the offset is unimportant.

The *ogee* offset (S-3), where no straight joins the two elbow portions, is used to accomplish a maximum throat-radius (see 2-P in Fig. 1-3). A heavy-gage transition offset is indicated in S-4 with mitered kink-lines so as to retain duct cross-section areas. Each throat segment has 3″ straight.

1. From points *A* and *B* as centers, draw arcs equal to each cheek width, 18″ and 24″, respectively.
2. From *A*, draw a tangent to the 24″ arc, and extend to *C*.
3. From *B*, draw a tangent to the 18″ arc, and extend to *D*.

An unequal cheek width offset with ogee curves (S-5) is developed as follows:

1. From sides of wide cheek, mark off points *A* and *B* equal to one-half narrow-cheek width or 7″.
2. Connect *AC* and *BC*.
3. 26″ offset (15″ set + 25″ cheek − 14″ cheek) ogee is developed from *AC*.
4. 15″ offset ogee is developed from *BC*.

Offsets, particularly in large ducts, can be conveniently made by the use of a reverse combination of bevel elbows. Two equal-cheek width 45° elbows with leg-length dimensions as indicated, are shown in S-6 to form a 22″ set with an 84″ running length.

An equal cheek 30° elbow and a transition 30° elbow are shown in Fig. 2-11, S-7 as a 15½″ set with a running length of 6′-4. Similarly, for all sets with unequal cheek widths, the larger set is equal to the sum of the smaller set and the difference of the duct widths (15½ + 45 − 35 = 25½″ offset).

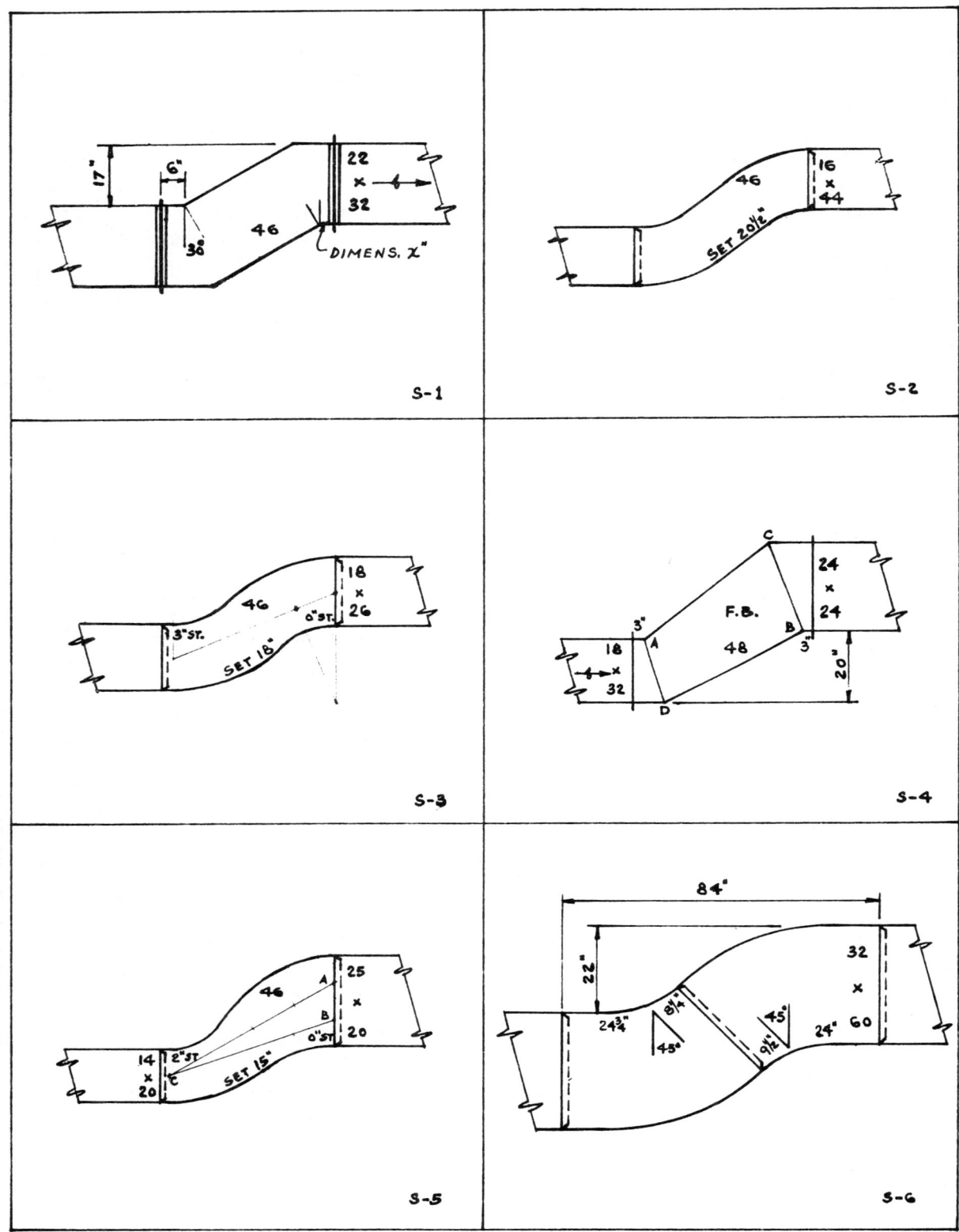

Fig. 2-10. Rectangular duct offsets.

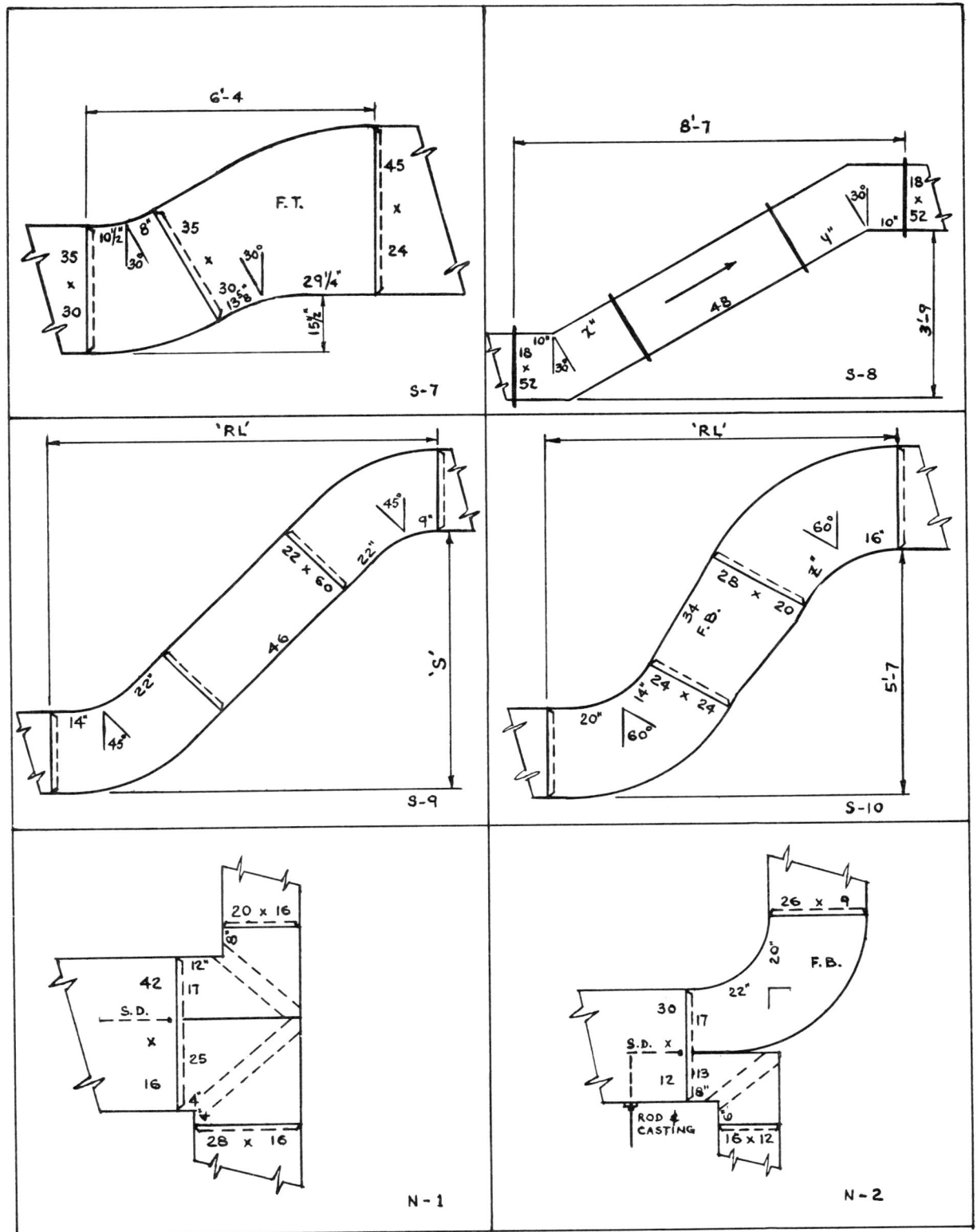

Fig. 2-11. Rectangular duct offsets and nesting 90° elbows.

Two 30° elbows and an intermediate straight-joint are required for a 3'-9 set of a heavy-gage duct run as in (S-8). Both elbows have short legs of 10" to complete a running length of 8'-7. Computations for *X* and *Y* are explained in Chapter 3.

Two 45° equal-cheek elbows and a straight-joint form offset (S-9). Although duct dimensions necessary for fabrication are shown, running length *RL*, and set *S*, must be calculated (see Chapter 3).

60° elbows of different cheek widths are connected to a transition for a set of 5'-7. Running length *RL* and leg length *Z* must be determined (see Chapter 3).

Construction of Nesting Fittings

A tee with turning vanes (Fig. 2-11, N-1) may be fabricated as two separate elbows, or as one fitting having an interior partition. The fabrication method need not be noted on the drawing unless field installation conditions require that the fitting be erected separately.

A supply duct splits into opposite directions with a 90° transition elbow and a square-throat vaned elbow (N-2). When a change in duct elevation is required at close proximity to a split, a horizontally nested connection (Fig. 2-12, N-3) can be used. For example, a 30" x 12" main at +9'-4 AFF is required to branch into opposite directions and one of the branches passes below a beam which is +9'-10 AFF. Clearance between the bottom of beam and bottom of the 30" x 12" main is 6", and a duct to this direction can be anticipated as a bottom *take-off*. Both elbows have flat, adjacent cheeks, as indicated by *FB* and *FT*, at +9'-9 AFF.

To acquire a sense of "fitting visualization," the draftsman should study transformations in duct sizes so that top and bottom elevations may be compared to each other. The 25" x 8" bottom elbow is 9'-9 *minus* 8" = +9'-1 AFF. As the bottom cheek *drops* 3", this is subtracted from 9'-4 which checks to the +9'-1 height. The 26" x 9" top elbow is 9'-9 *plus* 9" = +10'-6 AFF, for TOD (Top Of Duct). The TOD of the 30" x 12" main is +10'-4 and because the top cheek raises 2", this checks with +10'-6.

A vertical nest with a straight branch and a 90° elbow, as in N-4, requires careful thought before selection of the elbow is made. As the shop drawing is developed, specific elevations are determined by duct-space conditions and clearances for work of other trades. In this case, the elbow is costly to fabricate because seams are required for each cheek and the heel wrapper, due to their dimension relationships to standard sheet sizes. An alternate may be a square-throat elbow from 30" x 16" at the nest, to 32" x 16" at the slip connection, and then to a transformation fitting to the 46" x 12", as required.

A 14" x 34" return, or exhaust, main (N-5) at +10'-1 AFF divides into a 16" x 16" and a 30" x 10". The bottom branch is 18" at the main which locates the 30" x 10" at +10'-9 AFF. This may be calculated in two methods: A — (10'-1) + (18 − 10) = 10'-9; or B — the bottom of the transformation fitting drops 8" (the difference of 18 − 10); therefore (10'-1) + 8 = 10'-9. The plus height for the elbow is (10'-1) + 18 = 11'-7 AFF.

A 40" x 20" supply main branches three ways with flat-bottom fittings (N-6). Minus dimensions — bottom of slab to top of duct — can be calculated by subtracting the sum of the plus dimension (FIN.FL to BTM of duct) and the depth of each duct from the distance of floor to bottom of slab, with the following building conditions: floor-to-floor = 11'-6, concrete slabs = 5", and top of slab is finished floor; the minus dimension for the 40" x 20" duct at +8'-9 AFF is: (11'-6) − 5" − (8'-9 + 1'-8) = (11'-1) − (10'-5) = 8", called −8". N-6 in Fig. 2-12, indicates all fittings FB (flat bottom), so the minus dimensions will be the sum of the amount of drop in the top of each fitting and 8". The 25" x 13" would be (20 − 13 + 8) = 15", called 1'-3. A supply main from a heating and ventilating unit splits horizontally and vertically into three branch-ducts (N-7). Show plus-heights for the 23" x 12" and 36" x 12" ducts on the drawing.

Fabrication of Y Branch Fittings

This fitting (Fig. 2-12, Y-1) may be laid out with the bottom, straight side, and top in one piece,

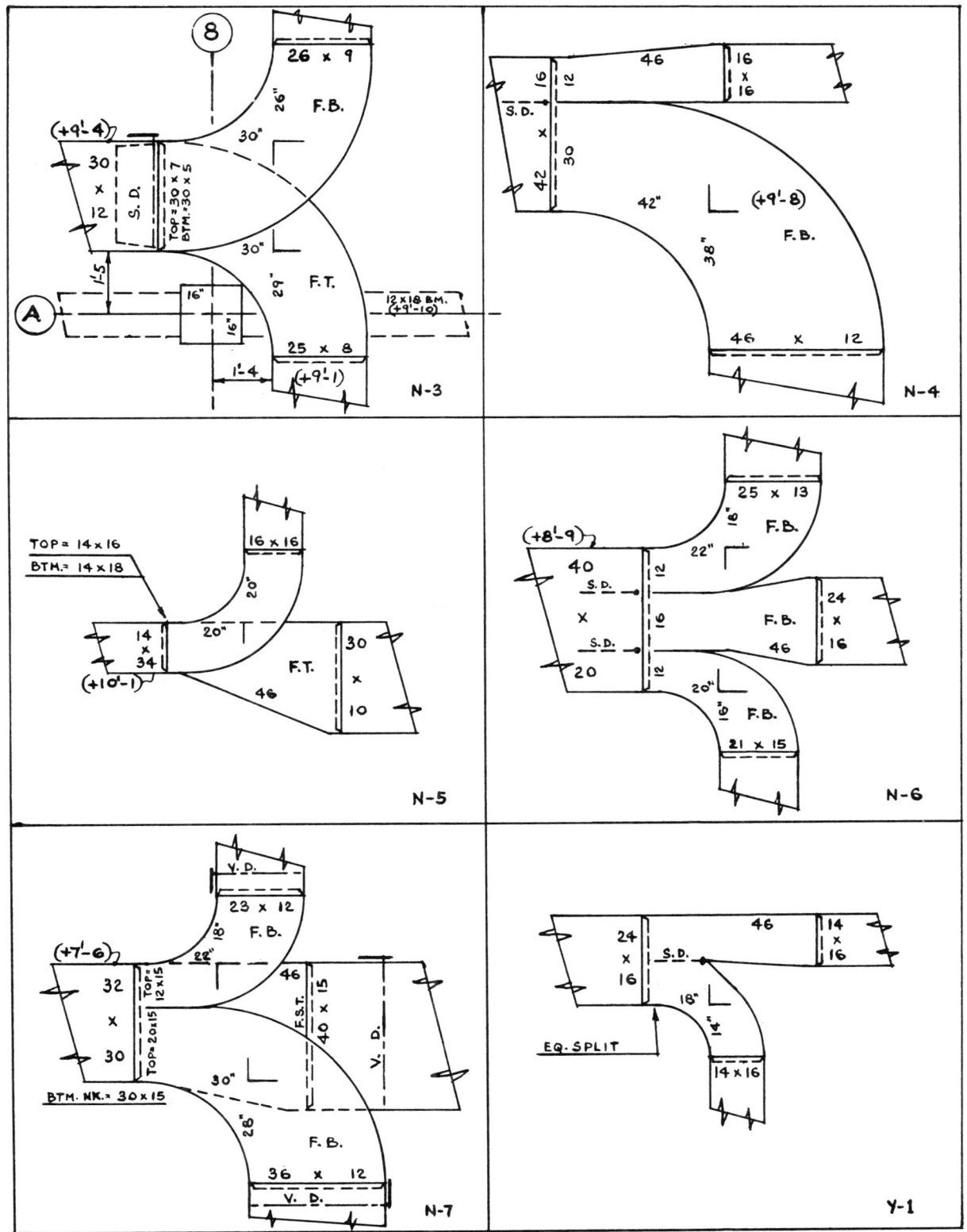

FIG. 2-12. Rectangular duct nested fittings and a Y-branch take-off.

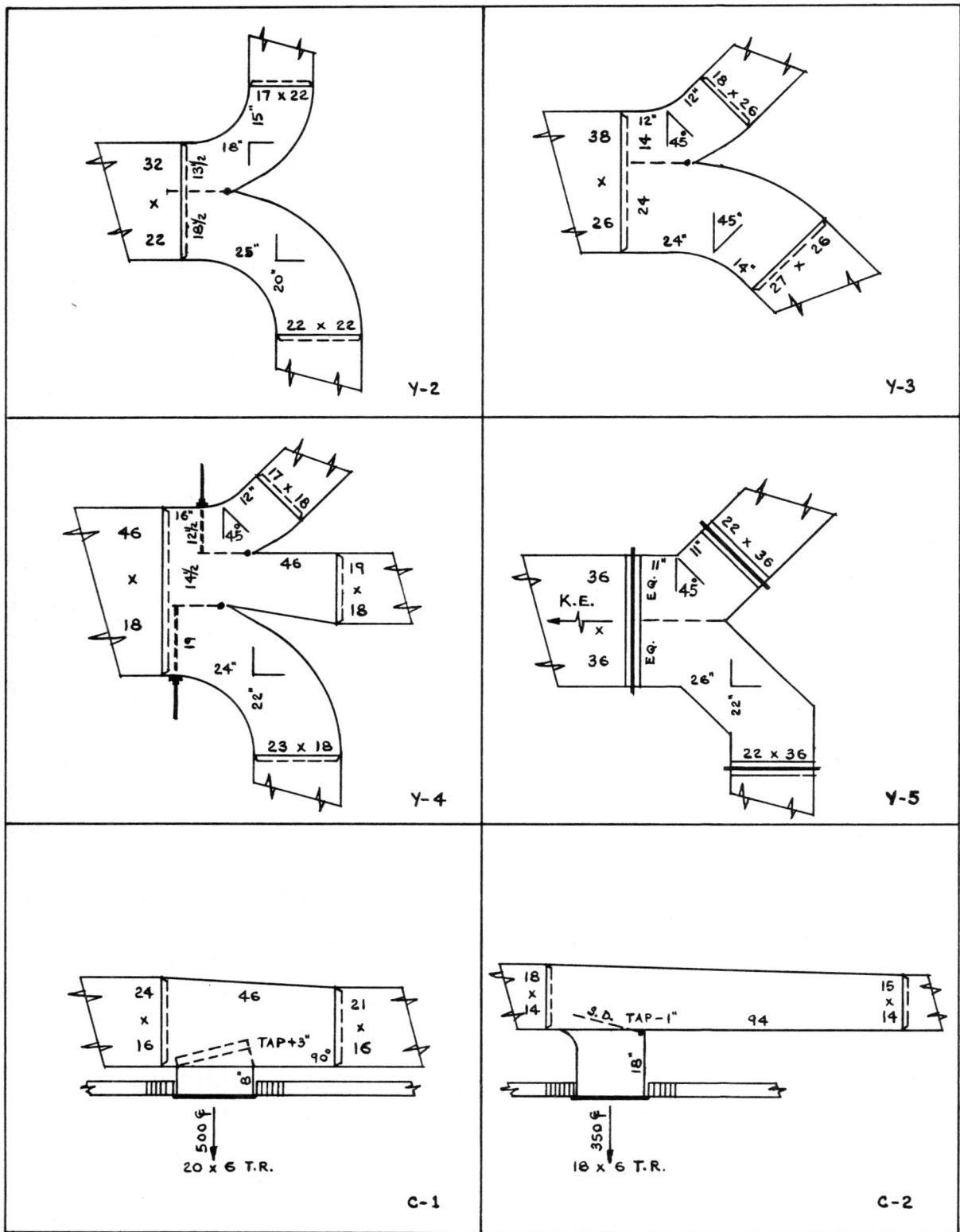

FIG. 2-13. Rectangular duct Y-branches and register collars.

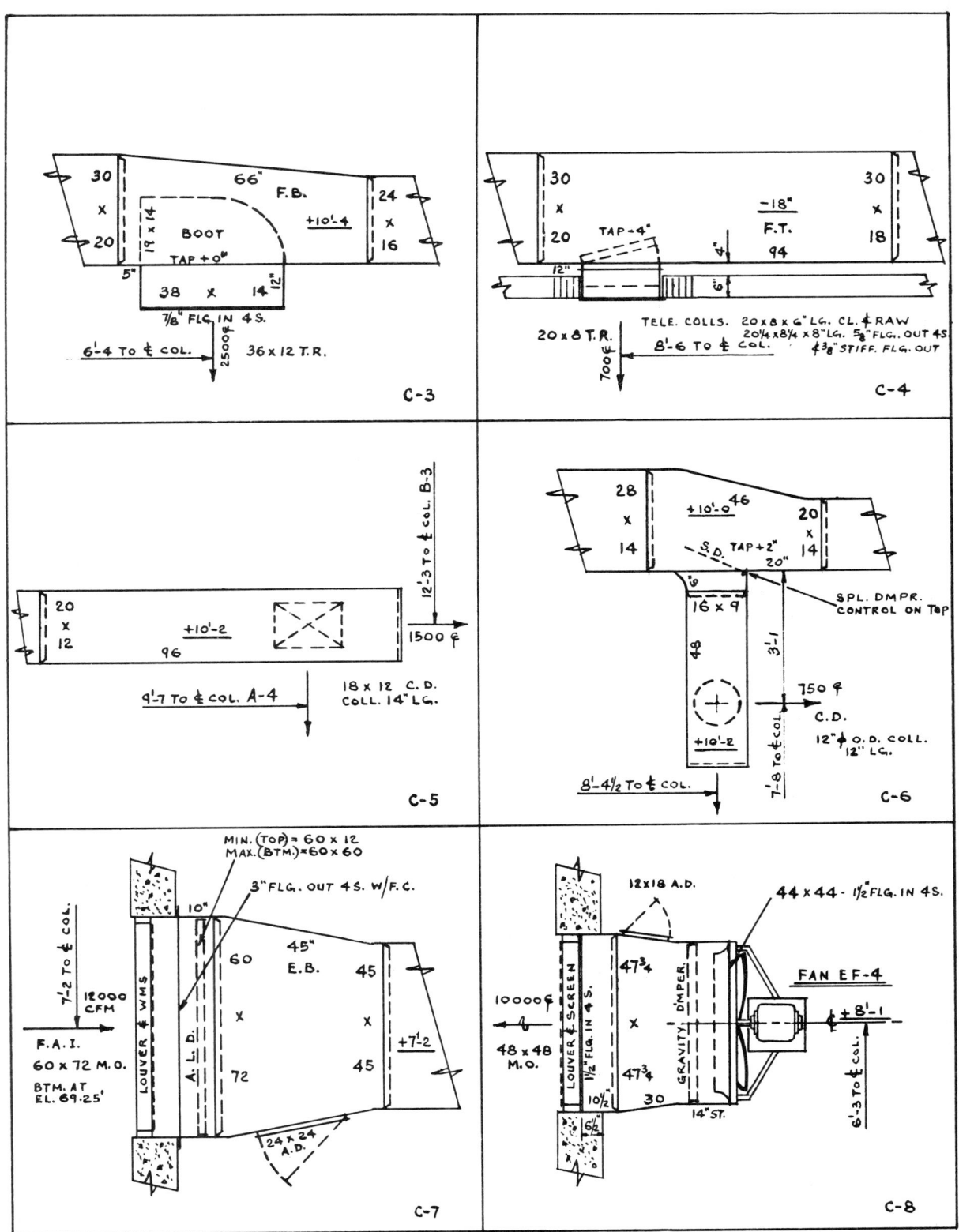

FIG. 2-14. Rectangular duct collars.

or with separate cheeks for the top and the bottom. However, the shop drawing should indicate a fabrication preference for erection requirements only, such as for installation in a restricted ceiling space. From the heel radius of the 90° branch, extend a 45° tangent to intersect a parallel line 12" from the straight side.

Three types of branch duct fittings (Fig. 2-13, Y-2, Y-3, and Y-4) each equal in height to their respective main, may have one-piece cheek layouts.

Two equivalent branch ducts (Y-5) are joined as a Y branch and are separated by an interior partition. The partition may be welded in a fixed position, or hinged at the corner as a splitter damper; the shop drawing should be noted accordingly, such as: (partition).

COLLARS

Collar Fabrication

A straight register collar with a factory manufactured volume-control-distribution device is shown in C-1, Fig. 2-13. (Note that the projection into the duct is on the downstream end of the collar.)

A splitter damper is shown in a throat-type register collar (C-2). The control should be indicated "on top" or "on BTM," whichever location is most accessible.

An internal positive-boot with a 19" x 14" neck opening is used here for a 36" x 12" register collar (C-3, Fig. 2-14) on an exposed duct. This type is often used on exposed ducts as the flanges turned in at the register connection provide a flush finish. Collars are increased 2" more than nominal register size.

Telescoping collars (C-4) are advantageous, especially for alteration jobs, as the slide-fit allows for length adjustment, thereby eliminating cutting, or adding to collar for register installation.

Ceiling-diffuser collar dimensions (C-5) are nominal *inside dimensions,* unless otherwise noted. Centerline arrows are for "tie-ins" and *do not* show air-distribution patterns. When internal baffles, or partial air-patterns are required, descriptive notes should be indicated, such as: (3) way blo, 90° baffle.

Ceiling-diffuser cuts generally indicate duct collars to fit *inside* the neck of the diffuser so that the connection lap is in the direction of airflow. As sheet metal dimensions always indicate nominal inside dimensions, the draftsman should note on the shop drawing (C-6) when an outside finished dimension is required, such as: (12" ϕ *OD* COLL 12" *LG*).

Minimum and maximum outside-air automatic dampers are located in a collar (C-7) which connects to the 60" x 72" masonry opening. This method is useful when the louver installation is not in the HVAC contract. An access door must always be provided for maintenance of equipment enclosed in a duct. Calculate the finished-floor elevation in the space where this duct shall be installed, and show it on the drawing.

A duct collar with 1½" flanges is shown (C-8) fastened to the inside of a louver frame. Note that ¼" is deducted from MO (Masonry Opening) for collar size.

While a shop drawing is being prepared, the sheet-metal draftsman must be constantly aware of construction requirements for the various duct systems because types of materials, connections, seams, bracing, and fitting design are determining factors in the layout. A duct system which requires special consideration is that one commonly called a "high-pressure system." It can be identified on the HVAC design drawings by acoustical duct lining and/or sound traps which are used to reduce sound levels and by pressure-reducing valves (PRV) and terminal boxes which are used for sound- and air-pressure control.

Duct tape, caulking material, gaskets, or welding are used to make all seams and connections completely airtight. Because of the pressures, rectangular ducts need adequate bracing and suitable material gages to resist pulsation or bowing. However, if duct space is available, the configuration of round ducts eliminates this problem.

Round Duct Fabrication

Round fittings and spiral conduit have fabrication and installation advantages because the connections and seams are simple to seal airtight; in addition straight sections in relatively long lengths may be used, thereby reducing the amount of connections needed. Drafting procedures for a round-duct system are similar to a plumbing or heating piping-layout, as standard fittings and straight pipe (spiral conduit) are fitted into the duct spaces. Calculations and dimensions are taken to duct centerlines UON (Unless Otherwise Noted), on the drawings.

A list of commonly used fittings is shown in Fig. 2-15. Dimensions indicated are not standard, as differences may occur depending on shop fabrication standards. Schedules of typical dimensions for 45° laterals and elbows are seen in Figs. 2-16 and 2-17, respectively.

Examples for Round Duct Fittings

Prepare a ½" = 1'-0 scale drawing of partial plans 1 to 4 (Fig. 2-18).

1. Use Figs. 2-15, 2-16, and 2-17 for fitting dimensions not shown.
2. Transformation segments are 12" long UON.
3. Show dimensions and notes required for fabrication and installation.
4. See Chapter 3 for computations to be calculated in Plans 3 and 4, Fig. 2-18.

Plan 1

Conditioned air from a high-pressure fan passes through a sound absorber (SA-4) which has 24" dia (ID) collars for entry and discharge. The first branch is a 12" dia, 90° tee and the 24" dia main continues as shown.

Plan 2

A 24" dia coupling is used to connect two sections of spiral conduit. The 45° lateral and reducer are fabricated as one fitting with the large end of the reducer (without lap L) welded to the lateral main. This is shown by XX on seam. The 8" dia branch, downstream, is from a conical tee take-off. The main duct continues to an 18" dia 90° elbow, and although it is a five-piece construction, the segments *need not be drawn* unless duct clearance is critical. The miter-line rise of an elbow = $\dfrac{\text{degrees of elbow}}{(\text{no. of elbow pieces} - 1)2}$. For a five-piece 90° elbow, the miter-line rise is 11.25°, $\left(\dfrac{90}{8}\right)$.

Plan 3

A 14" dia duct from the floor above is offset to avoid column *F-6* and continues to two branches. High-pressure duct systems *do not* have volume or splitter dampers.

Plan 4

A 16" dia duct is required to rise 2'-1 and offset 3'-4, using 45° elbows. These are called "rolling," or "rotating" fittings. Show tie-in of 12" dia 60° elbow which will connect to rigid conduit on another drawing (imaginary).

Rolling 90° Elbows

When elevation changes are anticipated in the vicinity of an elbow, the elbow may be "rolled" to a convenient degree and the duct-run leveled-off with a bevel fitting (see Fig. 1-2). A minimum amount of fittings is used in this method, therefore efficiency is improved in fabrication, erection, and system airflow.

Special Fittings

Some job conditions may require out-of-standard fittings. The draftsman should indicate in detail, on the shop drawing, all information necessary for shop and field. See Y branch in Plan 3, Fig. 2-18.

Transitions

Transitions should be equal-taper wherever possible, of sufficient length so that maximum side-slope is 20°. Unequal taper transitions must be noted on the drawing, as required, for example: FB, BD4", TU6½".

Acoustic Lining

The mechanical design drawings or job specifications will indicate thickness and lineal-feet required for acoustical duct linings. Duct dimensions on the plans generally indicate net clear sizes, therefore the ducts must be increased to allow for the required lining thickness (Fig. 2-19, part 1). When lining thickness varies for different systems or portions of a system, the draftsman shall indicate the type on the shop drawing by symbols or notes. Lining is generally ended by tape or metal nosing, or by a tight fit against the angle frame

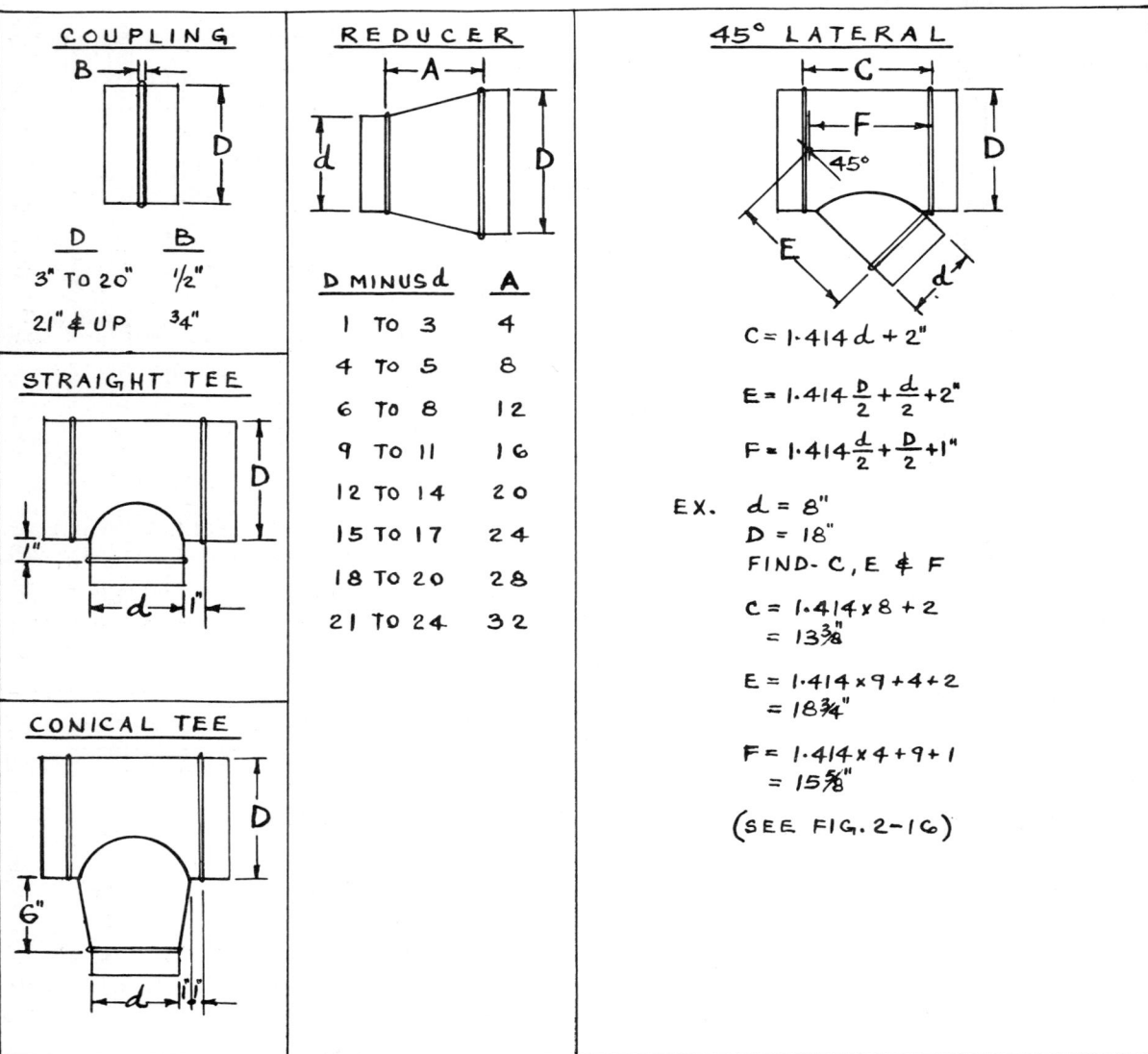

Fig. 2-15. Round fitting standards.

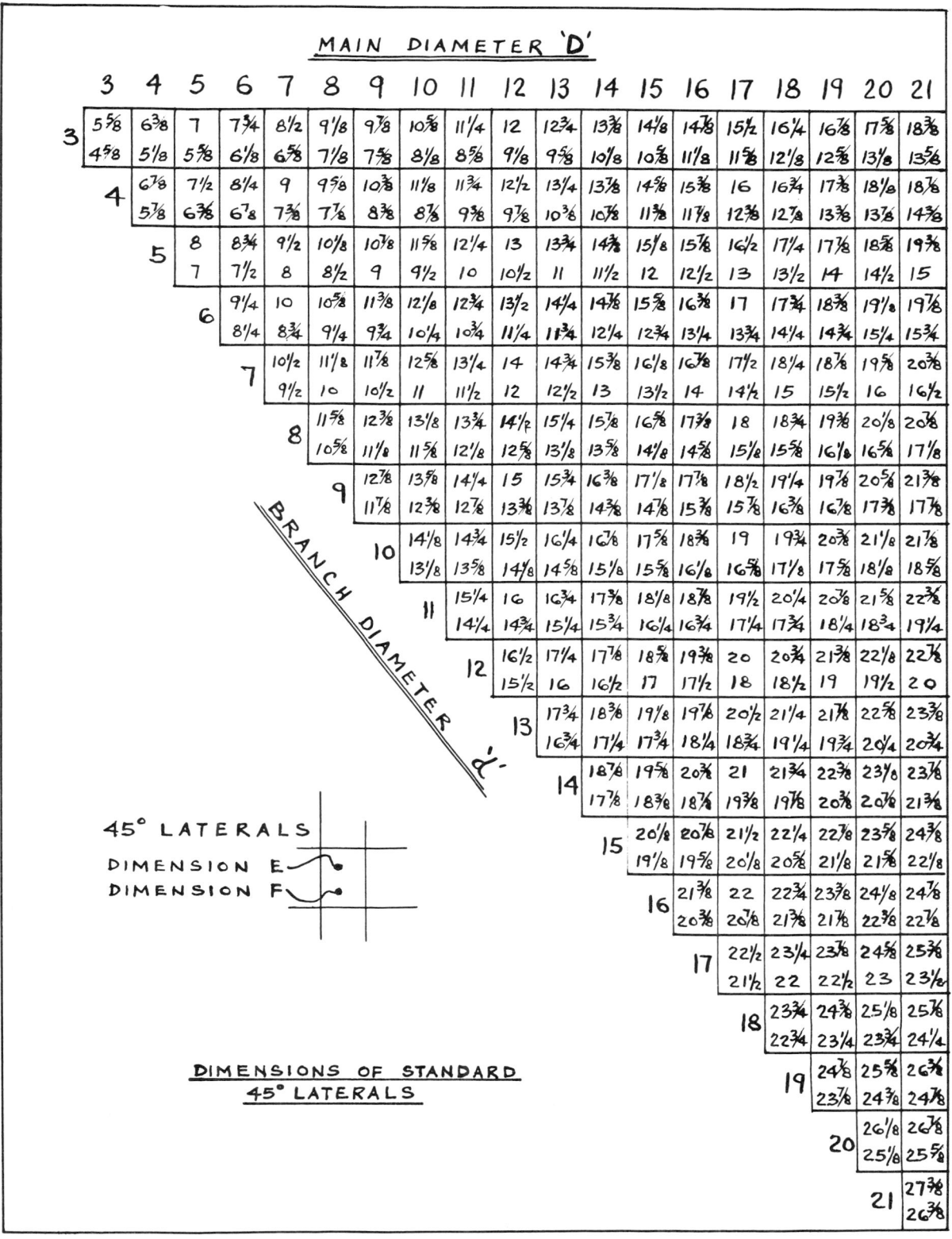

FIG. 2-16. Dimensions of standard 45° laterals.

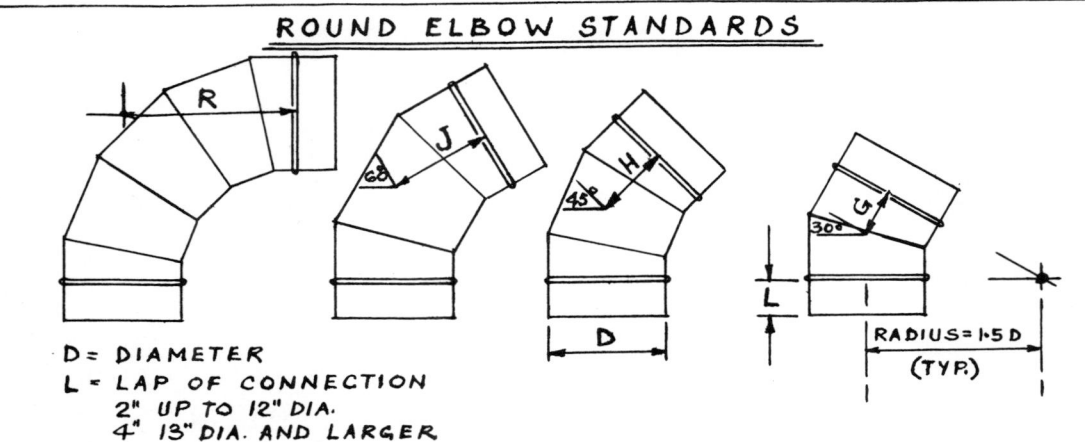

ROUND ELBOW STANDARDS

D = DIAMETER
L = LAP OF CONNECTION
2" UP TO 12" DIA.
4" 13" DIA. AND LARGER

D	R (90°)	J (60°)	H (45°)	G (30°)	D	R (90°)	J (60°)	H (45°)	G (30°)
4	6	3½	2½	1-5/8	26	39	22½	16-1/8	10½
5	7½	4-3/8	3-1/8	2	27	40½	23-3/8	16-3/4	10-7/8
6	9	5¼	3-3/4	2-3/8	28	42	24¼	17-3/8	11¼
7	10½	6	4-3/8	2-7/8	29	43½	25-1/8	18	11-5/8
8	12	6-7/8	5	3¼	30	45	26	18-5/8	12
9	13½	7-3/4	5-5/8	3-5/8	31	46½	26-7/8	19¼	12½
10	15	8-5/8	6¼	4	32	48	27-3/4	19-7/8	12-7/8
11	16½	9½	6-7/8	4-3/8	33	49½	28-5/8	20½	13¼
12	18	10-3/8	7½	4-7/8	34	51	29½	21-1/8	13-5/8
13	19½	11¼	8-1/8	5¼	35	52½	30¼	21-3/4	14-1/8
14	21	12-1/8	8-3/4	5-5/8	36	54	31-1/8	22-3/8	14½
15	22½	13	9-3/8	6	37	55½	32	23	14-7/8
16	24	13-7/8	9-7/8	6-3/8	38	57	32-7/8	23-5/8	15¼
17	25½	14-3/4	10½	6-7/8	39	58½	33-3/4	24¼	15-5/8
18	27	15-5/8	11-1/8	7¼	40	60	34-5/8	24-7/8	16-1/8
19	28½	16½	11-3/4	7-5/8	41	61½	35½	25½	16½
20	30	17-3/8	12-3/8	8	42	63	36-3/8	26	16-7/8
21	31½	18¼	13	8½	43	64½	37¼	26-3/4	17¼
22	33	19	13-5/8	8-7/8	44	66	38-1/8	27-3/8	17-5/8
23	34½	19-7/8	14¼	9¼	45	67½	39	28	18-1/8
24	36	20-3/4	14-7/8	9-5/8	46	69	39-7/8	28-5/8	18½
25	37½	21-5/8	15½	10	47	70½	40-3/4	29¼	18-7/8

FOR BEVEL ELBOWS WITH ₵ RADIUS OTHER THAN 1.5 × DIA.

ELBOW	MULTIPLY ₵ RADIUS BY
15°	.132
30°	.268
45°	.414
60°	.577

FIG. 2-17. Round elbow standards.

Ch. 2 DUCT CONSTRUCTION 41

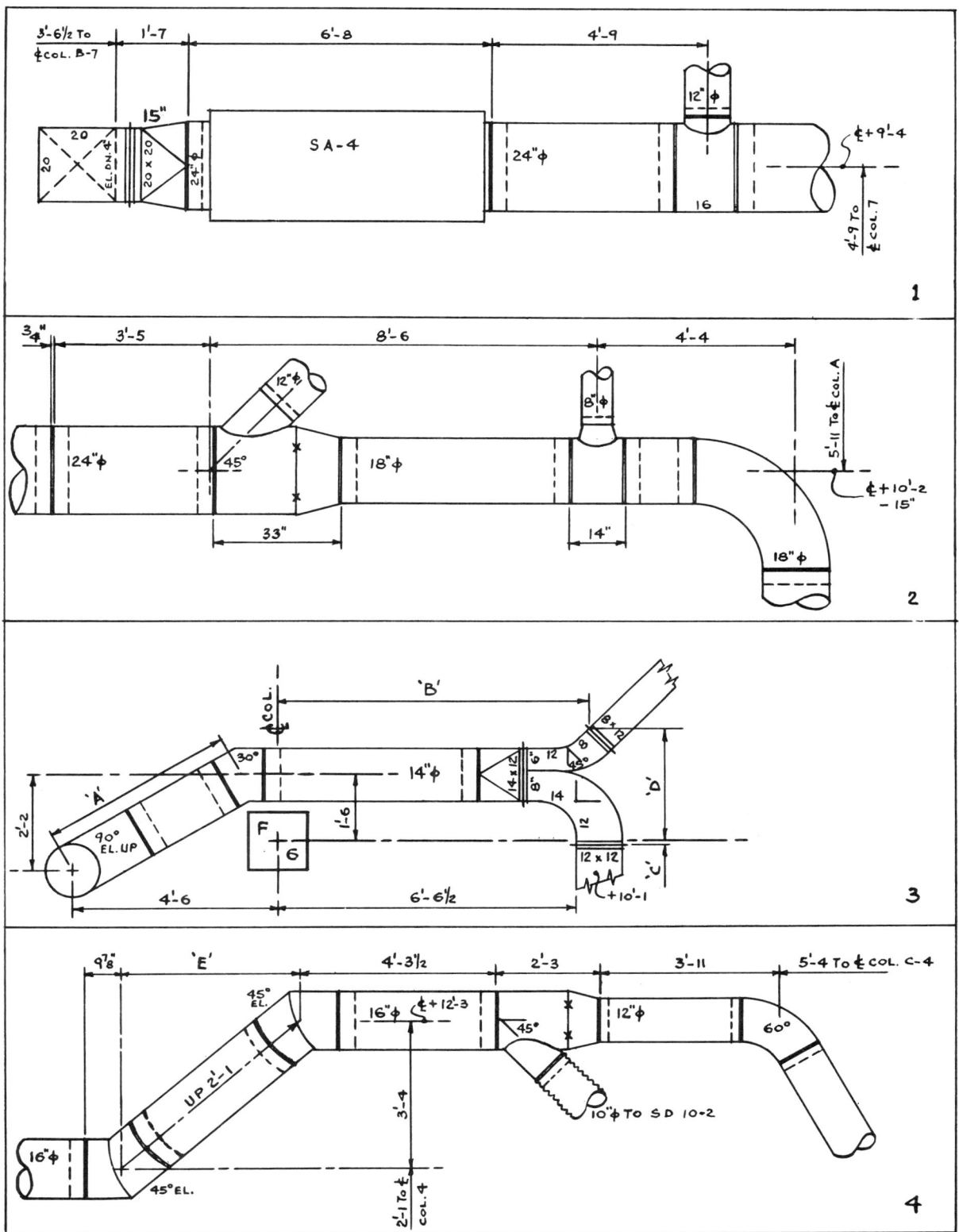

FIG. 2-18. Partial plans of round duct fittings.

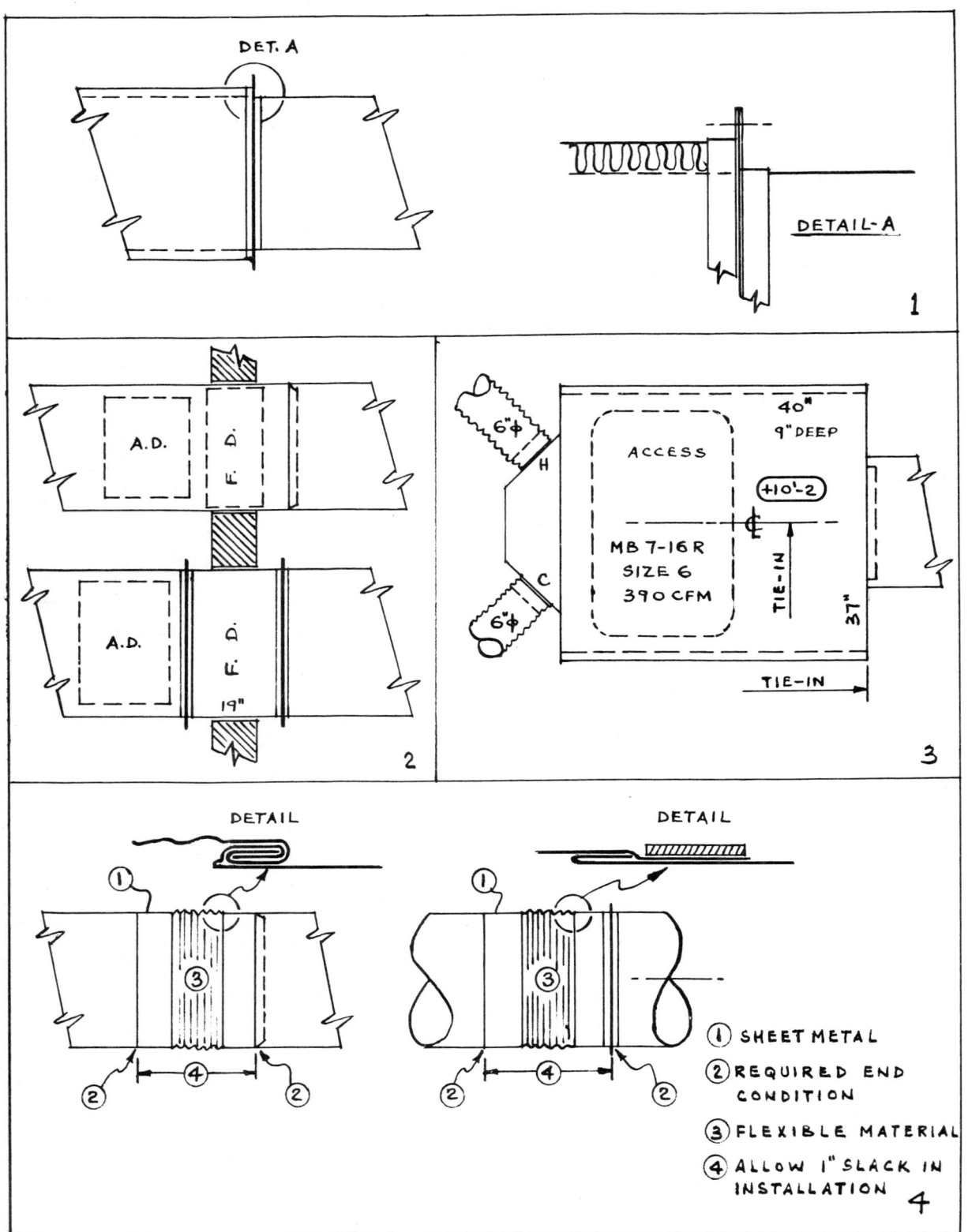

FIG. 2-19. Duct accessory constructions.

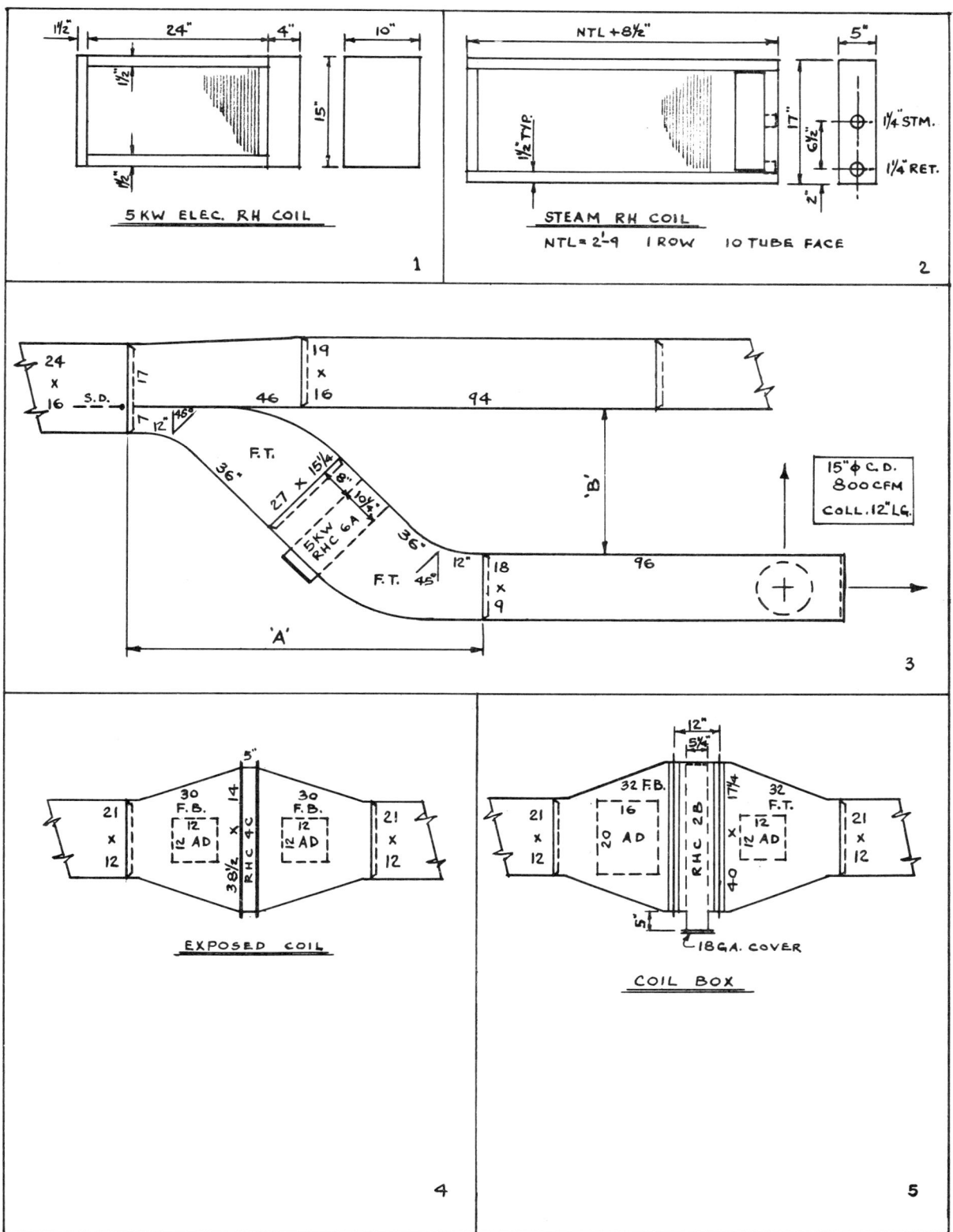

Fig. 2-20. Reheat coil cuts and duct connections.

on the unlined duct; both ducts using matched angles for the connection.

Terminal Boxes

Attenuation-box terminals should be positioned to allow adequate access for maintenance and adjustments. Indicate tie-in dimensions to the face of the discharge end, and to the longitudinal centerline (Fig. 2-19, part 3). Each box should be identified as per manufacturer's cuts and schedules for positive location. Rigid duct is usually terminated approximately 3'-0 from the box connection, and final connection is by flexible conduit. When flexible conduit is connected to a spiral-duct section, a coupling is required, which should be ordered and indicated on the shop drawing. Conventional, low-pressure duct construction is used for down-stream supply duct from terminal-box units.

Sound Absorbers

When manufactured sound-traps are used, the cuts should be checked carefully for duct-connection details to and from the unit. Material to be furnished by the shop, such as matched-angle frames, should be noted on the shop drawing.

Fire Dampers and Access Doors

Fire dampers of proper materials and gages must be used where specified, or to conform with local code-requirements. The sleeve-type fits into the duct and is designated on the shop drawing by a double dashed-line (Fig. 2-19, part 2). The spool-, or duct-type, has a matched angle connection on each end and is indicated as such. Access doors should be adequate for ease of inspection and maintenance. By using a perforated flat bar to hold the fusible link, an adjustment for volume control may be accomplished.

Flexible Duct Connections

Motor driven air-handling equipment is supported by vibration eliminator mountings to prevent the transmission of vibration to the building structure. Duct connections to and from this equipment are by flexible materials as the job specification may require (see Fig. 2-19, part 4).

Reheat Coils

Typical reheat coil cuts and duct connections are shown in Fig. 2-20, parts 1 and 2.

The 18" x 9" branch duct (Fig. 2-20, part 3) to a ceiling diffuser has an electric heating-coil. A complete frame to receive the coil is made by installing a 10¼" x 1½" sheet-metal channel fastened to the top, bottom, and far side of the duct, and with 1½" flanges turned in on the open end. The coil can then be installed and may be pulled out easily for maintenance.

1½" flanges are turned out on the transition fittings for the exposed coil (Fig. 2-20, part 4). 'U'-strips are placed over the duct and coil flanges and bolted for airtight closures.

The coil box (Fig. 2-20, part 5), while slightly more expensive to fabricate than that in part 4, completely encases the coil in an airtight duct. The duct run can also be completely erected prior to coil installation. This is advantageous, as it is not uncommon to install ductwork before coils are delivered to the job site.

CHAPTER 3

Practical Mathematics for Sheet Metal Draftsmen

Although a neat and elegant shop drawing is desirable, a sheet metal contractor is primarily interested in installing a duct system that can be efficiently fabricated, erected, and result in a profitable job. The draftsman is in a prime position to achieve this aim. To insure it, he must apply practical mathematical concepts and principles of arithmetic, algebra, plane geometry, solid geometry, and trigonometry in the process of making a drawing.

GENERAL ARITHMETIC REVIEW

Addition

When adding whole numbers, the corresponding digits of each number (units, tens, hundredths, etc.) should be placed in columns, one below the other. The numbers to be added are called addends and the result is the sum, or total.

Examples:

```
   17      748
  636       16
   42     1466    Addends
 1768      303
 ----     ----
 2463     2533    Sum, or Total
```

Whole numbers with decimal fractions are added similarly to whole numbers, locating the decimal points directly below each other.

Examples:

```
 16.125      46.0625
 42.25        7.6875
  8.875     119.375
 ------    --------
 67.250    173.1250
```

To add numbers in feet and inches, the *inch* column is added first and, dividing by 12, the total is converted to *feet* and any extra inches. The number of feet is then added to the *original* feet addends.

Examples:

```
  2'-10"                         9'-10"
  6'- 8"                         7'-10"
 12'- 9"                         3'-10"
 ------                         12'- 9"
   27" = 2'-3"                  -------
                                  39" = 3'-3"

  20'     Original addends, ft   31'
 + 2'                            + 3'
 ------                         ------
 22'-3"   Total                 34'-3"
```

Subtraction

Similar to addition, the proper digit columns must always be aligned.

Example:

```
 1978 cfm   Minuend       14.25  sq ft
 -1469      Subtrahend   - 6.875
 --------                -------
  509 cfm   Remainder     7.375 sq ft
```

In subtracting feet and inches, when the inches of the minuend are less than those of the subtrahend, the following method may be used:

12 inches (1 foot) are "borrowed" from the feet column in the minuend and then added to the number of inches in the inch column.

Example:

```
 14'-3"   (14' minus 12" = 13')
-11'-7"

          13'-15"   (3" + 12" = 15")
         -11'- 7"
         --------
           2'- 8"   Ans
```

Multiplication

Generally, multiplication is faster and easier when the multiplier contains the smallest number of digits, or the least value (discounting decimal points). For instance, the example below, on the left, is easier because it requires one step less:

764	Multiplicand	4.25
× 96	Multiplier	×16.5
4584		2 125
6876		25 50
		42 5
73344	Product	70.125

In all cases place the right-hand digit of each number group in line, disregarding the decimal point locations until the product is determined. The decimal point in the product is located by adding the number of digits to the *right* of the decimal points of the multiplicand and the multiplier. This amount is then counted from right to left in the product, and the decimal point placed to the left of that digit.

A good habit to acquire is "product approximation." By rounding off the multiplicand and the multiplier a quick check is accomplished. Using the previous examples: if 764 were rounded off to 800, and 96 to 100, the approximate result is 80,000; similarly, 4.25 is rounded off to 4, and 16.5 is rounded off to 17, and the approximate product is 68. Note that a number to be rounded off is increased only if the last series of digits is 50 percent or more of the unit increase.

Example:

128 to tens = 130; to hundreds = 100

56 to tens = 60; to hundreds = 100

When multiplying feet and inches, first determine the product of the inch column, convert this product to feet and inches, and then add the product of the feet column to get the final total.

Examples:

```
  3'-9       9" × 4 = 36" =  3'-0"
×  4         3' × 4      = 12'
                           15'-0"  Ans

  5'-7       7" × 6 = 42" =  3'-6"
×  6         5' × 6       = 30'
                            33'-6"  Ans
```

Division

Whole Numbers

```
                    194    Partial Quotient
Divisor   76) 14767        Dividend
              76
              ---
              716
              684
              ---
              327
              304
              ---
               23   Remainder
```

Decimals

```
              8.0
9.75.) 78.00.0
       78 00 0
```

Feet and Inches

```
          36 + 4 = 40
        5) 18'-4
           3'-8
```

Numbers with decimal fractions are divided as whole numbers. The decimal point is located by counting the number of digits to the right of the decimal point in the divisor, counting a similar amount to the right of the decimal point in the dividend, and placing the decimal point directly above — in the quotient. When dividing feet and inches, the feet in the remainder are converted to inches, added to the inches of the dividend, and then divided by the divisor.

Fractions

	Common	Improper		Mixed Number
Numerator	9	18		3
Denominator	16	5	=	3 $\frac{3}{5}$

The value of a fraction remains unchanged if the numerator and denominator are multiplied or divided by the same number.

$$\frac{3}{8} \times \frac{4}{4} = \frac{12}{32} \qquad \frac{14}{16} \div \frac{2}{2} = \frac{7}{8}$$

Addition and Subtraction of Fractions

When fractions are to be added or subtracted, the lowest common denominator (LCD) is selected, generally mentally.

In adding whole numbers and fractions:

```
10 3/8  | 3
14 1/4  | 2 = (1/4 = 2/8)
24      | 5
        | 8 (LCD)     = 24 5/8"  Ans
```

In adding feet and inches:

$$\begin{array}{r|l} 7'\text{-}9\frac{7}{8}'' & 7 \\ +6'\text{-}7\frac{3}{4}'' & 6 = (\frac{3}{4} = \frac{6}{8}) \\ \hline & \overline{13} \\ & \overline{8}\ (\text{LCD}) = 1\frac{5}{8}'' \end{array}$$

$$9'' + 7'' + 1\frac{5}{8}'' = 17\frac{5}{8}'' = 1'\text{-}5\frac{5}{8}''$$

$$7' + 6' + 1'\text{-}5\frac{5}{8}'' = 14'\text{-}5\frac{5}{8}''\quad Ans$$

In subtracting feet and inches:

$$\begin{array}{r} 27'\text{-}7\frac{1}{2}'' \\ -16'\text{-}9\frac{3}{4}'' \end{array} = \begin{array}{r} 26'\text{-}19\frac{1}{2}'' \\ 16'\text{-}\ 9\frac{3}{4}'' \end{array} = \begin{array}{r|l} 26'\text{-}18'' & 6 \\ 16'\text{-}\ 9'' & 3 \\ \hline 10'\text{-}\ 9'' & 3 \\ & \overline{4}\ (\text{LCD}) \end{array}$$

$$10'\text{-}9\frac{3}{4}''\quad Ans$$

Or:

$$\begin{array}{r} 27'\text{-}7\frac{1}{2}'' \\ -16'\text{-}9\frac{3}{4}'' \end{array} = \begin{array}{r} 27' \\ -17' \end{array}\ \begin{array}{l} 7\frac{1}{2}'' \\ +2\frac{1}{4}'' \end{array}\ (12''-9\frac{3}{4}''=2\frac{1}{4}'')$$

$$10'\text{-}9\frac{3}{4}''\quad Ans$$

Multiplication and Division of Fractions

Multiplying, by concellation:

$$10\frac{3}{4} \times 16$$

$$10\frac{3}{4} = \frac{10 \times 4 + 3}{4} = \frac{43}{4}$$

$$\frac{43}{\cancel{4}} \times \frac{\cancel{16}^{\,4}}{1} = 43 \times 4 = 172\quad Ans$$

Multiplying, feet and inches:

$$7'\text{-}9\frac{1}{2}'' \times 9$$

$$7' \times 9 = 63'$$

$$9\frac{1}{2}'' \times 9 = (9'' \times 9) + (\frac{1}{2}'' \times 9)$$
$$= 81'' + 4\frac{1}{2}'' = 85\frac{1}{2}''$$
$$= 7'\text{-}1\frac{1}{2}''$$

$$(63') + (7'\text{-}1\frac{1}{2}'') = 70'\text{-}1\frac{1}{2}''\quad Ans$$

Dividing:

$$\frac{3}{4} \div 12$$

Invert divisor, and multiply:

$$\frac{3}{4} \times \frac{1}{12} = \frac{1}{16}\quad Ans$$

Dividing, feet and inches:

$$9'\text{-}10\frac{1}{2}'' \div 3$$

$$9' \div 3 = 3'$$

$$10\frac{1}{2}'' = \frac{21}{2}$$

$$\frac{21}{2} \div 3 = \frac{21}{2} \times \frac{1}{3} = 3\frac{1}{2}''$$

$$3'\text{-}3\frac{1}{2}''\quad Ans$$

Fractions and Decimals

A common fraction may be converted to a decimal fraction by dividing the numerator by the denominator:

$$\frac{N}{D} = \text{Decimal}$$

For example:

$$\frac{1}{2} = \frac{1.00}{2} = .50$$

$$\frac{3}{4} = \frac{3.00}{4} = .75$$

To convert a decimal to a common fraction of any desired denominator:

$$\frac{N}{D} = \text{Decimal},\quad N = \text{Decimal} \times D$$

For example:

Change .4375 to 16ths.

$$N = .4375 \times 16$$

$$\begin{array}{r} .4375 \\ \times\quad 16 \\ \hline 2\ 6250 \\ 4\ 375\ \ \\ \hline 7.0000 = \frac{7}{16}''\quad Ans \end{array}$$

Example:

Convert $10\frac{3}{4}''$ to a decimal of a foot accurate to (4) places:

$$10\frac{3}{4}'' = 10.75''$$

Divide 10.75 by 12 (number of inches in one foot):

$$12\overline{)10.75000} = .8958'\quad Ans$$
$$.89583$$

Example:

Convert $7\frac{1}{2}''$ to a decimal of a foot:

$$\frac{N}{D} = \text{decimal} \qquad N = 7\frac{1}{2} = 7.5, \text{ or } \frac{15}{2}$$

$$D = 12 \text{ (inches per ft)}$$

$$\frac{15}{2} \div 12 = \frac{15}{2} \times \frac{1}{12} = \frac{5}{8} = .625' \quad Ans$$

or, $12 \overline{)7.500}$
         $.625 \quad Ans$

A decimal of a foot is converted to inches by multiplying the decimal by 12.

Example:

Convert .4375' to inches:

$$.4375 \times 12 = 5.25 = 5\frac{1}{4}'' \quad Ans$$

Table 3-1 is convenient for the conversion of decimals of a foot to inches.

For example:

What is the inches equivalent of .4375 foot? (1) Find .4375 under "Decimals of a Foot." (2) Read up the column from .4375 to get 5". (3) Read along the row from .4375 to the right to get $\frac{1}{4}''$. Thus, .4375 equals $5\frac{1}{4}''$.

A practical example:

The top of a duct 16" deep must be installed 12" below the centerline of a plumbing pipe. The pipe is noted on the design drawing at centerline elevation, 29.875'. What is the dimension of the bottom of the duct to the finished floor if the finished floor is at elevation 14.375'?

    29.875'
  − 14.375'
  ─────────
    15.500' = 15'-6''

(Centerline of pipe above finished floor)

    15'-6''
  − 1'-0''
  ─────────
    14'-6''

(Required height of top of duct to finished floor)

    14'-6''
  − 1'-4'' (16" Duct)
  ─────────
    13'-2'' *Ans*

POWERS, ROOTS, AND EXPONENTS

a to the first power $= a^1 = a \times 1$
a to the second power $= a^2 = a \times a$
a to the third power $= a^3 = a \times a \times a$

The small number at the upper right of a is called an *exponent* which indicates how many times a number is used as a factor. a^2 is read as "a squared," or "a to the second power," and a^3 is read as "a cubed" or "a to the third power."

For example:

$$6^2 = 6 \times 6 = 36$$
$$3^3 = 3 \times 3 \times 3 = 27$$

In an expression ab^2, the exponent "2" applies only to the letter b and would be: $a \times b \times b$.

When the *root* of a number is to be extracted, a radical sign $\sqrt{}$ is used. A number placed at the upper left corner of the radical sign indicates the index of the root. For example, $\sqrt[3]{27}$ means that the number to be found is the cube root of 27. If an index is not shown with a radical, the index (2) is understood.

Extracting a square root is simplified by factoring out exact squares such as:

$$\sqrt{256} = \sqrt{64 \times 4} = 8 \times 2 = 16$$
$$\sqrt{50} = \sqrt{25 \times 2} = 5\sqrt{2}$$
$$\sqrt{48} = \sqrt{16 \times 3} = 4\sqrt{3}$$

To find the exponent of similar numbers in a product, add the exponents of that number in the multiplicand and multiplier.

For example:

$$a^x \times a^y = a^{x+y}$$

let:

$$a = 5, \quad x = 2, \quad y = 3$$
$$a^{x+y} = 5^2 \times 5^3 = 5^5 = 3125$$

To find the exponent of similar numbers in a quotient, subtract the exponent in the divisor from the exponent in the dividend.

For example:

$$a^x \div a^y = a^{x-y}$$

let:

$$a = 15, \quad x = 3, \quad y = 2$$
$$15^3 \div 15^2 = 15^1 = 15$$

Table 3-1. Conversion from Decimals of a Foot to Inches

Inches												Fractions of an Inch	Decimals of an Inch
0	1	2	3	4	5	6	7	8	9	10	11		
Decimals of a Foot													
0	.0833	.1667	.2500	.3333	.4167	.500	.5833	.6667	.750	.8333	.9167	0	0
.0052	.0885	.1719	.2552	.3385	.4219	.5052	.5885	.6719	.7552	.8385	.9219	1/16	.0625
.0104	.0938	.1771	.2604	.3438	.4271	.5104	.5938	.6771	.7604	.8438	.9271	1/8	.1250
.0156	.0990	.1823	.2656	.3490	.4323	.5156	.5990	.6823	.7656	.8490	.9323	3/16	.1875
.0208	.1042	.1875	.2708	.3542	.4375	.5208	.6042	.6875	.7708	.8542	.9375	1/4	.250
.0260	.1094	.1927	.2760	.3594	.4427	.5260	.6094	.6927	.7760	.8594	.9427	5/16	.3125
.0313	.1146	.1979	.2812	.3646	.4479	.5313	.6146	.6979	.7813	.8646	.9479	3/8	.3750
.0365	.1198	.2031	.2865	.3698	.4531	.5365	.6198	.7031	.7865	.8698	.9531	7/16	.4375
.0417	.1250	.2083	.2917	.3750	.4583	.5417	.6250	.7083	.7917	.8750	.9583	1/2	.500
.0469	.1302	.2135	.2969	.3802	.4635	.5469	.6302	.7135	.7969	.8802	.9635	9/16	.5625
.0521	.1354	.2188	.3021	.3854	.4688	.5521	.6354	.7188	.8021	.8854	.9688	5/8	.6250
.0573	.1406	.2240	.3073	.3906	.4740	.5573	.6406	.7240	.8073	.8906	.9740	11/16	.6875
.0625	.1458	.2292	.3125	.3958	.4792	.5625	.6458	.7292	.8125	.8958	.9792	3/4	.750
.0677	.1510	.2344	.3177	.4010	.4844	.5677	.6510	.7344	.8177	.9010	.9844	13/16	.8125
.0729	.1563	.2396	.3229	.4063	.4896	.5729	.6563	.7396	.8229	.9063	.9896	7/8	.8750
.0781	.1615	.2448	.3281	.4115	.4948	.5781	.6615	.7448	.8281	.9115	.9948	15/16	.9375

To multiply unlike numbers with like exponents they must first be converted to whole numbers:

$$a^x \times b^x = (ab)^x$$

let:

$$a = 5, \quad b = 7, \quad x = 2$$

$$5^2 \times 7^2 = 35^2 = 1225$$

To divide unlike numbers with like exponents:

$$a^x \div b^x = \left(\frac{a}{b}\right)^x$$

let:

$$a = 18, \quad b = 6, \quad x = 3$$

$$18^3 \div 6^3 = \left(\frac{18}{6}\right)^3 = 3^3 = 27$$

INSTRUCTIONS FOR OPERATING THE MANNHEIM TYPE SLIDE RULE

With S, K, A, B, C1, C, D, L, and T

The slide rule is a simple and accurate device for solving mathematical problems. It involves multiplication, division, proportion, percentage, squares and square roots, cubes and cube roots, diameters and areas, reciprocals, logarithms and exponents, trigonometric formulae, and all sorts of combinations of these operations.

Description of the Slide Rule

The Mannheim type slide rule consists of three parts: a ruler, a slider, and a runner. The ruler (also called the body or the stock) carries six scales, marked S, K, A, D, L, T. The slider fits into and slides in grooves on the top side of the body. The slider carries the B, C1, and C scales. The runner (also called indicator or cursor) consists of a lens carrying a hairline and set in a frame which slides to the right or left, over the face of the rule.

Problems are worked (namely, the various operations of multiplication, division, taking square roots, and so on) by comparing two of the scales with each other. Since there are nine scales (marked S, K, A, B, C1, C, D, L, T,) it is easily

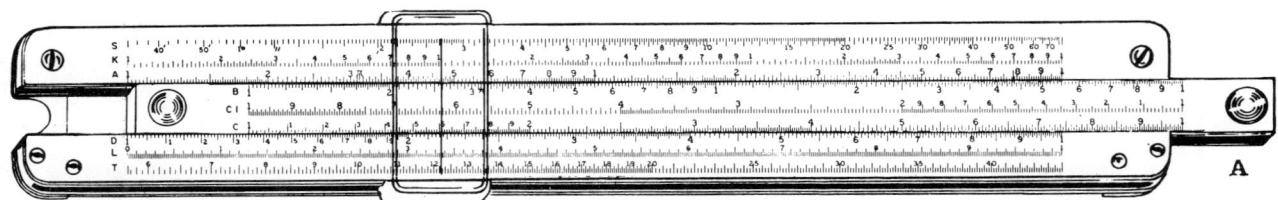

Rule showing the 9 scales, S, K, A, B, CI, C, D, L and T.

seen that there are numerous combinations taking two scales at a time. The manipulation of the slide rule consists in moving the slider along the body and in sliding the cursor to right or left over the face of body and slider.

It is important for the hairline on the cursor to be placed exactly at right angles to the direction in which the scales run. The setting of the hairline may be checked by centering it upon the reading A-1 (at the left or right end of the A scale); if the hairline is properly adjusted, it will also center upon the mark D-1. Any two readings (such as A-1 and D-1) which center upon the hairline when in proper adjustment are said to be **in register.**

The purpose of the cursor hairline is to enable one to read easily the figures on any one scale which lie in register with readings on any other scale.

Location of Numbers on the Scales

Experience proves that the beginner's first step is to learn how to locate numbers on the various scales. We cannot emphasize too strongly the necessity of learning how to read the scales. The various scales are not calibrated uniformly (except the L scale), and the marks on the scales do not measure lengths — they represent numbers. Since the reading of all the scales is done in much the same manner, it will be sufficient to illustrate the procedure with one scale. We use the C (and exactly similar D) scale for our example.

The C and D scales consist of nine **main** divisions, of steadily decreasing lengths as one proceeds to the right. The first line of each of these divisions is numbered: beginning at the left with 1 (called the left index), then 2, 3, 4, 5, 6, 7, 8, 9, and finally, 1 (which stands for 10, and is called the right index).

Dropping down to the center, we see that each of these main divisions is divided into 10 **secondary** divisions. Between the main divisions 1 and 2, the secondary divisions are numbered from 1 to 9 in smaller figures. The secondary divisions between the main divisions 2 and 3, 3 and 4, and so on, are not numbered, but they must be counted as 1, 2, 3, . . . 9.

Dropping down to the last line, we see that the secondary divisions are again subdivided into **tertiary** divisions. The secondary subdivisions between 1 and 2 are each divided into 10 tertiary parts. The shorter secondary divisions between main 2 and 3, and between main 3 and 4, are divided only into 5 tertiary parts. The still shorter secondary divisions between the main divisions beyond 4, are divided into two tertiary parts (of course, to avoid crowding the scale with marks).

To locate a three-digit number, say, on the C or D scale, one proceeds as follows: In the first place, certain technical terms must be explained. The first significant digit of a number is the first digit appearing on the left which is not zero: thus, 1 is the first significant digit in all the numbers, 125,

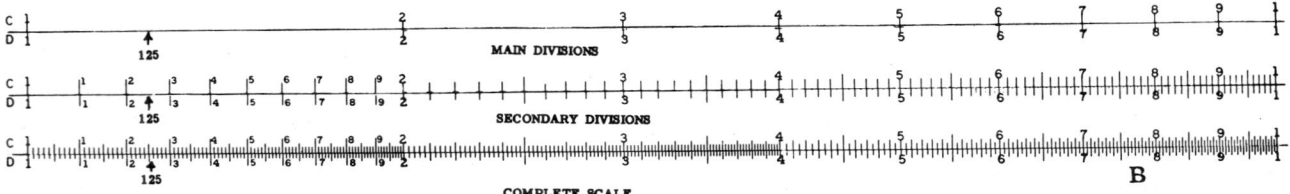

12.5, 1.25, 0.125, and 0.0125, etc. One locates on a slide rule only the sequence of numbers, 1-2-5; the decimal point has nothing to do with locating the number. As an example, let us locate this sequence 1-2-5 on the C scale.

Step 1. The first significant digit (1, for the number 125) locates the number as lying between the **main** divisions 1 and 2.

Step 2. The second significant digit (2, for the number 125) locates the number between the **secondary** divisions 2 and 3 of the **main** division 1 to 2.

Step 3. The third significant digit (5, for the number 125) locates the number as on the fifth of the ten **tertiary** divisions of the secondary range 2 to 3 of the **main** division 1 to 2. Had the number been 1257, we should then have located it 7/10 of the **tertiary** division between 1250 and 1260, this last shift being made by estimating the 7/10 by eye, since there are no fourth-order divisions.

The above procedure is to be followed for any number on the scale, except that it should be noted that the tertiary divisions toward the right end of the scale represent fifths (between main 2-3 and main 3-4) and halves (between main 4 and the right index). For example, the number 463 lies, first, on the main division between 4 and 5; second, on the secondary division between 460 and 470; and third, 3/5 of the first tertiary division between 460 and 465; this 3/5 being estimated by eye.

The decimal point is ignored in operating the slide rule. At the end of a calculation, the decimal point is located by estimating the answer from rounding off the factors and divisors. One quickly learns how to carry out such estimations. There is a method of keeping account of the decimal point in slide rule calculations; this method depends upon the theory of logarithms, but is universally ignored even by those who have a working knowledge of its use.

Description of the Scales

The C and D scales, which are exactly alike, are calibrated in proportion to the logarithms of the actual numbers which are marked on these scales. It may be a satisfaction to the reader to understand the theory of logarithms, but will not be necessary in using the slide rule. The C-D combination is used for multiplying, dividing, and in ratio and proportion.

The C1, or reciprocal, scale is calibrated in the same way as the C and D scales except that it reads from right to left, and is, namely, the C scale reversed in direction. The C1 scale may be used in reading off the reciprocal of a number as well as in multiplication and division.

Scales A and B, which are alike, are logarithmic scales just half as long as the C and D scales. If the left half of the A (and B) scale represents numbers from 1 to 10, the right half represents numbers 10 times as large, namely, from 10 to 100. Again, if the left half represents numbers from 100 to 1,000, the right half represents numbers from 1,000 to 10,000, and so on. These scales may be used in finding squares and square roots.

The K scale is a logarithmic scale 1/3 as long as the C scale. The second third of the scale represents numbers 10 times as large as those of the first third. The right third of the K scale represents numbers 100 times as large as the first third, and 10 times as large as the second third of the scale. Thus, the first, second, and third scales may represent numbers 1 to 10, 10 to 100, and 100 to 1,000, respectively.

The S scale is used in figuring the sines of angles.

The T scale is used in finding the tangents of angles.

The L scale is uniformly calibrated, and is used in combination with the C and D scales in finding the logarithms of numbers.

Slide Rule Operations

Multiplication. Use the C-D combination. To multiply two numbers, locate one factor on the D scale, set the index (either the left or right end) of the C scale in register with this factor on the D scale, locate the other factor on C, then, in register with this reading on C, the product will be found on the D scale.

For example, the left index of C is set on the number 1.35 of the D scale. If the cursor is moved over to 4 on C, then under this number and in register with it, the product of 1.35 × 4, which

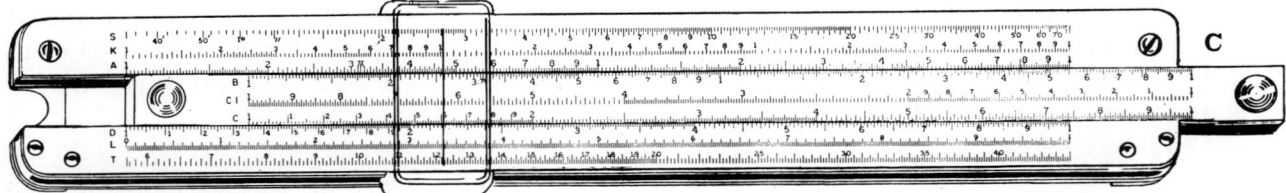

is 5.4, will be found on D. In register with C-6 and on D will be found the product 8.1 (1.35 × 6). How about 135 × 9? A glance will show that C-9 is off the D scale. In this and similar examples, the slider is to be pulled to the left in the ruler until the right index of C is in register with the first factor appearing on D (in the present example 1.35), and then the answer is again to be read on D in register with the second factor as read on C.

Division. Use the C-D combination. Division is the inverse of multiplication. Locate the divisor on the C scale and set this reading in register with the dividend on the D scale: the quotient will be found on the D scale in register with the index of the C scale (with whichever index of C, left or right, appears on the D scale). As an example, note that the divisor C-4 is in register with the dividend 5.4 on D, and that the quotient 1.35 lies on D in register with the left index of C. Note that with this setting of the slide rule, any number on the D scale, divided by the number in register on C, gives 1.35 as quotient.

Squares and Square Roots. Use the A-D combination of scales. To find the square of any number, locate this number on the D scale, set the hairline of the cursor upon this number, and read its square in register on the A scale.

To find the square root of any number, use the scales in reverse order. Thus, locate the given number whose square root is to be found upon the A scale (use the left half of A if this number has an odd number of digits, the right half if the number of digits is even), set the cursor upon this number, and read its square root in register on the D scale.

Cubes and Cube Roots. Use the D-K combination of scales. To find the cube of any number, locate this number on the D scale, set the cursor on this number, and read its cube in register, on the K scale.

To find the cube root of any number, use the D-K combination in the reverse order. Thus, locate the given number, whose cube root is desired, upon the K scale (use the left third of K if the number has 1, 4, 7, etc., digits, namely, if its number of digits is 1 plus a multiple of 3; use the middle third of K if the number of digits is 2 plus a multiple of 3, such as 2, 5, 8, etc., digits; and use the right third of K if the number of digits is an exact multiple of 3, such as 3, 6, 9, etc., digits), set the cursor upon this number and read its cube root in register, on D.

Proportion. Use the C-D combination of scales. Problems of proportion arise, for example, in the conversion of yards to feet, dollars to pounds, gallons to cubic feet, and so on. As an example of such use of the slide rule, let us set the rule for the yard-to-foot conversion. Place the index (either right or left) of the C scale upon the number 3, of the D scale. Note the number of yards and fraction thereof, on C, and read in register, on D, the corresponding number of feet. Conversely, locate the number of feet in any given distance, upon the D scale, and in register with this number, this distance will be found on C, expressed in yards.

Reciprocals. Use the C1-D combination of scales. To find the reciprocal of any number, locate this number on the D scale and set the cursor upon this reading: the reciprocal of this number will be found in register on the C1 scale. Alternatively, locate the number on the C1 scale, set the cursor upon this number, and read its reciprocal in register on the D scale.

Further Uses of the C1 Scale. Multiplication by use of the D-C1 combination: To multiply two numbers together, locate one factor on the D scale, the other on the C1 scale; set these two factors in register by use of the cursor, and read their product on D, in register with the index of the C1 scale. Since either the left or right index of C1 will always be found upon the scale, this method of multiplication never requires the reversal of the slider, which is frequently necessary when multiplying by use of the C-D combination.

Observe that this method of multiplication permits finding the product of three factors with one setting of the slider. As an example, let us calculate the volume of a wall 15.5 feet long, 8 feet high, and 0.55 foot thick. Solution: Set the cursor at 155 on D, draw the slider until 8 on C1 coincides with the hairline, move the cursor to 55, on C, then read the product 68.2 cubic feet, in register, on D.

Division by use of the D-C1 combination: To divide one number by another, locate the dividend on D, set the cursor on this number, draw the slider until its (right or left) index comes into register with the hairline, locate the divisor on C1, and set the cursor on this number; the quotient lies in register on the D scale.

Sine of an Angle. Use the A-S combination of scales. Set the indices of the S scale in exact register with those of the A scale. When using the No. 1200 and No. 1211 pocket rules, locate angle on S scale on rear of slide. Place in register with red indicator line. Then, read sine on the B scale in register with right index of A scale on face of rule.

To find the sine of any angle (greater than 30'), locate this angle on the S scale; its sine will be found in register on the A scale. Note that the left half of the A scale covers the angle range 30' to 5°45', and that the sines of angles in this range have one zero after the decimal point (for example, sin 4°30' is 0.0785); whereas the second half of the A scale covers the angle range, 5°45' to 90°, and that for sines of angles in this range, the significant digits begin immediately after the decimal point (for example, sin 35° is 0.574).

Given the sine of an angle, to find the angle, use the scales A and S in the reverse order. Thus, locate the numerical value of the sine on the A scale, and read off the angle in register on the S scale.

Tangent of an Angle. Use the D-T combination of scales. Set the indices of the T scale in exact register with those of the D scale. When using the No. 1200 and No. 1211 pocket rule, locate angle on T scale on rear of slide. Place in register with red indicator line. Then, read tangent on the C scale in register with right index of the D scale on the face of the rule.

To find the tangent of any angle (in the range from about 6° to 45°), locate the angle on the T scale, its tangent will be found in register on the D scale. Throughout the tangent scale, the significant digits begin immediately after the decimal point (for example, tan 19°20' is 0.351).

To find the tangent of an angle in the range 45° to 90°, find the tangent of 90° minus the given angle, and take the reciprocal of this value as explained in the paragraph on Reciprocals, above.

Given the tangent of an angle, to find the angle: If the given tangent lies in the range 0.1 to 1, locate this number on D, and read the corresponding angle in register, on the T scale: If the given tangent exceeds 1, find the reciprocal of the given number, locate this number on the D scale, read off the angle in register with this number on T, subtract this angle from 90° to find the required angle.

Logarithm of a Number. Use the D-L combination of scales. Set the indices of the L scale in exact register with those of the D scale.

To find the logarithm of a number, locate the significant digits on the D scale, the required logarithm will be found by use of the cursor, in register on the L scale. The characteristic of the logarithm is to be found from the position of the decimal point in the same way one determines it when using a logarithmic table.

Given the logarithm of a number, to find the number: locate the mantissa of the logarithm upon the L scale, and read off the significant digits of the number in register on the D scale. The decimal point for the number is fixed by the characteristic of the given logarithm, in the usual manner.

Extracting a Square Root

The draftsman often requires the square root of a number which does not contain an exact root. The following is an arithmetical method for that purpose:

Find the square root of 5207.0 to the closest hundredth.

54 PRACTICAL MATHEMATICS FOR SHEET METAL DRAFTSMEN Ch. 3

```
               6
       7   2 . 1  ↘
      ┌─────────────────────
      │ 52   07 . 00   00   00
      │ 49
  140 │  3   07
   2  │
  ─── │
  142 │  2   84
      │
 1440 │       23   00
   1  │
  ─── │
 1441 │       14   41
      │
14420 │        8   59   00
   5  │
 ──── │
14425 │        7   21   25
      │            1   37   75
```
Remainder is greater than half of 14425, therefore, (5) in the root is increased to (6)

Procedure:

1. Starting at the decimal point, separate the number both ways into groups of two digits.
2. Find the largest square (49) of the first pair of numbers (52) and subtract for a remainder of 3.
3. Place 7, the square root of 49, above 52, the original pair of numbers.
4. Place the next pair of numbers adjacent to the remainder of 3, to make 307.
5. A trial divisor is obtained by doubling the root found previously 7 to make 14, and adding a zero to make 140.
6. Divide the trial divisor of 140 into the first remainder of 307, this becomes the next number of the root, 2, and is placed above 07, the second pair of numbers.
7. Add this quotient (2) to the trial divisor, to make (142).
8. Multiply the divisor (142) by the new number of the root, 2, to make 284.
9. Subtract this from the first remainder to get 23.
10. Repeat procedures as from step 4, and so on, to obtain the square root of 72.16.

To check the answer, multiply:

```
       72.16
      ×72.16
      ──────
       432 96
       721 6
      14432
      50512
      ─────────
      5207.06 56
```

Example:
Solve $\sqrt{2}$ to three decimal places.

```
         1 . 4   1   4
      ┌─────────────────────
      │ 02 . 00   00   00
      │ 1
  20  │  1   00
   4  │
  ─── │
  24  │       96
      │
 280  │       4   00
   1  │
  ─── │
 281  │       2   81
      │
 2820 │       1   19   00
   4  │
 ──── │
 2824 │       1   12   96
      │            6   04
```
Remainder is less than half of 2824, so the last digit 4 in the root, remains.

To Find a Square Root from Tables (see Table 3-2)

Example: Find $\sqrt{5207}$ to two decimal places.

Check 'square' column (Table 3-2) for numbers closest to 5207.

$73^2 = 5329$
$72^2 = \underline{5184}$ (5207 is between 73^2 and 72^2)
         145

by interpolation:

```
       5207
      −5184
      ─────
         23
```

by dividing:

```
              .158 = .16
       145)23.000
           14 5
           ────
            8 50
            7 25
            ────
            1 250
            1 160
            ─────
               90
```

by adding:

$.16 + 72 = 72.16$

LOGARITHMS

Logarithms are exponents. Logarithms of numbers to the base 10 are *common logs.* The integral part of a log to the left of the decimal is the *characteristic,* and may be zero, positive, or negative.

(Text continued on page 65.)

Table 3-2. Powers, Roots, and Reciprocals of Numbers 1 to 500*

No.	Square	Cube	Sq. Root	Cube Root	Reciprocal	No.
1	1	1	1.00000	1.00000	1.0000000	1
2	4	8	1.41421	1.25992	0.5000000	2
3	9	27	1.73205	1.44225	0.3333333	3
4	16	64	2.00000	1.58740	0.2500000	4
5	25	125	2.23607	1.70998	0.2000000	5
6	36	216	2.44949	1.81712	0.1666667	6
7	49	343	2.64575	1.91293	0.1428571	7
8	64	512	2.82843	2.00000	0.1250000	8
9	81	729	3.00000	2.08008	0.1111111	9
10	100	1,000	3.16228	2.15443	0.1000000	10
11	121	1,331	3.31662	2.22398	0.0909091	11
12	144	1,728	3.46410	2.28943	0.0833333	12
13	169	2,197	3.60555	2.35133	0.0769231	13
14	196	2,744	3.74166	2.41014	0.0714286	14
15	225	3,375	3.87298	2.46621	0.0666667	15
16	256	4,096	4.00000	2.51984	0.0625000	16
17	289	4,913	4.12311	2.57128	0.0588235	17
18	324	5,832	4.24264	2.62074	0.0555556	18
19	361	6,859	4.35890	2.66840	0.0526316	19
20	400	8,000	4.47214	2.71442	0.0500000	20
21	441	9,261	4.58258	2.75892	0.0476190	21
22	484	10,648	4.69042	2.80204	0.0454545	22
23	529	12,167	4.79583	2.84387	0.0434783	23
24	576	13,824	4.89898	2.88450	0.0416667	24
25	625	15,625	5.00000	2.92402	0.0400000	25
26	676	17,576	5.09902	2.96250	0.0384615	26
27	729	19,683	5.19615	3.00000	0.0370370	27
28	784	21,952	5.29150	3.03659	0.0357143	28
29	841	24,389	5.38516	3.07232	0.0344828	29
30	900	27,000	5.47723	3.10723	0.0333333	30
31	961	29,791	5.56776	3.14138	0.0322581	31
32	1,024	32,768	5.65685	3.17480	0.0312500	32
33	1,089	35,937	5.74456	3.20753	0.0303030	33
34	1,156	39,304	5.83095	3.23961	0.0294118	34
35	1,225	42,875	5.91608	3.27107	0.0285714	35
36	1,296	46,656	6.00000	3.30193	0.0277778	36
37	1,369	50,653	6.08276	3.33222	0.0270270	37
38	1,444	54,872	6.16441	3.36198	0.0263158	38
39	1,521	59,319	6.24500	3.39121	0.0256410	39
40	1,600	64,000	6.32456	3.41995	0.0250000	40
41	1,681	68,921	6.40312	3.44822	0.0243902	41
42	1,764	74,088	6.48074	3.47603	0.0238095	42
43	1,849	79,507	6.55744	3.50340	0.0232558	43
44	1,936	85,184	6.63325	3.53035	0.0227273	44
45	2,025	91,125	6.70820	3.55689	0.0222222	45
46	2,116	97,336	6.78233	3.58305	0.0217391	46
47	2,209	103,823	6.85565	3.60883	0.0212766	47
48	2,304	110,592	6.92820	3.63424	0.0208333	48
49	2,401	117,649	7.00000	3.65931	0.0204082	49
50	2,500	125,000	7.07107	3.68403	0.0200000	50

* Reproduced from *Handbook of Air Conditioning, Heating and Ventilating*. Edited by Clifford Strock and Richard L. Koral. New York: Industrial Press Inc., 1965.

Table 3-2 (Continued). Powers, Roots, and Reciprocals of Numbers 1 to 500

No.	Square	Cube	Sq. Root	Cube Root	Reciprocal	No.
51	2,601	132,651	7.14143	3.70843	0.0196078	51
52	2,704	140,608	7.21110	3.73251	0.0192308	52
53	2,809	148,877	7.28011	3.75629	0.0188679	53
54	2,916	157,464	7.34847	3.77976	0.0185185	54
55	3,025	166,375	7.41620	3.80295	0.0181818	55
56	3,136	175,616	7.48331	3.82586	0.0178571	56
57	3,249	185,193	7.54983	3.84850	0.0175439	57
58	3,364	195,112	7.61577	3.87088	0.0172414	58
59	3,481	205,379	7.68115	3.89300	0.0169492	59
60	3,600	216,000	7.74597	3.91487	0.0166667	60
61	3,721	226,981	7.81025	3.93650	0.0163934	61
62	3,844	238,328	7.87401	3.95789	0.0161290	62
63	3,969	250,047	7.93725	3.97906	0.0158730	63
64	4,096	262,144	8.00000	4.00000	0.0156250	64
65	4,225	274,625	8.06226	4.02073	0.0153846	65
66	4,356	287,496	8.12404	4.04124	0.0151515	66
67	4,489	300,763	8.18535	4.06155	0.0149254	67
68	4,624	314,432	8.24621	4.08166	0.0147059	68
69	4,761	328,509	8.30662	4.10157	0.0144928	69
70	4,900	343,000	8.36660	4.12129	0.0142857	70
71	5,041	357,911	8.42615	4.14082	0.0140845	71
72	5,184	373,248	8.48528	4.16017	0.0138889	72
73	5,329	389,017	8.54400	4.17934	0.0136986	73
74	5,476	405,224	8.60233	4.19834	0.0135135	74
75	5,625	421,875	8.66025	4.21716	0.0133333	75
76	5,776	438,976	8.71780	4.23582	0.0131579	76
77	5,929	456,533	8.77496	4.25432	0.0129870	77
78	6,084	474,552	8.83176	4.27266	0.0128205	78
79	6,241	493,039	8.88819	4.29084	0.0126582	79
80	6,400	512,000	8.94427	4.30887	0.0125000	80
81	6,561	531,441	9.00000	4.32675	0.0123457	81
82	6,724	551,368	9.05539	4.34448	0.0121951	82
83	6,889	571,787	9.11043	4.36207	0.0120482	83
84	7,056	592,704	9.16515	4.37952	0.0119048	84
85	7,225	614,125	9.21954	4.39683	0.0117647	85
86	7,396	636,056	9.27362	4.41400	0.0116279	86
87	7,569	658,503	9.32738	4.43105	0.0114943	87
88	7,744	681,472	9.38083	4.44797	0.0113636	88
89	7,921	704,969	9.43398	4.46475	0.0112360	89
90	8,100	729,000	9.48683	4.48140	0.0111111	90
91	8,281	753,571	9.53939	4.49794	0.0109890	91
92	8,464	778,688	9.59166	4.51436	0.0108696	92
93	8,649	804,357	9.64365	4.53065	0.0107527	93
94	8,836	830,584	9.69536	4.54684	0.0106383	94
95	9,025	857,375	9.74679	4.56290	0.0105263	95
96	9,216	884,736	9.79796	4.57886	0.0104167	96
97	9,409	912,673	9.84886	4.59470	0.0103093	97
98	9,604	941,192	9.89949	4.61044	0.0102041	98
99	9,801	970,299	9.94987	4.62607	0.0101010	99
100	10,000	1,000,000	10.00000	4.64159	0.0100000	100

Table 3-2 (Continued). Powers, Roots, and Reciprocals of Numbers 1 to 500

No.	Square	Cube	Sq. Root	Cube Root	Reciprocal	No.
101	10,201	1,030,301	10.0499	4.65701	0.0099010	101
102	10,404	1,061,208	10.0995	4.67233	0.0098039	102
103	10,609	1,092,727	10.1489	4.68755	0.0097087	103
104	10,816	1,124,864	10.1980	4.70267	0.0096154	104
105	11,025	1,157,625	10.2470	4.71769	0.0095238	105
106	11,236	1,191,016	10.2956	4.73262	0.0094340	106
107	11,449	1,225,043	10.3441	4.74746	0.0093458	107
108	11,664	1,259,712	10.3923	4.76220	0.0092593	108
109	11,881	1,295,029	10.4403	4.77686	0.0091743	109
110	12,100	1,331,000	10.4881	4.79142	0.0090909	110
111	12,321	1,367,631	10.5357	4.80590	0.0090090	111
112	12,544	1,404,928	10.5830	4.82028	0.0089286	112
113	12,769	1,442,897	10.6301	4.83459	0.0088496	113
114	12,996	1,481,544	10.6771	4.84881	0.0087719	114
115	13,225	1,520,875	10.7238	4.86294	0.0086957	115
116	13,456	1,560,896	10.7703	4.87700	0.0086207	116
117	13,689	1,601,613	10.8167	4.89097	0.0085470	117
118	13,924	1,643,032	10.8628	4.90487	0.0084746	118
119	14,161	1,685,159	10.9087	4.91868	0.0084034	119
120	14,400	1,728,000	10.9545	4.93242	0.0083333	120
121	14,641	1,771,561	11.0000	4.94609	0.0082645	121
122	14,884	1,815,848	11.0454	4.95968	0.0081967	122
123	15,129	1,860,867	11.0905	4.97319	0.0081301	123
124	15,376	1,906,624	11.1355	4.98663	0.0080645	124
125	15,625	1,953,125	11.1803	5.00000	0.0080000	125
126	15,876	2,000,376	11.2250	5.01330	0.0079365	126
127	16,129	2,048,383	11.2694	5.02653	0.0078740	127
128	16,384	2,097,152	11.3137	5.03968	0.0078125	128
129	16,641	2,146,689	11.3578	5.05277	0.0077519	129
130	16,900	2,197,000	11.4018	5.06580	0.0076923	130
131	17,161	2,248,091	11.4455	5.07875	0.0076336	131
132	17,424	2,299,968	11.4891	5.09164	0.0075758	132
133	17,689	2,352,637	11.5326	5.10447	0.0075188	133
134	17,956	2,406,104	11.5758	5.11723	0.0074627	134
135	18,225	2,460,375	11.6190	5.12993	0.0074074	135
136	18,496	2,515,456	11.6619	5.14256	0.0073529	136
137	18,769	2,571,353	11.7047	5.15514	0.0072993	137
138	19,044	2,628,072	11.7473	5.16765	0.0072464	138
139	19,321	2,685,619	11.7898	5.18010	0.0071942	139
140	19,600	2,744,000	11.8322	5.19249	0.0071429	140
141	19,881	2,803,221	11.8743	5.20483	0.0070922	141
142	20,164	2,863,288	11.9164	5.21710	0.0070423	142
143	20,449	2,924,207	11.9583	5.22932	0.0069930	143
144	20,736	2,985,984	12.0000	5.24148	0.0069444	144
145	21,025	3,048,625	12.0416	5.25359	0.0068966	145
146	21,316	3,112,136	12.0830	5.26564	0.0068493	146
147	21,609	3,176,523	12.1244	5.27763	0.0068027	147
148	21,904	3,241,792	12.1655	5.28957	0.0067568	148
149	22,201	3,307,949	12.2066	5.30146	0.0067114	149
150	22,500	3,375,000	12.2474	5.31329	0.0066667	150

Table 3-2 (Continued). Powers, Roots, and Reciprocals of Numbers 1 to 500

No.	Square	Cube	Sq. Root	Cube Root	Reciprocal	No.
151	22,801	3,442,951	12.2882	5.32507	0.0066225	151
152	23,104	3,511,808	12.3288	5.33680	0.0065789	152
153	23,409	3,581,577	12.3693	5.34848	0.0065359	153
154	23,716	3,652,264	12.4097	5.36011	0.0064935	154
155	24,025	3,723,875	12.4499	5.37169	0.0064516	155
156	24,336	3,796,416	12.4900	5.38321	0.0064103	156
157	24,649	3,869,893	12.5300	5.39469	0.0063694	157
158	24,964	3,944,312	12.5698	5.40612	0.0063291	158
159	25,281	4,019,679	12.6095	5.41750	0.0062893	159
160	25,600	4,096,000	12.6491	5.42884	0.0062500	160
161	25,921	4,173,281	12.6886	5.44012	0.0062112	161
162	26,244	4,251,528	12.7279	5.45136	0.0061728	162
163	26,569	4,330,747	12.7671	5.46256	0.0061350	163
164	26,896	4,410,944	12.8062	5.47370	0.0060976	164
165	27,225	4,492,125	12.8452	5.48481	0.0060606	165
166	27,556	4,574,296	12.8841	5.49586	0.0060241	166
167	27,889	4,657,463	12.9228	5.50688	0.0059880	167
168	28,224	4,741,632	12.9615	5.51785	0.0059524	168
169	28,561	4,826,809	13.0000	5.52877	0.0059172	169
170	28,900	4,913,000	13.0384	5.53966	0.0058823	170
171	29,241	5,000,211	13.0767	5.55050	0.0058480	171
172	29,584	5,088,448	13.1149	5.56130	0.0058140	172
173	29,929	5,177,717	13.1529	5.57205	0.0057803	173
174	30,276	5,268,024	13.1909	5.58277	0.0057471	174
175	30,625	5,359,375	13.2288	5.59344	0.0057143	175
176	30,976	5,451,776	13.2665	5.60408	0.0056818	176
177	31,329	5,545,233	13.3041	5.61467	0.0056497	177
178	31,684	5,639,752	13.3417	5.62523	0.0056180	178
179	32,041	5,735,339	13.3791	5.63574	0.0055866	179
180	32,400	5,832,000	13.4164	5.64622	0.0055556	180
181	32,761	5,929,741	13.4536	5.65665	0.0055249	181
182	33,124	6,028,568	13.4907	5.66705	0.0054945	182
183	33,489	6,128,487	13.5277	5.67741	0.0054645	183
184	33,856	6,229,504	13.5647	5.68773	0.0054348	184
185	34,225	6,331,625	13.6015	5.69802	0.0054054	185
186	34,596	6,434,856	13.6382	5.70827	0.0053763	186
187	34,969	6,539,203	13.6748	5.71848	0.0053476	187
188	35,344	6,644,672	13.7113	5.72865	0.0053191	188
189	35,721	6,751,269	13.7477	5.73879	0.0052910	189
190	36,100	6,859,000	13.7840	5.74890	0.0052632	190
191	36,481	6,967,871	13.8203	5.75897	0.0052356	191
192	36,864	7,077,888	13.8564	5.76900	0.0052083	192
193	37,249	7,189,057	13.8924	5.77900	0.0051813	193
194	37,636	7,301,384	13.9284	5.78896	0.0051546	194
195	38,025	7,414,875	13.9642	5.79889	0.0051282	195
196	38,416	7,529,536	14.0000	5.80879	0.0051020	196
197	38,809	7,645,373	14.0357	5.81865	0.0050761	197
198	39,204	7,762,392	14.0712	5.82849	0.0050505	198
199	39,601	7,880,599	14.1067	5.83827	0.0050251	199
200	40,000	8,000,000	14.1421	5.84804	0.0050000	200

Table 3-2 (Continued). Powers, Roots, and Reciprocals of Numbers 1 to 500

No.	Square	Cube	Sq. Root	Cube Root	Reciprocal	No.
201	40,401	8,120,601	14.1774	5.85777	0.0049751	201
202	40,804	8,242,408	14.2127	5.86747	0.0049505	202
203	41,209	8,365,427	14.2478	5.87713	0.0049261	203
204	41,616	8,489,664	14.2829	5.88677	0.0049020	204
205	42,025	8,615,125	14.3178	5.89637	0.0048780	205
206	42,436	8,741,816	14.3527	5.90594	0.0048544	206
207	42,849	8,869,743	14.3875	5.91548	0.0048309	207
208	43,264	8,998,912	14.4222	5.92499	0.0048077	208
209	43,681	9,129,329	14.4568	5.93447	0.0047847	209
210	44,100	9,261,000	14.4914	5.94392	0.0047619	210
211	44,521	9,393,931	14.5258	5.95334	0.0047393	211
212	44,944	9,528,128	14.5602	5.96273	0.0047170	212
213	45,369	9,663,597	14.5945	5.97209	0.0046948	213
214	45,796	9,800,344	14.6287	5.98142	0.0046729	214
215	46,225	9,938,375	14.6629	5.99073	0.0046512	215
216	46,656	10,077,696	14.6969	6.00000	0.0046296	216
217	47,089	10,218,313	14.7309	6.00925	0.0046083	217
218	47,524	10,360,232	14.7648	6.01846	0.0045872	218
219	47,961	10,503,459	14.7986	6.02765	0.0045662	219
220	48,400	10,648,000	14.8324	6.03681	0.0045455	220
221	48,841	10,793,861	14.8661	6.04594	0.0045249	221
222	49,284	10,941,048	14.8997	6.05505	0.0045045	222
223	49,729	11,089,567	14.9332	6.06413	0.0044843	223
224	50,176	11,239,424	14.9666	6.07318	0.0044643	224
225	50,625	11,390,625	15.0000	6.08220	0.0044444	225
226	51,076	11,543,176	15.0333	6.09120	0.0044248	226
227	51,529	11,697,083	15.0665	6.10017	0.0044053	227
228	51,984	11,852,352	15.0997	6.10911	0.0043860	228
229	52,441	12,008,989	15.1327	6.11803	0.0043668	229
230	52,900	12,167,000	15.1658	6.12693	0.0043478	230
231	53,361	12,326,391	15.1987	6.13579	0.0043290	231
232	53,824	12,487,168	15.2315	6.14463	0.0043103	232
233	54,289	12,649,337	15.2643	6.15345	0.0042918	233
234	54,756	12,812,904	15.2971	6.16224	0.0042735	234
235	55,225	12,977,875	15.3297	6.17101	0.0042553	235
236	55,696	13,144,256	15.3623	6.17975	0.0042373	236
237	56,169	13,312,053	15.3948	6.18846	0.0042194	237
238	56,644	13,481,272	15.4272	6.19715	0.0042017	238
239	57,121	13,651,919	15.4596	6.20582	0.0041841	239
240	57,600	13,824,000	15.4919	6.21447	0.0041667	240
241	58,081	13,997,521	15.5242	6.22308	0.0041494	241
242	58,564	14,172,488	15.5563	6.23168	0.0041322	242
243	59,049	14,348,907	15.5885	6.24025	0.0041152	243
244	59,536	14,526,784	15.6205	6.24880	0.0040984	244
245	60,025	14,706,125	15.6525	6.25732	0.0040816	245
246	60,516	14,886,936	15.6844	6.26583	0.0040650	246
247	61,009	15,069,223	15.7162	6.27431	0.0040486	247
248	61,504	15,252,992	15.7480	6.28276	0.0040323	248
249	62,001	15,438,249	15.7797	6.29119	0.0040161	249
250	62,500	15,625,000	15.8114	6.29961	0.0040000	250

Table 3-2 (*Continued*).   Powers, Roots, and Reciprocals of Numbers 1 to 500

No.	Square	Cube	Sq. Root	Cube Root	Reciprocal	No.
251	63,001	15,813,251	15.8430	6.30799	0.0039841	251
252	63,504	16,003,008	15.8745	6.31636	0.0039683	252
253	64,009	16,194,277	15.9060	6.32470	0.0039526	253
254	64,516	16,387,064	15.9374	6.33303	0.0039370	254
255	65,025	16,581,375	15.9687	6.34133	0.0039216	255
256	65,536	16,777,216	16.0000	6.34960	0.0039063	256
257	66,049	16,974,593	16.0312	6.35786	0.0038911	257
258	66,564	17,173,512	16.0624	6.36610	0.0038760	258
259	67,081	17,373,979	16.0935	6.37431	0.0038610	259
260	67,600	17,576,000	16.1245	6.38250	0.0038462	260
261	68,121	17,779,581	16.1555	6.39068	0.0038314	261
262	68,644	17,984,728	16.1864	6.39883	0.0038168	262
263	69,169	18,191,447	16.2173	6.40696	0.0038023	263
264	69,696	18,399,744	16.2481	6.41507	0.0037879	264
265	70,225	18,609,625	16.2788	6.42316	0.0037736	265
266	70,756	18,821,096	16.3095	6.43123	0.0037594	266
267	71,289	19,034,163	16.3401	6.43928	0.0037453	267
268	71,824	19,248,832	16.3707	6.44731	0.0037313	268
269	72,361	19,465,109	16.4012	6.45531	0.0037175	269
270	72,900	19,683,000	16.4317	6.46330	0.0037037	270
271	73,441	19,902,511	16.4621	6.47127	0.0036900	271
272	73,984	20,123,648	16.4924	6.47922	0.0036765	272
273	74,529	20,346,417	16.5227	6.48715	0.0036630	273
274	75,076	20,570,824	16.5529	6.49507	0.0036496	274
275	75,625	20,796,875	16.5831	6.50296	0.0036364	275
276	76,176	21,024,576	16.6132	6.51083	0.0036232	276
277	76,729	21,253,933	16.6433	6.51868	0.0036101	277
278	77,284	21,484,952	16.6733	6.52652	0.0035971	278
279	77,841	21,717,639	16.7033	6.53434	0.0035842	279
280	78,400	21,952,000	16.7332	6.54213	0.0035714	280
281	78,961	22,188,041	16.7631	6.54991	0.0035587	281
282	79,524	22,425,768	16.7929	6.55767	0.0035461	282
283	80,089	22,665,187	16.8226	6.56541	0.0035336	283
284	80,656	22,906,304	16.8523	6.57314	0.0035211	284
285	81,225	23,149,125	16.8819	6.58084	0.0035088	285
286	81,796	23,393,656	16.9115	6.58853	0.0034965	286
287	82,369	23,639,903	16.9411	6.59620	0.0034843	287
288	82,944	23,887,872	16.9706	6.60385	0.0034722	288
289	83,521	24,137,569	17.0000	6.61149	0.0034602	289
290	84,100	24,389,000	17.0294	6.61911	0.0034483	290
291	84,681	24,642,171	17.0587	6.62671	0.0034364	291
292	85,264	24,897,088	17.0880	6.63429	0.0034247	292
293	85,849	25,153,757	17.1172	6.64185	0.0034130	293
294	86,436	25,412,184	17.1464	6.64940	0.0034014	294
295	87,025	25,672,375	17.1756	6.65693	0.0033898	295
296	87,616	25,934,336	17.2047	6.66444	0.0033784	296
297	88,209	26,198,073	17.2337	6.67194	0.0033670	297
298	88,804	26,463,592	17.2627	6.67942	0.0033557	298
299	89,401	26,730,899	17.2916	6.68688	0.0033445	299
300	90,000	27,000,000	17.3205	6.69433	0.0033333	300

Table 3-2 (Continued). Powers, Roots, and Reciprocals of Numbers 1 to 500

No.	Square	Cube	Sq. Root	Cube Root	Reciprocal	No.
301	90,601	27,270,901	17.3494	6.70176	0.0033223	301
302	91,204	27,543,608	17.3781	6.70917	0.0033113	302
303	91,809	27,818,127	17.4069	6.71657	0.0033003	303
304	92,416	28,094,464	17.4356	6.72395	0.0032895	304
305	93,025	28,372,625	17.4642	6.73132	0.0032787	305
306	93,636	28,652,616	17.4929	6.73866	0.0032680	306
307	94,249	28,934,443	17.5214	6.74600	0.0032573	307
308	94,864	29,218,112	17.5499	6.75331	0.0032468	308
309	95,481	29,503,629	17.5784	6.76061	0.0032362	309
310	96,100	29,791,000	17.6068	6.76790	0.0032258	310
311	96,721	30,080,231	17.6352	6.77517	0.0032154	311
312	97,344	30,371,328	17.6635	6.78242	0.0032051	312
313	97,969	30,664,297	17.6918	6.78966	0.0031949	313
314	98,596	30,959,144	17.7200	6.79688	0.0031847	314
315	99,225	31,255,875	17.7482	6.80409	0.0031746	315
316	99,856	31,554,496	17.7764	6.81128	0.0031646	316
317	100,489	31,855,013	17.8045	6.81846	0.0031546	317
318	101,124	32,157,432	17.8326	6.82562	0.0031447	318
319	101,761	32,461,759	17.8606	6.83277	0.0031348	319
320	102,400	32,768,000	17.8885	6.83990	0.0031250	320
321	103,041	33,076,161	17.9165	6.84702	0.0031153	321
322	103,684	33,386,248	17.9444	6.85412	0.0031056	322
323	104,329	33,698,267	17.9722	6.86121	0.0030960	323
324	104,976	34,012,224	18.0000	6.86829	0.0030864	324
325	105,625	34,328,125	18.0278	6.87534	0.0030769	325
326	106,276	34,645,976	18.0555	6.88239	0.0030675	326
327	106,929	34,965,783	18.0831	6.88942	0.0030581	327
328	107,584	35,287,552	18.1108	6.89643	0.0030488	328
329	108,241	35,611,289	18.1384	6.90344	0.0030395	329
330	108,900	35,937,000	18.1659	6.91042	0.0030303	330
331	109,561	36,264,691	18.1934	6.91740	0.0030211	331
332	110,224	36,594,368	18.2209	6.92436	0.0030120	332
333	110,889	36,926,037	18.2483	6.93131	0.0030030	333
334	111,556	37,259,704	18.2757	6.93823	0.0029940	334
335	112,225	37,595,375	18.3030	6.94515	0.0029851	335
336	112,896	37,933,056	18.3303	6.95205	0.0029762	336
337	113,569	38,272,753	18.3576	6.95894	0.0029674	337
338	114,244	38,614,472	18.3848	6.96582	0.0029586	338
339	114,921	38,958,219	18.4120	6.97268	0.0029499	339
340	115,600	39,304,000	18.4391	6.97953	0.0029412	340
341	116,281	39,651,821	18.4662	6.98637	0.0029326	341
342	116,964	40,001,688	18.4932	6.99319	0.0029240	342
343	117,649	40,353,607	18.5203	7.00000	0.0029155	343
344	118,336	40,707,584	18.5472	7.00680	0.0029070	344
345	119,025	41,063,625	18.5742	7.01358	0.0028986	345
346	119,716	41,421,736	18.6011	7.02035	0.0028902	346
347	120,409	41,781,923	18.6279	7.02711	0.0028818	347
348	121,104	42,144,192	18.6548	7.03385	0.0028736	348
349	121,801	42,508,549	18.6815	7.04059	0.0028653	349
350	122,500	42,875,000	18.7083	7.04730	0.0028571	350

Table 3-2 (Continued). Powers, Roots, and Reciprocals of Numbers 1 to 500

No.	Square	Cube	Sq. Root	Cube Root	Reciprocal	No.
351	123,201	43,243,551	18.7350	7.05400	0.0028490	351
352	123,904	43,614,208	18.7617	7.06070	0.0028409	352
353	124,609	43,986,977	18.7883	7.06738	0.0028329	353
354	125,316	44,361,864	18.8149	7.07404	0.0028249	354
355	126,025	44,738,875	18.8414	7.08070	0.0028169	355
356	126,736	45,118,016	18.8680	7.08734	0.0028090	356
357	127,449	45,499,293	18.8944	7.09397	0.0028011	357
358	128,164	45,882,712	18.9209	7.10059	0.0027933	358
359	128,881	46,268,279	18.9473	7.10719	0.0027855	359
360	129,600	46,656,000	18.9737	7.11379	0.0027778	360
361	130,321	47,045,881	19.0000	7.12037	0.0027701	361
362	131,044	47,437,928	19.0263	7.12694	0.0027624	362
363	131,769	47,832,147	19.0526	7.13349	0.0027548	363
364	132,496	48,228,544	19.0788	7.14004	0.0027473	364
365	133,225	48,627,125	19.1050	7.14657	0.0027397	365
366	133,956	49,027,896	19.1311	7.15309	0.0027322	366
367	134,689	49,430,863	19.1572	7.15960	0.0027248	367
368	135,424	49,836,032	19.1833	7.16610	0.0027174	368
369	136,161	50,243,409	19.2094	7.17258	0.0027100	369
370	136,900	50,653,000	19.2354	7.17905	0.0027027	370
371	137,641	51,064,811	19.2614	7.18552	0.0026954	371
372	138,384	51,478,848	19.2873	7.19197	0.0026882	372
373	139,129	51,895,117	19.3132	7.19841	0.0026810	373
374	139,876	52,313,624	19.3391	7.20483	0.0026738	374
375	140,625	52,734,375	19.3649	7.21125	0.0026667	375
376	141,376	53,157,376	19.3907	7.21765	0.0026596	376
377	142,129	53,582,633	19.4165	7.22405	0.0026525	377
378	142,884	54,010,152	19.4422	7.23043	0.0026455	378
379	143,641	54,439,939	19.4679	7.23680	0.0026385	379
380	144,400	54,872,000	19.4936	7.24316	0.0026316	380
381	145,161	55,306,341	19.5192	7.24950	0.0026247	381
382	145,924	55,742,968	19.5448	7.25584	0.0026178	382
383	146,689	56,181,887	19.5704	7.26217	0.0026110	383
384	147,456	56,623,104	19.5959	7.26848	0.0026042	384
385	148,225	57,066,625	19.6214	7.27479	0.0025974	385
386	148,996	57,512,456	19.6469	7.28108	0.0025907	386
387	149,769	57,960,603	19.6723	7.28736	0.0025840	387
388	150,544	58,411,072	19.6977	7.29363	0.0025773	388
389	151,321	58,863,869	19.7231	7.29989	0.0025707	389
390	152,100	59,319,000	19.7484	7.30614	0.0025641	390
391	152,881	59,776,471	19.7737	7.31238	0.0025575	391
392	153,664	60,236,288	19.7990	7.31861	0.0025510	392
393	154,449	60,698,457	19.8242	7.32483	0.0025445	393
394	155,236	61,162,984	19.8494	7.33104	0.0025381	394
395	156,025	61,629,875	19.8746	7.33723	0.0025316	395
396	156,816	62,099,136	19.8997	7.34342	0.0025253	396
397	157,609	62,570,773	19.9249	7.34960	0.0025189	397
398	158,404	63,044,792	19.9499	7.35576	0.0025126	398
399	159,201	63,521,199	19.9750	7.36192	0.0025063	399
400	160,000	64,000,000	20.0000	7.36806	0.0025000	400

Table 3-2 (Continued). Powers, Roots, and Reciprocals of Numbers 1 to 500

No.	Square	Cube	Sq. Root	Cube Root	Reciprocal	No.
401	160,801	64,481,201	20.0250	7.37420	0.0024938	401
402	161,604	64,964,808	20.0499	7.38032	0.0024876	402
403	162,409	65,450,827	20.0749	7.38644	0.0024814	403
404	163,216	65,939,264	20.0998	7.39254	0.0024752	404
405	164,025	66,430,125	20.1246	7.39864	0.0024691	405
406	164,836	66,923,416	20.1494	7.40472	0.0024631	406
407	165,649	67,419,143	20.1742	7.41080	0.0024570	407
408	166,464	67,917,312	20.1990	7.41686	0.0024510	408
409	167,281	68,417,929	20.2237	7.42291	0.0024450	409
410	168,100	68,921,000	20.2485	7.42896	0.0024390	410
411	168,921	69,426,531	20.2731	7.43499	0.0024331	411
412	169,744	69,934,528	20.2978	7.44102	0.0024272	412
413	170,569	70,444,997	20.3224	7.44703	0.0024213	413
414	171,396	70,957,944	20.3470	7.45304	0.0024155	414
415	172,225	71,473,375	20.3715	7.45904	0.0024096	415
416	173,056	71,991,296	20.3961	7.46502	0.0024038	416
417	173,889	72,511,713	20.4206	7.47100	0.0023981	417
418	174,724	73,034,632	20.4450	7.47697	0.0023923	418
419	175,561	73,560,059	20.4695	7.48292	0.0023866	419
420	176,400	74,088,000	20.4939	7.48887	0.0023810	420
421	177,241	74,618,461	20.5183	7.49481	0.0023753	421
422	178,084	75,151,448	20.5426	7.50074	0.0023697	422
423	178,929	75,686,967	20.5670	7.50666	0.0023641	423
424	179,776	76,225,024	20.5913	7.51257	0.0023585	424
425	180,625	76,765,625	20.6155	7.51847	0.0023529	425
426	181,476	77,308,776	20.6398	7.52437	0.0023474	426
427	182,329	77,854,483	20.6640	7.53025	0.0023419	427
428	183,184	78,402,752	20.6882	7.53612	0.0023364	428
429	184,041	78,953,589	20.7123	7.54199	0.0023310	429
430	184,900	79,507,000	20.7364	7.54784	0.0023256	430
431	185,761	80,062,991	20.7605	7.55369	0.0023202	431
432	186,624	80,621,568	20.7846	7.55953	0.0023148	432
433	187,489	81,182,737	20.8087	7.56535	0.0023095	433
434	188,356	81,746,504	20.8327	7.57117	0.0023041	434
435	189,225	82,312,875	20.8567	7.57698	0.0022989	435
436	190,096	82,881,856	20.8806	7.58279	0.0022936	436
437	190,969	83,453,453	20.9045	7.58858	0.0022883	437
438	191,844	84,027,672	20.9284	7.59436	0.0022831	438
439	192,721	84,604,519	20.9523	7.60014	0.0022779	439
440	193,600	85,184,000	20.9762	7.60590	0.0022727	440
441	194,481	85,766,121	21.0000	7.61166	0.0022676	441
442	195,364	86,350,888	21.0238	7.61741	0.0022624	442
443	196,249	86,938,307	21.0476	7.62315	0.0022573	443
444	197,136	87,528,384	21.0713	7.62888	0.0022523	444
445	198,025	88,121,125	21.0950	7.63461	0.0022472	445
446	198,916	88,716,536	21.1187	7.64032	0.0022422	446
447	199,809	89,314,623	21.1424	7.64603	0.0022371	447
448	200,704	89,915,392	21.1660	7.65172	0.0022321	448
449	201,601	90,518,849	21.1896	7.65741	0.0022272	449
450	202,500	91,125,000	21.2132	7.66309	0.0022222	450

Table 3-2 (Concluded). Powers, Roots, and Reciprocals of Numbers 1 to 500

No.	Square	Cube	Sq. Root	Cube Root	Reciprocal	No.
451	203,401	91,733,851	21.2368	7.66877	0.0022173	451
452	204,304	92,345,408	21.2603	7.67443	0.0022124	452
453	205,209	92,959,677	21.2838	7.68009	0.0022075	453
454	206,116	93,576,664	21.3073	7.68573	0.0022026	454
455	207,025	94,196,375	21.3307	7.69137	0.0021978	455
456	207,936	94,818,816	21.3542	7.69700	0.0021930	456
457	208,849	95,443,993	21.3776	7.70262	0.0021882	457
458	209,764	96,071,912	21.4009	7.70824	0.0021834	458
459	210,681	96,702,579	21.4243	7.71384	0.0021786	459
460	211,600	97,336,000	21.4476	7.71944	0.0021739	460
461	212,521	97,972,181	21.4709	7.72503	0.0021692	461
462	213,444	98,611,128	21.4942	7.73061	0.0021645	462
463	214,369	99,252,847	21.5174	7.73619	0.0021598	463
464	215,296	99,897,344	21.5407	7.74175	0.0021552	464
465	216,225	100,544,625	21.5639	7.74731	0.0021505	465
466	217,156	101,194,696	21.5870	7.75286	0.0021459	466
467	218,089	101,847,563	21.6102	7.75840	0.0021413	467
468	219,024	102,503,232	21.6333	7.76394	0.0021368	468
469	219,961	103,161,709	21.6564	7.76946	0.0021322	469
470	220,900	103,823,000	21.6795	7.77498	0.0021277	470
471	221,841	104,487,111	21.7025	7.78049	0.0021231	471
472	222,784	105,154,048	21.7256	7.78599	0.0021186	472
473	223,729	105,823,817	21.7486	7.79149	0.0021142	473
474	224,676	106,496,424	21.7715	7.79697	0.0021097	474
475	225,625	107,171,875	21.7945	7.80245	0.0021053	475
476	226,576	107,850,176	21.8174	7.80793	0.0021008	476
477	227,529	108,531,333	21.8403	7.81339	0.0020964	477
478	228,484	109,215,352	21.8632	7.81885	0.0020921	478
479	229,441	109,902,239	21.8861	7.82429	0.0020877	479
480	230,400	110,592,000	21.9089	7.82974	0.0020833	480
481	231,361	111,284,641	21.9317	7.83517	0.0020790	481
482	232,324	111,980,168	21.9545	7.84059	0.0020747	482
483	233,289	112,678,587	21.9773	7.84601	0.0020704	483
484	234,256	113,379,904	22.0000	7.85142	0.0020661	484
485	235,225	114,084,125	22.0227	7.85683	0.0020619	485
486	236,196	114,791,256	22.0454	7.86222	0.0020576	486
487	237,169	115,501,303	22.0681	7.86761	0.0020534	487
488	238,144	116,214,272	22.0907	7.87299	0.0020492	488
489	239,121	116,930,169	22.1133	7.87837	0.0020450	489
490	240,100	117,649,000	22.1359	7.88374	0.0020408	490
491	241,081	118,370,771	22.1585	7.88909	0.0020367	491
492	242,064	119,095,488	22.1811	7.89445	0.0020325	492
493	243,049	119,823,157	22.2036	7.89979	0.0020284	493
494	244,036	120,553,784	22.2261	7.90513	0.0020243	494
495	245,025	121,287,375	22.2486	7.91046	0.0020202	495
496	246,016	122,023,936	22.2711	7.91578	0.0020161	496
497	247,009	122,763,473	22.2935	7.92110	0.0020121	497
498	248,004	123,505,992	22.3159	7.92641	0.0020080	498
499	249,001	124,251,499	22.3383	7.93171	0.0020040	499
500	250,000	125,000,000	22.3607	7.93701	0.0020000	500

(Text continued from page 54.)

The decimal part of a log is the *mantissa,* which is found in *tables of logarithms.* (See Tables 3-3 and 3-4.)

The decimal point in a number is disregarded with respect to its mantissa.

For example: The mantissa of .95, 9.5, 95, 950, or .0095 is .97772.

The characteristic of a number is *one less* than the number of digits to the left of the decimal point.

For example: Log of 95 = 1.97772. Log of 9.5 = 0.97772

For numbers less than unity, the characteristic is *negative,* in numerical value *one more* than the number of *zeros* directly after the decimal point.

For example: Log of .0095 = −3.97772

To avoid the use of negative characteristics this may be written: 7.97772−10. Cumbersome calculations in multiplication, division, roots, and powers of numbers are simplified by using the system of logarithms.

For example: Log 259 = $10^{2.4133}$ = 2.4133. Log 25.9 = $10^{1.4133}$ = 1.4133

To multiply numbers, their logs are added:

$$a^x \times a^y = a^{x+y}$$

Conversely, to divide numbers, their logs are subtracted. As logs are exponents, a number may be raised to any power by multiplying the log of the number by its exponent. A root may be extracted by dividing the log of the number by the index of the root. Find: $\sqrt{6392}$.

Log 6392 = 3.80564 ÷ 2 = 1.90282 = 79.95 *Ans*

Interpolation

If the number for which the logarithm is required consists of five figures, it is possible by means of the small tables in the right-hand column of the logarithmic tables, headed "P.P." (proportional parts), to obtain the logarithm more accurately than by taking the nearest value for four figures. The logarithm of 1524.2, for example, is found as follows:

First find the difference between the nearest larger and the nearest smaller logarithms in the table. Log 1524 = 3.18298 and log 1525 = 3.18327. (See illustration.) The difference is

N.	L. 0	1	2	3	④	5	6	7	8	9	P. P.	
150	17 609	638	667	696	725	754	782	811	840	869		
151	898	926	955	984	*013	*041	*070	*099	*127	*156	㉙	28
⑮②	⑱184	213	241	270	⑳⑨⑧	㉗	355	384	412	441	1 | 2,9	2,8
153	469	498	526	554	583	611	639	667	696	724	② | 5,8	5,6
154	752	780	808	837	865	893	921	949	977	*005	3 | 8,7	8,4

0.00029. Then in the small table headed "29" in the right-hand column, find the figure opposite 2 (2 being the last or fifth figure in the given number). This figure is 5.8. Add this to the mantissa of the smaller of the two logarithms already found, disregarding the decimal point in the mantissa, and considering it, for the time being, as a whole number. Then, 18298 + 5.8 = 18303.8, or approximately, 18304. This is the mantissa of the logarithm of 1524.2 and the complete logarithm is 3.18304.

To find a number more accurately than to four figures, when the mantissa cannot be found exactly in the tables, find the mantissa which is nearest to, but less than, the given mantissa. Subtract this mantissa from the nearest larger mantissa in the tables and find in the right-hand column the small table headed by this difference. Then subtract the nearest smaller mantissa from the given logarithm and find the exact or approximate difference in the "proportional part" table. The corresponding figure in the left-hand column of the "proportional part" table is the fifth figure in the number sought, the other four figures being those corresponding to the logarithm next smaller than the given logarithm. In accordance with this rule, the number corresponding to the logarithm 4.46262 is found to be 29,015.

Table 3-3. Logarithms

N.	L.	0	1	2	3	4	5	6	7	8	9	P. P.		
100	00	000	043	087	130	173	217	260	303	346	389		44	43	42
101		432	475	518	561	604	647	689	732	775	817	1	4.4	4.3	4.2
102		860	903	945	988	*030	*072	*115	*157	*199	*242	2	8.8	8.6	8.4
103	01	284	326	368	410	452	494	536	578	620	662	3	13.2	12.9	12.6
104		703	745	787	828	870	912	953	995	*036	*078	4	17.6	17.2	16.8
105	02	119	160	202	243	284	325	366	407	449	490	5	22.0	21.5	21.0
106		531	572	612	653	694	735	776	816	857	898	6	26.4	25.8	25.2
107		938	979	*019	*060	*100	*141	*181	*222	*262	*302	7	30.8	30.1	29.4
108	03	342	383	423	463	503	543	583	623	663	703	8	35.2	34.4	33.6
109		743	782	822	862	902	941	981	*021	*060	*100	9	39.6	38.7	37.8
110	04	139	179	218	258	297	336	376	415	454	493		41	40	39
111		532	571	610	650	689	727	766	805	844	883	1	4.1	4.0	3.9
112		922	961	999	*038	*077	*115	*154	*192	*231	*269	2	8.2	8.0	7.8
113	05	308	346	385	423	461	500	538	576	614	652	3	12.3	12.0	11.7
114		690	729	767	805	843	881	918	956	994	*032	4	16.4	16.0	15.6
115	06	070	108	145	183	221	258	296	333	371	408	5	20.5	20.0	19.5
116		446	483	521	558	595	633	670	707	744	781	6	24.6	24.0	23.4
117		819	856	893	930	967	*004	*041	*078	*115	*151	7	28.7	28.0	27.3
118	07	188	225	262	298	335	372	408	445	482	518	8	32.8	32.0	31.2
119		555	591	628	664	700	737	773	809	846	882	9	36.9	36.0	35.1
120		918	954	990	*027	*063	*099	*135	*171	*207	*243		38	37	36
121	08	279	314	350	386	422	458	493	529	565	600	1	3.8	3.7	3.6
122		636	672	707	743	778	814	849	884	920	955	2	7.6	7.4	7.2
123		991	*026	*061	*096	*132	*167	*202	*237	*272	*307	3	11.4	11.1	10.8
124	09	342	377	412	447	482	517	552	587	621	656	4	15.2	14.8	14.4
125		691	726	760	795	830	864	899	934	968	*003	5	19.0	18.5	18.0
126	10	037	072	106	140	175	209	243	278	312	346	6	22.8	22.2	21.6
127		380	415	449	483	517	551	585	619	653	687	7	26.6	25.9	25.2
128		721	755	789	823	857	890	924	958	992	*025	8	30.4	29.6	28.8
129	11	059	093	126	160	193	227	261	294	327	361	9	34.2	33.3	32.4
130		394	428	461	494	528	561	594	628	661	694		35	34	33
131		727	760	793	826	860	893	926	959	992	*024	1	3.5	3.4	3.3
132	12	057	090	123	156	189	222	254	287	320	352	2	7.0	6.8	6.6
133		385	418	450	483	516	548	581	613	646	678	3	10.5	10.2	9.9
134		710	743	775	808	840	872	905	937	969	*001	4	14.0	13.6	13.2
135	13	033	066	098	130	162	194	226	258	290	322	5	17.5	17.0	16.5
136		354	386	418	450	481	513	545	577	609	640	6	21.0	20.4	19.8
137		672	704	735	767	799	830	862	893	925	956	7	24.5	23.8	23.1
138		988	*019	*051	*082	*114	*145	*176	*208	*239	*270	8	28.0	27.2	26.4
139	14	301	333	364	395	426	457	489	520	551	582	9	31.5	30.6	29.7
140		613	644	675	706	737	768	799	829	860	891		32	31	30
141		922	953	983	*014	*045	*076	*106	*137	*168	*198	1	3.2	3.1	3.0
142	15	229	259	290	320	351	381	412	442	473	503	2	6.4	6.2	6.0
143		534	564	594	625	655	685	715	746	776	806	3	9.6	9.3	9.0
144		836	866	897	927	957	987	*017	*047	*077	*107	4	12.8	12.4	12.0
145	16	137	167	197	227	256	286	316	346	376	406	5	16.0	15.5	15.0
146		435	465	495	524	554	584	613	643	673	702	6	19.2	18.6	18.0
147		732	761	791	820	850	879	909	938	967	997	7	22.4	21.7	21.0
148	17	026	056	085	114	143	173	202	231	260	289	8	25.6	24.8	24.0
149		319	348	377	406	435	464	493	522	551	580	9	28.8	27.9	27.0
150		609	638	667	696	725	754	782	811	840	869				

If the three last figures of the mantissa, as found in the table, are preceded by an (*), it indicates that these three figures belong to the group preceded by the two figures in the "L" column in the line next below. (Reproduced from *Handbook of Air Conditioning, Heating and Ventilating.* Edited by Clifford Strock and Richard L. Koral. New York: Industrial Press Inc., 1965.)

Table 3-3 (Continued). Logarithms

N.	L.	0	1	2	3	4	5	6	7	8	9	P. P.	
150	17	609	638	667	696	725	754	782	811	840	869	29	28
151		898	926	955	984	*013	*041	*070	*099	*127	*156	1 2.9	2.8
152	18	184	213	241	270	298	327	355	384	412	441	2 5.8	5.6
153		469	498	526	554	583	611	639	667	696	724	3 8.7	8.4
154		752	780	808	837	865	893	921	949	977	*005	4 11.6	11.2
155	19	033	061	089	117	145	173	201	229	257	285	5 14.5	14.0
156		312	340	368	396	424	451	479	507	535	562	6 17.4	16.8
157		590	618	645	673	700	728	756	783	811	838	7 20.3	19.6
158		866	893	921	948	976	*003	*030	*058	*085	*112	8 23.2	22.4
159	20	140	167	194	222	249	276	303	330	358	385	9 26.1	25.2
160		412	439	466	493	520	548	575	602	629	656	27	26
161		683	710	737	763	790	817	844	871	898	925	1 2.7	2.6
162		952	978	*005	*032	*059	*085	*112	*139	*165	*192	2 5.4	5.2
163	21	219	245	272	299	325	352	378	405	431	458	3 8.1	7.8
164		484	511	537	564	590	617	643	669	696	722	4 10.8	10.4
165		748	775	801	827	854	880	906	932	958	985	5 13.5	13.0
166	22	011	037	063	089	115	141	167	194	220	246	6 16.2	15.6
167		272	298	324	350	376	401	427	453	479	505	7 18.9	18.2
168		531	557	583	608	634	660	686	712	737	763	8 21.6	20.8
169		789	814	840	866	891	917	943	968	994	*019	9 24.3	23.4
170	23	045	070	096	121	147	172	198	223	249	274	25	
171		300	325	350	376	401	426	452	477	502	528	1 2.5	
172		553	578	603	629	654	679	704	729	754	779	2 5.0	
173		805	830	855	880	905	930	955	980	*005	*030	3 7.5	
174	24	055	080	105	130	155	180	204	229	254	279	4 10.0	
175		304	329	353	378	403	428	452	477	502	527	5 12.5	
176		551	576	601	625	650	674	699	724	748	773	6 15.0	
177		797	822	846	871	895	920	944	969	993	*018	7 17.5	
178	25	042	066	091	115	139	164	188	212	237	261	8 20.0	
179		285	310	334	358	382	406	431	455	479	503	9 22.5	
180		527	551	575	600	624	648	672	696	720	744	24	23
181		768	792	816	840	864	888	912	935	959	983	1 2.4	2.3
182	26	007	031	055	079	102	126	150	174	198	221	2 4.8	4.6
183		245	269	293	316	340	364	387	411	435	458	3 7.2	6.9
184		482	505	529	553	576	600	623	647	670	694	4 9.6	9.2
185		717	741	764	788	811	834	858	881	905	928	5 12.0	11.5
186		951	975	998	*021	*045	*068	*091	*114	*138	*161	6 14.4	13.8
187	27	184	207	231	254	277	300	323	346	370	393	7 16.8	16.1
188		416	439	462	485	508	531	554	577	600	623	8 19.2	18.4
189		646	669	692	715	738	761	784	807	830	852	9 21.6	20.7
190		875	898	921	944	967	989	*012	*035	*058	*081	22	21
191	28	103	126	149	171	194	217	240	262	285	307	1 2.2	2.1
192		330	353	375	398	421	443	466	488	511	533	2 4.4	4.2
193		556	578	601	623	646	668	691	713	735	758	3 6.6	6.3
194		780	803	825	847	870	892	914	937	959	981	4 8.8	8.4
195	29	003	026	048	070	092	115	137	159	181	203	5 11.0	10.5
196		226	248	270	292	314	336	358	380	403	425	6 13.2	12.6
197		447	469	491	513	535	557	579	601	623	645	7 15.4	14.7
198		667	688	710	732	754	776	798	820	842	863	8 17.6	16.8
199		885	907	929	951	973	994	*016	*038	*060	*081	9 19.8	18.9
200	30	103	125	146	168	190	211	233	255	276	298		

Table 3-3 (Continued). Logarithms

N.	L.	0	1	2	3	4	5	6	7	8	9	P. P.	
200	30	103	125	146	168	190	211	233	255	276	298	**22**	**21**
201		320	341	363	384	406	428	449	471	492	514	1 2.2	2.1
202		535	557	578	600	621	643	664	685	707	728	2 4.4	4.2
203		750	771	792	814	835	856	878	899	920	942	3 6.6	6.3
204		963	984	*006	*027	*048	*069	*091	*112	*133	*154	4 8.8	8.4
205	31	175	197	218	239	260	281	302	323	345	366	5 11.0	10.5
206		387	408	429	450	471	492	513	534	555	576	6 13.2	12.6
207		597	618	639	660	681	702	723	744	765	785	7 15.4	14.7
208		806	827	848	869	890	911	931	952	973	994	8 17.6	16.8
209	32	015	035	056	077	098	118	139	160	181	201	9 19.8	18.9
210		222	243	263	284	305	325	346	366	387	408	**20**	
211		428	449	469	490	510	531	552	572	593	613	1 2.0	
212		634	654	675	695	715	736	756	777	797	818	2 4.0	
213		838	858	879	899	919	940	960	980	*001	*021	3 6.0	
214	33	041	062	082	102	122	143	163	183	203	224	4 8.0	
215		244	264	284	304	325	345	365	385	405	425	5 10.0	
216		445	465	486	506	526	546	566	586	606	626	6 12.0	
217		646	666	686	706	726	746	766	786	806	826	7 14.0	
218		846	866	885	905	925	945	965	985	*005	*025	8 16.0	
219	34	044	064	084	104	124	143	163	183	203	223	9 18.0	
220		242	262	282	301	321	341	361	380	400	420	**19**	
221		439	459	479	498	518	537	557	577	596	616	1 1.9	
222		635	655	674	694	713	733	753	772	792	811	2 3.8	
223		830	850	869	889	908	928	947	967	986	*005	3 5.7	
224	35	025	044	064	083	102	122	141	160	180	199	4 7.6	
225		218	238	257	276	295	315	334	353	372	392	5 9.5	
226		411	430	449	468	488	507	526	545	564	583	6 11.4	
227		603	622	641	660	679	698	717	736	755	774	7 13.3	
228		793	813	832	851	870	889	908	927	946	965	8 15.2	
229		984	*003	*021	*040	*059	*078	*097	*116	*135	*154	9 17.1	
230	36	173	192	211	229	248	267	286	305	324	342	**18**	
231		361	380	399	418	436	455	474	493	511	530	1 1.8	
232		549	568	586	605	624	642	661	680	698	717	2 3.6	
233		736	754	773	791	810	829	847	866	884	903	3 5.4	
234		922	940	959	977	996	*014	*033	*051	*070	*088	4 7.2	
235	37	107	125	144	162	181	199	218	236	254	273	5 9.0	
236		291	310	328	346	365	383	401	420	438	457	6 10.8	
237		475	493	511	530	548	566	585	603	621	639	7 12.6	
238		658	676	694	712	731	749	767	785	803	822	8 14.4	
239		840	858	876	894	912	931	949	967	985	*003	9 16.2	
240	38	021	039	057	075	093	112	130	148	166	184	**17**	
241		202	220	238	256	274	292	310	328	346	364		
242		382	399	417	435	453	471	489	507	525	543	1 1.7	
243		561	578	596	614	632	650	668	686	703	721	2 3.4	
244		739	757	775	792	810	828	846	863	881	899	3 5.1	
245		917	934	952	970	987	*005	*023	*041	*058	*076	4 6.8	
246	39	094	111	129	146	164	182	199	217	235	252	5 8.5	
247		270	287	305	322	340	358	375	393	410	428	6 10.2	
248		445	463	480	498	515	533	550	568	585	602	7 11.9	
249		620	637	655	672	690	707	724	742	759	777	8 13.6	
250		794	811	829	846	863	881	898	915	933	950	9 15.3	

Table 3-3 (Continued). Logarithms

N.	L. 0	1	2	3	4	5	6	7	8	9	P. P.	
250	39 794	811	829	846	863	881	898	915	933	950	**18**	
251	967	985	*002	*019	*037	*054	*071	*088	*106	*123	1	1.8
252	40 140	157	175	192	209	226	243	261	278	295	2	3.6
253	312	329	346	364	381	398	415	432	449	466	3	5.4
254	483	500	518	535	552	569	586	603	620	637	4	7.2
255	654	671	688	705	722	739	756	773	790	807	5	9.0
256	824	841	858	875	892	909	926	943	960	976	6	10.8
257	993	*010	*027	*044	*061	*078	*095	*111	*128	*145	7	12.6
258	41 162	179	196	212	229	246	263	280	296	313	8	14.4
259	330	347	363	380	397	414	430	447	464	481	9	16.2
260	497	514	531	547	564	581	597	614	631	647	**17**	
261	664	681	697	714	731	747	764	780	797	814	1	1.7
262	830	847	863	880	896	913	929	946	963	979	2	3.4
263	996	*012	*029	*045	*062	*078	*095	*111	*127	*144	3	5.1
264	42 160	177	193	210	226	243	259	275	292	308	4	6.8
265	325	341	357	374	390	406	423	439	455	472	5	8.5
266	488	504	521	537	553	570	586	602	619	635	6	10.2
267	651	667	684	700	716	732	749	765	781	797	7	11.9
268	813	830	846	862	878	894	911	927	943	959	8	13.6
269	975	991	*008	*024	*040	*056	*072	*088	*104	*120	9	15.3
270	43 136	152	169	185	201	217	233	249	265	281	**16**	
271	297	313	329	345	361	377	393	409	425	441	1	1.6
272	457	473	489	505	521	537	553	569	584	600	2	3.2
273	616	632	648	664	680	696	712	727	743	759	3	4.8
274	775	791	807	823	838	854	870	886	902	917	4	6.4
275	933	949	965	981	996	*012	*028	*044	*059	*075	5	8.0
276	44 091	107	122	138	154	170	185	201	217	232	6	9.6
277	248	264	279	295	311	326	342	358	373	389	7	11.2
278	404	420	436	451	467	483	498	514	529	545	8	12.8
279	560	576	592	607	623	638	654	669	685	700	9	14.4
280	716	731	747	762	778	793	809	824	840	855	**15**	
281	871	886	902	917	932	948	963	979	994	*010	1	1.5
282	45 025	040	056	071	086	102	117	133	148	163	2	3.0
283	179	194	209	225	240	255	271	286	301	317	3	4.5
284	332	347	362	378	393	408	423	439	454	469	4	6.0
285	484	500	515	530	545	561	576	591	606	621	5	7.5
286	637	652	667	682	697	712	728	743	758	773	6	9.0
287	788	803	818	834	849	864	879	894	909	924	7	10.5
288	939	954	969	984	*000	*015	*030	*045	*060	*075	8	12.0
289	46 090	105	120	135	150	165	180	195	210	225	9	13.5
290	240	255	270	285	300	315	330	345	359	374	**14**	
291	389	404	419	434	449	464	479	494	509	523	1	1.4
292	538	553	568	583	598	613	627	642	657	672	2	2.8
293	687	702	716	731	746	761	776	790	805	820	3	4.2
294	835	850	864	879	894	909	923	938	953	967	4	5.6
295	982	997	*012	*026	*041	*056	*070	*085	*100	*114	5	7.0
296	47 129	144	159	173	188	202	217	232	246	261	6	8.4
297	276	290	305	319	334	349	363	378	392	407	7	9.8
298	422	436	451	465	480	494	509	524	538	553	8	11.2
299	567	582	596	611	625	640	654	669	683	698	9	12.6
300	712	727	741	756	770	784	799	813	828	842		

Table 3-3 (Continued). Logarithms

N.	L.	0	1	2	3	4	5	6	7	8	9	P. P.
300	47	712	727	741	756	770	784	799	813	828	842	
301		857	871	885	900	914	929	943	958	972	986	
302	48	001	015	029	044	058	073	087	101	116	130	
303		144	159	173	187	202	216	230	244	259	273	**15**
304		287	302	316	330	344	359	373	387	401	416	1 | 1.5
305		430	444	458	473	487	501	515	530	544	558	2 | 3.0
306		572	586	601	615	629	643	657	671	686	700	3 | 4.5
307		714	728	742	756	770	785	799	813	827	841	4 | 6.0
308		855	869	883	897	911	926	940	954	968	982	5 | 7.5
309		996	*010	*024	*038	*052	*066	*080	*094	*108	*122	6 | 9.0
310	49	136	150	164	178	192	206	220	234	248	262	7 | 10.5
311		276	290	304	318	332	346	360	374	388	402	8 | 12.0
312		415	429	443	457	471	485	499	513	527	541	9 | 13.5
313		554	568	582	596	610	624	638	651	665	679	
314		693	707	721	734	748	762	776	790	803	817	
315		831	845	859	872	886	900	914	927	941	955	**14**
316		969	982	996	*010	*024	*037	*051	*065	*079	*092	1 | 1.4
317	50	106	120	133	147	161	174	188	202	215	229	2 | 2.8
318		243	256	270	284	297	311	325	338	352	365	3 | 4.2
319		379	393	406	420	433	447	461	474	488	501	4 | 5.6
320		515	529	542	556	569	583	596	610	623	637	5 | 7.0
321		651	664	678	691	705	718	732	745	759	772	6 | 8.4
322		786	799	813	826	840	853	866	880	893	907	7 | 9.8
323		920	934	947	961	974	987	*001	*014	*028	*041	8 | 11.2
324	51	055	068	081	095	108	121	135	148	162	175	9 | 12.6
325		188	202	215	228	242	255	268	282	295	308	
326		322	335	348	362	375	388	402	415	428	441	**13**
327		455	468	481	495	508	521	534	548	561	574	1 | 1.3
328		587	601	614	627	640	654	667	680	693	706	2 | 2.6
329		720	733	746	759	772	786	799	812	825	838	3 | 3.9
330		851	865	878	891	904	917	930	943	957	970	4 | 5.2
331		983	996	*009	*022	*035	*048	*061	*075	*088	*101	5 | 6.5
332	52	114	127	140	153	166	179	192	205	218	231	6 | 7.8
333		244	257	270	284	297	310	323	336	349	362	7 | 9.1
334		375	388	401	414	427	440	453	466	479	492	8 | 10.4
335		504	517	530	543	556	569	582	595	608	621	9 | 11.7
336		634	647	660	673	686	699	711	724	737	750	
337		763	776	789	802	815	827	840	853	866	879	
338		892	905	917	930	943	956	969	982	994	*007	**12**
339	53	020	033	046	058	071	084	097	110	122	135	1 | 1.2
340		148	161	173	186	199	212	224	237	250	263	2 | 2.4
341		275	288	301	314	326	339	352	364	377	390	3 | 3.6
342		403	415	428	441	453	466	479	491	504	517	4 | 4.8
343		529	542	555	567	580	593	605	618	631	643	5 | 6.0
344		656	668	681	694	706	719	732	744	757	769	6 | 7.2
345		782	794	807	820	832	845	857	870	882	895	7 | 8.4
346		908	920	933	945	958	970	983	995	*008	*020	8 | 9.6
347	54	033	045	058	070	083	095	108	120	133	145	9 | 10.8
348		158	170	183	195	208	220	233	245	258	270	
349		283	295	307	320	332	345	357	370	382	394	
350		407	419	432	444	456	469	481	494	506	518	

Table 3-3 (Continued). Logarithms

N.	L. 0	1	2	3	4	5	6	7	8	9	P. P.	
350	54 407	419	432	444	456	469	481	494	506	518		
351	531	543	555	568	580	593	605	617	630	642		
352	654	667	679	691	704	716	728	741	753	765	**13**	
353	777	790	802	814	827	839	851	864	876	888		
354	900	913	925	937	949	962	974	986	998	*011	1	1.3
355	55 023	035	047	060	072	084	096	108	121	133	2	2.6
356	145	157	169	182	194	206	218	230	242	255	3	3.9
357	267	279	291	303	315	328	340	352	364	376	4	5.2
358	388	400	413	425	437	449	461	473	485	497	5	6.5
359	509	522	534	546	558	570	582	594	606	618	6	7.8
360	630	642	654	666	678	691	703	715	727	739	7	9.1
361	751	763	775	787	799	811	823	835	847	859	8	10.4
362	871	883	895	907	919	931	943	955	967	979	9	11.7
363	991	*003	*015	*027	*038	*050	*062	*074	*086	*098		
364	56 110	122	134	146	158	170	182	194	205	217		
365	229	241	253	265	277	289	301	312	324	336	**12**	
366	348	360	372	384	396	407	419	431	443	455	1	1.2
367	467	478	490	502	514	526	538	549	561	573	2	2.4
368	585	597	608	620	632	644	656	667	679	691	3	3.6
369	703	714	726	738	750	761	773	785	797	808	4	4.8
370	820	832	844	855	867	879	891	902	914	926	5	6.0
371	937	949	961	972	984	996	*008	*019	*031	*043	6	7.2
372	57 054	066	078	089	101	113	124	136	148	159	7	8.4
373	171	183	194	206	217	229	241	252	264	276	8	9.6
374	287	299	310	322	334	345	357	368	380	392	9	10.8
375	403	415	426	438	449	461	473	484	496	507		
376	519	530	542	553	565	576	588	600	611	623	**11**	
377	634	646	657	669	680	692	703	715	726	738	1	1.1
378	749	761	772	784	795	807	818	830	841	852	2	2.2
379	864	875	887	898	910	921	933	944	955	967	3	3.3
380	978	990	*001	*013	*024	*035	*047	*058	*070	*081	4	4.4
381	58 092	104	115	127	138	149	161	172	184	195	5	5.5
382	206	218	229	240	252	263	274	286	297	309	6	6.6
383	320	331	343	354	365	377	388	399	410	422	7	7.7
384	433	444	456	467	478	490	501	512	524	535	8	8.8
385	546	557	569	580	591	602	614	625	636	647	9	9.9
386	659	670	681	692	704	715	726	737	749	760		
387	771	782	794	805	816	827	838	850	861	872		
388	883	894	906	917	928	939	950	961	973	984	**10**	
389	995	*006	*017	*028	*040	*051	*062	*073	*084	*095	1	1.0
390	59 106	118	129	140	151	162	173	184	195	207	2	2.0
391	218	229	240	251	262	273	284	295	306	318	3	3.0
392	329	340	351	362	373	384	395	406	417	428	4	4.0
393	439	450	461	472	483	494	506	517	528	539	5	5.0
394	550	561	572	583	594	605	616	627	638	649	6	6.0
395	660	671	682	693	704	715	726	737	748	759	7	7.0
396	770	780	791	802	813	824	835	846	857	868	8	8.0
397	879	890	901	912	923	934	945	956	966	977	9	9.0
398	988	999	*010	*021	*032	*043	*054	*065	*076	*086		
399	60 097	108	119	130	141	152	163	173	184	195		
400	206	217	228	239	249	260	271	282	293	304		

Table 3-3 (Continued). Logarithms

N.	L.	0	1	2	3	4	5	6	7	8	9	P. P.
400	60	206	217	228	239	249	260	271	282	293	304	
401		314	325	336	347	358	369	379	390	401	412	
402		423	433	444	455	466	477	487	498	509	520	
403		531	541	552	563	574	584	595	606	617	627	
404		638	649	660	670	681	692	703	713	724	735	
405		746	756	767	778	788	799	810	821	831	842	
406		853	863	874	885	895	906	917	927	938	949	**11**
407		959	970	981	991	*002	*013	*023	*034	*045	*055	1 | 1.1
408	61	066	077	087	098	109	119	130	140	151	162	2 | 2.2
409		172	183	194	204	215	225	236	247	257	268	3 | 3.3
410		278	289	300	310	321	331	342	352	363	374	4 | 4.4
411		384	395	405	416	426	437	448	458	469	479	5 | 5.5
412		490	500	511	521	532	542	553	563	574	584	6 | 6.6
413		595	606	616	627	637	648	658	669	679	690	7 | 7.7
414		700	711	721	731	742	752	763	773	784	794	8 | 8.8
415		805	815	826	836	847	857	868	878	888	899	9 | 9.9
416		909	920	930	941	951	962	972	982	993	*003	
417	62	014	024	034	045	055	066	076	086	097	107	
418		118	128	138	149	159	170	180	190	201	211	
419		221	232	242	252	263	273	284	294	304	315	
420		325	335	346	356	366	377	387	397	408	418	**10**
421		428	439	449	459	469	480	490	500	511	521	1 | 1.0
422		531	542	552	562	572	583	593	603	613	624	2 | 2.0
423		634	644	655	665	675	685	696	706	716	726	3 | 3.0
424		737	747	757	767	778	788	798	808	818	829	4 | 4.0
425		839	849	859	870	880	890	900	910	921	931	5 | 5.0
426		941	951	961	972	982	992	*002	*012	*022	*033	6 | 6.0
427	63	043	053	063	073	083	094	104	114	124	134	7 | 7.0
428		144	155	165	175	185	195	205	215	225	236	8 | 8.0
429		246	256	266	276	286	296	306	317	327	337	9 | 9.0
430		347	357	367	377	387	397	407	417	428	438	
431		448	458	468	478	488	498	508	518	528	538	
432		548	558	568	579	589	599	609	619	629	639	
433		649	659	669	679	689	699	709	719	729	739	
434		749	759	769	779	789	799	809	819	829	839	
435		849	859	869	879	889	899	909	919	929	939	**9**
436		949	959	969	979	988	998	*008	*018	*028	*038	1 | 0.9
437	64	048	058	068	078	088	098	108	118	128	137	2 | 1.8
438		147	157	167	177	187	197	207	217	227	237	3 | 2.7
439		246	256	266	276	286	296	306	316	326	335	4 | 3.6
440		345	355	365	375	385	395	404	414	424	434	5 | 4.5
441		444	454	464	473	483	493	503	513	523	532	6 | 5.4
442		542	552	562	572	582	591	601	611	621	631	7 | 6.3
443		640	650	660	670	680	689	699	709	719	729	8 | 7.2
444		738	748	758	768	777	787	797	807	816	826	9 | 8.1
445		836	846	856	865	875	885	895	904	914	924	
446		933	943	953	963	972	982	992	*002	*011	*021	
447	65	031	040	050	060	070	079	089	099	108	118	
448		128	137	147	157	167	176	186	196	205	215	
449		225	234	244	254	263	273	283	292	302	312	
450		321	331	341	350	360	369	379	389	398	408	

Table 3-3 (Continued). Logarithms

N.	L. 0	1	2	3	4	5	6	7	8	9	P. P.
450	65 321	331	341	350	360	369	379	389	398	408	
451	418	427	437	447	456	466	475	485	495	504	
452	514	523	533	543	552	562	571	581	591	600	
453	610	619	629	639	648	658	667	677	686	696	
454	706	715	725	734	744	753	763	772	782	792	
455	801	811	820	830	839	849	858	868	877	887	
456	896	906	916	925	935	944	954	963	973	982	**10**
457	992	*001	*011	*020	*030	*039	*049	*058	*068	*077	1 | 1.0
458	66 087	096	106	115	124	134	143	153	162	172	2 | 2.0
459	181	191	200	210	219	229	238	247	257	266	3 | 3.0
460	276	285	295	304	314	323	332	342	351	361	4 | 4.0
461	370	380	389	398	408	417	427	436	445	455	5 | 5.0
462	464	474	483	492	502	511	521	530	539	549	6 | 6.0
463	558	567	577	586	596	605	614	624	633	642	7 | 7.0
464	652	661	671	680	689	699	708	717	727	736	8 | 8.0
465	745	755	764	773	783	792	801	811	820	829	9 | 9.0
466	839	848	857	867	876	885	894	904	913	922	
467	932	941	950	960	969	978	987	997	*006	*015	
468	67 025	034	043	052	062	071	080	089	099	108	
469	117	127	136	145	154	164	173	182	191	201	
470	210	219	228	237	247	256	265	274	284	293	**9**
471	302	311	321	330	339	348	357	367	376	385	1 | 0.9
472	394	403	413	422	431	440	449	459	468	477	2 | 1.8
473	486	495	504	514	523	532	541	550	560	569	3 | 2.7
474	578	587	596	605	614	624	633	642	651	660	4 | 3.6
475	669	679	688	697	706	715	724	733	742	752	5 | 4.5
476	761	770	779	788	797	806	815	825	834	843	6 | 5.4
477	852	861	870	879	888	897	906	916	925	934	7 | 6.3
478	943	952	961	970	979	988	997	*006	*015	*024	8 | 7.2
479	68 034	043	052	061	070	079	088	097	106	115	9 | 8.1
480	124	133	142	151	160	169	178	187	196	205	
481	215	224	233	242	251	260	269	278	287	296	
482	305	314	323	332	341	350	359	368	377	386	
483	395	404	413	422	431	440	449	458	467	476	
484	485	494	502	511	520	529	538	547	556	565	**8**
485	574	583	592	601	610	619	628	637	646	655	1 | 0.8
486	664	673	681	690	699	708	717	726	735	744	2 | 1.6
487	753	762	771	780	789	797	806	815	824	833	3 | 2.4
488	842	851	860	869	878	886	895	904	913	922	4 | 3.2
489	931	940	949	958	966	975	984	993	*002	*011	5 | 4.0
490	69 020	028	037	046	055	064	073	082	090	099	6 | 4.8
491	108	117	126	135	144	152	161	170	179	188	7 | 5.6
492	197	205	214	223	232	241	249	258	267	276	8 | 6.4
493	285	294	302	311	320	329	338	346	355	364	9 | 7.2
494	373	381	390	399	408	417	425	434	443	452	
495	461	469	478	487	496	504	513	522	531	539	
496	548	557	566	574	583	592	601	609	618	627	
497	636	644	653	662	671	679	688	697	705	714	
498	723	732	740	749	758	767	775	784	793	801	
499	810	819	827	836	845	854	862	871	880	888	
500	897	906	914	923	932	940	949	958	966	975	

Table 3-3 (Continued).   Logarithms

N.	L. 0	1	2	3	4	5	6	7	8	9	P. P.
500	69 897	906	914	923	932	940	949	958	966	975	
501	984	992	*001	*010	*018	*027	*036	*044	*053	*062	
502	70 070	079	088	096	105	114	122	131	140	148	
503	157	165	174	183	191	200	209	217	226	234	
504	243	252	260	269	278	286	295	303	312	321	
505	329	338	346	355	364	372	381	389	398	406	
506	415	424	432	441	449	458	467	475	484	492	**9**
507	501	509	518	526	535	544	552	561	569	578	1 | 0.9
508	586	595	603	612	621	629	638	646	655	663	2 | 1.8
509	672	680	689	697	706	714	723	731	740	749	3 | 2.7
510	757	766	774	783	791	800	808	817	825	834	4 | 3.6
511	842	851	859	868	876	885	893	902	910	919	5 | 4.5
512	927	935	944	952	961	969	978	986	995	*003	6 | 5.4
513	71 012	020	029	037	046	054	063	071	079	088	7 | 6.3
514	096	105	113	122	130	139	147	155	164	172	8 | 7.2
515	181	189	198	206	214	223	231	240	248	257	9 | 8.1
516	265	273	282	290	299	307	315	324	332	341	
517	349	357	366	374	383	391	399	408	416	425	
518	433	441	450	458	466	475	483	492	500	508	
519	517	525	533	542	550	559	567	575	584	592	
520	600	609	617	625	634	642	650	659	667	675	**8**
521	684	692	700	709	717	725	734	742	750	759	1 | 0.8
522	767	775	784	792	800	809	817	825	834	842	2 | 1.6
523	850	858	867	875	883	892	900	908	917	925	3 | 2.4
524	933	941	950	958	966	975	983	991	999	*008	4 | 3.2
525	72 016	024	032	041	049	057	066	074	082	090	5 | 4.0
526	099	107	115	123	132	140	148	156	165	173	6 | 4.8
527	181	189	198	206	214	222	230	239	247	255	7 | 5.6
528	263	272	280	288	296	304	313	321	329	337	8 | 6.4
529	346	354	362	370	378	387	395	403	411	419	9 | 7.2
530	428	436	444	452	460	469	477	485	493	501	
531	509	518	526	534	542	550	558	567	575	583	
532	591	599	607	616	624	632	640	648	656	665	
533	673	681	689	697	705	713	722	730	738	746	
534	754	762	770	779	787	795	803	811	819	827	
535	835	843	852	860	868	876	884	892	900	908	**7**
536	916	925	933	941	949	957	965	973	981	989	1 | 0.7
537	997	*006	*014	*022	*030	*038	*046	*054	*062	*070	2 | 1.4
538	73 078	086	094	102	111	119	127	135	143	151	3 | 2.1
539	159	167	175	183	191	199	207	215	223	231	4 | 2.8
540	239	247	255	263	272	280	288	296	304	312	5 | 3.5
541	320	328	336	344	352	360	368	376	384	392	6 | 4.2
542	400	408	416	424	432	440	448	456	464	472	7 | 4.9
543	480	488	496	504	512	520	528	536	544	552	8 | 5.6
544	560	568	576	584	592	600	608	616	624	632	9 | 6.3
545	640	648	656	664	672	679	687	695	703	711	
546	719	727	735	743	751	759	767	775	783	791	
547	799	807	815	823	830	838	846	854	862	870	
548	878	886	894	902	910	918	926	933	941	949	
549	957	965	973	981	989	997	*005	*013	*020	*028	
550	74 036	044	052	060	068	076	084	092	099	107	

Table 3-3 (Continued). Logarithms

N.	L.	0	1	2	3	4	5	6	7	8	9	P. P.
550	74	036	044	052	060	068	076	084	092	099	107	
551		115	123	131	139	147	155	162	170	178	186	
552		194	202	210	218	225	233	241	249	257	265	
553		273	280	288	296	304	312	320	327	335	343	
554		351	359	367	374	382	390	398	406	414	421	
555		429	437	445	453	461	468	476	484	492	500	
556		507	515	523	531	539	547	554	562	570	578	
557		586	593	601	609	617	624	632	640	648	656	
558		663	671	679	687	695	702	710	718	726	733	
559		741	749	757	764	772	780	788	796	803	811	
560		819	827	834	842	850	858	865	873	881	889	**8**
561		896	904	912	920	927	935	943	950	958	966	1 | 0.8
562		974	981	989	997	*005	*012	*020	*028	*035	*043	2 | 1.6
563	75	051	059	066	074	082	089	097	105	113	120	3 | 2.4
564		128	136	143	151	159	166	174	182	189	197	4 | 3.2
565		205	213	220	228	236	243	251	259	266	274	5 | 4.0
566		282	289	297	305	312	320	328	335	343	351	6 | 4.8
567		358	366	374	381	389	397	404	412	420	427	7 | 5.6
568		435	442	450	458	465	473	481	488	496	504	8 | 6.4
569		511	519	526	534	542	549	557	565	572	580	9 | 7.2
570		587	595	603	610	618	626	633	641	648	656	
571		664	671	679	686	694	702	709	717	724	732	
572		740	747	755	762	770	778	785	793	800	808	
573		815	823	831	838	846	853	861	868	876	884	
574		891	899	906	914	921	929	937	944	952	959	
575		967	974	982	989	997	*005	*012	*020	*027	*035	
576	76	042	050	057	065	072	080	087	095	103	110	
577		118	125	133	140	148	155	163	170	178	185	
578		193	200	208	215	223	230	238	245	253	260	
579		268	275	283	290	298	305	313	320	328	335	
580		343	350	358	365	373	380	388	395	403	410	**7**
581		418	425	433	440	448	455	462	470	477	485	1 | 0.7
582		492	500	507	515	522	530	537	545	552	559	2 | 1.4
583		567	574	582	589	597	604	612	619	626	634	3 | 2.1
584		641	649	656	664	671	678	686	693	701	708	4 | 2.8
585		716	723	730	738	745	753	760	768	775	782	5 | 3.5
586		790	797	805	812	819	827	834	842	849	856	6 | 4.2
587		864	871	879	886	893	901	908	916	923	930	7 | 4.9
588		938	945	953	960	967	975	982	989	997	*004	8 | 5.6
589	77	012	019	026	034	041	048	056	063	070	078	9 | 6.3
590		085	093	100	107	115	122	129	137	144	151	
591		159	166	173	181	188	195	203	210	217	225	
592		232	240	247	254	262	269	276	283	291	298	
593		305	313	320	327	335	342	349	357	364	371	
594		379	386	393	401	408	415	422	430	437	444	
595		452	459	466	474	481	488	495	503	510	517	
596		525	532	539	546	554	561	568	576	583	590	
597		597	605	612	619	627	634	641	648	656	663	
598		670	677	685	692	699	706	714	721	728	735	
599		743	750	757	764	772	779	786	793	801	808	
600		815	822	830	837	844	851	859	866	873	880	

Table 3-3 (Continued). Logarithms

N.	L.	0	1	2	3	4	5	6	7	8	9	P. P.
600	77	815	822	830	837	844	851	859	866	873	880	
601		887	895	902	909	916	924	931	938	945	952	
602		960	967	974	981	988	996	*003	*010	*017	*025	
603	78	032	039	046	053	061	068	075	082	089	097	
604		104	111	118	125	132	140	147	154	161	168	
605		176	183	190	197	204	211	219	226	233	240	**8**
606		247	254	262	269	276	283	290	297	305	312	1|0.8
607		319	326	333	340	347	355	362	369	376	383	2|1.6
608		390	398	405	412	419	426	433	440	447	455	3|2.4
609		462	469	476	483	490	497	504	512	519	526	4|3.2
610		533	540	547	554	561	569	576	583	590	597	5|4.0
611		604	611	618	625	633	640	647	654	661	668	6|4.8
612		675	682	689	696	704	711	718	725	732	739	7|5.6
613		746	753	760	767	774	781	789	796	803	810	8|6.4
614		817	824	831	838	845	852	859	866	873	880	9|7.2
615		888	895	902	909	916	923	930	937	944	951	
616		958	965	972	979	986	993	*000	*007	*014	*021	
617	79	029	036	043	050	057	064	071	078	085	092	
618		099	106	113	120	127	134	141	148	155	162	
619		169	176	183	190	197	204	211	218	225	232	
620		239	246	253	260	267	274	281	288	295	302	**7**
621		309	316	323	330	337	344	351	358	365	372	1|0.7
622		379	386	393	400	407	414	421	428	435	442	2|1.4
623		449	456	463	470	477	484	491	498	505	511	3|2.1
624		518	525	532	539	546	553	560	567	574	581	4|2.8
625		588	595	602	609	616	623	630	637	644	650	5|3.5
626		657	664	671	678	685	692	699	706	713	720	6|4.2
627		727	734	741	748	754	761	768	775	782	789	7|4.9
628		796	803	810	817	824	831	837	844	851	858	8|5.6
629		865	872	879	886	893	900	906	913	920	927	9|6.3
630		934	941	948	955	962	969	975	982	989	996	
631	80	003	010	017	024	030	037	044	051	058	065	
632		072	079	085	092	099	106	113	120	127	134	
633		140	147	154	161	168	175	182	188	195	202	
634		209	216	223	229	236	243	250	257	264	271	
635		277	284	291	298	305	312	318	325	332	339	**6**
636		346	353	359	366	373	380	387	393	400	407	1|0.6
637		414	421	428	434	441	448	455	462	468	475	2|1.2
638		482	489	496	502	509	516	523	530	536	543	3|1.8
639		550	557	564	570	577	584	591	598	604	611	4|2.4
640		618	625	632	638	645	652	659	665	672	679	5|3.0
641		686	693	699	706	713	720	726	733	740	747	6|3.6
642		754	760	767	774	781	787	794	801	808	814	7|4.2
643		821	828	835	841	848	855	862	868	875	882	8|4.8
644		889	895	902	909	916	922	929	936	943	949	9|5.4
645		956	963	969	976	983	990	996	*003	*010	*017	
646	81	023	030	037	043	050	057	064	070	077	084	
647		090	097	104	111	117	124	131	137	144	151	
648		158	164	171	178	184	191	198	204	211	218	
649		224	231	238	245	251	258	265	271	278	285	
650		291	298	305	311	318	325	331	338	345	351	

Table 3-3 (Continued). Logarithms

N.	L.	0	1	2	3	4	5	6	7	8	9	P. P.
650	81	291	298	305	311	318	325	331	338	345	351	
651		358	365	371	378	385	391	398	405	411	418	
652		425	431	438	445	451	458	465	471	478	485	
653		491	498	505	511	518	525	531	538	544	551	
654		558	564	571	578	584	591	598	604	611	617	
655		624	631	637	644	651	657	664	671	677	684	
656		690	697	704	710	717	723	730	737	743	750	
657		757	763	770	776	783	790	796	803	809	816	
658		823	829	836	842	849	856	862	869	875	882	
659		889	895	902	908	915	921	928	935	941	948	
660		954	961	968	974	981	987	994	*000	*007	*014	**7**
661	82	020	027	033	040	046	053	060	066	073	079	1 | 0.7
662		086	092	099	105	112	119	125	132	138	145	2 | 1.4
663		151	158	164	171	178	184	191	197	204	210	3 | 2.1
664		217	223	230	236	243	249	256	263	269	276	4 | 2.8
665		282	289	295	302	308	315	321	328	334	341	5 | 3.5
666		347	354	360	367	373	380	387	393	400	406	6 | 4.2
667		413	419	426	432	439	445	452	458	465	471	7 | 4.9
668		478	484	491	497	504	510	517	523	530	536	8 | 5.6
669		543	549	556	562	569	575	582	588	595	601	9 | 6.3
670		607	614	620	627	633	640	646	653	659	666	
671		672	679	685	692	698	705	711	718	724	730	
672		737	743	750	756	763	769	776	782	789	795	
673		802	808	814	821	827	834	840	847	853	860	
674		866	872	879	885	892	898	905	911	918	924	
675		930	937	943	950	956	963	969	975	982	988	
676		995	*001	*008	*014	*020	*027	*033	*040	*046	*052	
677	83	059	065	072	078	085	091	097	104	110	117	
678		123	129	136	142	149	155	161	168	174	181	
679		187	193	200	206	213	219	225	232	238	245	
680		251	257	264	270	276	283	289	296	302	308	**6**
681		315	321	327	334	340	347	353	359	366	372	1 | 0.6
682		378	385	391	398	404	410	417	423	429	436	2 | 1.2
683		442	448	455	461	467	474	480	487	493	499	3 | 1.8
684		506	512	518	525	531	537	544	550	556	563	4 | 2.4
685		569	575	582	588	594	601	607	613	620	626	5 | 3.0
686		632	639	645	651	658	664	670	677	683	689	6 | 3.6
687		696	702	708	715	721	727	734	740	746	753	7 | 4.2
688		759	765	771	778	784	790	797	803	809	816	8 | 4.8
689		822	828	835	841	847	853	860	866	872	879	9 | 5.4
690		885	891	897	904	910	916	923	929	935	942	
691		948	954	960	967	973	979	985	992	998	*004	
692	84	011	017	023	029	036	042	048	055	061	067	
693		073	080	086	092	098	105	111	117	123	130	
694		136	142	148	155	161	167	173	180	186	192	
695		198	205	211	217	223	230	236	242	248	255	
696		261	267	273	280	286	292	298	305	311	317	
697		323	330	336	342	348	354	361	367	373	379	
698		386	392	398	404	410	417	423	429	435	442	
699		448	454	460	466	473	479	485	491	497	504	
700		510	516	522	528	535	541	547	553	559	566	

Table 3-3 (*Continued*). Logarithms

N.	L.	0	1	2	3	4	5	6	7	8	9	P. P.
700	84	510	516	522	528	535	541	547	553	559	566	
701		572	578	584	590	597	603	609	615	621	628	
702		634	640	646	652	658	665	671	677	683	689	
703		696	702	708	714	720	726	733	739	745	751	
704		757	763	770	776	782	788	794	800	807	813	
705		819	825	831	837	844	850	856	862	868	874	
706		880	887	893	899	905	911	917	924	930	936	**7**
707		942	948	954	960	967	973	979	985	991	997	1|0.7
708	85	003	009	016	022	028	034	040	046	052	058	2|1.4
709		065	071	077	083	089	095	101	107	114	120	3|2.1
710		126	132	138	144	150	156	163	169	175	181	4|2.8
711		187	193	199	205	211	217	224	230	236	242	5|3.5
712		248	254	260	266	272	278	285	291	297	303	6|4.2
713		309	315	321	327	333	339	345	352	358	364	7|4.9
714		370	376	382	388	394	400	406	412	418	425	8|5.6
715		431	437	443	449	455	461	467	473	479	485	9|6.3
716		491	497	503	509	516	522	528	534	540	546	
717		552	558	564	570	576	582	588	594	600	606	
718		612	618	625	631	637	643	649	655	661	667	
719		673	679	685	691	697	703	709	715	721	727	
720		733	739	745	751	757	763	769	775	781	788	**6**
721		794	800	806	812	818	824	830	836	842	848	1|0.6
722		854	860	866	872	878	884	890	896	902	908	2|1.2
723		914	920	926	932	938	944	950	956	962	968	3|1.8
724		974	980	986	992	998	*004	*010	*016	*022	*028	4|2.4
725	86	034	040	046	052	058	064	070	076	082	088	5|3.0
726		094	100	106	112	118	124	130	136	141	147	6|3.6
727		153	159	165	171	177	183	189	195	201	207	7|4.2
728		213	219	225	231	237	243	249	255	261	267	8|4.8
729		273	279	285	291	297	303	308	314	320	326	9|5.4
730		332	338	344	350	356	362	368	374	380	386	
731		392	398	404	410	415	421	427	433	439	445	
732		451	457	463	469	475	481	487	493	499	504	
733		510	516	522	528	534	540	546	552	558	564	
734		570	576	581	587	593	599	605	611	617	623	
735		629	635	641	646	652	658	664	670	676	682	**5**
736		688	694	700	705	711	717	723	729	735	741	1|0.5
737		747	753	759	764	770	776	782	788	794	800	2|1.0
738		806	812	817	823	829	835	841	847	853	859	3|1.5
739		864	870	876	882	888	894	900	906	911	917	4|2.0
740		923	929	935	941	947	953	958	964	970	976	5|2.5
741		982	988	994	999	*005	*011	*017	*023	*029	*035	6|3.0
742	87	040	046	052	058	064	070	075	081	087	093	7|3.5
743		099	105	111	116	122	128	134	140	146	151	8|4.0
744		157	163	169	175	181	186	192	198	204	210	9|4.5
745		216	221	227	233	239	245	251	256	262	268	
746		274	280	286	291	297	303	309	315	320	326	
747		332	338	344	349	355	361	367	373	379	384	
748		390	396	402	408	413	419	425	431	437	442	
749		448	454	460	466	471	477	483	489	495	500	
750		506	512	518	523	529	535	541	547	552	558	

Table 3-3 (Continued). Logarithms

N.	L.	0	1	2	3	4	5	6	7	8	9	P. P.
750	87	506	512	518	523	529	535	541	547	552	558	
751		564	570	576	581	587	593	599	604	610	616	
752		622	628	633	639	645	651	656	662	668	674	
753		679	685	691	697	703	708	714	720	726	731	
754		737	743	749	754	760	766	772	777	783	789	
755		795	800	806	812	818	823	829	835	841	846	
756		852	858	864	869	875	881	887	892	898	904	
757		910	915	921	927	933	938	944	950	955	961	
758		967	973	978	984	990	996	*001	*007	*013	*018	
759	88	024	030	036	041	047	053	058	064	070	076	
760		081	087	093	098	104	110	116	121	127	133	6
761		138	144	150	156	161	167	173	178	184	190	1 | 0.6
762		195	201	207	213	218	224	230	235	241	247	2 | 1.2
763		252	258	264	270	275	281	287	292	298	304	3 | 1.8
764		309	315	321	326	332	338	343	349	355	360	4 | 2.4
765		366	372	377	383	389	395	400	406	412	417	5 | 3.0
766		423	429	434	440	446	451	457	463	468	474	6 | 3.6
767		480	485	491	497	502	508	513	519	525	530	7 | 4.2
768		536	542	547	553	559	564	570	576	581	587	8 | 4.8
769		593	598	604	610	615	621	627	632	638	643	9 | 5.4
770		649	655	660	666	672	677	683	689	694	700	
771		705	711	717	722	728	734	739	745	750	756	
772		762	767	773	779	784	790	795	801	807	812	
773		818	824	829	835	840	846	852	857	863	868	
774		874	880	885	891	897	902	908	913	919	925	
775		930	936	941	947	953	958	964	969	975	981	
776		986	992	997	*003	*009	*014	*020	*025	*031	*037	
777	89	042	048	053	059	064	070	076	081	087	092	
778		098	104	109	115	120	126	131	137	143	148	
779		154	159	165	170	176	182	187	193	198	204	
780		209	215	221	226	232	237	243	248	254	260	5
781		265	271	276	282	287	293	298	304	310	315	1 | 0.5
782		321	326	332	337	343	348	354	360	365	371	2 | 1.0
783		376	382	387	393	398	404	409	415	421	426	3 | 1.5
784		432	437	443	448	454	459	465	470	476	481	4 | 2.0
785		487	492	498	504	509	515	520	526	531	537	5 | 2.5
786		542	548	553	559	564	570	575	581	586	592	6 | 3.0
787		597	603	609	614	620	625	631	636	642	647	7 | 3.5
788		653	658	664	669	675	680	686	691	697	702	8 | 4.0
789		708	713	719	724	730	735	741	746	752	757	9 | 4.5
790		763	768	774	779	785	790	796	801	807	812	
791		818	823	829	834	840	845	851	856	862	867	
792		873	878	883	889	894	900	905	911	916	922	
793		927	933	938	944	949	955	960	966	971	977	
794		982	988	993	998	*004	*009	*015	*020	*026	*031	
795	90	037	042	048	053	059	064	069	075	080	086	
796		091	097	102	108	113	119	124	129	135	140	
797		146	151	157	162	168	173	179	184	189	195	
798		200	206	211	217	222	227	233	238	244	249	
799		255	260	266	271	276	282	287	293	298	304	
800		309	314	320	325	331	336	342	347	352	358	

Table 3-3 (Continued). Logarithms

N.	L.	0	1	2	3	4	5	6	7	8	9	P. P.
800	90	309	314	320	325	331	336	342	347	352	358	
801		363	369	374	380	385	390	396	401	407	412	
802		417	423	428	434	439	445	450	455	461	466	
803		472	477	482	488	493	499	504	509	515	520	
804		526	531	536	542	547	553	558	563	569	574	
805		580	585	590	596	601	607	612	617	623	628	
806		634	639	644	650	655	660	666	671	677	682	
807		687	693	698	703	709	714	720	725	730	736	
808		741	747	752	757	763	768	773	779	784	789	
809		795	800	806	811	816	822	827	832	838	843	**6**
810		849	854	859	865	870	875	881	886	891	897	
811		902	907	913	918	924	929	934	940	945	950	1 | 0.6
812		956	961	966	972	977	982	988	993	998	*004	2 | 1.2
813	91	009	014	020	025	030	036	041	046	052	057	3 | 1.8
814		062	068	073	078	084	089	094	100	105	110	4 | 2.4
815		116	121	126	132	137	142	148	153	158	164	5 | 3.0
816		169	174	180	185	190	196	201	206	212	217	6 | 3.6
817		222	228	233	238	243	249	254	259	265	270	7 | 4.2
818		275	281	286	291	297	302	307	312	318	323	8 | 4.8
819		328	334	339	344	350	355	360	365	371	376	9 | 5.4
820		381	387	392	397	403	408	413	418	424	429	
821		434	440	445	450	455	461	466	471	477	482	
822		487	492	498	503	508	514	519	524	529	535	
823		540	545	551	556	561	566	572	577	582	587	
824		593	598	603	609	614	619	624	630	635	640	
825		645	651	656	661	666	672	677	682	687	693	
826		698	703	709	714	719	724	730	735	740	745	
827		751	756	761	766	772	777	782	787	793	798	
828		803	808	814	819	824	829	834	840	845	850	
829		855	861	866	871	876	882	887	892	897	903	
830		908	913	918	924	929	934	939	944	950	955	**5**
831		960	965	971	976	981	986	991	997	*002	*007	1 | 0.5
832	92	012	018	023	028	033	038	044	049	054	059	2 | 1.0
833		065	070	075	080	085	091	096	101	106	111	3 | 1.5
834		117	122	127	132	137	143	148	153	158	163	4 | 2.0
835		169	174	179	184	189	195	200	205	210	215	5 | 2.5
836		221	226	231	236	241	247	252	257	262	267	6 | 3.0
837		273	278	283	288	293	298	304	309	314	319	7 | 3.5
838		324	330	335	340	345	350	355	361	366	371	8 | 4.0
839		376	381	387	392	397	402	407	412	418	423	9 | 4.5
840		428	433	438	443	449	454	459	464	469	474	
841		480	485	490	495	500	505	511	516	521	526	
842		531	536	542	547	552	557	562	567	572	578	
843		583	588	593	598	603	609	614	619	624	629	
844		634	639	645	650	655	660	665	670	675	681	
845		686	691	696	701	706	711	716	722	727	732	
846		737	742	747	752	758	763	768	773	778	783	
847		788	793	799	804	809	814	819	824	829	834	
848		840	845	850	855	860	865	870	875	881	886	
849		891	896	901	906	911	916	921	927	932	937	
850		942	947	952	957	962	967	973	978	983	988	

Table 3-3 (Continued). Logarithms

N.	L. 0	1	2	3	4	5	6	7	8	9	P. P.
850	92 942	947	952	957	962	967	973	978	983	988	
851	993	998	*003	*008	*013	*018	*024	*029	*034	*039	
852	93 044	049	054	059	064	069	075	080	085	090	
853	095	100	105	110	115	120	125	131	136	141	
854	146	151	156	161	166	171	176	181	186	192	
855	197	202	207	212	217	222	227	232	237	242	**6**
856	247	252	258	263	268	273	278	283	288	293	1|0.6
857	298	303	308	313	318	323	328	334	339	344	2|1.2
858	349	354	359	364	369	374	379	384	389	394	3|1.8
859	399	404	409	414	420	425	430	435	440	445	4|2.4
860	450	455	460	465	470	475	480	485	490	495	5|3.0
861	500	505	510	515	520	526	531	536	541	546	6|3.6
862	551	556	561	566	571	576	581	586	591	596	7|4.2
863	601	606	611	616	621	626	631	636	641	646	8|4.8
864	651	656	661	666	671	676	682	687	692	697	9|5.4
865	702	707	712	717	722	727	732	737	742	747	
866	752	757	762	767	772	777	782	787	792	797	
867	802	807	812	817	822	827	832	837	842	847	
868	852	857	862	867	872	877	882	887	892	897	
869	902	907	912	917	922	927	932	937	942	947	
870	952	957	962	967	972	977	982	987	992	997	
871	94 002	007	012	017	022	027	032	037	042	047	**5**
872	052	057	062	067	072	077	082	086	091	096	1|0.5
873	101	106	111	116	121	126	131	136	141	146	2|1.0
874	151	156	161	166	171	176	181	186	191	196	3|1.5
875	201	206	211	216	221	226	231	236	240	245	4|2.0
876	250	255	260	265	270	275	280	285	290	295	5|2.5
877	300	305	310	315	320	325	330	335	340	345	6|3.0
878	349	354	359	364	369	374	379	384	389	394	7|3.5
879	399	404	409	414	419	424	429	433	438	443	8|4.0
880	448	453	458	463	468	473	478	483	488	493	9|4.5
881	498	503	507	512	517	522	527	532	537	542	
882	547	552	557	562	567	571	576	581	586	591	
883	596	601	606	611	616	621	626	630	635	640	
884	645	650	655	660	665	670	675	680	685	689	
885	694	699	704	709	714	719	724	729	734	738	**4**
886	743	748	753	758	763	768	773	778	783	787	
887	792	797	802	807	812	817	822	827	832	836	1|0.4
888	841	846	851	856	861	866	871	876	880	885	2|0.8
889	890	895	900	905	910	915	919	924	929	934	3|1.2
890	939	944	949	954	959	963	968	973	978	983	4|1.6
891	988	993	998	*002	*007	*012	*017	*022	*027	*032	5|2.0
892	95 036	041	046	051	056	061	066	071	075	080	6|2.4
893	085	090	095	100	105	109	114	119	124	129	7|2.8
894	134	139	143	148	153	158	163	168	173	177	8|3.2
895	182	187	192	197	202	207	211	216	221	226	9|3.6
896	231	236	240	245	250	255	260	265	270	274	
897	279	284	289	294	299	303	308	313	318	323	
898	328	332	337	342	347	352	357	361	366	371	
899	376	381	386	390	395	400	405	410	415	419	
900	424	429	434	439	444	448	453	458	463	468	

Table 3-3 (Continued). Logarithms

N.	L.	0	1	2	3	4	5	6	7	8	9	P. P.
900	95	424	429	434	439	444	448	453	458	463	468	
901		472	477	482	487	492	497	501	506	511	516	
902		521	525	530	535	540	545	550	554	559	564	
903		569	574	578	583	588	593	598	602	607	612	
904		617	622	626	631	636	641	646	650	655	660	
905		665	670	674	679	684	689	694	698	703	708	
906		713	718	722	727	732	737	742	746	751	756	
907		761	766	770	775	780	785	789	794	799	804	
908		809	813	818	823	828	832	837	842	847	852	
909		856	861	866	871	875	880	885	890	895	899	
910		904	909	914	918	923	928	933	938	942	947	**5**
911		952	957	961	966	971	976	980	985	990	995	1 | 0.5
912		999	*004	*009	*014	*019	*023	*028	*033	*038	*042	2 | 1.0
913	96	047	052	057	061	066	071	076	080	085	090	3 | 1.5
914		095	099	104	109	114	118	123	128	133	137	4 | 2.0
915		142	147	152	156	161	166	171	175	180	185	5 | 2.5
916		190	194	199	204	209	213	218	223	227	232	6 | 3.0
917		237	242	246	251	256	261	265	270	275	280	7 | 3.5
918		284	289	294	298	303	308	313	317	322	327	8 | 4.0
919		332	336	341	346	350	355	360	365	369	374	9 | 4.5
920		379	384	388	393	398	402	407	412	417	421	
921		426	431	435	440	445	450	454	459	464	468	
922		473	478	483	487	492	497	501	506	511	515	
923		520	525	530	534	539	544	548	553	558	562	
924		567	572	577	581	586	591	595	600	605	609	
925		614	619	624	628	633	638	642	647	652	656	
926		661	666	670	675	680	685	689	694	699	703	
927		708	713	717	722	727	731	736	741	745	750	
928		755	759	764	769	774	778	783	788	792	797	
929		802	806	811	816	820	825	830	834	839	844	
930		848	853	858	862	867	872	876	881	886	890	**4**
931		895	900	904	909	914	918	923	928	932	937	1 | 0.4
932		942	946	951	956	960	965	970	974	979	984	2 | 0.8
933		988	993	997	*002	*007	*011	*016	*021	*025	*030	3 | 1.2
934	97	035	039	044	049	053	058	063	067	072	077	4 | 1.6
935		081	086	090	095	100	104	109	114	118	123	5 | 2.0
936		128	132	137	142	146	151	155	160	165	169	6 | 2.4
937		174	179	183	188	192	197	202	206	211	216	7 | 2.8
938		220	225	230	234	239	243	248	253	257	262	8 | 3.2
939		267	271	276	280	285	290	294	299	304	308	9 | 3.6
940		313	317	322	327	331	336	340	345	350	354	
941		359	364	368	373	377	382	387	391	396	400	
942		405	410	414	419	424	428	433	437	442	447	
943		451	456	460	465	470	474	479	483	488	493	
944		497	502	506	511	516	520	525	529	534	539	
945		543	548	552	557	562	566	571	575	580	585	
946		589	594	598	603	607	612	617	621	626	630	
947		635	640	644	649	653	658	663	667	672	676	
948		681	685	690	695	699	704	708	713	717	722	
949		727	731	736	740	745	749	754	759	763	768	
950		772	777	782	786	791	795	800	804	809	813	

Table 3-3 (Concluded). Logarithms

N.	L.	0	1	2	3	4	5	6	7	8	9	P. P.
950	97	772	777	782	786	791	795	800	804	809	813	
951		818	823	827	832	836	841	845	850	855	859	
952		864	868	873	877	882	886	891	896	900	905	
953		909	914	918	923	928	932	937	941	946	950	
954		955	959	964	968	973	978	982	987	991	996	
955	98	000	005	009	014	019	023	028	032	037	041	
956		046	050	055	059	064	068	073	078	082	087	
957		091	096	100	105	109	114	118	123	127	132	
958		137	141	146	150	155	159	164	168	173	177	
959		182	186	191	195	200	204	209	214	218	223	
960		227	232	236	241	245	250	254	259	263	268	**5**
961		272	277	281	286	290	295	299	304	308	313	1 | 0.5
962		318	322	327	331	336	340	345	349	354	358	2 | 1.0
963		363	367	372	376	381	385	390	394	399	403	3 | 1.5
964		408	412	417	421	426	430	435	439	444	448	4 | 2.0
965		453	457	462	466	471	475	480	484	489	493	5 | 2.5
966		498	502	507	511	516	520	525	529	534	538	6 | 3.0
967		543	547	552	556	561	565	570	574	579	583	7 | 3.5
968		588	592	597	601	605	610	614	619	623	628	8 | 4.0
969		632	637	641	646	650	655	659	664	668	673	9 | 4.5
970		677	682	686	691	695	700	704	709	713	717	
971		722	726	731	735	740	744	749	753	758	762	
972		767	771	776	780	784	789	793	798	802	807	
973		811	816	820	825	829	834	838	843	847	851	
974		856	860	865	869	874	878	883	887	892	896	
975		900	905	909	914	918	923	927	932	936	941	
976		945	949	954	958	963	967	972	976	981	985	
977		989	994	998	*003	*007	*012	*016	*021	*025	*029	
978	99	034	038	043	047	052	056	061	065	069	074	
979		078	083	087	092	096	100	105	109	114	118	
980		123	127	131	136	140	145	149	154	158	162	
981		167	171	176	180	185	189	193	198	202	207	**4**
982		211	216	220	224	229	233	238	242	247	251	1 | 0.4
983		255	260	264	269	273	277	282	286	291	295	2 | 0.8
984		300	304	308	313	317	322	326	330	335	339	3 | 1.2
985		344	348	352	357	361	366	370	374	379	383	4 | 1.6
986		388	392	396	401	405	410	414	419	423	427	5 | 2.0
987		432	436	441	445	449	454	458	463	467	471	6 | 2.4
988		476	480	484	489	493	498	502	506	511	515	7 | 2.8
989		520	524	528	533	537	542	546	550	555	559	8 | 3.2
990		564	568	572	577	581	585	590	594	599	603	9 | 3.6
991		607	612	616	621	625	629	634	638	642	647	
992		651	656	660	664	669	673	677	682	686	691	
993		695	699	704	708	712	717	721	726	730	734	
994		739	743	747	752	756	760	765	769	774	778	
995		782	787	791	795	800	804	808	813	817	822	
996		826	830	835	839	843	848	852	856	861	865	
997		870	874	878	883	887	891	896	900	904	909	
998		913	917	922	926	930	935	939	944	948	952	
999		957	961	965	970	974	978	983	987	991	996	
1000	∞	000	004	009	013	017	022	026	030	035	039	

Table 3-4. Condensed Table of Common Logarithms*

N	0	1	2	3	4	5	6	7	8	9
10	0000	0043	0086	0128	0170	0212	0253	0294	0334	0374
11	0414	0453	0492	0531	0569	0607	0645	0682	0719	0755
12	0792	0828	0864	0899	0934	0969	1004	1038	1072	1106
13	1139	1173	1206	1239	1271	1303	1335	1367	1399	1430
14	1461	1492	1523	1553	1584	1614	1644	1673	1703	1732
15	1761	1790	1818	1847	1875	1903	1931	1959	1987	2014
16	2041	2068	2095	2122	2148	2175	2201	2227	2253	2279
17	2304	2330	2355	2380	2405	2430	2455	2480	2504	2529
18	2553	2577	2601	2625	2648	2672	2695	2718	2742	2765
19	2788	2810	2833	2856	2878	2900	2923	2945	2967	2989
20	3010	3032	3054	3075	3096	3118	3139	3160	3181	3201
21	3222	3243	3263	3284	3304	3324	3345	3365	3385	3404
22	3424	3444	3464	3483	3502	3522	3541	3560	3579	3598
23	3617	3636	3655	3674	3692	3711	3729	3747	3766	3784
24	3802	3820	3838	3856	3874	3892	3909	3927	3945	3962
25	3979	3997	4014	4031	4048	4065	4082	4099	4116	4133
26	4150	4166	4183	4200	4216	4232	4249	4265	4281	4298
27	4314	4330	4346	4362	4378	4393	4409	4425	4440	4456
28	4472	4487	4502	4518	4533	4548	4564	4579	4594	4609
29	4624	4639	4654	4669	4683	4698	4713	4728	4742	4757
30	4771	4786	4800	4814	4829	4843	4857	4871	4886	4900
31	4914	4928	4942	4955	4969	4983	4997	5011	5024	5038
32	5051	5065	5079	5092	5105	5119	5132	5145	5159	5172
33	5185	5198	5211	5224	5237	5250	5263	5276	5289	5302
34	5315	5328	5340	5353	5366	5378	5391	5403	5416	5428
35	5441	5453	5465	5478	5490	5502	5514	5527	5539	5551
36	5563	5575	5587	5599	5611	5623	5635	5647	5658	5670
37	5682	5694	5705	5717	5729	5740	5752	5763	5775	5786
38	5798	5809	5821	5832	5843	5855	5866	5877	5888	5899
39	5911	5922	5933	5944	5955	5966	5977	5988	5999	6010
40	6021	6031	6042	6053	6064	6075	6085	6096	6107	6117
41	6128	6138	6149	6160	6170	6180	6191	6201	6212	6222
42	6232	6243	6253	6263	6274	6284	6294	6304	6314	6325
43	6335	6345	6355	6365	6375	6385	6395	6405	6415	6425
44	6435	6444	6454	6464	6474	6484	6493	6503	6513	6522
45	6532	6542	6551	6561	6571	6580	6590	6599	6609	6618
46	6628	6637	6646	6656	6665	6675	6684	6693	6702	6712
47	6721	6730	6739	6749	6758	6767	6776	6785	6794	6803
48	6812	6821	6830	6839	6848	6857	6866	6875	6884	6893
49	6902	6911	6920	6928	6937	6946	6955	6964	6972	6981
50	6990	6998	7007	7016	7024	7033	7042	7050	7059	7067
51	7076	7084	7093	7101	7110	7118	7126	7135	7143	7152
52	7160	7168	7177	7185	7193	7202	7210	7218	7226	7235
53	7243	7251	7259	7267	7275	7284	7292	7300	7308	7316
54	7324	7332	7340	7348	7356	7364	7372	7380	7388	7396
55	7404	7412	7419	7427	7435	7443	7451	7459	7466	7474
56	7482	7490	7497	7505	7513	7520	7528	7536	7543	7551
57	7559	7566	7574	7582	7589	7597	7604	7612	7619	7627
58	7634	7642	7649	7657	7664	7672	7679	7686	7694	7701
59	7709	7716	7723	7731	7738	7745	7752	7760	7767	7774
60	7782	7789	7796	7803	7810	7818	7825	7832	7839	7846
61	7853	7860	7868	7875	7882	7889	7896	7903	7910	7917
62	7924	7931	7938	7945	7952	7959	7966	7973	7980	7987
63	7993	8000	8007	8014	8021	8028	8035	8041	8048	8055
64	8062	8069	8075	8082	8089	8096	8102	8109	8116	8122
65	8129	8136	8142	8149	8156	8162	8169	8176	8182	8189
66	8195	8202	8209	8215	8222	8228	8235	8241	8248	8254
67	8261	8267	8274	8280	8287	8293	8299	8306	8312	8319
68	8325	8331	8338	8344	8351	8357	8363	8370	8376	8382
69	8388	8395	8401	8407	8414	8420	8426	8432	8439	8445
70	8451	8457	8463	8470	8476	8482	8488	8494	8500	8506
71	8513	8519	8525	8531	8537	8543	8549	8555	8561	8567
72	8573	8579	8585	8591	8597	8603	8609	8615	8621	8627
73	8633	8639	8645	8651	8657	8663	8669	8675	8681	8686
74	8692	8698	8704	8710	8716	8722	8727	8733	8739	8745
75	8751	8756	8762	8768	8774	8779	8785	8791	8797	8802
76	8808	8814	8820	8825	8831	8837	8842	8848	8854	8859
77	8865	8871	8876	8882	8887	8893	8899	8904	8910	8915
78	8921	8927	8932	8938	8943	8949	8954	8960	8965	8971
79	8976	8982	8987	8993	8998	9004	9009	9015	9020	9025
80	9031	9036	9042	9047	9053	9058	9063	9069	9074	9079
81	9085	9090	9096	9101	9106	9112	9117	9122	9128	9133
82	9138	9143	9149	9154	9159	9165	9170	9175	9180	9186
83	9191	9196	9201	9206	9212	9217	9222	9227	9232	9238
84	9243	9248	9253	9258	9263	9269	9274	9279	9284	9289
85	9294	9299	9304	9309	9315	9320	9325	9330	9335	9340
86	9345	9350	9355	9360	9365	9370	9375	9380	9385	9390
87	9395	9400	9405	9410	9415	9420	9425	9430	9435	9440
88	9445	9450	9455	9460	9465	9469	9474	9479	9484	9489
89	9494	9499	9504	9509	9513	9518	9523	9528	9533	9538
90	9542	9547	9552	9557	9562	9566	9571	9576	9581	9586
91	9590	9595	9600	9605	9609	9614	9619	9624	9628	9633
92	9638	9643	9647	9652	9657	9661	9666	9671	9675	9680
93	9685	9689	9694	9699	9703	9708	9713	9717	9722	9727
94	9731	9736	9741	9745	9750	9754	9759	9763	9768	9773
95	9777	9782	9786	9791	9795	9800	9805	9809	9814	9818
96	9823	9827	9832	9836	9841	9845	9850	9854	9859	9863
97	9868	9872	9877	9881	9886	9890	9894	9899	9903	9908
98	9912	9917	9921	9926	9930	9934	9939	9943	9948	9952
99	9956	9961	9965	9969	9974	9978	9983	9987	9991	9996

* This table gives the mantissas of numbers with the decimal point omitted in each case. Characteristics are determined by inspection from the numbers.

The sheet metal draftsman must often calculate lengths, areas, or volumes of triangles, circles, and other geometrical figures by application of equations and formulas. An equation is a mathematical expression of equality. When an equation states a proven rule, it is called a formula. The following procedure is helpful for solving equations:

1. Substitute values of the given letters.
2. Combine similar terms.
3. Simplify terms.
4. When both sides of an equation are divided, multiplied, added to, or subtracted by the same number, the equality will remain unchanged.
5. Solve for the unknown.
6. Prove the result by substitution.

MATHEMATICAL ABBREVIATIONS AND SYMBOLS

A	Area	∡	Angle
a	Vertical side of plane figure	⊙	Circle
b	Base of plane figure	ϕ	Diameter
c	Hypotenuse of right triangle	÷	Divided by
C	Circumference of circle	:	Is to (in proportion, or ratio)
d	Diameter of circle	=	Is equal to
D	Large diameter	×	Multiplied by
h	Vertical height	π	Pi
P	Perimeter	+	Plus, sign of addition
r	Radius of circle	±	Plus or minus
R	Large radius	▭	Rectangle
sh	Slant height	√	Square root
V	Volume	a^2	a to the second power
W	Width	▢	Square
x,y,z	Unknowns	△	Triangle
		∴	Therefore

Right Triangle

$$a^2 + b^2 = c^2 \quad \text{when, } a = 3$$
$$a = \sqrt{c^2 - b^2} \quad b = 4$$
$$b = \sqrt{c^2 - a^2} \quad c = 5$$
$$c = \sqrt{a^2 + b^2} \quad \text{or proportional parts}$$

$$A = \frac{ab}{2}$$

$$P = a + b + c$$

Example 1:

Find the true length of a line to the nearest ⅛" which is 66" long in plan view and has a slope of 2" per foot.

Let:

a = Distance of difference in height of slope line

b = Line in plan view 66" long

b = 66" = 5.5'

a = 2" slope per foot at 5.5' = 11"

$c = \sqrt{66^2 + 11^2}$
$= \sqrt{4477} = 66⅞''$ *Ans*

Example 2:

Compute $\sqrt{4477}$ (in Example 1) by logs: The characteristic of 4477 is (3), which is one less than the number of digits (4).

From log tables, the mantissa is .65099. To find the square root, the log of a number is divided by two, therefore,

$$3.65099 \div 2 = 1.82549$$

From tables, log 1.82549 = 6691. As the characteristic is (1), the decimal point is to the right of the second digit or 66.91. To the nearest ⅛", 66.91" = 66⅞" *Ans*

Example 3:

In a right triangle find c, to the nearest 16th of an inch, when $a = 64''$ and $b = 47''$.

$c = \sqrt{a^2 + b^2}$ Log 6305 = 3.79969
$= \sqrt{64^2 + 47^2}$ $3.79969 \div 2 = 1.89984$
$= \sqrt{4096 + 2209}$
$= \sqrt{6305}$ $c = 79.4''$

$c = 79⅜''$ *Ans*

to find .4" to the nearest 16th of an inch:

$$.4 \times 16 = 6.4$$
$$= \tfrac{6}{16} = \tfrac{3}{8}''$$

Example 4:

What is the area of the triangle in Example 1, in square feet?

$a = 11''$, $b = 66''$. One square foot = 144 sq in.

$$A = \frac{ab}{2} = \frac{11 \times 66}{2 \times 144} = 2.52 \text{ sq ft}$$

or, from Table 3-1, convert inches to decimals of a foot:

$$a = 11'' = .9167'$$
$$b = 66'' \div 2 = 33'' = 2.75'$$
$$.9167 \times 2.75 = 2.52 \text{ sq ft}$$

Circle

In a circle, the ratio of the circumference to the diameter is indicated by the Greek letter π (Pi). The value of pi to four decimal places is 3.1416.

Equivalent Functions of π

$\pi = 3.1416$

$\dfrac{1}{\pi} = 0.3183$

$2\pi = 6.2832$

$\dfrac{1}{2\pi} = 0.15915$

$\dfrac{\pi}{2} = 1.5708$

$\dfrac{\pi}{4} = 0.7854$

$\pi^2 = 9.8696$

$\dfrac{1}{\pi^2} = 0.1013$

$\sqrt{\pi} = 1.772$

$\dfrac{1}{\sqrt{\pi}} = 0.564$

$\pi^3 = 31.0063$

$\dfrac{4\pi}{3} = 4.1888$

$$A = \pi r^2$$
$$= 3.1416 r^2$$
$$= .7854 d^2$$

$$C = 2\pi r$$
$$= 6.283 r$$
$$= \pi d$$
$$= 3.1416 d$$

$$r = \dfrac{C}{2\pi}$$
$$= \dfrac{C}{6.283}$$
$$= C \times .15915$$

$$d = \dfrac{C}{\pi}$$
$$= \dfrac{C}{3.1416}$$
$$= C \times .3183$$

Example 1:

What is the square inch area, to the nearest inch, of a circle 11 inches in diameter?

$$\text{Use } A = .7854 d^2$$
$$= (.7854) 121$$
$$= 95.0334$$
$$A = 95 \text{ sq in.}$$

Example 2:

Find the diameter, to the nearest quarter of an inch, of a circle having a circumference of 135''.

$$\text{Use } d = C \times 0.3183$$
$$= 135 \times .3183$$
$$= 42.97$$
$$d = 43''$$

Circular Sector

$$l = \dfrac{r \times \angle \times \pi}{180} \qquad A = \dfrac{\angle \text{ of sector}}{360} \pi r^2$$
$$= .01745 r \angle$$

Example:

A shop drawing indicates that a 90° elbow, 26'' in diameter is to be rotated clockwise 25° down from the horizontal. Show how an accurate rotation may be accomplished for the field installation.

$$l = .01745 r \angle$$
$$= .01745 \times 25 \times 13$$
$$= 5.67125$$
$$l = 5^{11}\!/_{16}''$$

$$r = \dfrac{26}{2} = 13''$$
$$\angle = 25°$$

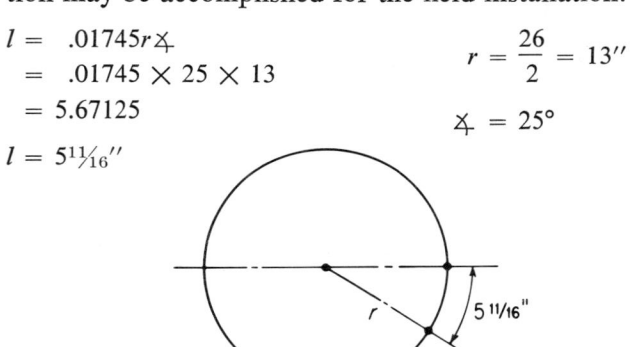

Flat Oval

$$A = .7854 d^2 + bd \qquad P = \pi d + 2b$$

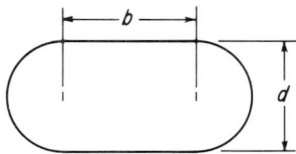

Example 1:

Find the area of a 30'' x 16'' flat oval to the nearest tenth of a square inch.

$A = .7854d^2 + bd$
$= 201.062 + (14)(16)$
$A = 425.1$ sq in.

$b = 30 - 16 = 14$
$d = 16$

Example 1:

Find the cross-section area, to the nearest tenth of a sq ft of a 66″ x 30″ duct.

$b = 66'' = 5.5'$
$h = 30'' = 2.5'$
$A = bh$
$= 5.5 \times 2.5$
$A = 13.75$ sq ft

Square

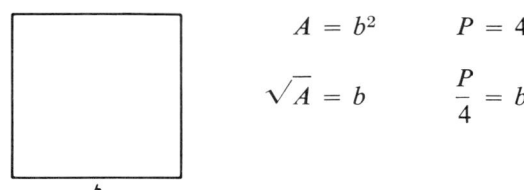

$A = b^2 \qquad P = 4b$
$\sqrt{A} = b \qquad \dfrac{P}{4} = b$

Example 1:

Find the cross-section area, to the nearest tenth of a square foot, of an 18″ x 18″ duct.

$A = b^2 \qquad b = 18'' = 1.5'$
$A = 1.5 \times 1.5$
$ = 2.25$ sq ft

Example 2:

Calculate the size of a square duct with a cross-section area of 729 sq in.

$b = \sqrt{A}$
$ = \sqrt{729}$
$b = 27''$

Example 3:

What is the size of a square duct with a perimeter of 92″?

$b = \dfrac{P}{4}$
$b = \dfrac{92}{4} = 23''$

Rectangle

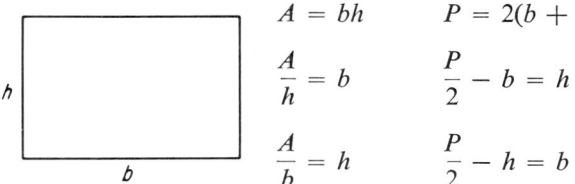

$A = bh \qquad P = 2(b + h)$
$\dfrac{A}{h} = b \qquad \dfrac{P}{2} - b = h$
$\dfrac{A}{b} = h \qquad \dfrac{P}{2} - h = b$

Example 2:

Determine the width and perimeter of a duct which has an area of 4.25 sq ft and is 18″ high.

$\dfrac{A}{h} = b$
$\dfrac{4.25 \times 144}{18} = b = 34''$
$P = 2(b + h)$
$ = 2(34 + 18)$
$P = 104''$

Square Prism

$V = bhw$

Surface $A_1 = (bh + bw + hw)2$

Surface A_2 of duct $hw = 2b(h + w)$

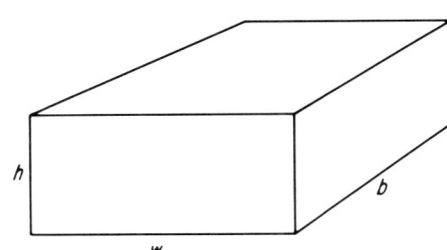

Example 1:

Find the square foot volume of a space which has an 8′-6″ ceiling height and with floor dimensions of 36′9″ x 22′10″.

$V = bhw$
$ = 36.75 \times 8.5 \times 22.8333$
$V = 7132.6$ cu ft   *Ans*

$b = 36'\text{-}9'' = 36.75'$
$h = 8'\text{-}6'' = 8.5'$
$w = 22'\text{-}10'' = 22.8333'$

Example 2:

Calculate the surface area, in square feet, of a 42" x 16" plenum, 8'-0" long.

$$b = 8'$$
$$h = 16'' = 1'\text{-}4'' = 1.33'$$
$$w = 42'' = 3'\text{-}6'' = 3.5'$$
$$A_1 = (bh + bw + hw)2$$
$$= [8(1.33 + 3.5) + (1.33)3.5]2$$
$$A_1 = 86.6 \text{ sq ft} \quad Ans$$

Example 3:

Determine the surface area in square feet of a 66" x 30" duct, 45 feet long.

$$A_2 = 2b(h + w)$$
$$= 2 \times 45(2.5 + 5.5)$$
$$A_2 = 720 \text{ sq ft} \quad Ans$$

$$h = 30'' = 2'\text{-}6'' = 2.5'$$
$$w = 66'' = 5'\text{-}6'' = 5.5'$$
$$b = 45'$$

where:
$Wt = (w + h)bF$
$Wt = $ weight of galvanized duct, in pounds
$w = $ width, inches
$h = $ height, inches
$b = $ length, lineal feet
$F = $ factor from table (allows for seams)

Values of 'F'

Duct Gage	F
26	.16
24	.21
22	.25
20	.30
18	.39
16	.47
14	.57
12	.80
10	1.00

Example 4:

A duct run is estimated at 245 feet long.

(a) Find the weight of a 78" x 12" duct of 20-gage galvanized steel.
(b) What would be the saving in weight if a 30" x 27" duct of 24-gage galvanized steel was used?

$$w = 78 \text{ and } 30 \quad b = 245$$
$$h = 12 \text{ and } 27 \quad F = .30 \text{ and } .21$$
$$Wt = (w + h)bF$$
$$Wt = (78 + 12)245 \times .30$$
$$= 6{,}615 \text{ lbs} \quad Ans \text{ (a)}$$
$$Wt = (30 + 27)245 \times .21$$
$$= 2{,}932.65 \text{ lbs for } 30'' \times 27''$$

$$\begin{array}{r} 6{,}615 \\ -2{,}933 \\ \hline 3{,}682 \text{ lbs difference} \end{array} \quad Ans \text{ (b)}$$

Cylinder

Volume, $V = .7854d^2h$
Area of cylindrical surface $= S = \pi dh$

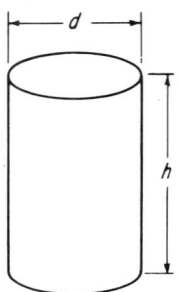

Example 1:

A 10-gage stainless steel tank has a diameter of 30 inches and a capacity of 275 gallons. How long must the tank be?

One gallon $= 231$ cu in.

$$V = 275 \text{ gal} \times \frac{231 \text{ cu in.}}{1 \text{ gal}} = 63{,}525 \text{ cu in.}$$
$$d = 30''$$
$$V = .7854d^2h$$
$$63{,}525 = .7854 \times 30^2 h$$
$$\frac{63{,}525}{706.9} = h$$
$$90'' = h \quad Ans$$

Example 2:

How many square feet of material are in a duct 26" in diameter and 30 feet long?

$$d = 26'' = 2'\text{-}2'' = 2.1667'$$
$$h = 36'$$
$$S = \pi dh$$
$$= 3.14 \times 2.167 \times 36$$
$$S = 245 \text{ sq ft} \quad Ans$$

PROPORTION

When the ratio of two numbers is equal to the ratio of two other numbers, the four are said to be proportional.

A proportion is a simple fractional equation.

Consider: $\quad \dfrac{3}{5} = \dfrac{9}{15}$

As a fraction, it is read: three-fifths equals nine-fifteenths.

As a proportion, it is read: three is to five as nine is to fifteen.

It may also be written: $3:5 = 9:15$

where, 3 and 15 are the extremes,
and 5 and 9 are the means.

The sheet metal draftsman often applies the rules of proportioning as in the following relationships with sides of similar triangles.

In the triangles ADE, ABC

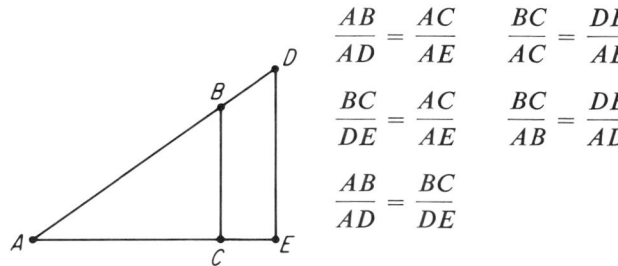

$$\frac{AB}{AD} = \frac{AC}{AE} \qquad \frac{BC}{AC} = \frac{DE}{AE}$$

$$\frac{BC}{DE} = \frac{AC}{AE} \qquad \frac{BC}{AB} = \frac{DE}{AD}$$

$$\frac{AB}{AD} = \frac{BC}{DE}$$

Example 1:

Assume the low point of the bottom of a supply duct is 8'-2" above the finished floor and slopes upward 4'-2" within a distance of 10'-10" from the low point. Find the height of the duct from the finished floor at a horizontal distance of 7'-7" from the low point.

In triangles ABC and ADE

Let:

$BC = X$ $\qquad \dfrac{BC}{DE} = \dfrac{AC}{AE}$

$DE = 4'\text{-}2''$

$AC = 7'\text{-}7''$ $\qquad \dfrac{X}{4'\text{-}2''} = \dfrac{7'\text{-}7''}{10'\text{-}10''}$

$AE = 10'\text{-}10''$

$$\dfrac{X}{50} = \dfrac{91}{130}$$

$$X = \left(\dfrac{91}{130}\right) 50 = 35 = 2'\text{-}11''$$

Height at point B is 2'-11" above height at point A, or $2'\text{-}11'' + 8'\text{-}2'' = 11'\text{-}1''$ *Ans.*

Example 2:

The top of a central-system apparatus casing is 11'-2" above the finished floor at a bank of reheat coils. Starting 6" from the leaving-air side of the coils, the top slopes downward 2'-8" to a distance 8'-10" from the coils. Find the height of the casing top at a distance of 6'-9" from the coil face.

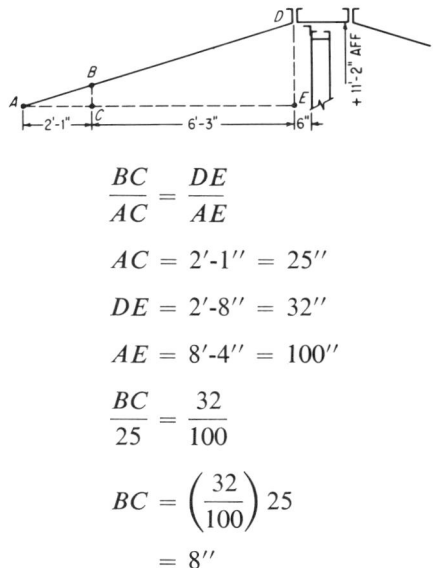

$$\frac{BC}{AC} = \frac{DE}{AE}$$

$AC = 2'\text{-}1'' = 25''$

$DE = 2'\text{-}8'' = 32''$

$AE = 8'\text{-}4'' = 100''$

$$\frac{BC}{25} = \frac{32}{100}$$

$$BC = \left(\frac{32}{100}\right) 25$$

$$= 8''$$

Point $B =$
$(11'\text{-}2'') - (2'\text{-}8'') + 8'' = 9'\text{-}2''$ AFF *Ans*

TRIGONOMETRY OF THE RIGHT TRIANGLE

A right angle is equal to 90°. When one of the angles of a triangle is 90° it is called a right triangle. The legs of a right triangle are the sides which form the 90° angle and the hypotenuse is the side opposite the 90° angle. The other two angles of the triangle are acute and their sum is 90°. Trigonometry tables are based on the ratio of two sides of the right triangle, length-to-length as in the following formulas:

$$\text{Sine} = \frac{\text{Side Opposite}}{\text{Hypotenuse}} = \text{Sine } A = \frac{a}{c};$$

$$\text{Sine } B = \frac{b}{c}$$

$$\text{Cosine} = \frac{\text{Side Adjacent}}{\text{Hypotenuse}} = \text{Cosine } A = \frac{b}{c};$$

$$\text{Cosine } B = \frac{a}{c}$$

(Text continued on page 105.)

Table 3-5. Values of the Natural Trigonometric Functions

0° — **179°**

M	Sine	Cosine	Tan.	Cotan.	Secant	Cosec.	M
0	0.00000	1.0000	0.00000	Infinite	1.0000	Infinite	60
1	.00029	.0000	.00029	3437.7	.0000	3437.7	59
2	.00058	.0000	.00058	1718.9	.0000	1718.9	58
3	.00087	.0000	.00087	1145.9	.0000	1145.9	57
4	.00116	.0000	.00116	859.44	.0000	859.44	56
5	0.00145	1.0000	0.00145	687.55	1.0000	687.55	55
6	.00174	.0000	.00174	572.96	.0000	572.96	54
7	.00204	.0000	.00204	491.11	.0000	491.11	53
8	.00233	.0000	.00233	429.72	.0000	429.72	52
9	.00262	.0000	.00262	381.97	.0000	381.97	51
10	0.00291	0.99999	0.00291	343.77	1.0000	343.77	50
11	.00320	.99999	.00320	312.52	.0000	312.52	49
12	.00349	.99999	.00349	286.48	.0000	286.48	48
13	.00378	.99999	.00378	264.44	.0000	264.44	47
14	.00407	.99999	.00407	245.55	.0000	245.55	46
15	0.00436	0.99999	0.00436	229.18	1.0000	229.18	45
16	.00465	.99999	.00465	214.86	.0000	214.86	44
17	.00494	.99999	.00494	202.22	.0000	202.22	43
18	.00524	.99999	.00524	190.98	.0000	190.99	42
19	.00553	.99998	.00553	180.93	.0000	180.93	41
20	0.00582	0.99998	0.00582	171.88	1.0000	171.89	40
21	.00611	.99998	.00611	163.70	.0000	163.70	39
22	.00640	.99998	.00640	156.26	.0000	156.26	38
23	.00669	.99998	.00669	149.46	.0000	149.47	37
24	.00698	.99997	.00698	143.24	.0000	143.24	36
25	0.00727	0.99997	0.00727	137.51	1.0000	137.51	35
26	.00756	.99997	.00756	132.22	.0000	132.22	34
27	.00785	.99997	.00785	127.32	.0000	127.32	33
28	.00814	.99997	.00814	122.77	.0000	122.78	32
29	.00843	.99996	.00844	118.54	.0000	118.54	31
30	0.00873	0.99996	0.00873	114.59	1.0000	114.59	30
31	.00902	.99996	.00902	110.89	.0000	110.90	29
32	.00931	.99996	.00931	107.43	.0000	107.43	28
33	.00960	.99995	.00960	104.17	.0000	104.17	27
34	.00989	.99995	.00989	101.11	.0000	101.11	26
35	0.01018	0.99995	0.01018	98.218	1.0000	98.223	25
36	.01047	.99994	.01047	95.489	.0000	95.495	24
37	.01076	.99994	.01076	92.908	.0000	92.914	23
38	.01105	.99994	.01105	90.463	.0001	90.469	22
39	.01134	.99993	.01134	88.143	.0001	88.149	21
40	0.01163	0.99993	0.01164	85.940	1.0001	85.946	20
41	.01193	.99993	.01193	83.843	.0001	83.849	19
42	.01222	.99992	.01222	81.847	.0001	81.853	18
43	.01251	.99992	.01251	79.943	.0001	79.950	17
44	.01280	.99992	.01280	78.126	.0001	78.133	16
45	0.01309	0.99991	0.01309	76.390	1.0001	76.396	15
46	.01338	.99991	.01338	74.729	.0001	74.736	14
47	.01367	.99991	.01367	73.139	.0001	73.146	13
48	.01396	.99990	.01396	71.615	.0001	71.622	12
49	.01425	.99990	.01425	70.153	.0001	70.160	11
50	0.01454	0.99989	0.01454	68.750	1.0001	68.757	10
51	.01483	.99989	.01484	67.402	.0001	67.409	9
52	.01512	.99988	.01513	66.105	.0001	66.113	8
53	.01542	.99988	.01542	64.858	.0001	64.866	7
54	.01571	.99988	.01571	63.657	.0001	63.664	6
55	0.01600	0.99987	0.01600	62.499	1.0001	62.507	5
56	.01629	.99987	.01629	61.383	.0001	61.391	4
57	.01658	.99987	.01658	60.306	.0001	60.314	3
58	.01687	.99986	.01687	59.266	.0001	59.274	2
59	.01716	.99985	.01716	58.261	.0001	58.270	1
60	0.01745	0.99985	0.01745	57.290	1.0001	57.299	0
M	Cosine	Sine	Cotan.	Tan.	Cosec.	Secant	M

90° — **89°**

1° — **178°**

M	Sine	Cosine	Tan.	Cotan.	Secant	Cosec.	M
0	0.01745	0.99985	0.01745	57.290	1.0001	57.299	60
1	.01774	.99984	.01775	56.350	.0001	56.359	59
2	.01803	.99984	.01804	55.441	.0001	55.450	58
3	.01832	.99983	.01833	54.561	.0002	54.570	57
4	.01861	.99983	.01862	53.708	.0002	53.718	56
5	0.01891	0.99982	0.01891	52.882	1.0002	52.891	55
6	.01920	.99981	.01920	52.081	.0002	52.090	54
7	.01949	.99981	.01949	51.303	.0002	51.313	53
8	.01978	.99980	.01978	50.548	.0002	50.558	52
9	.02007	.99980	.02007	49.816	.0002	49.826	51
10	0.02036	0.99979	0.02036	49.104	1.0002	49.114	50
11	.02065	.99979	.02066	48.412	.0002	48.422	49
12	.02094	.99978	.02095	47.739	.0002	47.750	48
13	.02123	.99977	.02124	47.085	.0002	47.096	47
14	.02152	.99977	.02153	46.449	.0002	46.460	46
15	0.02181	0.99976	0.02182	45.829	1.0002	45.840	45
16	.02210	.99975	.02211	45.226	.0002	45.237	44
17	.02240	.99975	.02240	44.638	.0002	44.650	43
18	.02269	.99974	.02269	44.066	.0002	44.077	42
19	.02298	.99974	.02298	43.508	.0003	43.520	41
20	0.02326	0.99973	0.02327	42.964	1.0003	42.976	40
21	.02356	.99972	.02357	42.433	.0003	42.445	39
22	.02385	.99971	.02386	41.916	.0003	41.928	38
23	.02414	.99971	.02415	41.410	.0003	41.423	37
24	.02443	.99970	.02444	40.917	.0003	40.930	36
25	0.02472	0.99969	0.02473	40.436	1.0003	40.448	35
26	.02501	.99969	.02502	39.965	.0003	39.978	34
27	.02530	.99968	.02531	39.506	.0003	39.518	33
28	.02559	.99967	.02560	39.057	.0003	39.069	32
29	.02589	.99966	.02589	38.618	.0003	38.631	31
30	0.02618	0.99966	0.02618	38.188	1.0003	38.201	30
31	.02647	.99965	.02648	37.769	.0003	37.782	29
32	.02676	.99964	.02677	37.358	.0003	37.371	28
33	.02705	.99963	.02706	36.956	.0004	36.969	27
34	.02734	.99963	.02735	36.563	.0004	36.576	26
35	0.02763	0.99962	0.02764	36.177	1.0004	36.191	25
36	.02792	.99961	.02793	35.800	.0004	35.814	24
37	.02821	.99960	.02822	35.431	.0004	35.445	23
38	.02850	.99959	.02851	35.069	.0004	35.084	22
39	.02879	.99958	.02881	34.715	.0004	34.729	21
40	0.02908	0.99958	0.02910	34.368	1.0004	34.382	20
41	.02937	.99957	.02939	34.027	.0004	34.042	19
42	.02967	.99956	.02968	33.693	.0004	33.708	18
43	.02996	.99955	.02997	33.366	.0004	33.381	17
44	.03025	.99954	.03026	33.045	.0004	33.060	16
45	0.03054	0.99953	0.03055	32.730	1.0005	32.745	15
46	.03083	.99952	.03084	32.421	.0005	32.437	14
47	.03112	.99951	.03113	32.118	.0005	32.134	13
48	.03141	.99951	.03143	31.820	.0005	31.836	12
49	.03170	.99950	.03172	31.528	.0005	31.544	11
50	0.03199	0.99949	0.03201	31.241	1.0005	31.257	10
51	.03228	.99948	.03230	30.960	.0005	30.976	9
52	.03257	.99947	.03259	30.683	.0005	30.699	8
53	.03286	.99946	.03288	30.411	.0005	30.428	7
54	.03315	.99945	.03317	30.145	.0005	30.161	6
55	0.03344	0.99944	0.03346	29.882	1.0005	29.899	5
56	.03374	.99943	.03375	29.624	.0006	29.641	4
57	.03403	.99942	.03405	29.371	.0006	29.388	3
58	.03432	.99941	.03434	29.122	.0006	29.139	2
59	.03461	.99940	.03463	28.877	.0006	28.894	1
60	0.03490	0.99939	0.03492	28.636	1.0006	28.654	0
M	Cosine	Sine	Cotan.	Tan.	Cosec.	Secant	M

91° — **88°**

2° — **177°**

M	Sine	Cosine	Tan.	Cotan.	Secant	Cosec.	M
0	0.03490	0.99939	0.03492	28.636	1.0006	28.654	60
1	.03519	.99938	.03521	28.399	.0006	28.417	59
2	.03548	.99937	.03550	28.166	.0006	28.184	58
3	.03577	.99936	.03579	27.937	.0006	27.955	57
4	.03606	.99935	.03608	27.712	.0006	27.730	56
5	0.03635	0.99934	0.03638	27.490	1.0007	27.508	55
6	.03664	.99933	.03667	27.271	.0007	27.290	54
7	.03693	.99932	.03696	27.056	.0007	27.075	53
8	.03722	.99931	.03725	26.845	.0007	26.864	52
9	.03751	.99930	.03754	26.637	.0007	26.655	51
10	0.03781	0.99928	0.03783	26.432	1.0007	26.450	50
11	.03810	.99927	.03812	26.230	.0007	26.249	49
12	.03839	.99926	.03842	26.031	.0007	26.050	48
13	.03868	.99925	.03871	25.835	.0007	25.854	47
14	.03897	.99924	.03900	25.642	.0008	25.661	46
15	0.03926	0.99923	0.03929	25.452	1.0008	25.471	45
16	.03955	.99922	.03958	25.264	.0008	25.284	44
17	.03984	.99921	.03987	25.080	.0008	25.100	43
18	.04013	.99919	.04016	24.898	.0008	24.918	42
19	.04042	.99918	.04045	24.718	.0008	24.739	41
20	0.04071	0.99917	0.04075	24.542	1.0008	24.562	40
21	.04100	.99916	.04104	24.367	.0008	24.388	39
22	.04129	.99915	.04133	24.196	.0008	24.216	38
23	.04158	.99913	.04162	24.026	.0009	24.047	37
24	.04187	.99912	.04191	23.859	.0009	23.880	36
25	0.04217	0.99911	0.04220	23.694	1.0009	23.716	35
26	.04246	.99910	.04249	23.532	.0009	23.553	34
27	.04275	.99908	.04279	23.372	.0009	23.393	33
28	.04304	.99907	.04308	23.214	.0009	23.235	32
29	.04333	.99906	.04337	23.058	.0009	23.079	31
30	0.04362	0.99905	0.04366	22.904	1.0009	22.925	30
31	.04391	.99903	.04395	22.752	.0010	22.774	29
32	.04420	.99902	.04424	22.602	.0010	22.624	28
33	.04449	.99901	.04453	22.454	.0010	22.476	27
34	.04478	.99900	.04483	22.308	.0010	22.330	26
35	0.04507	0.99898	0.04512	22.164	1.0010	22.186	25
36	.04536	.99897	.04541	22.022	.0010	22.044	24
37	.04565	.99896	.04570	21.881	.0010	21.904	23
38	.04594	.99894	.04599	21.742	.0010	21.765	22
39	.04623	.99893	.04628	21.606	.0011	21.629	21
40	0.04652	0.99892	0.04657	21.470	1.0011	21.494	20
41	.04681	.99890	.04687	21.337	.0011	21.360	19
42	.04711	.99889	.04716	21.205	.0011	21.228	18
43	.04740	.99888	.04745	21.075	.0011	21.098	17
44	.04769	.99886	.04774	20.946	.0011	20.970	16
45	0.04798	0.99885	0.04803	20.819	1.0011	20.843	15
46	.04827	.99883	.04832	20.693	.0012	20.717	14
47	.04856	.99882	.04862	20.569	.0012	20.593	13
48	.04885	.99881	.04891	20.446	.0012	20.471	12
49	.04914	.99879	.04920	20.325	.0012	20.350	11
50	0.04943	0.99878	0.04949	20.205	1.0012	20.230	10
51	.04972	.99876	.04978	20.087	.0012	20.112	9
52	.05001	.99875	.05007	19.970	.0012	19.995	8
53	.05030	.99873	.05037	19.854	.0013	19.880	7
54	.05059	.99872	.05066	19.740	.0013	19.766	6
55	0.05088	0.99870	0.05095	19.627	1.0013	19.653	5
56	.05117	.99869	.05124	19.515	.0013	19.541	4
57	.05146	.99867	.05153	19.405	.0013	19.431	3
58	.05175	.99866	.05182	19.296	.0013	19.322	2
59	05204	.99864	.05212	19.188	.0013	19.214	1
60	0.05234	0.99863	0.05241	19.081	1.0014	19.107	0
M	Cosine	Sine	Cotan.	Tan.	Cosec.	Secant	M

92° — **87°**

For angles from 0° to 45°, read degrees and function at top and minutes at left; for 45° to 90°, read degrees and function at bottom and minutes at right; for 90° to 135°, read degrees and function at bottom and minutes at left; and for 135° to 180°, read degrees and function at top and minutes at right.

Table 3-5 (*Continued*). Values of the Natural Trigonometric Functions

3° — **176°**

M	Sine	Cosine	Tan.	Cotan.	Secant	Cosec.	M
0	0.05234	0.99863	0.05241	19.081	1.0014	19.107	60
1	.05263	.99861	.05270	18.975	.0014	19.002	59
2	.05292	.99860	.05299	18.871	.0014	18.897	58
3	.05321	.99858	.05328	18.768	.0014	18.794	57
4	.05350	.99857	.05357	18.665	.0014	18.692	56
5	0.05379	0.99855	0.05387	18.564	1.0014	18.591	55
6	.05408	.99854	.05416	18.464	.0015	18.491	54
7	.05437	.99852	.05445	18.365	.0015	18.393	53
8	.05466	.99850	.05474	18.268	.0015	18.295	52
9	.05495	.99849	.05503	18.171	.0015	18.198	51
10	0.05524	0.99847	0.05532	18.075	1.0015	18.103	50
11	.05553	.99846	.05562	17.980	.0015	18.008	49
12	.05582	.99844	.05591	17.886	.0016	17.914	48
13	.05611	.99842	.05620	17.793	.0016	17.821	47
14	.05640	.99841	.05649	17.701	.0016	17.730	46
15	0.05669	0.99839	0.05678	17.610	1.0016	17.639	45
16	.05698	.99837	.05707	17.520	.0016	17.549	44
17	.05727	.99836	.05737	17.431	.0016	17.460	43
18	.05756	.99834	.05766	17.343	.0017	17.372	42
19	.05785	.99832	.05795	17.256	.0017	17.285	41
20	0.05814	0.99831	0.05824	17.169	1.0017	17.198	40
21	.05843	.99829	.05853	17.084	.0017	17.113	39
22	.05872	.99827	.05883	16.999	.0017	17.028	38
23	.05902	.99826	.05912	16.915	.0017	16.944	37
24	.05931	.99824	.05941	16.832	.0018	16.861	36
25	0.05960	0.99822	0.05970	16.750	1.0018	16.779	35
26	.05989	.99820	.05999	16.668	.0018	16.698	34
27	.06018	.99819	.06029	16.587	.0018	16.617	33
28	.06047	.99817	.06058	16.507	.0018	16.538	32
29	.06076	.99815	.06087	16.428	.0018	16.459	31
30	0.06105	0.99813	0.06116	16.350	1.0019	16.380	30
31	.06134	.99812	.06145	16.272	.0019	16.303	29
32	.06163	.99810	.06175	16.195	.0019	16.226	28
33	.06192	.99808	.06204	16.119	.0019	16.150	27
34	.06221	.99806	.06233	16.043	.0019	16.075	26
35	0.06250	0.99804	0.06262	15.969	1.0019	16.000	25
36	.06279	.99803	.06291	15.894	.0020	15.926	24
37	.06308	.99801	.06321	15.821	.0020	15.853	23
38	.06337	.99799	.06350	15.748	.0020	15.780	22
39	.06366	.99797	.06379	15.676	.0020	15.708	21
40	0.06395	0.99795	0.06408	15.605	1.0020	15.637	20
41	.06424	.99793	.06437	15.534	.0021	15.566	19
42	.06453	.99791	.06467	15.464	.0021	15.496	18
43	.06482	.99790	.06496	15.394	.0021	15.427	17
44	.06511	.99788	.06525	15.325	.0021	15.358	16
45	0.06540	0.99786	0.06554	15.257	1.0021	15.290	15
46	.06569	.99784	.06583	15.189	.0022	15.222	14
47	.06598	.99782	.06613	15.122	.0022	15.155	13
48	.06627	.99780	.06642	15.056	.0022	15.089	12
49	.06656	.99778	.06671	14.990	.0022	15.023	11
50	0.06685	0.99776	0.06700	14.924	1.0022	14.958	10
51	.06714	.99774	.06730	14.860	.0023	14.893	9
52	.06743	.99772	.06759	14.795	.0023	14.829	8
53	.06772	.99770	.06788	14.732	.0023	14.765	7
54	.06801	.99768	.06817	14.668	.0023	14.702	6
55	0.06830	0.99766	0.06846	14.606	1.0023	14.640	5
56	.06859	.99764	.06876	14.544	.0024	14.578	4
57	.06888	.99762	.06905	14.482	.0024	14.517	3
58	.06918	.99760	.06934	14.421	.0024	14.456	2
59	.06947	.99758	.06963	14.361	.0024	14.395	1
60	0.06976	0.99756	0.06993	14.301	1.0024	14.335	0
M	Cosine	Sine	Cotan.	Tan.	Cosec.	Secant	M

93° — **86°**

4° — **175°**

M	Sine	Cosine	Tan.	Cotan.	Secant	Cosec.	M
0	0.06976	0.99756	0.06993	14.301	1.0024	14.335	60
1	.07005	.99754	.07022	14.241	.0025	14.276	59
2	.07034	.99752	.07051	14.182	.0025	14.217	58
3	.07063	.99750	.07080	14.123	.0025	14.159	57
4	.07092	.99748	.07110	14.065	.0025	14.101	56
5	0.07121	0.99746	0.07139	14.008	1.0025	14.043	55
6	.07150	.99744	.07168	13.951	.0026	13.986	54
7	.07179	.99742	.07197	13.894	.0026	13.930	53
8	.07208	.99740	.07226	13.838	.0026	13.874	52
9	.07237	.99738	.07256	13.782	.0026	13.818	51
10	0.07266	0.99736	0.07285	13.727	1.0026	13.763	50
11	.07295	.99733	.07314	13.672	.0027	13.708	49
12	.07324	.99731	.07343	13.617	.0027	13.654	48
13	.07353	.99729	.07373	13.563	.0027	13.600	47
14	.07382	.99727	.07402	13.510	.0027	13.547	46
15	0.07411	0.99725	0.07431	13.457	1.0027	13.494	45
16	.07440	.99723	.07460	13.404	.0028	13.441	44
17	.07469	.99721	.07490	13.351	.0028	13.389	43
18	.07498	.99718	.07519	13.299	.0028	13.337	42
19	.07527	.99716	.07548	13.248	.0028	13.286	41
20	0.07556	0.99714	0.07577	13.197	1.0029	13.235	40
21	.07585	.99712	.07607	13.146	.0029	13.184	39
22	.07614	.99710	.07636	13.096	.0029	13.134	38
23	.07643	.99707	.07665	13.046	.0029	13.084	37
24	.07672	.99705	.07694	12.996	.0029	13.034	36
25	0.07701	0.99703	0.07724	12.947	1.0030	12.985	35
26	.07730	.99701	.07753	12.898	.0030	12.937	34
27	.07759	.99698	.07782	12.849	.0030	12.888	33
28	.07788	.99696	.07812	12.801	.0030	12.840	32
29	.07817	.99694	.07841	12.754	.0031	12.793	31
30	0.07846	0.99692	0.07870	12.706	1.0031	12.745	30
31	.07875	.99689	.07899	12.659	.0031	12.698	29
32	.07904	.99687	.07929	12.612	.0031	12.652	28
33	.07933	.99685	.07958	12.566	.0032	12.606	27
34	.07962	.99682	.07987	12.520	.0032	12.560	26
35	0.07991	0.99680	0.08016	12.474	1.0032	12.514	25
36	.08020	.99678	.08046	12.429	.0032	12.469	24
37	.08049	.99675	.08075	12.384	.0032	12.424	23
38	.08078	.99673	.08104	12.339	.0033	12.379	22
39	.08107	.99671	.08134	12.295	.0033	12.335	21
40	0.08136	0.99668	0.08163	12.250	1.0033	12.291	20
41	.08165	.99666	.08192	12.207	.0033	12.248	19
42	.08194	.99664	.08221	12.163	.0034	12.204	18
43	.08223	.99661	.08251	12.120	.0034	12.161	17
44	.08252	.99659	.08280	12.077	.0034	12.118	16
45	0.08281	0.99656	0.08309	12.035	1.0034	12.076	15
46	.08310	.99654	.08339	11.992	.0035	12.034	14
47	.08339	.99652	.08368	11.950	.0035	11.992	13
48	.08368	.99649	.08397	11.909	.0035	11.950	12
49	.08397	.99647	.08426	11.867	.0035	11.909	11
50	0.08426	0.99644	0.08456	11.826	1.0036	11.868	10
51	.08455	.99642	.08485	11.785	.0036	11.828	9
52	.08484	.99639	.08514	11.745	.0036	11.787	8
53	.08513	.99637	.08544	11.704	.0036	11.747	7
54	.08542	.99634	.08573	11.664	.0037	11.707	6
55	0.08571	0.99632	0.08602	11.625	1.0037	11.668	5
56	.08600	.99629	.08632	11.585	.0037	11.628	4
57	.08629	.99627	.08661	11.546	.0037	11.589	3
58	.08658	.99624	.08690	11.507	.0038	11.550	2
59	.08687	.99622	.08719	11.468	.0038	11.512	1
60	0.08715	0.99619	0.08749	11.430	1.0038	11.474	0
M	Cosine	Sine	Cotan.	Tan.	Cosec.	Secant	M

94° — **85°**

5° — **174°**

M	Sine	Cosine	Tan.	Cotan.	Secant	Cosec.	M
0	0.08715	0.99619	0.08749	11.430	1.0038	11.474	60
1	.08744	.99617	.08778	11.392	.0038	11.436	59
2	.08773	.99614	.08807	11.354	.0039	11.398	58
3	.08802	.99612	.08837	11.316	.0039	11.360	57
4	.08831	.99609	.08866	11.279	.0039	11.323	56
5	0.08860	0.99607	0.08895	11.242	1.0039	11.286	55
6	.08889	.99604	.08925	11.205	.0040	11.249	54
7	.08918	.99601	.08954	11.168	.0040	11.213	53
8	.08947	.99599	.08983	11.132	.0040	11.176	52
9	.08976	.99596	.09013	11.095	.0040	11.140	51
10	0.09005	0.99594	0.09042	11.059	1.0041	11.104	50
11	.09034	.99591	.09071	11.024	.0041	11.069	49
12	.09063	.99588	.09101	10.988	.0041	11.033	48
13	.09092	.99586	.09130	10.953	.0041	10.998	47
14	.09121	.99583	.09159	10.918	.0042	10.963	46
15	0.09150	0.99580	0.09189	10.883	1.0042	10.929	45
16	.09179	.99578	.09218	10.848	.0042	10.894	44
17	.09208	.99575	.09247	10.814	.0043	10.860	43
18	.09237	.99572	.09277	10.780	.0043	10.826	42
19	.09266	.99570	.09306	10.746	.0043	10.792	41
20	0.09295	0.99567	0.09335	10.712	1.0043	10.758	40
21	.09324	.99564	.09365	10.678	.0044	10.725	39
22	.09353	.99562	.09394	10.645	.0044	10.692	38
23	.09382	.99559	.09423	10.612	.0044	10.659	37
24	.09411	.99556	.09453	10.579	.0044	10.626	36
25	0.09440	0.99553	0.09482	10.546	1.0045	10.593	35
26	.09469	.99551	.09511	10.514	.0045	10.561	34
27	.09498	.99548	.09541	10.481	.0045	10.529	33
28	.09527	.99545	.09570	10.449	.0046	10.497	32
29	.09556	.99542	.09599	10.417	.0046	10.465	31
30	0.09584	0.99540	0.09629	10.385	1.0046	10.433	30
31	.09613	.99537	.09658	10.354	.0046	10.402	29
32	.09642	.99534	.09688	10.322	.0047	10.371	28
33	.09671	.99531	.09717	10.291	.0047	10.340	27
34	.09700	.99528	.09746	10.260	.0047	10.309	26
35	0.09729	0.99525	0.09776	10.229	1.0048	10.278	25
36	.09758	.99523	.09805	10.199	.0048	10.248	24
37	.09787	.99520	.09834	10.168	.0048	10.217	23
38	.09816	.99517	.09864	10.138	.0048	10.187	22
39	.09845	.99514	.09893	10.108	.0049	10.157	21
40	0.09874	0.99511	0.09922	10.078	1.0049	10.127	20
41	.09903	.99508	.09952	10.048	.0049	10.098	19
42	.09932	.99505	.09981	10.019	.0050	10.068	18
43	.09961	.99503	.10011	9.9893	.0050	10.039	17
44	.09990	.99500	.10040	9.9601	.0050	10.010	16
45	0.10019	0.99497	0.10069	9.9310	1.0050	9.9812	15
46	.10048	.99494	.10099	9.9021	.0051	9.9525	14
47	.10077	.99491	.10128	9.8734	.0051	9.9239	13
48	.10106	.99488	.10158	9.8448	.0051	9.8955	12
49	.10134	.99485	.10187	9.8164	.0052	9.8672	11
50	0.10163	0.99482	0.10216	9.7882	1.0052	9.8391	10
51	.10192	.99479	.10246	9.7601	.0052	9.8112	9
52	.10221	.99476	.10275	9.7322	.0053	9.7834	8
53	.10250	.99473	.10305	9.7044	.0053	9.7558	7
54	.10279	.99470	.10334	9.6768	.0053	9.7283	6
55	0.10308	0.99467	0.10363	9.6493	1.0053	9.7010	5
56	.10337	.99464	.10393	9.6220	.0054	9.6739	4
57	.10366	.99461	.10422	9.5949	.0054	9.6469	3
58	.10395	.99458	.10452	9.5679	.0054	9.6200	2
59	.10424	.99455	.10481	9.5411	.0055	9.5933	1
60	0.10453	0.99452	0.10510	9.5144	1.0055	9.5668	0
M	Cosine	Sine	Cotan.	Tan.	Cosec.	Secant	M

95° — **84°**

For angles from 0° to 45°, read degrees and function at top and minutes at left; for 45° to 90°, read degrees and function at bottom and minutes at right; for 90° to 135°, read degrees and function at bottom and minutes at left; and for 135° to 180° read degrees and function at top and minutes at right.

Table 3-5 (Continued). Values of the Natural Trigonometric Functions

Table 3-5 (*Continued*). Values of the Natural Trigonometric Functions

9° — **170°**

M	Sine	Cosine	Tan.	Cotan.	Secant	Cosec.	M
0	0.15643	0.98769	0.15838	6.3137	1.0125	6.3924	60
1	.15672	.98764	.15868	.3019	.0125	.3807	59
2	.15701	.98760	.15898	.2901	.0125	.3690	58
3	.15730	.98755	.15928	.2783	.0126	.3574	57
4	.15758	.98750	.15958	.2665	.0126	.3458	56
5	0.15787	0.98746	0.15987	6.2548	1.0127	6.3343	55
6	.15816	.98741	.16017	.2432	.0127	.3228	54
7	.15844	.98737	.16047	.2316	.0128	.3113	53
8	.15873	.98732	.16077	.2200	.0128	.2999	52
9	.15902	.98727	.16107	.2085	.0129	.2885	51
10	0.15931	0.98723	0.16137	6.1970	1.0129	6.2772	50
11	.15959	.98718	.16167	.1856	.0130	.2659	49
12	.15988	.98714	.16196	.1742	.0130	.2546	48
13	.16017	.98709	.16226	.1628	.0131	.2434	47
14	.16045	.98704	.16256	.1515	.0131	.2322	46
15	0.16074	0.98700	0.16286	6.1402	1.0132	6.2211	45
16	.16103	.98695	.16316	.1290	.0132	.2100	44
17	.16132	.98690	.16346	.1178	.0133	.1990	43
18	.16160	.98685	.16376	.1066	.0133	.1880	42
19	.16189	.98681	.16405	.0955	.0134	.1770	41
20	0.16218	0.98676	0.16435	6.0844	1.0134	6.1661	40
21	.16246	.98671	.16465	.0734	.0135	.1552	39
22	.16275	.98667	.16495	.0624	.0135	.1443	38
23	.16304	.98662	.16525	.0514	.0136	.1335	37
24	.16333	.98657	.16555	.0405	.0136	.1227	36
25	0.16361	0.98652	0.16585	6.0296	1.0136	6.1120	35
26	.16390	.98648	.16615	.0188	.0137	.1013	34
27	.16419	.98643	.16644	.0080	.0137	.0906	33
28	.16447	.98638	.16674	5.9972	.0138	.0800	32
29	.16476	.98633	.16704	.9865	.0138	.0694	31
30	0.16505	0.98628	0.16734	5.9758	1.0139	6.0588	30
31	.16533	.98624	.16764	.9651	.0139	.0483	29
32	.16562	.98619	.16794	.9545	.0140	.0379	28
33	.16591	.98614	.16824	.9439	.0140	.0274	27
34	.16619	.98609	.16854	.9333	.0141	.0170	26
35	0.16648	0.98604	0.16884	5.9228	1.0141	6.0066	25
36	.16677	.98600	.16914	.9123	.0142	5.9963	24
37	.16705	.98595	.16944	.9019	.0142	.9860	23
38	.16734	.98590	.16973	.8915	.0143	.9758	22
39	.16763	.98585	.17003	.8811	.0143	.9655	21
40	0.16791	0.98580	0.17033	5.8708	1.0144	5.9554	20
41	.16820	.98575	.17063	.8605	.0144	.9452	19
42	.16849	.98570	.17093	.8502	.0145	.9351	18
43	.16878	.98565	.17123	.8400	.0145	.9250	17
44	.16906	.98560	.17153	.8298	.0146	.9150	16
45	0.16935	0.98556	0.17183	5.8196	1.0146	5.9049	15
46	.16964	.98551	.17213	.8095	.0147	.8950	14
47	.16992	.98546	.17243	.7994	.0147	.8850	13
48	.17021	.98541	.17273	.7894	.0148	.8751	12
49	.17050	.98536	.17303	.7794	.0148	.8652	11
50	0.17078	0.98531	0.17333	5.7694	1.0149	5.8554	10
51	.17107	.98526	.17363	.7594	.0150	.8456	9
52	.17136	.98521	.17393	.7495	.0150	.8358	8
53	.17164	.98516	.17423	.7396	.0151	.8261	7
54	.17193	.98511	.17453	.7297	.0151	.8163	6
55	0.17221	0.98506	0.17483	5.7199	1.0152	5.8067	5
56	.17250	.98501	.17513	.7101	.0152	.7970	4
57	.17279	.98496	.17543	.7004	.0153	.7874	3
58	.17307	.98491	.17573	.6906	.0153	.7778	2
59	.17336	.98486	.17603	.6809	.0154	.7683	1
60	0.17365	0.98481	0.17633	5.6713	1.0154	5.7588	0
M	Cosine	Sine	Cotan.	Tan.	Cosec.	Secant	M

99° — **80°**

10° — **169°**

M	Sine	Cosine	Tan.	Cotan.	Secant	Cosec.	M
0	0.17365	0.98481	0.17633	5.6713	1.0154	5.7588	60
1	.17393	.98476	.17663	.6616	.0155	.7493	59
2	.17422	.98471	.17693	.6520	.0155	.7398	58
3	.17451	.98465	.17723	.6425	.0156	.7304	57
4	.17479	.98460	.17753	.6329	.0156	.7210	56
5	0.17508	0.98455	0.17783	5.6234	1.0157	5.7117	55
6	.17537	.98450	.17813	.6140	.0157	.7023	54
7	.17565	.98445	.17843	.6045	.0158	.6930	53
8	.17594	.98440	.17873	.5951	.0158	.6838	52
9	.17622	.98435	.17903	.5857	.0159	.6745	51
10	0.17651	0.98430	0.17933	5.5764	1.0159	5.6653	50
11	.17680	.98425	.17963	.5670	.0160	.6561	49
12	.17708	.98419	.17993	.5578	.0160	.6470	48
13	.17737	.98414	.18023	.5485	.0161	.6379	47
14	.17766	.98409	.18053	.5393	.0162	.6288	46
15	0.17794	0.98404	0.18083	5.5301	1.0162	5.6197	45
16	.17823	.98399	.18113	.5209	.0163	.6107	44
17	.17852	.98394	.18143	.5117	.0163	.6017	43
18	.17880	.98388	.18173	.5026	.0164	.5928	42
19	.17909	.98383	.18203	.4936	.0164	.5838	41
20	0.17937	0.98378	0.18233	5.4845	1.0165	5.5749	40
21	.17966	.98373	.18263	.4755	.0165	.5660	39
22	.17995	.98368	.18293	.4665	.0166	.5572	38
23	.18023	.98362	.18323	.4575	.0166	.5484	37
24	.18052	.98357	.18353	.4486	.0167	.5396	36
25	0.18080	0.98352	0.18383	5.4396	1.0167	5.5308	35
26	.18109	.98347	.18413	.4308	.0168	.5221	34
27	.18138	.98341	.18444	.4219	.0169	.5134	33
28	.18166	.98336	.18474	.4131	.0169	.5047	32
29	.18195	.98331	.18504	.4043	.0170	.4960	31
30	0.18223	0.98325	0.18534	5.3955	1.0170	5.4874	30
31	.18252	.98320	.18564	.3868	.0171	.4788	29
32	.18281	.98315	.18594	.3780	.0171	.4702	28
33	.18309	.98309	.18624	.3694	.0172	.4617	27
34	.18338	.98304	.18654	.3607	.0172	.4532	26
35	0.18366	0.98299	0.18684	5.3521	1.0173	5.4447	25
36	.18395	.98293	.18714	.3434	.0174	.4362	24
37	.18424	.98288	.18745	.3349	.0174	.4278	23
38	.18452	.98283	.18775	.3263	.0175	.4194	22
39	.18481	.98277	.18805	.3178	.0175	.4110	21
40	0.18509	0.98272	0.18835	5.3093	1.0176	5.4026	20
41	.18538	.98267	.18865	.3008	.0176	.3943	19
42	.18567	.98261	.18895	.2923	.0177	.3860	18
43	.18595	.98256	.18925	.2839	.0177	.3777	17
44	.18624	.98250	.18955	.2755	.0178	.3695	16
45	0.18652	0.98245	0.18985	5.2671	1.0179	5.3612	15
46	.18681	.98240	.19016	.2588	.0179	.3530	14
47	.18709	.98234	.19046	.2505	.0180	.3449	13
48	.18738	.98229	.19076	.2422	.0180	.3367	12
49	.18767	.98223	.19106	.2339	.0181	.3286	11
50	0.18795	0.98218	0.19136	5.2257	1.0181	5.3205	10
51	.18824	.98212	.19166	.2174	.0182	.3124	9
52	.18852	.98207	.19197	.2092	.0182	.3044	8
53	.18881	.98201	.19227	.2011	.0183	.2963	7
54	.18909	.98196	.19257	.1929	.0184	.2883	6
55	0.18938	0.98190	0.19287	5.1848	1.0184	5.2803	5
56	.18967	.98185	.19317	.1767	.0185	.2724	4
57	.18995	.98179	.19347	.1686	.0185	.2645	3
58	.19024	.98174	.19378	.1606	.0186	.2566	2
59	.19052	.98168	.19408	.1525	.0186	.2487	1
60	0.19081	0.98163	0.19438	5.1445	1.0187	5.2408	0
M	Cosine	Sine	Cotan.	Tan.	Cosec.	Secant	M

100° — **79°**

11° — **168°**

M	Sine	Cosine	Tan.	Cotan.	Secant	Cosec.	M
0	0.19081	0.98163	0.19438	5.1445	1.0187	5.2408	60
1	.19109	.98157	.19468	.1366	.0188	.2330	59
2	.19138	.98152	.19498	.1286	.0188	.2252	58
3	.19166	.98146	.19529	.1207	.0189	.2174	57
4	.19195	.98140	.19559	.1128	.0189	.2097	56
5	0.19224	0.98135	0.19589	5.1049	1.0190	5.2019	55
6	.19252	.98129	.19619	.0970	.0191	.1942	54
7	.19281	.98124	.19649	.0892	.0191	.1865	53
8	.19309	.98118	.19680	.0814	.0192	.1788	52
9	.19338	.98112	.19710	.0736	.0192	.1712	51
10	0.19366	0.98107	0.19740	5.0658	1.0193	5.1636	50
11	.19395	.98101	.19770	.0581	.0193	.1560	49
12	.19423	.98095	.19800	.0504	.0194	.1484	48
13	.19452	.98090	.19831	.0427	.0195	.1409	47
14	.19480	.98084	.19861	.0350	.0195	.1333	46
15	0.19509	0.98078	0.19891	5.0273	1.0196	5.1258	45
16	.19537	.98073	.19921	.0197	.0196	.1183	44
17	.19566	.98067	.19952	.0121	.0197	.1109	43
18	.19595	.98061	.19982	.0045	.0198	.1034	42
19	.19623	.98056	.20012	4.9969	.0198	.0960	41
20	0.19652	0.98050	0.20042	4.9894	1.0199	5.0886	40
21	.19680	.98044	.20073	.9819	.0199	.0812	39
22	.19709	.98039	.20103	.9744	.0200	.0739	38
23	.19737	.98033	.20133	.9669	.0201	.0666	37
24	.19766	.98027	.20163	.9594	.0201	.0593	36
25	0.19794	0.98021	0.20194	4.9520	1.0202	5.0520	35
26	.19823	.98016	.20224	.9446	.0202	.0447	34
27	.19851	.98010	.20254	.9372	.0203	.0375	33
28	.19880	.98004	.20285	.9298	.0204	.0302	32
29	.19908	.97998	.20315	.9225	.0204	.0230	31
30	0.19937	0.97992	0.20345	4.9151	1.0205	5.0158	30
31	.19965	.97987	.20375	.9078	.0205	.0087	29
32	.19994	.97981	.20406	.9006	.0206	.0015	28
33	.20022	.97975	.20436	.8933	.0207	4.9944	27
34	.20051	.97969	.20466	.8860	.0207	.9873	26
35	0.20079	0.97963	0.20497	4.8788	1.0208	4.9802	25
36	.20108	.97957	.20527	.8716	.0208	.9732	24
37	.20136	.97952	.20557	.8644	.0209	.9661	23
38	.20165	.97946	.20588	.8573	.0210	.9591	22
39	.20193	.97940	.20618	.8501	.0210	.9521	21
40	0.20222	0.97934	0.20648	4.8430	1.0211	4.9452	20
41	.20250	.97928	.20679	.8359	.0211	.9382	19
42	.20279	.97922	.20709	.8288	.0212	.9313	18
43	.20307	.97916	.20739	.8217	.0213	.9243	17
44	.20336	.97910	.20770	.8147	.0213	.9175	16
45	0.20364	0.97904	0.20800	4.8077	1.0214	4.9106	15
46	.20393	.97899	.20830	.8007	.0215	.9037	14
47	.20421	.97893	.20861	.7937	.0215	.8969	13
48	.20450	.97887	.20891	.7867	.0216	.8901	12
49	.20478	.97881	.20921	.7798	.0216	.8833	11
50	0.20506	0.97875	0.20952	4.7728	1.0217	4.8765	10
51	.20535	.97869	.20982	.7659	.0218	.8697	9
52	.20563	.97863	.21012	.7591	.0218	.8630	8
53	.20592	.97857	.21043	.7522	.0219	.8563	7
54	.20620	.97851	.21073	.7453	.0220	.8496	6
55	0.20649	0.97845	0.21104	4.7385	1.0220	4.8429	5
56	.20677	.97839	.21134	.7317	.0221	.8362	4
57	.20706	.97833	.21164	.7249	.0221	.8296	3
58	.20734	.97827	.21195	.7181	.0222	.8229	2
59	.20763	.97821	.21225	.7114	.0223	.8163	1
60	0.20791	0.97815	0.21256	4.7046	1.0223	4.8097	0
M	Cosine	Sine	Cotan.	Tan.	Cosec.	Secant	M

101° — **78°**

For angles from 0° to 45°, read degrees and function at top and minutes at left; for 45° to 90°, read degrees and function at bottom and minutes at right; for 90° to 135°, read degrees and function at bottom and minutes at left; and for 135° to 180°, read degrees and function at top and minutes at right.

Table 3-5 (*Continued*). Values of the Natural Trigonometric Functions

12°							167°
M	Sine	Cosine	Tan.	Cotan.	Secant	Cosec.	M
0	0.20791	0.97815	0.21256	4.7046	1.0223	4.8097	60
1	.20820	.97809	.21286	.6979	.0224	.8032	59
2	.20848	.97803	.21316	.6912	.0225	.7966	58
3	.20876	.97797	.21347	.6845	.0225	.7901	57
4	.20905	.97790	.21377	.6778	.0226	.7835	56
5	0.20933	0.97784	0.21408	4.6712	1.0226	4.7770	55
6	.20962	.97778	.21438	.6646	.0227	.7706	54
7	.20990	.97772	.21468	.6580	.0228	.7641	53
8	.21019	.97766	.21499	.6514	.0228	.7576	52
9	.21047	.97760	.21529	.6448	.0229	.7512	51
10	0.21076	0.97754	0.21560	4.6382	1.0230	4.7448	50
11	.21104	.97748	.21590	.6317	.0230	.7384	49
12	.21132	.97741	.21621	.6252	.0231	.7320	48
13	.21161	.97735	.21651	.6187	.0232	.7257	47
14	.21189	.97729	.21682	.6122	.0232	.7193	46
15	0.21218	0.97723	0.21712	4.6057	1.0233	4.7130	45
16	.21246	.97717	.21742	.5993	.0234	.7067	44
17	.21275	.97711	.21773	.5928	.0234	.7004	43
18	.21303	.97704	.21803	.5864	.0235	.6942	42
19	.21331	.97698	.21834	.5800	.0235	.6879	41
20	0.21360	0.97692	0.21864	4.5736	1.0236	4.6817	40
21	.21388	.97686	.21895	.5673	.0237	.6754	39
22	.21417	.97680	.21925	.5609	.0237	.6692	38
23	.21445	.97673	.21956	.5546	.0238	.6631	37
24	.21473	.97667	.21986	.5483	.0239	.6569	36
25	0.21502	0.97661	0.22017	4.5420	1.0239	4.6507	35
26	.21530	.97655	.22047	.5357	.0240	.6446	34
27	.21559	.97648	.22078	.5294	.0241	.6385	33
28	.21587	.97642	.22108	.5232	.0241	.6324	32
29	.21615	.97636	.22139	.5169	.0242	.6263	31
30	0.21644	0.97630	0.22169	4.5107	1.0243	4.6201	30
31	.21672	.97623	.22200	.5045	.0243	.6142	29
32	.21701	.97617	.22230	.4983	.0244	.6081	28
33	21729	.97611	.22261	.4921	.0245	.6021	27
34	.21757	.97604	.22291	.4860	.0245	.5961	26
35	0.21786	0.97598	0.22322	4.4799	1.0246	4.5901	25
36	.21814	.97592	.22353	.4737	.0247	.5841	24
37	.21843	.97585	.22383	.4676	.0247	.5782	23
38	.21871	.97579	.22414	.4615	.0248	.5722	22
39	.21899	.97573	.22444	.4555	.0249	.5663	21
40	0.21928	0.97566	0.22475	4.4494	1.0249	4.5604	20
41	.21956	.97560	.22505	.4434	.0250	.5545	19
42	.21985	.97553	.22536	.4373	.0251	.5486	18
43	.22013	.97547	.22566	.4313	.0251	.5428	17
44	.22041	.97541	.22597	.4253	.0252	.5369	16
45	0.22070	0.97534	0.22628	4.4194	1.0253	4.5311	15
46	.22098	.97528	.22658	.4134	.0253	.5253	14
47	.22126	.97521	.22689	.4074	.0254	.5195	13
48	.22155	.97515	.22719	.4015	.0255	.5137	12
49	.22183	.97508	.22750	.3956	.0255	.5079	11
50	0.22211	0.97502	0.22781	4.3897	1.0256	4.5021	10
51	.22240	.97495	.22811	.3838	.0257	.4964	9
52	.22268	.97489	.22842	.3779	.0257	.4907	8
53	.22297	.97483	.22872	.3721	.0258	.4850	7
54	.22325	.97476	.22903	.3662	.0259	.4793	6
55	0.22353	0.97470	0.22934	4.3604	1.0260	4.4736	5
56	.22382	.97463	.22964	.3546	.0260	.4679	4
57	.22410	.97457	.22995	.3488	.0261	.4623	3
58	.22438	.97450	.23025	.3430	.0262	.4566	2
59	.22467	.97443	.23056	.3372	.0262	.4510	1
60	0.22495	0.97437	0.23087	4.3315	1.0263	4.4454	0
M	Cosine	Sine	Cotan.	Tan.	Cosec.	Secant	M
102°							77°

13°							166°
M	Sine	Cosine	Tan.	Cotan.	Secant	Cosec.	M
0	0.22495	0.97437	0.23087	4.3315	1.0263	4.4454	60
1	.22523	.97430	.23117	.3257	.0264	.4398	59
2	.22552	.97424	.23148	.3200	.0264	.4342	58
3	.22580	.97417	.23179	.3143	.0265	.4287	57
4	.22608	.97411	.23209	.3086	.0266	.4231	56
5	0.22637	0.97404	0.23240	4.3029	1.0266	4.4176	55
6	.22665	.97398	.23270	.2972	.0267	.4121	54
7	.22693	.97391	.23301	.2916	.0268	.4065	53
8	.22722	.97384	.23332	.2859	.0268	.4011	52
9	.22750	.97378	.23363	.2803	.0269	.3956	51
10	0.22778	0.97371	0.23393	4.2747	1.0270	4.3901	50
11	.22807	.97364	.23424	.2691	.0271	.3847	49
12	.22835	.97358	.23455	.2635	.0271	.3792	48
13	.22863	.97351	.23485	.2579	.0272	.3738	47
14	.22892	.97344	.23516	.2524	.0273	.3684	46
15	0.22920	0.97338	0.23547	4.2468	1.0273	4.3630	45
16	.22948	.97331	.23577	.2413	.0274	.3576	44
17	.22977	.97324	.23608	.2358	.0275	.3522	43
18	.23005	.97318	.23639	.2303	.0276	.3469	42
19	.23033	.97311	.23670	.2248	.0276	.3415	41
20	0.23061	0.97304	0.23700	4.2193	1.0277	4.3362	40
21	.23090	.97298	.23731	.2139	.0278	.3309	39
22	.23118	.97291	.23762	.2084	.0278	.3256	38
23	.23146	.97284	.23793	.2030	.0279	.3203	37
24	.23175	.97277	.23823	.1976	.0280	.3150	36
25	0.23203	0.97271	0.23854	4.1921	1.0280	4.3098	35
26	.23231	.97264	.23885	.1867	.0281	.3045	34
27	.23260	.97257	.23916	.1814	.0282	.2993	33
28	.23288	.97250	.23946	.1760	.0283	.2941	32
29	.23316	.97244	.23977	.1706	.0283	.2888	31
30	0.23344	0.97237	0.24008	4.1653	1.0284	4.2836	30
31	.23373	.97230	.24039	.1600	.0285	.2785	29
32	.23401	.97223	.24069	.1546	.0285	.2733	28
33	.23429	.97216	.24100	.1493	.0286	.2681	27
34	.23458	.97210	.24131	.1440	.0287	.2630	26
35	0.23486	0.97203	0.24162	4.1388	1.0288	4.2579	25
36	.23514	.97196	.24192	.1335	.0288	.2527	24
37	.23542	.97189	.24223	.1282	.0289	.2476	23
38	.23571	.97182	.24254	.1230	.0290	.2425	22
39	.23599	.97175	.24285	.1178	.0291	.2375	21
40	0.23627	0.97169	0.24316	4.1126	1.0291	4.2324	20
41	.23655	.97162	.24346	.1073	.0292	.2273	19
42	.23684	.97155	.24377	.1022	.0293	.2223	18
43	.23712	.97148	.24408	.0970	.0293	.2173	17
44	.23740	.97141	.24439	.0918	.0294	.2122	16
45	0.23768	0.97134	0.24470	4.0867	1.0295	4.2072	15
46	.23797	.97127	.24501	.0815	.0296	.2022	14
47	.23825	.97120	.24531	.0764	.0296	.1972	13
48	.23853	.97113	.24562	.0713	.0297	.1923	12
49	.23881	.97106	.24593	.0662	.0298	.1873	11
50	0.23910	0.97099	0.24624	4.0611	1.0299	4.1824	10
51	.23938	.97092	.24655	.0560	.0299	.1774	9
52	.23966	.97086	.24686	.0509	.0300	.1725	8
53	.23994	.97079	.24717	.0458	.0301	.1676	7
54	.24023	.97072	.24747	.0408	.0302	.1627	6
55	0.24051	0.97065	0.24778	4.0358	1.0302	4.1578	5
56	.24079	.97058	.24809	.0307	.0303	.1529	4
57	.24107	.97051	.24840	.0257	.0304	.1481	3
58	.24136	.97044	.24871	.0207	.0305	.1432	2
59	.24164	.97037	.24902	.0157	.0305	.1384	1
60	0.24192	0.97029	0.24933	4.0108	1.0306	4.1336	0
M	Cosine	Sine	Cotan.	Tan.	Cosec.	Secant	M
103°							76°

14°							165°
M	Sine	Cosine	Tan.	Cotan.	Secant	Cosec.	M
0	0.24192	0.97029	0.24933	4.0108	1.0306	4.1336	60
1	.24220	.97022	.24964	.0058	.0307	.1287	59
2	.24249	.97015	.24995	.0009	.0308	.1239	58
3	.24277	.97008	.25025	3.9959	.0308	.1191	57
4	.24305	.97001	.25056	.9910	.0309	.1144	56
5	0.24333	0.96994	0.25087	3.9861	1.0310	4.1096	55
6	.24361	.96987	.25118	.9812	.0311	.1048	54
7	.24390	.96980	.25149	.9763	.0311	.1001	53
8	.24418	.96973	.25180	.9714	.0312	.0953	52
9	.24446	.96966	.25211	.9665	.0313	.0906	51
10	0.24474	0.96959	0.25242	3.9616	1.0314	4.0859	50
11	.24502	.96952	.25273	.9568	.0314	.0812	49
12	.24531	.96944	.25304	.9520	.0315	.0765	48
13	.24559	.96937	.25335	.9471	.0316	.0718	47
14	.24587	.96930	.25366	.9423	.0317	.0672	46
15	0.24615	0.96923	0.25397	3.9375	1.0317	4.0625	45
16	.24643	.96916	.25428	.9327	.0318	.0579	44
17	.24672	.96909	.25459	.9279	.0319	.0532	43
18	.24700	.96901	.25490	.9231	.0320	.0486	42
19	.24728	.96894	.25521	.9184	.0320	.0440	41
20	0.24756	0.96887	0.25552	3.9136	1.0321	4.0394	40
21	.24784	.96880	.25583	.9089	.0322	.0348	39
22	.24813	.96873	.25614	.9042	.0323	.0302	38
23	.24841	.96865	.25645	.8994	.0323	.0256	37
24	.24869	.96858	.25676	.8947	.0324	.0211	36
25	0.24897	0.96851	0.25707	3.8900	1.0325	4.0165	35
26	.24925	.96844	.25738	.8853	.0326	.0120	34
27	.24953	.96836	.25769	.8807	.0327	.0074	33
28	.24982	.96829	.25800	.8760	.0327	.0029	32
29	.25010	.96822	.25831	.8713	.0328	3.9984	31
30	0.25038	0.96815	0.25862	3.8667	1.0329	3.9939	30
31	.25066	.96807	.25893	.8621	.0330	.9894	29
32	.25094	.96800	.25924	.8574	.0330	.9850	28
33	.25122	.96793	.25955	.8528	.0331	.9805	27
34	.25151	.96785	.25986	.8482	.0332	.9760	26
35	0.25179	0.96778	0.26017	3.8436	1.0333	3.9716	25
36	.25207	.96771	.26048	.8390	.0334	.9672	24
37	.25235	.96763	.26079	.8345	.0334	.9627	23
38	.25263	.96756	.26110	.8299	.0335	.9583	22
39	.25291	.96749	.26141	.8254	.0336	.9539	21
40	0.25319	0.96741	0.26172	3.8208	1.0337	3.9495	20
41	.25348	.96734	.26203	.8163	.0338	.9451	19
42	.25376	.96727	.26234	.8118	.0338	.9408	18
43	.25404	.96719	.26266	.8073	.0339	.9364	17
44	.25432	.96712	.26297	.8027	.0340	.9320	16
45	0.25460	0.96704	0.26328	3.7983	1.0341	3.9277	15
46	.25488	.96697	.26359	.7938	.0341	.9234	14
47	.25516	.96690	.26390	.7893	.0342	.9190	13
48	.25544	.96682	.26421	.7848	.0343	.9147	12
49	.25573	.96675	.26452	.7804	.0344	.9104	11
50	0.25601	0.96667	0.26483	3.7759	1.0345	3.9061	10
51	.25629	.96660	.26514	.7715	.0345	.9018	9
52	.25657	.96652	.26546	.7671	.0346	.8976	8
53	.25685	.96645	.26577	.7627	.0347	.8933	7
54	.25713	.96638	.26608	.7583	.0348	.8890	6
55	0.25741	0.96630	0.26639	3.7539	1.0349	3.8848	5
56	.25769	.96623	.26670	.7495	.0349	.8805	4
57	.25798	.96615	.26701	.7451	.0350	.8763	3
58	.25826	.96608	.26732	.7407	.0351	.8721	2
59	.25854	.96600	.26764	.7364	.0352	.8679	1
60	0.25882	0.96593	0.26795	3.7320	1.0353	3.8637	0
M	Cosine	Sine	Cotan.	Tan.	Cosec.	Secant	M
104°							75°

For angles from 0° to 45°, read degrees and function at top and minutes at left; for 45° to 90°, read degrees and function at bottom and minutes at right; for 90° to 135°, read degrees and function at bottom and minutes at left; and for 135° to 180° read degrees and function at top and minutes at right.

Table 3-5 (Continued). Values of the Natural Trigonometric Functions

For angles from 0° to 45°, read degrees and function at top and minutes at left; for 45° to 90°, read degrees and function at bottom and minutes at left; and for 135° to 180°, read degrees and function at top and minutes at right; for 90° to 135°, read degrees and function at bottom and minutes at right.

Table 3-5 (*Continued*). Values of the Natural Trigonometric Functions

18° — **161°**

M	Sine	Cosine	Tan.	Cotan.	Secant	Cosec.	M
0	0.30902	0.95106	0.32492	3.0777	1.0515	3.2361	60
1	.30929	.95097	.32524	.0746	.0516	.2332	59
2	.30957	.95088	.32556	.0716	.0517	.2303	58
3	.30985	.95079	.32588	.0686	.0518	.2274	57
4	.31012	.95070	.32621	.0655	.0519	.2245	56
5	0.31040	0.95061	0.32653	3.0625	1.0520	3.2216	55
6	.31068	.95051	.32685	.0595	.0521	.2188	54
7	.31095	.95042	.32717	.0565	.0522	.2159	53
8	.31123	.95033	.32749	.0535	.0523	.2131	52
9	.31150	.95024	.32782	.0505	.0524	.2102	51
10	0.31178	0.95015	0.32814	3.0475	1.0525	3.2074	50
11	.31206	.95006	.32846	.0445	.0526	.2045	49
12	.31233	.94997	.32878	.0415	.0527	.2017	48
13	.31261	.94988	.32910	.0385	.0528	.1989	47
14	.31289	.94979	.32943	.0356	.0529	.1960	46
15	0.31316	0.94970	0.32975	3.0326	1.0530	3.1932	45
16	.31344	.94961	.33007	.0296	.0531	.1904	44
17	.31372	.94952	.33039	.0267	.0532	.1876	43
18	.31399	.94942	.33072	.0237	.0533	.1848	42
19	.31427	.94933	.33104	.0208	.0534	.1820	41
20	0.31454	0.94924	0.33136	3.0178	1.0535	3.1792	40
21	.31482	.94915	.33169	.0149	.0536	.1764	39
22	.31510	.94906	.33201	.0120	.0537	.1736	38
23	.31537	.94897	.33233	.0090	.0538	.1708	37
24	.31565	.94888	.33265	.0061	.0539	.1681	36
25	0.31592	0.94878	0.33298	3.0032	1.0540	3.1653	35
26	.31620	.94869	.33330	.0003	.0541	.1625	34
27	.31648	.94860	.33362	2.9974	.0542	.1598	33
28	.31675	.94851	.33395	.9945	.0543	.1570	32
29	.31703	.94841	.33427	.9916	.0544	.1543	31
30	0.31730	0.94832	0.33459	2.9887	1.0545	3.1515	30
31	.31758	.94823	.33492	.9858	.0546	.1488	29
32	.31786	.94814	.33524	.9829	.0547	.1461	28
33	.31813	.94805	.33557	.9800	.0548	.1433	27
34	.31841	.94795	.33589	.9772	.0549	.1406	26
35	0.31868	0.94786	0.33621	2.9743	1.0550	3.1379	25
36	.31896	.94777	.33654	.9714	.0551	.1352	24
37	.31923	.94767	.33686	.9686	.0552	.1325	23
38	.31951	.94758	.33718	.9657	.0553	.1298	22
39	.31978	.94749	.33751	.9629	.0554	.1271	21
40	0.32006	0.94740	0.33783	2.9600	1.0555	3.1244	20
41	.32034	.94730	.33816	.9572	.0556	.1217	19
42	.32061	.94721	.33848	.9544	.0557	.1190	18
43	.32089	.94712	.33880	.9515	.0558	.1163	17
44	.32116	.94702	.33913	.9487	.0559	.1137	16
45	0.32144	0.94693	0.33945	2.9459	1.0560	3.1110	15
46	.32171	.94684	.33978	.9431	.0561	.1083	14
47	.32199	.94674	.34010	.9403	.0562	.1057	13
48	.32226	.94665	.34043	.9375	.0563	.1030	12
49	.32254	.94655	.34075	.9347	.0565	.1004	11
50	0.32282	0.94646	0.34108	2.9319	1.0566	3.0977	10
51	.32309	.94637	.34140	.9291	.0567	.0951	9
52	.32337	.94627	.34173	.9263	.0568	.0925	8
53	.32364	.94618	.34205	.9235	.0569	.0898	7
54	.32392	.94608	.34238	.9208	.0570	.0872	6
55	0.32419	0.94599	0.34270	2.9180	1.0571	3.0846	5
56	.32447	.94590	.34303	.9152	.0572	.0820	4
57	.32474	.94580	.34335	.9125	.0573	.0793	3
58	.32502	.94571	.34368	.9097	.0574	.0767	2
59	.32529	.94561	.34400	.9069	.0575	.0741	1
60	0.32557	0.94552	0.34433	2.9042	1.0576	3.0715	0
M	Cosine	Sine	Cotan.	Tan.	Cosec.	Secant	M

108° — **71°**

19° — **160°**

M	Sine	Cosine	Tan.	Cotan.	Secant	Cosec.	M
0	0.32557	0.94552	0.34433	2.9042	1.0576	3.0715	60
1	.32584	.94542	.34465	.9015	.0577	.0690	59
2	.32612	.94533	.34498	.8987	.0578	.0664	58
3	.32639	.94523	.34530	.8960	.0579	.0638	57
4	.32667	.94514	.34563	.8933	.0580	.0612	56
5	0.32694	0.94504	0.34595	2.8905	1.0581	3.0586	55
6	.32722	.94495	.34628	.8878	.0582	.0561	54
7	.32749	.94485	.34661	.8851	.0584	.0535	53
8	.32777	.94476	.34693	.8824	.0585	.0509	52
9	.32804	.94466	.34726	.8797	.0586	.0484	51
10	0.32832	0.94457	0.34758	2.8770	1.0587	3.0458	50
11	.32859	.94447	.34791	.8743	.0588	.0433	49
12	.32887	.94438	.34824	.8716	.0589	.0407	48
13	.32914	.94428	.34856	.8689	.0590	.0382	47
14	.32942	.94418	.34889	.8662	.0591	.0357	46
15	0.32969	0.94409	0.34921	2.8636	1.0592	3.0331	45
16	.32996	.94399	.34954	.8609	.0593	.0306	44
17	.33024	.94390	.34987	.8582	.0594	.0281	43
18	.33051	.94380	.35019	.8555	.0595	.0256	42
19	.33079	.94370	.35052	.8529	.0596	.0231	41
20	0.33106	0.94361	0.35085	2.8502	1.0598	3.0206	40
21	.33134	.94351	.35117	.8476	.0599	.0181	39
22	.33161	.94341	.35150	.8449	.0600	.0156	38
23	.33189	.94332	.35183	.8423	.0601	.0131	37
24	.33216	.94322	.35215	.8396	.0602	.0106	36
25	0.33243	0.94313	0.35248	2.8370	1.0603	3.0081	35
26	.33271	.94303	.35281	.8344	.0604	.0056	34
27	.33298	.94293	.35314	.8318	.0605	.0031	33
28	.33326	.94283	.35346	.8291	.0606	.0007	32
29	.33353	.94274	.35379	.8265	.0607	2.9982	31
30	0.33381	0.94264	0.35412	2.8239	1.0608	2.9957	30
31	.33408	.94254	.35445	.8213	.0609	.9933	29
32	.33435	.94245	.35477	.8187	.0611	.9908	28
33	.33463	.94235	.35510	.8161	.0612	.9884	27
34	.33490	.94225	.35543	.8135	.0613	.9859	26
35	0.33518	0.94215	0.35576	2.8109	1.0614	2.9835	25
36	.33545	.94206	.35608	.8083	.0615	.9810	24
37	.33572	.94196	.35641	.8057	.0616	.9786	23
38	.33600	.94186	.35674	.8032	.0617	.9762	22
39	.33627	.94176	.35707	.8006	.0618	.9738	21
40	0.33655	0.94167	0.35739	2.7980	1.0619	2.9713	20
41	.33682	.94157	.35772	.7954	.0620	.9689	19
42	.33709	.94147	.35805	.7929	.0622	.9665	18
43	.33737	.94137	.35838	.7903	.0623	.9641	17
44	.33764	.94127	.35871	.7878	.0624	.9617	16
45	0.33792	0.94118	0.35904	2.7852	1.0625	2.9593	15
46	.33819	.94108	.35936	.7827	.0626	.9569	14
47	.33846	.94098	.35969	.7801	.0627	.9545	13
48	.33874	.94088	.36002	.7776	.0628	.9521	12
49	.33901	.94078	.36035	.7751	.0629	.9497	11
50	0.33928	0.94068	0.36068	2.7725	1.0630	2.9474	10
51	.33956	.94058	.36101	.7700	.0632	.9450	9
52	.33983	.94049	.36134	.7675	.0633	.9426	8
53	.34011	.94039	.36167	.7650	.0634	.9402	7
54	.34038	.94029	.36199	.7625	.0635	.9379	6
55	0.34065	0.94019	0.36232	2.7600	1.0636	2.9355	5
56	.34093	.94009	.36265	.7575	.0637	.9332	4
57	.34120	.93999	.36298	.7550	.0638	.9308	3
58	.34147	.93989	.36331	.7525	.0639	.9285	2
59	.34175	.93979	.36364	.7500	.0641	.9261	1
60	0.34202	0.93969	0.36397	2.7475	1.0642	2.9238	0
M	Cosine	Sine	Cotan.	Tan.	Cosec.	Secant	M

109° — **70°**

20° — **159°**

M	Sine	Cosine	Tan.	Cotan.	Secant	Cosec.	M
0	0.34202	0.93969	0.36397	2.7475	1.0642	2.9238	60
1	.34229	.93959	.36430	.7450	.0643	.9215	59
2	.34257	.93949	.36463	.7425	.0644	.9191	58
3	.34284	.93939	.36496	.7400	.0645	.9168	57
4	.34311	.93929	.36529	.7376	.0646	.9145	56
5	0.34339	0.93919	0.36562	2.7351	1.0647	2.9122	55
6	.34366	.93909	.36595	.7326	.0648	.9098	54
7	.34393	.93899	.36628	.7302	.0650	.9075	53
8	.34421	.93889	.36661	.7277	.0651	.9052	52
9	.34448	.93879	.36694	.7252	.0652	.9029	51
10	0.34475	0.93869	0.36727	2.7228	1.0653	2.9006	50
11	.34502	.93859	.36760	.7204	.0654	.8983	49
12	.34530	.93849	.36793	.7179	.0655	.8960	48
13	.34557	.93839	.36826	.7155	.0656	.8937	47
14	.34584	.93829	.36859	.7130	.0658	.8915	46
15	0.34612	0.93819	0.36892	2.7106	1.0659	2.8892	45
16	.34639	.93809	.36925	.7082	.0660	.8869	44
17	.34666	.93799	.36958	.7058	.0661	.8846	43
18	.34693	.93789	.36991	.7033	.0662	.8824	42
19	.34721	.93779	.37024	.7009	.0663	.8801	41
20	0.34748	0.93769	0.37057	2.6985	1.0664	2.8778	40
21	.34775	.93758	.37090	.6961	.0666	.8756	39
22	.34803	.93748	.37123	.6937	.0667	.8733	38
23	.34830	.93738	.37156	.6913	.0668	.8711	37
24	.34857	.93728	.37190	.6889	.0669	.8688	36
25	0.34884	0.93718	0.37223	2.6865	1.0670	2.8666	35
26	.34912	.93708	.37256	.6841	.0671	.8644	34
27	.34939	.93698	.37289	.6817	.0673	.8621	33
28	.34966	.93687	.37322	.6794	.0674	.8599	32
29	.34993	.93677	.37355	.6770	.0675	.8577	31
30	0.35021	0.93667	0.37388	2.6746	1.0676	2.8554	30
31	.35048	.93657	.37422	.6722	.0677	.8532	29
32	.35075	.93647	.37455	.6699	.0678	.8510	28
33	.35102	.93637	.37488	.6675	.0679	.8488	27
34	.35130	.93626	.37521	.6652	.0681	.8466	26
35	0.35157	0.93616	0.37554	2.6628	1.0682	2.8444	25
36	.35184	.93606	.37587	.6604	.0683	.8422	24
37	.35211	.93596	.37621	.6581	.0684	.8400	23
38	.35239	.93585	.37654	.6558	.0685	.8378	22
39	.35266	.93575	.37687	.6534	.0686	.8356	21
40	0.35293	0.93565	0.37720	2.6511	1.0688	2.8334	20
41	.35320	.93555	.37754	.6487	.0689	.8312	19
42	.35347	.93544	.37787	.6464	.0690	.8290	18
43	.35375	.93534	.37820	.6441	.0691	.8269	17
44	.35402	.93524	.37853	.6418	.0692	.8247	16
45	0.35429	0.93513	0.37887	2.6394	1.0694	2.8225	15
46	.35456	.93503	.37920	.6371	.0695	.8204	14
47	.35483	.93493	.37953	.6348	.0696	.8182	13
48	.35511	.93482	.37986	.6325	.0697	.8160	12
49	.35538	.93472	.38020	.6302	.0698	.8139	11
50	0.35565	0.93462	0.38053	2.6279	1.0699	2.8117	10
51	.35592	.93451	.38086	.6256	.0701	.8096	9
52	.35619	.93441	.38120	.6233	.0702	.8074	8
53	.35647	.93431	.38153	.6210	.0703	.8053	7
54	.35674	.93420	.38186	.6187	.0704	.8032	6
55	0.35701	0.93410	0.38220	2.6164	1.0705	2.8010	5
56	.35728	.93400	.38253	.6142	.0707	.7989	4
57	.35755	.93389	.38286	.6119	.0708	.7968	3
58	.35782	.93379	.38320	.6096	.0709	.7947	2
59	.35810	.93368	.38353	.6073	.0710	.7925	1
60	0.35837	0.93358	0.38386	2.6051	1.0711	2.7904	0
M	Cosine	Sine	Cotan.	Tan.	Cosec.	Secant	M

110° — **69°**

For angles from 0° to 45°, read degrees and function at top and minutes at left; for 45° to 90°, read degrees and function at bottom and minutes at right; for 90° to 135°, read degrees and function at bottom and minutes at left; and for 135° to 180°, read degrees and function at top and minutes at right.

Table 3-5 (*Continued*). Values of the Natural Trigonometric Functions

21°							158°
M	Sine	Cosine	Tan.	Cotan.	Secant	Cosec.	M
0	0.35837	0.93358	0.38386	2.6051	1.0711	2.7904	60
1	.35864	.93348	.38420	.6028	.0713	.7883	59
2	.35891	.93337	.38453	.6006	.0714	.7862	58
3	.35918	.93327	.38486	.5983	.0715	.7841	57
4	.35945	.93316	.38520	.5960	.0716	.7820	56
5	0.35972	0.93306	0.38553	2.5938	1.0717	2.7799	55
6	.36000	.93295	.38587	.5916	.0719	.7778	54
7	.36027	.93285	.38620	.5893	.0720	.7757	53
8	.36054	.93274	.38654	.5871	.0721	.7736	52
9	.36081	.93264	.38687	.5848	.0722	.7715	51
10	0.36108	0.93253	0.38720	2.5826	1.0723	2.7694	50
11	.36135	.93243	.38754	.5804	.0725	.7674	49
12	.36162	.93232	.38787	.5781	.0726	.7653	48
13	.36189	.93222	.38821	.5759	.0727	.7632	47
14	.36217	.93211	.38854	.5737	.0728	.7611	46
15	0.36244	0.93201	0.38888	2.5715	1.0729	2.7591	45
16	.36271	.93190	.38921	.5693	.0731	.7570	44
17	.36298	.93180	.38955	.5671	.0732	.7550	43
18	.36325	.93169	.38988	.5649	.0733	.7529	42
19	.36352	.93158	.39022	.5627	.0734	.7509	41
20	0.36379	0.93148	0.39055	2.5605	1.0736	2.7488	40
21	.36406	.93137	.39089	.5583	.0737	.7468	39
22	.36433	.93127	.39122	.5561	.0738	.7447	38
23	.36460	.93116	.39156	.5539	.0739	.7427	37
24	.36488	.93105	.39189	.5517	.0740	.7406	36
25	0.36515	0.93095	0.39223	2.5495	1.0742	2.7386	35
26	.36542	.93084	.39257	.5473	.0743	.7366	34
27	.36569	.93074	.39290	.5451	.0744	.7346	33
28	.36596	.93063	.39324	.5430	.0745	.7325	32
29	.36623	.93052	.39357	.5408	.0747	.7305	31
30	0.36650	0.93042	0.39391	2.5386	1.0748	2.7285	30
31	.36677	.93031	.39425	.5365	.0749	.7265	29
32	.36704	.93020	.39458	.5343	.0750	.7245	28
33	.36731	.93010	.39492	.5322	.0751	.7225	27
34	.36758	.92999	.39525	.5300	.0753	.7205	26
35	0.36785	0.92988	0.39559	2.5278	1.0754	2.7185	25
36	.36812	.92978	.39593	.5257	.0755	.7165	24
37	.36839	.92967	.39626	.5236	.0756	.7145	23
38	.36866	.92956	.39660	.5214	.0758	.7125	22
39	.36893	.92945	.39694	.5193	.0759	.7105	21
40	0.36921	0.92935	0.39727	2.5171	1.0760	2.7085	20
41	.36948	.92924	.39761	.5150	.0761	.7065	19
42	.36975	.92913	.39795	.5129	.0763	.7045	18
43	.37002	.92902	.39828	.5108	.0764	.7026	17
44	.37029	.92892	.39862	.5086	.0765	.7006	16
45	0.37056	0.92881	0.39896	2.5065	1.0766	2.6986	15
46	.37083	.92870	.39930	.5044	.0768	.6967	14
47	.37110	.92859	.39963	.5023	.0769	.6947	13
48	.37137	.92848	.39997	.5002	.0770	.6927	12
49	.37164	.92838	.40031	.4981	.0771	.6908	11
50	0.37191	0.92827	0.40065	2.4960	1.0773	2.6888	10
51	.37218	.92816	.40098	.4939	.0774	.6869	9
52	.37245	.92805	.40132	.4918	.0775	.6849	8
53	.37272	.92794	.40166	.4897	.0776	.6830	7
54	.37299	.92784	.40200	.4876	.0778	.6810	6
55	0.37326	0.92773	0.40233	2.4855	1.0779	2.6791	5
56	.37353	.92762	.40267	.4834	.0780	.6772	4
57	.37380	.92751	.40301	.4813	.0781	.6752	3
58	.37407	.92740	.40335	.4792	.0783	.6733	2
59	.37434	.92729	.40369	.4772	.0784	.6714	1
60	0.37461	0.92718	0.40403	2.4751	1.0785	2.6695	0
M	Cosine	Sine	Cotan.	Tan.	Cosec.	Secant	M
111°							68°

22°							157°
M	Sine	Cosine	Tan.	Cotan.	Secant	Cosec.	M
0	0.37461	0.92718	0.40403	2.4751	1.0785	2.6695	60
1	.37488	.92707	.40436	.4730	.0787	.6675	59
2	.37514	.92696	.40470	.4709	.0788	.6656	58
3	.37541	.92686	.40504	.4689	.0789	.6637	57
4	.37568	.92675	.40538	.4668	.0790	.6618	56
5	0.37595	0.92664	0.40572	2.4647	1.0792	2.6599	55
6	.37622	.92653	.40606	.4627	.0793	.6580	54
7	.37649	.92642	.40640	.4606	.0794	.6561	53
8	.37676	.92631	.40673	.4586	.0795	.6542	52
9	.37703	.92620	.40707	.4565	.0797	.6523	51
10	0.37730	0.92609	0.40741	2.4545	1.0798	2.6504	50
11	.37757	.92598	.40775	.4525	.0799	.6485	49
12	.37784	.92587	.40809	.4504	.0801	.6466	48
13	.37811	.92576	.40843	.4484	.0802	.6447	47
14	.37838	.92565	.40877	.4463	.0803	.6428	46
15	0.37865	0.92554	0.40911	2.4443	1.0804	2.6410	45
16	.37892	.92543	.40945	.4423	.0806	.6391	44
17	.37919	.92532	.40979	.4403	.0807	.6372	43
18	.37946	.92521	.41013	.4382	.0808	.6353	42
19	.37972	.92510	.41047	.4362	.0810	.6335	41
20	0.37999	0.92499	0.41081	2.4342	1.0811	2.6316	40
21	.38026	.92488	.41115	.4322	.0812	.6297	39
22	.38053	.92477	.41149	.4302	.0813	.6279	38
23	.38080	.92466	.41183	.4282	.0815	.6260	37
24	.38107	.92455	.41217	.4262	.0816	.6242	36
25	0.38134	0.92443	0.41251	2.4242	1.0817	2.6223	35
26	.38151	.92432	.41285	.4222	.0819	.6205	34
27	.38188	.92421	.41319	.4202	.0820	.6186	33
28	.38214	.92410	.41353	.4182	.0821	.6168	32
29	.38241	.92399	.41387	.4162	.0823	.6150	31
30	0.38268	0.92388	0.41421	2.4142	1.0824	2.6131	30
31	.38295	.92377	.41455	.4122	.0825	.6113	29
32	.38322	.92366	.41489	.4102	.0826	.6095	28
33	.38349	.92354	.41524	.4083	.0828	.6076	27
34	.38376	.92343	.41558	.4063	.0829	.6058	26
35	0.38403	0.92332	0.41592	2.4043	1.0830	2.6040	25
36	.38429	.92321	.41626	.4023	.0832	.6022	24
37	.38456	.92310	.41660	.4004	.0833	.6003	23
38	.38483	.92299	.41694	.3984	.0834	.5985	22
39	.38510	.92287	.41728	.3964	.0836	.5967	21
40	0.38537	0.92276	0.41762	2.3945	1.0837	2.5949	20
41	.38564	.92265	.41797	.3925	.0838	.5931	19
42	.38591	.92254	.41831	.3906	.0840	.5913	18
43	.38617	.92242	.41865	.3886	.0841	.5895	17
44	.38644	.92231	.41899	.3867	.0842	.5877	16
45	0.38671	0.92220	0.41933	2.3847	1.0844	2.5859	15
46	.38698	.92209	.41968	.3828	.0845	.5841	14
47	.38725	.92197	.42002	.3808	.0846	.5823	13
48	.38751	.92186	.42036	.3789	.0847	.5805	12
49	.38778	.92175	.42070	.3770	.0849	.5787	11
50	0.38805	0.92164	0.42105	2.3750	1.0850	2.5770	10
51	.38832	.92152	.42139	.3731	.0851	.5752	9
52	.38859	.92141	.42173	.3712	.0853	.5734	8
53	.38886	.92130	.42207	.3692	.0854	.5716	7
54	.38912	.92118	.42242	.3673	.0855	.5699	6
55	0.38939	0.92107	0.42276	2.3654	1.0857	2.5681	5
56	.38966	.92096	.42310	.3635	.0858	.5663	4
57	.38993	.92084	.42344	.3616	.0859	.5646	3
58	.39019	.92073	.42379	.3597	.0861	.5628	2
59	.39046	.92062	.42413	.3577	.0862	.5610	1
60	0.39073	0.92050	0.42447	2.3558	1.0864	2.5593	0
M	Cosine	Sine	Cotan.	Tan.	Cosec.	Secant	M
112°							67°

23°							156°
M	Sine	Cosine	Tan.	Cotan.	Secant	Cosec.	M
0	0.39073	0.92050	0.42447	2.3558	1.0864	2.5593	60
1	.39100	.92039	.42482	.3539	.0865	.5575	59
2	.39126	.92028	.42516	.3520	.0866	.5558	58
3	.39153	.92016	.42550	.3501	.0868	.5540	57
4	.39180	.92005	.42585	.3482	.0869	.5523	56
5	0.39207	0.91993	0.42619	2.3463	1.0870	2.5506	55
6	.39234	.91982	.42654	.3445	.0872	.5488	54
7	.39260	.91971	.42688	.3426	.0873	.5471	53
8	.39287	.91959	.42722	.3407	.0874	.5453	52
9	.39314	.91948	.42757	.3388	.0876	.5436	51
10	0.39341	0.91936	0.42791	2.3369	1.0877	2.5419	50
11	.39367	.91925	.42826	.3350	.0878	.5402	49
12	.39394	.91913	.42860	.3332	.0880	.5384	48
13	.39421	.91902	.42894	.3313	.0881	.5367	47
14	.39448	.91891	.42929	.3294	.0882	.5350	46
15	0.39474	0.91879	0.42963	2.3276	1.0884	2.5333	45
16	.39501	.91868	.42998	.3257	.0885	.5316	44
17	.39528	.91856	.43032	.3238	.0886	.5299	43
18	.39554	.91845	.43067	.3220	.0888	.5281	42
19	.39581	.91833	.43101	.3201	.0889	.5264	41
20	0.39608	0.91822	0.43136	2.3183	1.0891	2.5247	40
21	.39635	.91810	.43170	.3164	.0892	.5230	39
22	.39661	.91798	.43205	.3145	.0893	.5213	38
23	.39688	.91787	.43239	.3127	.0895	.5196	37
24	.39715	.91775	.43274	.3109	.0896	.5179	36
25	0.39741	0.91764	0.43308	2.3090	1.0897	2.5163	35
26	.39768	.91752	.43343	.3072	.0899	.5146	34
27	.39795	.91741	.43377	.3053	.0900	.5129	33
28	.39821	.91729	.43412	.3035	.0902	.5112	32
29	.39848	.91718	.43447	.3017	.0903	.5095	31
30	0.39875	0.91706	0.43481	2.2998	1.0904	2.5078	30
31	.39901	.91694	.43516	.2980	.0906	.5062	29
32	.39928	.91683	.43550	.2962	.0907	.5045	28
33	.39955	.91671	.43585	.2944	.0908	.5028	27
34	.39981	.91659	.43620	.2925	.0910	.5011	26
35	0.40008	0.91648	0.43654	2.2907	1.0911	2.4995	25
36	.40035	.91636	.43689	.2889	.0913	.4978	24
37	.40061	.91625	.43723	.2871	.0914	.4961	23
38	.40088	.91613	.43758	.2853	.0915	.4945	22
39	.40115	.91601	.43793	.2835	.0917	.4928	21
40	0.40141	0.91590	0.43827	2.2817	1.0918	2.4912	20
41	.40168	.91578	.43862	.2799	.0920	.4895	19
42	.40195	.91566	.43897	.2781	.0921	.4879	18
43	.40221	.91554	.43932	.2763	.0922	.4862	17
44	.40248	.91543	.43966	.2745	.0924	.4846	16
45	0.40275	0.91531	0.44001	2.2727	1.0925	2.4829	15
46	.40301	.91519	.44036	.2709	.0927	.4813	14
47	.40328	.91508	.44070	.2691	.0928	.4797	13
48	.40354	.91496	.44105	.2673	.0929	.4780	12
49	.40381	.91484	.44140	.2655	.0931	.4764	11
50	0.40408	0.91472	0.44175	2.2637	1.0932	2.4748	10
51	.40434	.91461	.44209	.2619	.0934	.4731	9
52	.40461	.91449	.44244	.2602	.0935	.4715	8
53	.40487	.91437	.44279	.2584	.0936	.4699	7
54	.40514	.91425	.44314	.2566	.0938	.4683	6
55	0.40541	0.91414	0.44349	2.2548	1.0939	2.4666	5
56	.40567	.91402	.44383	.2531	.0941	.4650	4
57	.40594	.91390	.44418	.2513	.0942	.4634	3
58	.40620	.91378	.44453	.2495	.0943	.4618	2
59	.40647	.91366	.44488	.2478	.0945	.4602	1
60	0.40674	0.91354	0.44523	2.2460	1.0946	2.4586	0
M	Cosine	Sine	Cotan.	Tan.	Cosec.	Secant	M
113°							66°

For angles from 0° to 45°, read degrees and function at top and minutes at left; for 45° to 90°, read degrees and function at bottom and minutes at right; for 90° to 135°, read degrees and function at bottom and minutes at left; and for 135° to 180° read degrees and function at top and minutes at right.

Table 3-5 (*Continued*). Values of the Natural Trigonometric Functions

24° **155°**

M	Sine	Cosine	Tan.	Cotan.	Secant	Cosec.	M
0	0.40674	0.91354	0.44523	2.2460	1.0946	2.4586	60
1	.40700	.91343	.44558	.2443	.0948	.4570	59
2	.40727	.91331	.44593	.2425	.0949	.4554	58
3	.40753	.91319	.44627	.2408	.0951	.4538	57
4	.40780	.91307	.44662	.2390	.0952	.4522	56
5	0.40806	0.91295	0.44697	2.2373	1.0953	2.4506	55
6	.40833	.91283	.44732	.2355	.0955	.4490	54
7	.40860	.91271	.44767	.2338	.0956	.4474	53
8	.40886	.91260	.44802	.2320	.0958	.4458	52
9	.40913	.91248	.44837	.2303	.0959	.4442	51
10	0.40939	0.91236	0.44872	2.2286	1.0961	2.4426	50
11	.40966	.91224	.44907	.2268	.0962	.4411	49
12	.40992	.91212	.44942	.2251	.0963	.4395	48
13	.41019	.91200	.44977	.2234	.0965	.4379	47
14	.41045	.91188	.45012	.2216	.0966	.4363	46
15	0.41072	0.91176	0.45047	2.2199	1.0968	2.4347	45
16	.41098	.91164	.45082	.2182	.0969	.4332	44
17	.41125	.91152	.45117	.2165	.0971	.4316	43
18	.41151	.91140	.45152	.2147	.0972	.4300	42
19	.41178	.91128	.45187	.2130	.0973	.4285	41
20	0.41204	0.91116	0.45222	2.2113	1.0975	2.4269	40
21	.41231	.91104	.45257	.2096	.0976	.4254	39
22	.41257	.91092	.45292	.2079	.0978	.4238	38
23	.41284	.91080	.45327	.2062	.0979	.4222	37
24	.41310	.91068	.45362	.2045	.0981	.4207	36
25	0.41337	0.91056	0.45397	2.2028	1.0982	2.4191	35
26	.41363	.91044	.45432	.2011	.0984	.4176	34
27	.41390	.91032	.45467	.1994	.0985	.4160	33
28	.41416	.91020	.45502	.1977	.0986	.4145	32
29	.41443	.91008	.45537	.1960	.0988	.4130	31
30	0.41469	0.90996	0.45573	2.1943	1.0989	2.4114	30
31	.41496	.90984	.45608	.1926	.0991	.4099	29
32	.41522	.90972	.45643	.1909	.0992	.4083	28
33	.41549	.90960	.45678	.1892	.0994	.4068	27
34	.41575	.90948	.45713	.1875	.0995	.4053	26
35	0.41602	0.90936	0.45748	2.1859	1.0997	2.4037	25
36	.41628	.90924	.45783	.1842	.0998	.4022	24
37	.41654	.90911	.45819	.1825	.1000	.4007	23
38	.41681	.90899	.45854	.1808	.1001	.3992	22
39	.41707	.90887	.45889	.1792	.1003	.3976	21
40	0.41734	0.90875	0.45924	2.1775	1.1004	2.3961	20
41	.41760	.90863	.45960	.1758	.1005	.3946	19
42	.41787	.90851	.45995	.1741	.1007	.3931	18
43	.41813	.90839	.46030	.1725	.1008	.3916	17
44	.41839	.90826	.46065	.1708	.1010	.3901	16
45	0.41866	0.90814	0.46101	2.1692	1.1011	2.3886	15
46	.41892	.90802	.46136	.1675	.1013	.3871	14
47	.41919	.90790	.46171	.1658	.1014	.3856	13
48	.41945	.90778	.46206	.1642	.1016	.3841	12
49	.41972	.90765	.46242	.1625	.1017	.3826	11
50	0.41998	0.90753	0.46277	2.1609	1.1019	2.3811	10
51	.42024	.90741	.46312	.1592	.1020	.3796	9
52	.42051	.90729	.46348	.1576	.1022	.3781	8
53	.42077	.90717	.46383	.1559	.1023	.3766	7
54	.42103	.90704	.46418	.1543	.1025	.3751	6
55	0.42130	0.90692	0.46454	2.1527	1.1026	2.3736	5
56	.42156	.90680	.46489	.1510	.1028	.3721	4
57	.42183	.90668	.46524	.1494	.1029	.3706	3
58	.42209	.90655	.46560	.1478	.1031	.3691	2
59	.42235	.90643	.46595	.1461	.1032	.3677	1
60	0.42262	0.90631	0.46631	2.1445	1.1034	2.3662	0
M	Cosine	Sine	Cotan.	Tan.	Cosec.	Secant	M

114° **65°**

25° **154°**

M	Sine	Cosine	Tan.	Cotan.	Secant	Cosec.	M
0	0.42262	0.90631	0.46631	2.1445	1.1034	2.3662	60
1	.42288	.90618	.46666	.1429	.1035	.3647	59
2	.42314	.90606	.46702	.1412	.1037	.3632	58
3	.42341	.90594	.46737	.1396	.1038	.3618	57
4	.42367	.90581	.46772	.1380	.1040	.3603	56
5	0.42394	0.90569	0.46808	2.1364	1.1041	2.3588	55
6	.42420	.90557	.46843	.1348	.1043	.3574	54
7	.42446	.90544	.46879	.1331	.1044	.3559	53
8	.42473	.90532	.46914	.1315	.1046	.3544	52
9	.42499	.90520	.46950	.1299	.1047	.3530	51
10	0.42525	0.90507	0.46985	2.1283	1.1049	2.3515	50
11	.42552	.90495	.47021	.1267	.1050	.3501	49
12	.42578	.90483	.47056	.1251	.1052	.3486	48
13	.42604	.90470	.47092	.1235	.1053	.3472	47
14	.42630	.90458	.47127	.1219	.1055	.3457	46
15	0.42657	0.90445	0.47163	2.1203	1.1056	2.3443	45
16	.42683	.90433	.47199	.1187	.1058	.3428	44
17	.42709	.90421	.47234	.1171	.1059	.3414	43
18	.42736	.90408	.47270	.1155	.1061	.3399	42
19	.42762	.90396	.47305	.1139	.1062	.3385	41
20	0.42788	0.90383	0.47341	2.1123	1.1064	2.3371	40
21	.42815	.90371	.47376	.1107	.1065	.3356	39
22	.42841	.90358	.47412	.1092	.1067	.3342	38
23	.42867	.90346	.47448	.1076	.1068	.3328	37
24	.42893	.90333	.47483	.1060	.1070	.3313	36
25	0.42920	0.90321	0.47519	2.1044	1.1072	2.3299	35
26	.42946	.90308	.47555	.1028	.1073	.3285	34
27	.42972	.90296	.47590	.1013	.1075	.3271	33
28	.42998	.90283	.47626	.0997	.1076	.3256	32
29	.43025	.90271	.47662	.0981	.1078	.3242	31
30	0.43051	0.90258	0.47697	2.0965	1.1079	2.3228	30
31	.43077	.90246	.47733	.0950	.1081	.3214	29
32	.43104	.90233	.47769	.0934	.1082	.3200	28
33	.43130	.90221	.47805	.0918	.1084	.3186	27
34	.43156	.90208	.47840	.0903	.1085	.3172	26
35	0.43182	0.90196	0.47876	2.0887	1.1087	2.3158	25
36	.43208	.90183	.47912	.0872	.1088	.3143	24
37	.43235	.90171	.47948	.0856	.1090	.3129	23
38	.43261	.90158	.47983	.0840	.1092	.3115	22
39	.43287	.90145	.48019	.0825	.1093	.3101	21
40	0.43313	0.90133	0.48055	2.0809	1.1095	2.3087	20
41	.43340	.90120	.48091	.0794	.1096	.3073	19
42	.43366	.90108	.48127	.0778	.1098	.3059	18
43	.43392	.90095	.48162	.0763	.1099	.3046	17
44	.43418	.90082	.48198	.0747	.1101	.3032	16
45	0.43444	0.90070	0.48234	2.0732	1.1102	2.3018	15
46	.43471	.90057	.48270	.0717	.1104	.3004	14
47	.43497	.90044	.48306	.0701	.1106	.2990	13
48	.43523	.90032	.48342	.0686	.1107	.2976	12
49	.43549	.90019	.48378	.0671	.1109	.2962	11
50	0.43575	0.90006	0.48414	2.0655	1.1110	2.2949	10
51	.43602	.89994	.48449	.0640	.1112	.2935	9
52	.43628	.89981	.48485	.0625	.1113	.2921	8
53	.43654	.89968	.48521	.0609	.1115	.2907	7
54	.43680	.89956	.48557	.0594	.1116	.2894	6
55	0.43706	0.89943	0.48593	2.0579	1.1118	2.2880	5
56	.43732	.89930	.48629	.0564	.1120	.2866	4
57	.43759	.89918	.48665	.0548	.1121	.2853	3
58	.43785	.89905	.48701	.0533	.1123	.2839	2
59	.43811	.89892	.48737	.0518	.1124	.2825	1
60	0.43837	0.89879	0.48773	2.0503	1.1126	2.2812	0
M	Cosine	Sine	Cotan.	Tan.	Cosec.	Secant	M

115° **64°**

26° **153°**

M	Sine	Cosine	Tan.	Cotan.	Secant	Cosec.	M
0	0.43837	0.89879	0.48773	2.0503	1.1126	2.2812	60
1	.43863	.89867	.48809	.0488	.1127	.2798	59
2	.43889	.89854	.48845	.0473	.1129	.2784	58
3	.43915	.89841	.48881	.0458	.1131	.2771	57
4	.43942	.89828	.48917	.0443	.1132	.2757	56
5	0.43968	0.89815	0.48953	2.0427	1.1134	2.2744	55
6	.43994	.89803	.48989	.0412	.1135	.2730	54
7	.44020	.89790	.49025	.0397	.1137	.2717	53
8	.44046	.89777	.49062	.0382	.1139	.2703	52
9	.44072	.89764	.49098	.0367	.1140	.2690	51
10	0.44098	0.89751	0.49134	2.0352	1.1142	2.2676	50
11	.44124	.89739	.49170	.0338	.1143	.2663	49
12	.44150	.89726	.49206	.0323	.1145	.2650	48
13	.44177	.89713	.49242	.0308	.1147	.2636	47
14	.44203	.89700	.49278	.0293	.1148	.2623	46
15	0.44229	0.89687	0.49314	2.0278	1.1150	2.2610	45
16	.44255	.89674	.49351	.0263	.1151	.2596	44
17	.44281	.89661	.49387	.0248	.1153	.2583	43
18	.44307	.89649	.49423	.0233	.1155	.2570	42
19	.44333	.89636	.49459	.0219	.1156	.2556	41
20	0.44359	0.89623	0.49495	2.0204	1.1158	2.2543	40
21	.44385	.89610	.49532	.0189	.1159	.2530	39
22	.44411	.89597	.49568	.0174	.1161	.2517	38
23	.44437	.89584	.49604	.0159	.1163	.2503	37
24	.44463	.89571	.49640	.0145	.1164	.2490	36
25	0.44489	0.89558	0.49677	2.0130	1.1166	2.2477	35
26	.44516	.89545	.49713	.0115	.1167	.2464	34
27	.44542	.89532	.49749	.0101	.1169	.2451	33
28	.44568	.89519	.49785	.0086	.1171	.2438	32
29	.44594	.89506	.49822	.0071	.1172	.2425	31
30	0.44620	0.89493	0.49858	2.0057	1.1174	2.2411	30
31	.44646	.89480	.49894	.0042	.1176	.2398	29
32	.44672	.89467	.49931	.0028	.1177	.2385	28
33	.44698	.89454	.49967	.0013	.1179	.2372	27
34	.44724	.89441	.50003	1.9998	.1180	.2359	26
35	0.44750	0.89428	0.50040	1.9984	1.1182	2.2346	25
36	.44776	.89415	.50076	.9969	.1184	.2333	24
37	.44802	.89402	.50113	.9955	.1185	.2320	23
38	.44828	.89389	.50149	.9940	.1187	.2307	22
39	.44854	.89376	.50185	.9926	.1189	.2294	21
40	0.44880	0.89363	0.50222	1.9912	1.1190	2.2282	20
41	.44906	.89350	.50258	.9897	.1192	.2269	19
42	.44932	.89337	.50295	.9883	.1193	.2256	18
43	.44958	.89324	.50331	.9868	.1195	.2243	17
44	.44984	.89311	.50368	.9854	.1197	.2230	16
45	0.45010	0.89298	0.50404	1.9840	1.1198	2.2217	15
46	.45036	.89285	.50441	.9825	.1200	.2204	14
47	.45062	.89272	.50477	.9811	.1202	.2192	13
48	.45088	.89258	.50514	.9797	.1203	.2179	12
49	.45114	.89245	.50550	.9782	.1205	.2166	11
50	0.45140	0.89232	0.50587	1.9768	1.1207	2.2153	10
51	.45166	.89219	.50623	.9754	.1208	.2141	9
52	.45191	.89206	.50660	.9739	.1210	.2128	8
53	.45217	.89193	.50696	.9725	.1212	.2115	7
54	.45243	.89180	.50733	.9711	.1213	.2103	6
55	0.45269	0.89166	0.50769	1.9697	1.1215	2.2090	5
56	.45295	.89153	.50806	.9683	.1217	.2077	4
57	.45321	.89140	.50843	.9668	.1218	.2065	3
58	.45347	.89127	.50879	.9654	.1220	.2052	2
59	.45373	.89114	.50916	.9640	.1222	.2039	1
60	0.45399	0.89101	0.50952	1.9626	1.1223	2.2027	0
M	Cosine	Sine	Cotan.	Tan.	Cosec.	Secant	M

116° **63°**

For angles from 0° to 45°, read degrees and function at top and minutes at left; for 45° to 90°, read degrees and function at bottom and minutes at right; for 90° to 135°, read degrees and function at bottom and minutes at left; and for 135° to 180°, read degrees and function at top and minutes at right.

Table 3-5 (*Continued*). Values of the Natural Trigonometric Functions

27°							152°
M	Sine	Cosine	Tan.	Cotan.	Secant	Cosec.	M
0	0.45399	0.89101	0.50952	1.9626	1.1223	2.2027	60
1	.45425	.89087	.50989	.9612	.1225	.2014	59
2	.45451	.89074	.51026	.9598	.1226	.2002	58
3	.45477	.89061	.51062	.9584	.1228	.1989	57
4	.45503	.89048	.51099	.9570	.1230	.1977	56
5	0.45528	0.89034	0.51136	1.9556	1.1231	2.1964	55
6	.45554	.89021	.51172	.9542	.1233	.1952	54
7	.45580	.89008	.51209	.9528	.1235	.1939	53
8	.45606	.88995	.51246	.9514	.1237	.1927	52
9	.45632	.88981	.51283	.9500	.1238	.1914	51
10	0.45658	0.88968	0.51319	1.9486	1.1240	2.1902	50
11	.45684	.88955	.51356	.9472	.1242	.1889	49
12	.45710	.88942	.51393	.9458	.1243	.1877	48
13	.45736	.88928	.51430	.9444	.1245	.1865	47
14	.45761	.88915	.51466	.9430	.1247	.1852	46
15	0.45787	0.88902	0.51503	1.9416	1.1248	2.1840	45
16	.45813	.88888	.51540	.9402	.1250	.1828	44
17	.45839	.88875	.51577	.9388	.1252	.1815	43
18	.45865	.88862	.51614	.9375	.1253	.1803	42
19	.45891	.88848	.51651	.9361	.1255	.1791	41
20	0.45917	0.88835	0.51687	1.9347	1.1257	2.1778	40
21	.45942	.88822	.51724	.9333	.1258	.1766	39
22	.45968	.88808	.51761	.9319	.1260	.1754	38
23	.45994	.88795	.51798	.9306	.1262	.1742	37
24	.46020	.88781	.51835	.9292	.1264	.1730	36
25	0.46046	0.88768	0.51872	1.9278	1.1265	2.1717	35
26	.46072	.88755	.51909	.9264	.1267	.1705	34
27	.46097	.88741	.51946	.9251	.1269	.1693	33
28	.46123	.88728	.51983	.9237	.1270	.1681	32
29	.46149	.88714	.52020	.9223	.1272	.1669	31
30	0.46175	0.88701	0.52057	1.9210	1.1274	2.1657	30
31	.46201	.88688	.52094	.9196	.1275	.1645	29
32	.46226	.88674	.52131	.9182	.1277	.1633	28
33	.46252	.88661	.52168	.9169	.1279	.1620	27
34	.46278	.88647	.52205	.9155	.1281	.1608	26
35	0.46304	0.88634	0.52242	1.9142	1.1282	2.1596	25
36	.46330	.88620	.52279	.9128	.1284	.1584	24
37	.46355	.88607	.52316	.9115	.1286	.1572	23
38	.46381	.88593	.52353	.9101	.1287	.1560	22
39	.46407	.88580	.52390	.9088	.1289	.1548	21
40	0.46433	0.88566	0.52427	1.9074	1.1291	2.1536	20
41	.46458	.88553	.52464	.9061	.1293	.1525	19
42	.46484	.88539	.52501	.9047	.1294	.1513	18
43	.46510	.88526	.52538	.9034	.1296	.1501	17
44	.46536	.88512	.52575	.9020	.1298	.1489	16
45	0.46561	0.88499	0.52612	1.9007	1.1299	2.1477	15
46	.46587	.88485	.52650	.8993	.1301	.1465	14
47	.46613	.88472	.52687	.8980	.1303	.1453	13
48	.46639	.88458	.52724	.8967	.1305	.1441	12
49	.46664	.88444	.52761	.8953	.1306	.1430	11
50	0.46690	0.88431	0.52798	1.8940	1.1308	2.1418	10
51	.46716	.88417	.52836	.8927	.1310	.1406	9
52	.46741	.88404	.52873	.8913	.1312	.1394	8
53	.46767	.88390	.52910	.8900	.1313	.1382	7
54	.46793	.88376	.52947	.8887	.1315	.1371	6
55	0.46819	0.88363	0.52984	1.8873	1.1317	2.1359	5
56	.46844	.88349	.53022	.8860	.1319	.1347	4
57	.46870	.88336	.53059	.8847	.1320	.1335	3
58	.46896	.88322	.53096	.8834	.1322	.1324	2
59	.46921	.88308	.53134	.8820	.1324	.1312	1
60	0.46947	0.88295	0.53171	1.8807	1.1326	2.1300	0
M	Cosine	Sine	Cotan.	Tan.	Cosec.	Secant	M
117°							62°

28°							151°
M	Sine	Cosine	Tan.	Cotan.	Secant	Cosec.	M
0	0.46947	0.88295	0.53171	1.8807	1.1326	2.1300	60
1	.46973	.88281	.53208	.8794	.1327	.1289	59
2	.46998	.88267	.53245	.8781	.1329	.1277	58
3	.47024	.88254	.53283	.8768	.1331	.1266	57
4	.47050	.88240	.53320	.8754	.1333	.1254	56
5	0.47075	0.88226	0.53358	1.8741	1.1334	2.1242	55
6	.47101	.88213	.53395	.8728	.1336	.1231	54
7	.47127	.88199	.53432	.8715	.1338	.1219	53
8	.47152	.88185	.53470	.8702	.1340	.1208	52
9	.47178	.88171	.53507	.8689	.1341	.1196	51
10	0.47204	0.88158	0.53545	1.8676	1.1343	2.1185	50
11	.47229	.88144	.53582	.8663	.1345	.1173	49
12	.47255	.88130	.53619	.8650	.1347	.1162	48
13	.47281	.88117	.53657	.8637	.1349	.1150	47
14	.47306	.88103	.53694	.8624	.1350	.1139	46
15	0.47332	0.88089	0.53732	1.8611	1.1352	2.1127	45
16	.47357	.88075	.53769	.8598	.1354	.1116	44
17	.47383	.88061	.53807	.8585	.1356	.1104	43
18	.47409	.88048	.53844	.8572	.1357	.1093	42
19	.47434	.88034	.53882	.8559	.1359	.1082	41
20	0.47460	0.88020	0.53919	1.8546	1.1361	2.1070	40
21	.47486	.88006	.53957	.8533	.1363	.1059	39
22	.47511	.87992	.53995	.8520	.1365	.1048	38
23	.47537	.87979	.54032	.8507	.1366	.1036	37
24	.47562	.87965	.54070	.8495	.1368	.1025	36
25	0.47588	0.87951	0.54107	1.8482	1.1370	2.1014	35
26	.47613	.87937	.54145	.8469	.1372	.1002	34
27	.47639	.87923	.54183	.8456	.1373	.0991	33
28	.47665	.87909	.54220	.8443	.1375	.0980	32
29	.47690	.87895	.54258	.8430	.1377	.0969	31
30	0.47716	0.87882	0.54295	1.8418	1.1379	2.0957	30
31	.47741	.87868	.54333	.8405	.1381	.0946	29
32	.47767	.87854	.54371	.8392	.1382	.0935	28
33	.47792	.87840	.54409	.8379	.1384	.0924	27
34	.47818	.87826	.54446	.8367	.1386	.0912	26
35	0.47844	0.87812	0.54484	1.8354	1.1388	2.0901	25
36	.47869	.87798	.54522	.8341	.1390	.0890	24
37	.47895	.87784	.54559	.8329	.1391	.0879	23
38	.47920	.87770	.54597	.8316	.1393	.0868	22
39	.47946	.87756	.54635	.8303	.1395	.0857	21
40	0.47971	0.87742	0.54673	1.8291	1.1397	2.0846	20
41	.47997	.87728	.54711	.8278	.1399	.0835	19
42	.48022	.87715	.54748	.8265	.1401	.0824	18
43	.48048	.87701	.54786	.8253	.1402	.0812	17
44	.48073	.87687	.54824	.8240	.1404	.0801	16
45	0.48099	0.87673	0.54862	1.8227	1.1406	2.0790	15
46	.48124	.87659	.54900	.8215	.1408	.0779	14
47	.48150	.87645	.54937	.8202	.1410	.0768	13
48	.48175	.87631	.54975	.8190	.1411	.0757	12
49	.48201	.87617	.55013	.8177	.1413	.0746	11
50	0.48226	0.87603	0.55051	1.8165	1.1415	2.0735	10
51	.48252	.87588	.55089	.8152	.1417	.0725	9
52	.48277	.87574	.55127	.8140	.1419	.0714	8
53	.48303	.87560	.55165	.8127	.1421	.0703	7
54	.48328	.87546	.55203	.8115	.1422	.0692	6
55	0.48354	0.87532	0.55241	1.8102	1.1424	2.0681	5
56	.48379	.87518	.55279	.8090	.1426	.0670	4
57	.48405	.87504	.55317	.8078	.1428	.0659	3
58	.48430	.87490	.55355	.8065	.1430	.0648	2
59	.48455	.87476	.55393	.8053	.1432	.0637	1
60	0.48481	0.87462	0.55431	1.8040	1.1433	2.0627	0
M	Cosine	Sine	Cotan.	Tan.	Cosec.	Secant	M
118°							61°

29°							150°
M	Sine	Cosine	Tan.	Cotan.	Secant	Cosec.	M
0	0.48481	0.87462	0.55431	1.8040	1.1433	2.0627	60
1	.48506	.87448	.55469	.8028	.1435	.0616	59
2	.48532	.87434	.55507	.8016	.1437	.0605	58
3	.48557	.87420	.55545	.8003	.1439	.0594	57
4	.48583	.87405	.55583	.7991	.1441	.0583	56
5	0.48608	0.87391	0.55621	1.7979	1.1443	2.0573	55
6	.48633	.87377	.55659	.7966	.1445	.0562	54
7	.48659	.87363	.55697	.7954	.1446	.0551	53
8	.48684	.87349	.55735	.7942	.1448	.0540	52
9	.48710	.87335	.55774	.7930	.1450	.0530	51
10	0.48735	0.87320	0.55812	1.7917	1.1452	2.0519	50
11	.48760	.87306	.55850	.7905	.1454	.0508	49
12	.48786	.87292	.55888	.7893	.1456	.0498	48
13	.48811	.87278	.55926	.7881	.1458	.0487	47
14	.48837	.87264	.55964	.7868	.1459	.0476	46
15	0.48862	0.87250	0.56003	1.7856	1.1461	2.0466	45
16	.48887	.87235	.56041	.7844	.1463	.0455	44
17	.48913	.87221	.56079	.7832	.1465	.0444	43
18	.48938	.87207	.56117	.7820	.1467	.0434	42
19	.48964	.87193	.56156	.7808	.1469	.0423	41
20	0.48989	0.87178	0.56194	1.7795	1.1471	2.0413	40
21	.49014	.87164	.56232	.7783	.1473	.0402	39
22	.49040	.87150	.56270	.7771	.1474	.0392	38
23	.49065	.87136	.56309	.7759	.1476	.0381	37
24	.49090	.87121	.56347	.7747	.1478	.0370	36
25	0.49116	0.87107	0.56385	1.7735	1.1480	2.0360	35
26	.49141	.87093	.56424	.7723	.1482	.0349	34
27	.49166	.87078	.56462	.7711	.1484	.0339	33
28	.49192	.87064	.56500	.7699	.1486	.0329	32
29	.49217	.87050	.56539	.7687	.1488	.0318	31
30	0.49242	0.87035	0.56577	1.7675	1.1489	2.0308	30
31	.49268	.87021	.56616	.7663	.1491	.0297	29
32	.49293	.87007	.56654	.7651	.1493	.0287	28
33	.49318	.86992	.56692	.7639	.1495	.0276	27
34	.49343	.86978	.56731	.7627	.1497	.0266	26
35	0.49369	0.86964	0.56769	1.7615	1.1499	2.0256	25
36	.49394	.86949	.56808	.7603	.1501	.0245	24
37	.49419	.86935	.56846	.7591	.1503	.0235	23
38	.49445	.86921	.56885	.7579	.1505	.0224	22
39	.49470	.86906	.56923	.7567	.1507	.0214	21
40	0.49495	0.86892	0.56962	1.7555	1.1508	2.0204	20
41	.49521	.86877	.57000	.7544	.1510	.0194	19
42	.49546	.86863	.57039	.7532	.1512	.0183	18
43	.49571	.86849	.57077	.7520	.1514	.0173	17
44	.49596	.86834	.57116	.7508	.1516	.0163	16
45	0.49622	0.86820	0.57155	1.7496	1.1518	2.0152	15
46	.49647	.86805	.57193	.7484	.1520	.0142	14
47	.49672	.86791	.57232	.7473	.1522	.0132	13
48	.49697	.86776	.57270	.7461	.1524	.0122	12
49	.49723	.86762	.57309	.7449	.1526	.0111	11
50	0.49748	0.86748	0.57348	1.7437	1.1528	2.0101	10
51	.49773	.86733	.57386	.7426	.1530	.0091	9
52	.49798	.86719	.57425	.7414	.1531	.0081	8
53	.49823	.86704	.57464	.7402	.1533	.0071	7
54	.49849	.86690	.57503	.7390	.1535	.0061	6
55	0.49874	0.86675	0.57541	1.7379	1.1537	2.0050	5
56	.49899	.86661	.57580	.7367	.1539	.0040	4
57	.49924	.86646	.57619	.7355	.1541	.0030	3
58	.49950	.86632	.57657	.7344	.1543	.0020	2
59	.49975	.86617	.57696	.7332	.1545	.0010	1
60	0.50000	0.86603	0.57735	1.7320	1.1547	2.0000	0
M	Cosine	Sine	Cotan.	Tan.	Cosec.	Secant	M
119°							60°

For angles from 0° to 45°, read degrees and function at top and minutes at left; for 45° to 90°, read degrees and function at bottom and minutes at right; for 90° to 135°, read degrees and function at bottom and minutes at left; and for 135° to 180°, read degrees and function at top and minutes at right.

Table 3-5 (*Continued*). Values of the Natural Trigonometric Functions

30° — **149°**

M	Sine	Cosine	Tan.	Cotan.	Secant	Cosec.	M
0	0.50000	0.86603	0.57735	1.7320	1.1547	2.0000	60
1	.50025	.86588	.57774	.7309	.1549	1.9990	59
2	.50050	.86573	.57813	.7297	.1551	.9980	58
3	.50075	.86559	.57851	.7286	.1553	.9970	57
4	.50101	.86544	.57890	.7274	.1555	.9960	56
5	0.50126	0.86530	0.57929	1.7262	1.1557	1.9950	55
6	.50151	.86515	.57968	.7251	.1559	.9940	54
7	.50176	.86500	.58007	.7239	.1561	.9930	53
8	.50201	.86486	.58046	.7228	.1562	.9920	52
9	.50226	.86471	.58085	.7216	.1564	.9910	51
10	0.50252	0.86457	0.58123	1.7205	1.1566	1.9900	50
11	.50277	.86442	.58162	.7193	.1568	.9890	49
12	.50302	.86427	.58201	.7182	.1570	.9880	48
13	.50327	.86413	.58240	.7170	.1572	.9870	47
14	.50352	.86398	.58279	.7159	.1574	.9860	46
15	0.50377	0.86383	0.58318	1.7147	1.1576	1.9850	45
16	.50402	.86369	.58357	.7136	.1578	.9840	44
17	.50428	.86354	.58396	.7124	.1580	.9830	43
18	.50453	.86339	.58435	.7113	.1582	.9820	42
19	.50478	.86325	.58474	.7101	.1584	.9811	41
20	0.50503	0.86310	0.58513	1.7090	1.1586	1.9801	40
21	.50528	.86295	.58552	.7079	.1588	.9791	39
22	.50553	.86281	.58591	.7067	.1590	.9781	38
23	.50578	.86266	.58630	.7056	.1592	.9771	37
24	.50603	.86251	.58670	.7044	.1594	.9761	36
25	0.50628	0.86237	0.58709	1.7033	1.1596	1.9752	35
26	.50653	.86222	.58748	.7022	.1598	.9742	34
27	.50679	.86207	.58787	.7010	.1600	.9732	33
28	.50704	.86192	.58826	.6999	.1602	.9722	32
29	.50729	.86178	.58865	.6988	.1604	.9713	31
30	0.50754	0.86163	0.58904	1.6977	1.1606	1.9703	30
31	.50779	.86148	.58944	.6965	.1608	.9693	29
32	.50804	.86133	.58983	.6954	.1610	.9683	28
33	.50829	.86118	.59022	.6943	.1612	.9674	27
34	.50854	.86104	.59061	.6931	.1614	.9664	26
35	0.50879	0.86089	0.59100	1.6920	1.1616	1.9654	25
36	.50904	.86074	.59140	.6909	.1618	.9645	24
37	.50929	.86059	.59179	.6898	.1620	.9635	23
38	.50954	.86044	.59218	.6887	.1622	.9625	22
39	.50979	.86030	.59258	.6875	.1624	.9616	21
40	0.51004	0.86015	0.59297	1.6864	1.1626	1.9606	20
41	.51029	.86000	.59336	.6853	.1628	.9596	19
42	.51054	.85985	.59376	.6842	.1630	.9587	18
43	.51079	.85970	.59415	.6831	.1632	.9577	17
44	.51104	.85955	.59454	.6820	.1634	.9568	16
45	0.51129	0.85941	0.59494	1.6808	1.1636	1.9558	15
46	.51154	.85926	.59533	.6797	.1638	.9549	14
47	.51179	.85911	.59572	.6786	.1640	.9539	13
48	.51204	.85896	.59612	.6775	.1642	.9530	12
49	.51229	.85881	.59651	.6764	.1644	.9520	11
50	0.51254	0.85866	0.59691	1.6753	1.1646	1.9510	10
51	.51279	.85851	.59730	.6742	.1648	.9501	9
52	.51304	.85836	.59770	.6731	.1650	.9491	8
53	.51329	.85821	.59809	.6720	.1652	.9482	7
54	.51354	.85806	.59849	.6709	.1654	.9473	6
55	0.51379	0.85791	0.59888	1.6698	1.1656	1.9463	5
56	.51404	.85777	.59928	.6687	.1658	.9454	4
57	.51429	.85762	.59967	.6676	.1660	.9444	3
58	.51454	.85747	.60007	.6665	.1662	.9435	2
59	.51479	.85732	.60046	.6654	.1664	.9425	1
60	0.51504	0.85717	0.60086	1.6643	1.1666	1.9416	0
M	Cosine	Sine	Cotan.	Tan.	Cosec.	Secant	M

120° — **59°**

31° — **148°**

M	Sine	Cosine	Tan.	Cotan.	Secant	Cosec.	M
0	0.51504	0.85717	0.60086	1.6643	1.1666	1.9416	60
1	.51529	.85702	.60126	.6632	.1668	.9407	59
2	.51554	.85687	.60165	.6621	.1670	.9397	58
3	.51578	.85672	.60205	.6610	.1672	.9388	57
4	.51603	.85657	.60244	.6599	.1674	.9378	56
5	0.51628	0.85642	0.60284	1.6588	1.1676	1.9369	55
6	.51653	.85627	.60324	.6577	.1678	.9360	54
7	.51678	.85612	.60363	.6566	.1681	.9350	53
8	.51703	.85597	.60403	.6555	.1683	.9341	52
9	.51728	.85582	.60443	.6544	.1685	.9332	51
10	0.51753	0.85566	0.60483	1.6534	1.1687	1.9322	50
11	.51778	.85551	.60522	.6523	.1689	.9313	49
12	.51803	.85536	.60562	.6512	.1691	.9304	48
13	.51827	.85521	.60602	.6501	.1693	.9295	47
14	.51852	.85506	.60642	.6490	.1695	.9285	46
15	0.51877	0.85491	0.60681	1.6479	1.1697	1.9276	45
16	.51902	.85476	.60721	.6469	.1699	.9267	44
17	.51927	.85461	.60761	.6458	.1701	.9258	43
18	.51952	.85446	.60801	.6447	.1703	.9248	42
19	.51977	.85431	.60841	.6436	.1705	.9239	41
20	0.52002	0.85416	0.60881	1.6425	1.1707	1.9230	40
21	.52026	.85400	.60920	.6415	.1709	.9221	39
22	.52051	.85385	.60960	.6404	.1712	.9212	38
23	.52076	.85370	.61000	.6393	.1714	.9203	37
24	.52101	.85355	.61040	.6383	.1716	.9193	36
25	0.52126	0.85340	0.61080	1.6372	1.1718	1.9184	35
26	.52151	.85325	.61120	.6361	.1720	.9175	34
27	.52175	.85309	.61160	.6350	.1722	.9166	33
28	.52200	.85294	.61200	.6340	.1724	.9157	32
29	.52225	.85279	.61240	.6329	.1726	.9148	31
30	0.52250	0.85264	0.61280	1.6318	1.1728	1.9139	30
31	.52275	.85249	.61320	.6308	.1730	.9130	29
32	.52299	.85234	.61360	.6297	.1732	.9121	28
33	.52324	.85218	.61400	.6286	.1734	.9112	27
34	.52349	.85203	.61440	.6276	.1737	.9102	26
35	0.52374	0.85188	0.61480	1.6265	1.1739	1.9093	25
36	.52398	.85173	.61520	.6255	.1741	.9084	24
37	.52423	.85157	.61560	.6244	.1743	.9075	23
38	.52448	.85142	.61601	.6233	.1745	.9066	22
39	.52473	.85127	.61641	.6223	.1747	.9057	21
40	0.52498	0.85112	0.61681	1.6212	1.1749	1.9048	20
41	.52522	.85096	.61721	.6202	.1751	.9039	19
42	.52547	.85081	.61761	.6191	.1753	.9030	18
43	.52572	.85066	.61801	.6181	.1756	.9021	17
44	.52597	.85050	.61842	.6170	.1758	.9013	16
45	0.52621	0.85035	0.61882	1.6160	1.1760	1.9004	15
46	.52646	.85020	.61922	.6149	.1762	.8995	14
47	.52671	.85004	.61962	.6139	.1764	.8986	13
48	.52695	.84989	.62003	.6128	.1766	.8977	12
49	.52720	.84974	.62043	.6118	.1768	.8968	11
50	0.52745	0.84959	0.62083	1.6107	1.1770	1.8959	10
51	.52770	.84943	.62123	.6097	.1772	.8950	9
52	.52794	.84928	.62164	.6086	.1775	.8941	8
53	.52819	.84912	.62204	.6076	.1777	.8932	7
54	.52844	.84897	.62244	.6066	.1779	.8924	6
55	0.52868	0.84882	0.62285	1.6055	1.1781	1.8915	5
56	.52893	.84866	.62325	.6045	.1783	.8906	4
57	.52918	.84851	.62366	.6034	.1785	.8897	3
58	.52942	.84836	.62406	.6024	.1787	.8888	2
59	.52967	.84820	.62446	.6014	.1790	.8879	1
60	0.52992	0.84805	0.62487	1.6003	1.1792	1.8871	0
M	Cosine	Sine	Cotan.	Tan.	Cosec.	Secant	M

121° — **58°**

32° — **147°**

M	Sine	Cosine	Tan.	Cotan.	Secant	Cosec.	M
0	0.52992	0.84805	0.62487	1.6003	1.1792	1.8871	60
1	.53016	.84789	.62527	.5993	.1794	.8862	59
2	.53041	.84774	.62568	.5983	.1796	.8853	58
3	.53066	.84758	.62608	.5972	.1798	.8844	57
4	.53090	.84743	.62649	.5962	.1800	.8836	56
5	0.53115	0.84728	0.62689	1.5952	1.1802	1.8827	55
6	.53140	.84712	.62730	5941	.1805	.8818	54
7	.53164	.84697	62770	.5931	.1807	.8809	53
8	.53189	.84681	.62811	.5921	.1809	.8801	52
9	.53214	.84666	.62851	.5910	.1811	.8792	51
10	0.53238	0.84650	0.62892	1.5900	1.1813	1.8783	50
11	.53263	.84635	.62933	.5890	.1815	.8775	49
12	.53288	.84619	.62973	.5880	.1818	.8766	48
13	.53312	.84604	.63014	.5869	.1820	.8757	47
14	.53337	.84588	.63055	.5859	.1822	.8749	46
15	0.53361	0.84573	0.63095	1.5849	1.1824	1.8740	45
16	.53386	.84557	.63136	.5839	.1826	.8731	44
17	.53411	.84542	.63177	.5829	.1828	.8723	43
18	.53435	.84526	.63217	.5818	.1831	.8714	42
19	.53460	.84511	.63258	.5808	.1833	.8706	41
20	0.53484	0.84495	0.63299	1.5798	1.1835	1.8697	40
21	.53509	.84479	.63339	.5788	.1837	.8688	39
22	.53533	.84464	.63380	.5778	.1839	.8680	38
23	.53558	.84448	.63421	.5768	.1841	.8671	37
24	.53583	.84433	.63462	.5757	.1844	.8663	36
25	0.53607	0.84417	0.63503	1.5747	1.1846	1.8654	35
26	.53632	.84402	.63543	.5737	.1848	.8646	34
27	.53656	.84386	.63584	.5727	.1850	.8637	33
28	.53681	.84370	.63625	.5717	.1852	.8629	32
29	.53705	.84355	.63666	.5707	.1855	.8620	31
30	0.53730	0.84339	0.63707	1.5697	1.1857	1.8611	30
31	.53754	.84323	.63748	.5687	.1859	.8603	29
32	.53779	.84308	.63789	.5677	.1861	.8595	28
33	.53803	.84292	.63830	.5667	.1863	.8586	27
34	.53828	.84276	.63871	.5657	.1866	.8578	26
35	0.53852	0.84261	0.63912	1.5646	1.1868	1.8569	25
36	.53877	.84245	.63953	.5636	.1870	.8561	24
37	.53901	.84229	.63994	.5626	.1872	.8552	23
38	.53926	.84214	.64035	.5616	.1874	.8544	22
39	.53950	.84198	.64076	.5606	.1877	.8535	21
40	0.53975	0.84182	0.64117	1.5596	1.1879	1.8527	20
41	.53999	.84167	.64158	.5586	.1881	.8519	19
42	.54024	.84151	.64199	.5577	.1883	.8510	18
43	.54048	.84135	.64240	.5567	.1886	.8502	17
44	.54073	.84120	.64281	.5557	.1888	.8493	16
45	0.54097	0.84104	0.64322	1.5547	1.1890	1.8485	15
46	.54122	.84088	.64363	.5537	.1892	.8477	14
47	.54146	.84072	.64404	.5527	.1894	.8468	13
48	.54171	.84057	.64446	.5517	.1897	.8460	12
49	.54195	.84041	.64487	.5507	.1899	.8452	11
50	0.54220	0.84025	0.64528	1.5497	1.1901	1.8443	10
51	.54244	.84009	.64569	.5487	.1903	.8435	9
52	.54268	.83993	.64610	.5477	.1906	.8427	8
53	.54293	.83978	.64652	.5467	.1908	.8418	7
54	.54317	.83962	.64693	.5458	.1910	.8410	6
55	0.54342	0.83946	0.64734	1.5448	1.1912	1.8402	5
56	.54366	.83930	.64775	.5438	.1915	.8394	4
57	.54391	.83914	.64817	.5428	.1917	.8385	3
58	.54415	.83899	.64858	.5418	.1919	.8377	2
59	.54439	.83883	.64899	.5408	.1921	.8369	1
60	0.54464	0.83867	0.64941	1.5399	1.1924	1.8361	0
M	Cosine	Sine	Cotan.	Tan.	Cosec.	Secant	M

122° — **57°**

For angles from 0° to 45°, read degrees and function at top and minutes at left; for 45° to 90°, read degrees and function at bottom and minutes at right; for 90° to 135°, read degrees and function at bottom and minutes at left; and for 135° to 180° read degrees and function at top and minutes at right.

Table 3-5 (*Continued*). Values of the Natural Trigonometric Functions

33° — **146°**

M	Sine	Cosine	Tan.	Cotan.	Secant	Cosec.	M
0	0.54464	0.83867	0.64941	1.5399	1.1924	1.8361	60
1	.54488	.83851	.64982	.5389	.1926	.8352	59
2	.54513	.83835	.65023	.5379	.1928	.8344	58
3	.54537	.83819	.65065	.5369	.1930	.8336	57
4	.54561	.83804	.65106	.5359	.1933	.8328	56
5	0.54586	0.83788	0.65148	1.5350	1.1935	1.8320	55
6	.54610	.83772	.65189	.5340	.1937	.8311	54
7	.54634	.83756	.65231	.5330	.1939	.8303	53
8	.54659	.83740	.65272	.5320	.1942	.8295	52
9	.54683	.83724	.65314	.5311	.1944	.8287	51
10	0.54708	0.83708	0.65355	1.5301	1.1946	1.8279	50
11	.54732	.83692	.65397	.5291	.1948	.8271	49
12	.54756	.83676	.65438	.5282	.1951	.8263	48
13	.54781	.83660	.65480	.5272	.1953	.8255	47
14	.54805	.83644	.65521	.5262	.1955	.8246	46
15	0.54829	0.83629	0.65563	1.5252	1.1958	1.8238	45
16	.54854	.83613	.65604	.5243	.1960	.8230	44
17	.54878	.83597	.65646	.5233	.1962	.8222	43
18	.54902	.83581	.65688	.5223	.1964	.8214	42
19	.54926	.83565	.65729	.5214	.1967	.8206	41
20	0.54951	0.83549	0.65771	1.5204	1.1969	1.8198	40
21	.54975	.83533	.65813	.5195	.1971	.8190	39
22	.54999	.83517	.65854	.5185	.1974	.8182	38
23	.55024	.83501	.65896	.5175	.1976	.8174	37
24	.55048	.83485	.65938	.5166	.1978	.8166	36
25	0.55072	0.83469	0.65980	1.5156	1.1980	1.8158	35
26	.55097	.83453	.66021	.5147	.1983	.8150	34
27	.55121	.83437	.66063	.5137	.1985	.8142	33
28	.55145	.83421	.66105	.5127	.1987	.8134	32
29	.55169	.83405	.66147	.5118	.1990	.8126	31
30	0.55194	0.83388	0.66188	1.5108	1.1992	1.8118	30
31	.55218	.83372	.66230	.5099	.1994	.8110	29
32	.55242	.83356	.66272	.5089	.1997	.8102	28
33	.55266	.83340	.66314	.5080	.1999	.8094	27
34	.55291	.83324	.66356	.5070	.2001	.8086	26
35	0.55315	0.83308	0.66398	1.5061	1.2004	1.8078	25
36	.55339	.83292	.66440	.5051	.2006	.8070	24
37	.55363	.83276	.66482	.5042	.2008	.8062	23
38	.55388	.83260	.66524	.5032	.2010	.8054	22
39	.55412	.83244	.66566	.5023	.2013	.8047	21
40	0.55436	0.83228	0.66608	1.5013	1.2015	1.8039	20
41	.55460	.83211	.66650	.5004	.2017	.8031	19
42	.55484	.83195	.66692	.4994	.2020	.8023	18
43	.55509	.83179	.66734	.4985	.2022	.8015	17
44	.55533	.83163	.66776	.4975	.2024	.8007	16
45	0.55557	0.83147	0.66818	1.4966	1.2027	1.7999	15
46	.55581	.83131	.66860	.4957	.2029	.7992	14
47	.55605	.83115	.66902	.4947	.2031	.7984	13
48	.55629	.83098	.66944	.4938	.2034	.7976	12
49	.55654	.83082	.66986	.4928	.2036	.7968	11
50	0.55678	0.83066	0.67028	1.4919	1.2039	1.7960	10
51	.55702	.83050	.67071	.4910	.2041	.7953	9
52	.55726	.83034	.67113	.4900	.2043	.7945	8
53	.55750	.83017	.67155	.4891	.2046	.7937	7
54	.55774	.83001	.67197	.4881	.2048	.7929	6
55	0.55799	0.82985	0.67239	1.4872	1.2050	1.7921	5
56	.55823	.82969	.67282	.4863	.2053	.7914	4
57	.55847	.82952	.67324	.4853	.2055	.7906	3
58	.55871	.82936	.67366	.4844	.2057	.7898	2
59	.55895	.82920	.67408	.4835	.2060	.7891	1
60	0.55919	0.82904	0.67451	1.4826	1.2062	1.7883	0
M	Cosine	Sine	Cotan.	Tan.	Cosec.	Secant	M

123° — **56°**

34° — **145°**

M	Sine	Cosine	Tan.	Cotan.	Secant	Cosec.	M
0	0.55919	0.82904	0.67451	1.4826	1.2062	1.7883	60
1	.55943	.82887	.67493	.4816	.2064	.7875	59
2	.55967	.82871	.67535	.4807	.2067	.7867	58
3	.55992	.82855	.67578	.4798	.2069	.7860	57
4	.56016	.82839	.67620	.4788	.2072	.7852	56
5	0.56040	0.82822	0.67663	1.4779	1.2074	1.7844	55
6	.56064	.82806	.67705	.4770	.2076	.7837	54
7	.56088	.82790	.67747	.4761	.2079	.7829	53
8	.56112	.82773	.67790	.4751	.2081	.7821	52
9	.56136	.82757	.67832	.4742	.2083	.7814	51
10	0.56160	0.82741	0.67875	1.4733	1.2086	1.7806	50
11	.56184	.82724	.67917	.4724	.2088	.7798	49
12	.56208	.82708	.67960	.4714	.2091	.7791	48
13	.56232	.82692	.68002	.4705	.2093	.7783	47
14	.56256	.82675	.68045	.4696	.2095	.7776	46
15	0.56280	0.82659	0.68087	1.4687	1.2098	1.7768	45
16	.56304	.82643	.68130	.4678	.2100	.7760	44
17	.56328	.82626	.68173	.4669	.2103	.7753	43
18	.56353	.82610	.68215	.4659	.2105	.7745	42
19	.56377	.82593	.68258	.4650	.2107	.7738	41
20	0.56401	0.82577	0.68301	1.4641	1.2110	1.7730	40
21	.56425	.82561	.68343	.4632	.2112	.7723	39
22	.56449	.82544	.68386	.4623	.2115	.7715	38
23	.56473	.82528	.68429	.4614	.2117	.7708	37
24	.56497	.82511	.68471	.4605	.2119	.7700	36
25	0.56521	0.82495	0.68514	1.4595	1.2122	1.7693	35
26	.56545	.82478	.68557	.4586	.2124	.7685	34
27	.56569	.82462	.68600	.4577	.2127	.7678	33
28	.56593	.82445	.68642	.4568	.2129	.7670	32
29	.56617	.82429	.68685	.4559	.2132	.7663	31
30	0.56641	0.82413	0.68728	1.4550	1.2134	1.7655	30
31	.56664	.82396	.68771	.4541	.2136	.7648	29
32	.56688	.82380	.68814	.4532	.2139	.7640	28
33	.56712	.82363	.68857	.4523	.2141	.7633	27
34	.56736	.82347	.68899	.4514	.2144	.7625	26
35	0.56760	0.82330	0.68942	1.4505	1.2146	1.7618	25
36	.56784	.82314	.68985	.4496	.2149	.7610	24
37	.56808	.82297	.69028	.4487	.2151	.7603	23
38	.56832	.82280	.69071	.4478	.2153	.7596	22
39	.56856	.82264	.69114	.4469	.2156	.7588	21
40	0.56880	0.82247	0.69157	1.4460	1.2158	1.7581	20
41	.56904	.82231	.69200	.4451	.2161	.7573	19
42	.56928	.82214	.69243	.4442	.2163	.7566	18
43	.56952	.82198	.69286	.4433	.2166	.7559	17
44	.56976	.82181	.69329	.4424	.2168	.7551	16
45	0.57000	0.82165	0.69372	1.4415	1.2171	1.7544	15
46	.57023	.82148	.69415	.4406	.2173	.7537	14
47	.57047	.82131	.69459	.4397	.2175	.7529	13
48	.57071	.82115	.69502	.4388	.2178	.7522	12
49	.57095	.82098	.69545	.4379	.2180	.7514	11
50	0.57119	0.82082	0.69588	1.4370	1.2183	1.7507	10
51	.57143	.82065	.69631	.4361	.2185	.7500	9
52	.57167	.82048	.69674	.4352	.2188	.7493	8
53	.57191	.82032	.69718	.4343	.2190	.7485	7
54	.57214	.82015	.69761	.4335	.2193	.7478	6
55	0.57238	0.81998	0.69804	1.4326	1.2195	1.7471	5
56	.57262	.81982	.69847	.4317	.2198	.7463	4
57	.57286	.81965	.69891	.4308	.2200	.7456	3
58	.57310	.81948	.69934	.4299	.2203	.7449	2
59	.57334	.81932	.69977	.4290	.2205	.7442	1
60	0.57358	0.81915	0.70021	1.4281	1.2208	1.7434	0
M	Cosine	Sine	Cotan.	Tan.	Cosec.	Secant	M

124° — **55°**

35° — **144°**

M	Sine	Cosine	Tan.	Cotan.	Secant	Cosec.	M
0	0.57358	0.81915	0.70021	1.4281	1.2208	1.7434	60
1	.57381	.81898	.70064	.4273	.2210	.7427	59
2	.57405	.81882	.70107	.4264	.2213	.7420	58
3	.57429	.81865	.70151	.4255	.2215	.7413	57
4	.57453	.81848	.70194	.4246	.2218	.7405	56
5	0.57477	0.81832	0.70238	1.4237	1.2220	1.7398	55
6	.57500	.81815	.70281	.4228	.2223	.7391	54
7	.57524	.81798	.70325	.4220	.2225	.7384	53
8	.57548	.81781	.70368	.4211	.2228	.7377	52
9	.57572	.81765	.70412	.4202	.2230	.7369	51
10	0.57596	0.81748	0.70455	1.4193	1.2233	1.7362	50
11	.57619	.81731	.70499	.4185	.2235	.7355	49
12	.57643	.81714	.70542	.4176	.2238	.7348	48
13	.57667	.81698	.70586	.4167	.2240	.7341	47
14	.57691	.81681	.70629	.4158	.2243	.7334	46
15	0.57714	0.81664	0.70673	1.4150	1.2245	1.7327	45
16	.57738	.81647	.70717	.4141	.2248	.7319	44
17	.57762	.81630	.70760	.4132	.2250	.7312	43
18	.57786	.81614	.70804	.4123	.2253	.7305	42
19	.57809	.81597	.70848	.4115	.2255	.7298	41
20	0.57833	0.81580	0.70891	1.4106	1.2258	1.7291	40
21	.57857	.81563	.70935	.4097	.2260	.7284	39
22	.57881	.81546	.70979	.4089	.2263	.7277	38
23	.57904	.81530	.71022	.4080	.2265	.7270	37
24	.57928	.81513	.71066	.4071	.2268	.7263	36
25	0.57952	0.81496	0.71110	1.4063	1.2270	1.7256	35
26	.57975	.81479	.71154	.4054	.2273	.7249	34
27	.57999	.81462	.71198	.4045	.2276	.7242	33
28	.58023	.81445	.71241	.4037	.2278	.7234	32
29	.58047	.81428	.71285	.4028	.2281	.7227	31
30	0.58070	0.81411	0.71329	1.4019	1.2283	1.7220	30
31	.58094	.81395	.71373	.4011	.2286	.7213	29
32	.58118	.81378	.71417	.4002	.2288	.7206	28
33	.58141	.81361	.71461	.3994	.2291	.7199	27
34	.58165	.81344	.71505	.3985	.2293	.7192	26
35	0.58189	0.81327	0.71549	1.3976	1.2296	1.7185	25
36	.58212	.81310	.71593	.3968	.2298	.7178	24
37	.58236	.81293	.71637	.3959	.2301	.7171	23
38	.58259	.81276	.71681	.3951	.2304	.7164	22
39	.58283	.81259	.71725	.3942	.2306	.7157	21
40	0.58307	0.81242	0.71769	1.3933	1.2309	1.7151	20
41	.58330	.81225	.71813	.3925	.2311	.7144	19
42	.58354	.81208	.71857	.3916	.2314	.7137	18
43	.58378	.81191	.71901	.3908	.2316	.7130	17
44	.58401	.81174	.71945	.3899	.2319	.7123	16
45	0.58425	0.81157	0.71990	1.3891	1.2322	1.7116	15
46	.58448	.81140	.72034	.3882	.2324	.7109	14
47	.58472	.81123	.72078	.3874	.2327	.7102	13
48	.58496	.81106	.72122	.3865	.2329	.7095	12
49	.58519	.81089	.72166	.3857	.2332	.7088	11
50	0.58543	0.81072	0.72211	1.3848	1.2335	1.7081	10
51	.58566	.81055	.72255	.3840	.2337	.7075	9
52	.58590	.81038	.72299	.3831	.2340	.7068	8
53	.58614	.81021	.72344	.3823	.2342	.7061	7
54	.58637	.81004	.72388	.3814	.2345	.7054	6
55	0.58661	0.80987	0.72432	1.3806	1.2348	1.7047	5
56	.58684	.80970	.72477	.3797	.2350	.7040	4
57	.58708	.80953	.72521	.3789	.2353	.7033	3
58	.58731	.80936	.72565	.3781	.2355	.7027	2
59	.58755	.80919	.72610	.3772	.2358	.7020	1
60	0.58778	0.80902	0.72654	1.3764	1.2361	1.7013	0
M	Cosine	Sine	Cotan.	Tan.	Cosec.	Secant	M

125° — **54°**

For angles from 0° to 45°, read degrees and function at top and minutes at left; for 45° to 90°, read degrees and function at bottom and minutes at right; for 90° to 135°, read degrees and function at bottom and minutes at left; and for 135° to 180° read degrees and function at top and minutes at right.

Table 3-5 (*Continued*). Values of the Natural Trigonometric Functions

36°							143°
M	Sine	Cosine	Tan.	Cotan.	Secant	Cosec.	M
0	0.58778	0.80902	0.72654	1.3764	1.2361	1.7013	60
1	.58802	.80885	.72699	.3755	.2363	.7006	59
2	.58825	.80867	.72743	.3747	.2366	.6999	58
3	.58849	.80850	.72788	.3738	.2368	.6993	57
4	.58873	.80833	.72832	.3730	.2371	.6986	56
5	0.58896	0.80816	0.72877	1.3722	1.2374	1.6979	55
6	.58920	.80799	.72921	.3713	.2376	.6972	54
7	.58943	.80782	.72966	.3705	.2379	.6965	53
8	.58967	.80765	.73010	.3697	.2382	.6959	52
9	.58990	.80747	.73055	.3688	.2384	.6952	51
10	0.59014	0.80730	0.73100	1.3680	1.2387	1.6945	50
11	.59037	.80713	.73144	.3672	.2389	.6938	49
12	.59060	.80696	.73189	.3663	.2392	.6932	48
13	.59084	.80679	.73234	.3655	.2395	.6925	47
14	.59107	.80662	.73278	.3647	.2397	.6918	46
15	0.59131	0.80644	0.73323	1.3638	1.2400	1.6912	45
16	.59154	.80627	.73368	.3630	.2403	.6905	44
17	.59178	.80610	.73412	.3622	.2405	.6898	43
18	.59201	.80593	.73457	.3613	.2408	.6891	42
19	.59225	.80576	.73502	.3605	.2411	.6885	41
20	0.59248	0.80558	0.73547	1.3597	1.2413	1.6878	40
21	.59272	.80541	.73592	.3588	.2416	.6871	39
22	.59295	.80524	.73637	.3580	.2419	.6865	38
23	.59318	.80507	.73681	.3572	.2421	.6858	37
24	.59342	.80489	.73726	.3564	.2424	.6851	36
25	0.59365	0.80472	0.73771	1.3555	1.2427	1.6845	35
26	.59389	.80455	.73816	.3547	.2429	.6838	34
27	.59412	.80437	.73861	.3539	.2432	.6831	33
28	.59435	.80420	.73906	.3531	.2435	.6825	32
29	.59459	.80403	.73951	.3522	.2437	.6818	31
30	0.59482	0.80386	0.73996	1.3514	1.2440	1.6812	30
31	.59506	.80368	.74041	.3506	.2443	.6805	29
32	.59529	.80351	.74086	.3498	.2445	.6798	28
33	.59552	.80334	.74131	.3489	.2448	.6792	27
34	.59576	.80316	.74176	.3481	.2451	.6785	26
35	0.59599	0.80299	0.74221	1.3473	1.2453	1.6779	25
36	.59622	.80282	.74266	.3465	.2456	.6772	24
37	.59646	.80264	.74312	.3457	.2459	.6766	23
38	.59669	.80247	.74357	.3449	.2461	.6759	22
39	.59692	.80230	.74402	.3440	.2464	.6752	21
40	0.59716	0.80212	0.74447	1.3432	1.2467	1.6746	20
41	.59739	.80195	.74492	.3424	.2470	.6739	19
42	.59762	.80177	.74538	.3416	.2472	.6733	18
43	.59786	.80160	.74583	.3408	.2475	.6726	17
44	.59809	.80143	.74628	.3400	.2478	.6720	16
45	0.59832	0.80125	0.74673	1.3392	1.2480	1.6713	15
46	.59856	.80108	.74719	.3383	.2483	.6707	14
47	.59879	.80090	.74764	.3375	.2486	.6700	13
48	.59902	.80073	.74809	.3367	.2488	.6694	12
49	.59926	.80056	.74855	.3359	.2491	.6687	11
50	0.59949	0.80038	0.74900	1.3351	1.2494	1.6681	10
51	.59972	.80021	.74946	.3343	.2497	.6674	9
52	.59995	.80003	.74991	.3335	.2499	.6668	8
53	.60019	.79986	.75037	.3327	.2502	.6661	7
54	.60042	.79968	.75082	.3319	.2505	.6655	6
55	0.60065	0.79951	0.75128	1.3311	1.2508	1.6648	5
56	.60088	.79933	.75173	.3303	.2510	.6642	4
57	.60112	.79916	.75219	.3294	.2513	.6636	3
58	.60135	.79898	.75264	.3286	.2516	.6629	2
59	.60158	.79881	.75310	.3278	.2519	.6623	1
60	0.60181	0.79863	0.75355	1.3270	1.2521	1.6616	0
M	Cosine	Sine	Cotan.	Tan.	Cosec.	Secant	M
126°							53°

37°							142°
M	Sine	Cosine	Tan.	Cotan.	Secant	Cosec.	M
0	0.60181	0.79863	0.75355	1.3270	1.2521	1.6616	60
1	.60205	.79846	.75401	.3262	.2524	.6610	59
2	.60228	.79828	.75447	.3254	.2527	.6603	58
3	.60251	.79811	.75492	.3246	.2530	.6597	57
4	.60274	.79793	.75538	.3238	.2532	.6591	56
5	0.60298	0.79776	0.75584	1.3230	1.2535	1.6584	55
6	.60320	.79758	.75629	.3222	.2538	.6578	54
7	.60344	.79741	.75675	.3214	.2541	.6572	53
8	.60367	.79723	.75721	.3206	.2543	.6565	52
9	.60390	.79706	.75767	.3198	.2546	.6559	51
10	0.60413	0.79688	0.75812	1.3190	1.2549	1.6552	50
11	.60437	.79670	.75858	.3182	.2552	.6546	49
12	.60460	.79653	.75904	.3174	.2554	.6540	48
13	.60483	.79635	.75950	.3166	.2557	.6533	47
14	.60506	.79618	.75996	.3159	.2560	.6527	46
15	0.60529	0.79600	0.76042	1.3151	1.2563	1.6521	45
16	.60552	.79582	.76088	.3143	.2565	.6514	44
17	.60576	.79565	.76134	.3135	.2568	.6508	43
18	.60599	.79547	.76179	.3127	.2571	.6502	42
19	.60622	.79530	.76225	.3119	.2574	.6496	41
20	0.60645	0.79512	0.76271	1.3111	1.2577	1.6489	40
21	.60668	.79494	.76317	.3103	.2579	.6483	39
22	.60691	.79477	.76364	.3095	.2582	.6477	38
23	.60714	.79459	.76410	.3087	.2585	.6470	37
24	.60737	.79441	.76456	.3079	.2588	.6464	36
25	0.60761	0.79424	0.76502	1.3071	1.2591	1.6458	35
26	.60784	.79406	.76548	.3064	.2593	.6452	34
27	.60807	.79388	.76594	.3056	.2596	.6445	33
28	.60830	.79371	.76640	.3048	.2599	.6439	32
29	.60853	.79353	.76686	.3040	.2602	.6433	31
30	0.60876	0.79335	0.76733	1.3032	1.2605	1.6427	30
31	.60899	.79318	.76779	.3024	.2607	.6420	29
32	.60922	.79300	.76825	.3016	.2610	.6414	28
33	.60945	.79282	.76871	.3009	.2613	.6408	27
34	.60968	.79264	.76918	.3001	.2616	.6402	26
35	0.60991	0.79247	0.76964	1.2993	1.2619	1.6396	25
36	.61014	.79229	.77010	.2985	.2622	.6389	24
37	.61037	.79211	.77057	.2977	.2624	.6383	23
38	.61061	.79193	.77103	.2970	.2627	.6377	22
39	.61084	.79176	.77149	.2962	.2630	.6371	21
40	0.61107	0.79158	0.77196	1.2954	1.2633	1.6365	20
41	.61130	.79140	.77242	.2946	.2636	.6359	19
42	.61153	.79122	.77289	.2938	.2639	.6352	18
43	.61176	.79104	.77335	.2931	.2641	.6346	17
44	.61199	.79087	.77382	.2923	.2644	.6340	16
45	0.61222	0.79069	0.77428	1.2915	1.2647	1.6334	15
46	.61245	.79051	.77475	.2907	.2650	.6328	14
47	.61268	.79033	.77521	.2900	.2653	.6322	13
48	.61290	.79015	.77568	.2892	.2656	.6316	12
49	.61314	.78998	.77614	.2884	.2659	.6309	11
50	0.61337	0.78980	0.77661	1.2876	1.2661	1.6303	10
51	.61360	.78962	.77708	.2869	.2664	.6297	9
52	.61383	.78944	.77754	.2861	.2667	.6291	8
53	.61405	.78926	.77801	.2853	.2670	.6285	7
54	.61428	.78908	.77848	.2845	.2673	.6279	6
55	0.61451	0.78890	0.77895	1.2838	1.2676	1.6273	5
56	.61474	.78873	.77941	.2830	.2679	.6267	4
57	.61497	.78855	.77988	.2822	.2681	.6261	3
58	.61520	.78837	.78035	.2815	.2684	.6255	2
59	.61543	.78819	.78082	.2807	.2687	.6249	1
60	0.61566	0.78801	0.78128	1.2799	1.2690	1.6243	0
M	Cosine	Sine	Cotan.	Tan.	Cosec.	Secant	M
127°							52°

38°							141°
M	Sine	Cosine	Tan.	Cotan.	Secant	Cosec.	M
0	0.61566	0.78801	0.78128	1.2799	1.2690	1.6243	60
1	.61589	.78783	.78175	.2792	.2693	.6237	59
2	.61612	.78765	.78222	.2784	.2696	.6231	58
3	.61635	.78747	.78269	.2776	.2699	.6224	57
4	.61658	.78729	.78316	.2769	.2702	.6218	56
5	0.61681	0.78711	0.78363	1.2761	1.2705	1.6212	55
6	.61703	.78693	.78410	.2753	.2707	.6206	54
7	.61726	.78675	.78457	.2746	.2710	.6200	53
8	.61749	.78657	.78504	.2738	.2713	.6194	52
9	.61772	.78640	.78551	.2730	.2716	.6188	51
10	0.61795	0.78622	0.78598	1.2723	1.2719	1.6182	50
11	.61818	.78604	.78645	.2715	.2722	.6176	49
12	.61841	.78586	.78692	.2708	.2725	.6170	48
13	.61864	.78568	.78739	.2700	.2728	.6164	47
14	.61886	.78550	.78786	.2692	.2731	.6159	46
15	0.61909	0.78532	0.78834	1.2685	1.2734	1.6153	45
16	.61932	.78514	.78881	.2677	.2737	.6147	44
17	.61955	.78496	.78928	.2670	.2739	.6141	43
18	.61978	.78478	.78975	.2662	.2742	.6135	42
19	.62001	.78460	.79022	.2655	.2745	.6129	41
20	0.62023	0.78441	0.79070	1.2647	1.2748	1.6123	40
21	.62046	.78423	.79117	.2639	.2751	.6117	39
22	.62069	.78405	.79164	.2632	.2754	.6111	38
23	.62092	.78387	.79212	.2624	.2757	.6105	37
24	.62115	.78369	.79259	.2617	.2760	.6099	36
25	0.62137	0.78351	0.79306	1.2609	1.2763	1.6093	35
26	.62160	.78333	.79354	.2602	.2766	.6087	34
27	.62183	.78315	.79401	.2594	.2769	.6081	33
28	.62206	.78297	.79449	.2587	.2772	.6077	32
29	.62229	.78279	.79496	.2579	.2775	.6070	31
30	0.62251	0.78261	0.79543	1.2572	1.2778	1.6064	30
31	.62274	.78243	.79591	.2564	.2781	.6058	29
32	.62297	.78224	.79639	.2557	.2784	.6052	28
33	.62320	.78206	.79686	.2549	.2787	.6046	27
34	.62342	.78188	.79734	.2542	.2790	.6040	26
35	0.62365	0.78170	0.79781	1.2534	1.2793	1.6034	25
36	.62388	.78152	.79829	.2527	.2795	.6029	24
37	.62411	.78134	.79876	.2519	.2798	.6023	23
38	.62433	.78116	.79924	.2512	.2801	.6017	22
39	.62456	.78097	.79972	.2504	.2804	.6011	21
40	0.62479	0.78079	0.80020	1.2497	1.2807	1.6005	20
41	.62501	.78061	.80067	.2489	.2810	.6000	19
42	.62524	.78043	.80115	.2482	.2813	.5994	18
43	.62547	.78025	.80163	.2475	.2816	.5988	17
44	.62570	.78007	.80211	.2467	.2819	.5982	16
45	0.62592	0.77988	0.80258	1.2460	1.2822	1.5976	15
46	.62615	.77970	.80306	.2452	.2825	.5971	14
47	.62638	.77952	.80354	.2445	.2828	.5965	13
48	.62660	.77934	.80402	.2437	.2831	.5959	12
49	.62683	.77915	.80450	.2430	.2834	.5953	11
50	0.62706	0.77897	0.80498	1.2423	1.2837	1.5947	10
51	.62728	.77879	.80546	.2415	.2840	.5942	9
52	.62751	.77861	.80594	.2408	.2843	.5936	8
53	.62774	.77842	.80642	.2400	.2846	.5930	7
54	.62796	.77824	.80690	.2393	.2849	.5924	6
55	0.62819	0.77806	0.80738	1.2386	1.2852	1.5919	5
56	.62841	.77788	.80786	.2378	.2855	.5913	4
57	.62864	.77769	.80834	.2371	.2858	.5907	3
58	.62887	.77751	.80882	.2364	.2861	.5901	2
59	.62909	.77733	.80930	.2356	.2864	.5896	1
60	0.62932	0.77715	0.80978	1.2349	1.2867	1.5890	0
M	Cosine	Sine	Cotan.	Tan.	Cosec.	Secant	M
128°							51°

For angles from 0° to 45°, read degrees and function at top and minutes at left; for 45° to 90°, read degrees and function at bottom and minutes at right; for 90° to 135°, read degrees and function at bottom and minutes at left; and for 135° to 180° read degrees and function at top and minutes at right.

Table 3-5 (*Continued*). Values of the Natural Trigonometric Functions

39° — **140°**

M	Sine	Cosine	Tan.	Cotan.	Secant	Cosec.	M
0	0.62932	0.77715	0.80978	1.2349	1.2867	1.5890	60
1	.62955	.77696	.81026	.2342	.2871	.5884	59
2	.62977	.77678	.81075	.2334	.2874	.5879	58
3	.63000	.77660	.81123	.2327	.2877	.5873	57
4	.63022	.77641	.81171	.2320	.2880	.5867	56
5	0.63045	0.77623	0.81219	1.2312	1.2883	1.5862	55
6	.63067	.77605	.81268	.2305	.2886	.5856	54
7	.63090	.77586	.81316	.2297	.2889	.5850	53
8	.63113	.77568	.81364	.2290	.2892	.5845	52
9	.63135	.77549	.81413	.2283	.2895	.5839	51
10	0.63158	0.77531	0.81461	1.2276	1.2898	1.5833	50
11	.63180	.77513	.81509	.2268	.2901	.5828	49
12	.63203	.77494	.81558	.2261	.2904	.5822	48
13	.63225	.77476	.81606	.2254	.2907	.5816	47
14	.63248	.77458	.81655	.2247	.2910	.5811	46
15	0.63270	0.77439	0.81703	1.2239	1.2913	1.5805	45
16	.63293	.77421	.81752	.2232	.2916	.5799	44
17	.63315	.77402	.81800	.2225	.2919	.5794	43
18	.63338	.77384	.81849	.2218	.2922	.5788	42
19	.63360	.77365	.81898	.2210	.2926	.5783	41
20	0.63383	0.77347	0.81946	1.2203	1.2929	1.5777	40
21	.63405	.77329	.81995	.2196	.2932	.5771	39
22	.63428	.77310	.82043	.2189	.2935	.5766	38
23	.63450	.77292	.82092	.2181	.2938	.5760	37
24	.63473	.77273	.82141	.2174	.2941	.5755	36
25	0.63495	0.77255	0.82190	1.2167	1.2944	1.5749	35
26	.63518	.77236	.82238	.2160	.2947	.5743	34
27	.63540	.77218	.82287	.2152	.2950	.5738	33
28	.63563	.77199	.82336	.2145	.2953	.5732	32
29	.63585	.77181	.82385	.2138	.2956	.5727	31
30	0.63608	0.77162	0.82434	1.2131	1.2960	1.5721	30
31	.63630	.77144	.82482	.2124	.2963	.5716	29
32	.63653	.77125	.82531	.2117	.2966	.5710	28
33	.63675	.77107	.82580	.2109	.2969	.5705	27
34	.63697	.77088	.82629	.2102	.2972	.5699	26
35	0.63720	0.77070	0.82678	1.2095	1.2975	1.5694	25
36	.63742	.77051	.82727	.2088	.2978	.5688	24
37	.63765	.77033	.82776	.2081	.2981	.5683	23
38	.63787	.77014	.82825	.2074	.2985	.5677	22
39	.63810	.76996	.82874	.2066	.2988	.5672	21
40	0.63832	0.76977	0.82923	1.2059	1.2991	1.5666	20
41	.63854	.76958	.82972	.2052	.2994	.5661	19
42	.63877	.76940	.83022	.2045	.2997	.5655	18
43	.63899	.76921	.83071	.2038	.3000	.5650	17
44	.63921	.76903	.83120	.2031	.3003	.5644	16
45	0.63944	0.76884	0.83169	1.2024	1.3006	1.5639	15
46	.63966	.76865	.83218	.2016	.3010	.5633	14
47	.63989	.76847	.83267	.2009	.3013	.5628	13
48	.64011	.76828	.83317	.2002	.3016	.5622	12
49	.64033	.76810	.83366	.1995	.3019	.5617	11
50	0.64056	0.76791	0.83415	1.1988	1.3022	1.5611	10
51	.64078	.76772	.83465	.1981	.3025	.5606	9
52	.64100	.76754	.83514	.1974	.3029	.5600	8
53	.64123	.76735	.83563	.1967	.3032	.5595	7
54	.64145	.76716	.83613	.1960	.3035	.5590	6
55	0.64167	0.76698	0.83662	1.1953	1.3038	1.5584	5
56	.64189	.76679	.83712	.1946	.3041	.5579	4
57	.64212	.76660	.83761	.1939	.3044	.5573	3
58	.64234	.76642	.83811	.1932	.3048	.5568	2
59	.64256	.76623	.83860	.1924	.3051	.5563	1
60	0.64279	0.76604	0.83910	1.1917	1.3054	1.5557	0
M	Cosine	Sine	Cotan.	Tan.	Cosec.	Secant	M

129° — **50°**

40° — **139°**

M	Sine	Cosine	Tan.	Cotan.	Secant	Cosec.	M
0	0.64279	0.76604	0.83910	1.1917	1.3054	1.5557	60
1	.64301	.76586	.83959	.1910	.3057	.5552	59
2	.64323	.76567	.84009	.1903	.3060	.5546	58
3	.64345	.76548	.84059	.1896	.3064	.5541	57
4	.64368	.76530	.84108	.1889	.3067	.5536	56
5	0.64390	0.76511	0.84158	1.1882	1.3070	1.5530	55
6	.64412	.76492	.84208	.1875	.3073	.5525	54
7	.64435	.76473	.84257	.1868	.3076	.5520	53
8	.64457	.76455	.84307	.1861	.3080	.5514	52
9	.64479	.76436	.84357	.1854	.3083	.5509	51
10	0.64501	0.76417	0.84407	1.1847	1.3086	1.5503	50
11	.64523	.76398	.84457	.1840	.3089	.5498	49
12	.64546	.76380	.84506	.1833	.3092	.5493	48
13	.64568	.76361	.84556	.1826	.3096	.5487	47
14	.64590	.76342	.84606	.1819	.3099	.5482	46
15	0.64612	0.76323	0.84656	1.1812	1.3102	1.5477	45
16	.64635	.76304	.84706	.1805	.3105	.5471	44
17	.64657	.76286	.84756	.1798	.3109	.5466	43
18	.64679	.76267	.84806	.1791	.3112	.5461	42
19	.64701	.76248	.84856	.1785	.3115	.5456	41
20	0.64723	0.76229	0.84906	1.1778	1.3118	1.5450	40
21	.64745	.76210	.84956	.1771	.3121	.5445	39
22	.64768	.76191	.85006	.1764	.3125	.5440	38
23	.64790	.76173	.85056	.1757	.3128	.5434	37
24	.64812	.76154	.85107	.1750	.3131	.5429	36
25	0.64834	0.76135	0.85157	1.1743	1.3134	1.5424	35
26	.64856	.76116	.85207	.1736	.3138	.5419	34
27	.64878	.76097	.85257	.1729	.3141	.5413	33
28	.64900	.76078	.85307	.1722	.3144	.5408	32
29	.64923	.76059	.85358	.1715	.3148	.5403	31
30	0.64945	0.76041	0.85408	1.1708	1.3151	1.5398	30
31	.64967	.76022	.85458	.1702	.3154	.5392	29
32	.64989	.76003	.85509	.1695	.3157	.5387	28
33	.65011	.75984	.85559	.1688	.3161	.5382	27
34	.65033	.75965	.85609	.1681	.3164	.5377	26
35	0.65055	0.75946	0.85660	1.1674	1.3167	1.5371	25
36	.65077	.75927	.85710	.1667	.3170	.5366	24
37	.65100	.75908	.85761	.1660	.3174	.5361	23
38	.65121	.75889	.85811	.1653	.3177	.5356	22
39	.65144	.75870	.85862	.1647	.3180	.5351	21
40	0.65166	0.75851	0.85912	1.1640	1.3184	1.5345	20
41	.65188	.75832	.85963	.1633	.3187	.5340	19
42	.65210	.75813	.86013	.1626	.3190	.5335	18
43	.65232	.75794	.86064	.1619	.3193	.5330	17
44	.65254	.75775	.86115	.1612	.3197	.5325	16
45	0.65276	0.75756	0.86165	1.1605	1.3200	1.5319	15
46	.65298	.75737	.86216	.1599	.3203	.5314	14
47	.65320	.75718	.86267	.1592	.3207	.5309	13
48	.65342	.75700	.86318	.1585	.3210	.5304	12
49	.65364	.75680	.86368	.1578	.3213	.5299	11
50	0.65386	0.75661	0.86419	1.1571	1.3217	1.5294	10
51	.65408	.75642	.86470	.1565	.3220	.5289	9
52	.65430	.75623	.86521	.1558	.3223	.5283	8
53	.65452	.75604	.86572	.1551	.3227	.5278	7
54	.65474	.75585	.86623	.1544	.3230	.5273	6
55	0.65496	0.75566	0.86674	1.1537	1.3233	1.5268	5
56	.65518	.75547	.86725	.1531	.3237	.5263	4
57	.65540	.75528	.86775	.1524	.3240	.5258	3
58	.65562	.75509	.86826	.1517	.3243	.5253	2
59	.65584	.75490	.86878	.1510	.3247	.5248	1
60	0.65606	0.75471	0.86929	1.1504	1.3250	1.5242	0
M	Cosine	Sine	Cotan.	Tan.	Cosec.	Secant	M

130° — **49°**

41° — **138°**

M	Sine	Cosine	Tan.	Cotan.	Secant	Cosec.	M
0	0.65606	0.75471	0.86929	1.1504	1.3250	1.5242	60
1	.65628	.75452	.86980	.1497	.3253	.5237	59
2	.65650	.75433	.87031	.1490	.3257	.5232	58
3	.65672	.75414	.87082	.1483	.3260	.5227	57
4	.65694	.75394	.87133	.1477	.3263	.5222	56
5	0.65716	0.75375	0.87184	1.1470	1.3267	1.5217	55
6	.65737	.75356	.87235	.1463	.3270	.5212	54
7	.65759	.75337	.87287	.1456	.3274	.5207	53
8	.65781	.75318	.87338	.1450	.3277	.5202	52
9	.65803	.75299	.87389	.1443	.3280	.5197	51
10	0.65825	0.75280	0.87441	1.1436	1.3284	1.5192	50
11	.65847	.75261	.87492	.1430	.3287	.5187	49
12	.65869	.75241	.87543	.1423	.3290	.5182	48
13	.65891	.75222	.87595	.1416	.3294	.5177	47
14	.65913	.75203	.87646	.1409	.3297	.5171	46
15	0.65934	0.75184	0.87698	1.1403	1.3301	1.5166	45
16	.65956	.75165	.87749	.1396	.3304	.5161	44
17	.65978	.75146	.87801	.1389	.3307	.5156	43
18	.66000	.75126	.87852	.1383	.3311	.5151	42
19	.66022	.75107	.87904	.1376	.3314	.5146	41
20	0.66044	0.75088	0.87955	1.1369	1.3318	1.5141	40
21	.66066	.75069	.88007	.1363	.3321	.5136	39
22	.66087	.75049	.88058	.1356	.3324	.5131	38
23	.66109	.75030	.88110	.1349	.3328	.5126	37
24	.66131	.75011	.88162	.1343	.3331	.5121	36
25	0.66153	0.74992	0.88213	1.1336	1.3335	1.5116	35
26	.66175	.74973	.88265	.1329	.3338	.5111	34
27	.66197	.74953	.88317	.1323	.3342	.5106	33
28	.66218	.74934	.88369	.1316	.3345	.5101	32
29	.66240	.74915	.88421	.1309	.3348	.5096	31
30	0.66262	0.74895	0.88472	1.1303	1.3352	1.5092	30
31	.66284	.74876	.88524	.1296	.3355	.5087	29
32	.66305	.74857	.88576	.1290	.3359	.5082	28
33	.66327	.74838	.88628	.1283	.3362	.5077	27
34	.66349	.74818	.88680	.1276	.3366	.5072	26
35	0.66371	0.74799	0.88732	1.1270	1.3369	1.5067	25
36	.66393	.74780	.88784	.1263	.3372	.5062	24
37	.66414	.74760	.88836	.1257	.3376	.5057	23
38	.66436	.74741	.88888	.1250	.3379	.5052	22
39	.66458	.74722	.88940	.1243	.3383	.5047	21
40	0.66479	0.74702	0.88992	1.1237	1.3386	1.5042	20
41	.66501	.74683	.89044	.1230	.3390	.5037	19
42	.66523	.74664	.89097	.1224	.3393	.5032	18
43	.66545	.74644	.89149	.1217	.3397	.5027	17
44	.66566	.74625	.89201	.1211	.3400	.5022	16
45	0.66588	0.74606	0.89253	1.1204	1.3404	1.5018	15
46	.66610	.74586	.89306	.1197	.3407	.5013	14
47	.66631	.74567	.89358	.1191	.3411	.5008	13
48	.66653	.74548	.89410	.1184	.3414	.5003	12
49	.66675	.74528	.89463	.1178	.3418	.4998	11
50	0.66697	0.74509	0.89515	1.1171	1.3421	1.4993	10
51	.66718	.74489	.89567	.1165	.3425	.4988	9
52	.66740	.74470	.89620	.1158	.3428	.4983	8
53	.66762	.74450	.89672	.1152	.3432	.4979	7
54	.66783	.74431	.89725	.1145	.3435	.4974	6
55	0.66805	0.74412	0.89777	1.1139	1.3439	1.4969	5
56	.66826	.74392	.89830	.1132	.3442	.4964	4
57	.66848	.74373	.89882	.1126	.3446	.4959	3
58	.66870	.74353	.89935	.1119	.3449	.4954	2
59	.66891	.74334	.89988	.1113	.3453	.4949	1
60	0.66913	0.74314	0.90040	1.1106	1.3456	1.4945	0
M	Cosine	Sine	Cotan.	Tan.	Cosec.	Secant	M

131° — **48°**

For angles from 0° to 45°, read degrees and function at top and minutes at left; for 45° to 90°, read degrees and function at bottom and minutes at right; for 90° to 135°, read degrees and function at bottom and minutes at left; and for 135° to 180° read degrees and function at top and minutes at right.

Table 3-5 (*Concluded*). Values of the Natural Trigonometric Functions

42°							137°
M	Sine	Cosine	Tan.	Cotan.	Secant	Cosec.	M
0	0.66913	0.74314	0.90040	1.1106	1.3456	1.4945	60
1	.66935	.74295	.90093	.1100	.3460	.4940	59
2	.66956	.74275	.90146	.1093	.3463	.4935	58
3	.66978	.74256	.90198	.1086	.3467	.4930	57
4	.66999	.74236	.90251	.1080	.3470	.4925	56
5	0.67021	0.74217	0.90304	1.1074	1.3474	1.4921	55
6	.67043	.74197	.90357	.1067	.3477	.4916	54
7	.67064	.74178	.90410	.1061	.3481	.4911	53
8	.67086	.74158	.90463	.1054	.3485	.4906	52
9	.67107	.74139	.90515	.1048	.3488	.4901	51
10	0.67129	0.74119	0.90568	1.1041	1.3492	1.4897	50
11	.67150	.74100	.90621	.1035	.3495	.4892	49
12	.67172	.74080	.90674	.1028	.3499	.4887	48
13	.67194	.74061	.90727	.1022	.3502	.4882	47
14	.67215	.74041	.90780	.1015	.3506	.4877	46
15	0.67237	0.74022	0.90834	1.1009	1.3509	1.4873	45
16	.67258	.74002	.90887	.1003	.3513	.4868	44
17	.67280	.73983	.90940	.0996	.3517	.4863	43
18	.67301	.73963	.90993	.0990	.3520	.4858	42
19	.67323	.73943	.91046	.0983	.3524	.4854	41
20	0.67344	0.73924	0.91099	1.0977	1.3527	1.4849	40
21	.67366	.73904	.91153	.0971	.3531	.4844	39
22	.67387	.73885	.91206	.0964	.3534	.4839	38
23	.67409	.73865	.91259	.0958	.3538	.4835	37
24	.67430	.73845	.91312	.0951	.3542	.4830	36
25	0.67452	0.73826	0.91366	1.0945	1.3545	1.4825	35
26	.67473	.73806	.91419	.0939	.3549	.4821	34
27	.67495	.73787	.91473	.0932	.3552	.4816	33
28	.67516	.73767	.91526	.0926	.3556	.4811	32
29	.67537	.73747	.91580	.0919	.3560	.4806	31
30	0.67559	0.73728	0.91633	1.0913	1.3563	1.4802	30
31	.67580	.73708	.91687	.0907	.3567	.4797	29
32	.67602	.73688	.91740	.0900	.3571	.4792	28
33	.67623	.73669	.91794	.0894	.3574	.4788	27
34	.67645	.73649	.91847	.0888	.3578	.4783	26
35	0.67666	0.73629	0.91901	1.0881	1.3581	1.4778	25
36	.67688	.73610	.91955	.0875	.3585	.4774	24
37	.67709	.73590	.92008	.0868	.3589	.4769	23
38	.67730	.73570	.92062	.0862	.3592	.4764	22
39	.67752	.73551	.92116	.0856	.3596	.4760	21
40	0.67773	0.73531	0.92170	1.0849	1.3600	1.4755	20
41	.67794	.73511	.92223	.0843	.3603	.4750	19
42	.67816	.73491	.92277	.0837	.3607	.4746	18
43	.67837	.73472	.92331	.0830	.3611	.4741	17
44	.67859	.73452	.92385	.0824	.3614	.4736	16
45	0.67880	0.73432	0.92439	1.0818	1.3618	1.4732	15
46	.67901	.73412	.92493	.0812	.3622	.4727	14
47	.67923	.73393	.92547	.0805	.3625	.4723	13
48	.67944	.73373	.92601	.0799	.3629	.4718	12
49	.67965	.73353	.92655	.0793	.3633	.4713	11
50	0.67987	0.73333	0.92709	1.0786	1.3636	1.4709	10
51	.68008	.73314	.92763	.0780	.3640	.4704	9
52	.68029	.73294	.92817	.0774	.3644	.4699	8
53	.68051	.73274	.92871	.0767	.3647	.4695	7
54	.68072	.73254	.92926	.0761	.3651	.4690	6
55	0.68093	0.73234	0.92980	1.0755	1.3655	1.4686	5
56	.68115	.73215	.93034	.0749	.3658	.4681	4
57	.68136	.73195	.93088	.0742	.3662	.4676	3
58	.68157	.73175	.93143	.0736	.3666	.4672	2
59	.68178	.73155	.93197	.0730	.3669	.4667	1
60	0.68200	0.73135	0.93251	1.0724	1.3673	1.4663	0
M	Cosine	Sine	Cotan.	Tan.	Cosec.	Secant	M
132°							47°

43°							136°
M	Sine	Cosine	Tan.	Cotan.	Secant	Cosec.	M
0	0.68200	0.73135	0.93251	1.0724	1.3673	1.4663	60
1	.68221	.73115	.93306	.0717	.3677	.4658	59
2	.68242	.73096	.93360	.0711	.3681	.4654	58
3	.68264	.73076	.93415	.0705	.3684	.4649	57
4	.68285	.73056	.93469	.0699	.3688	.4644	56
5	0.68306	0.73036	0.93524	1.0692	1.3692	1.4640	55
6	.68327	.73016	.93578	.0686	.3695	.4635	54
7	.68349	.72996	.93633	.0680	.3699	.4631	53
8	.68370	.72976	.93687	.0674	.3703	.4626	52
9	.68391	.72956	.93742	.0667	.3707	.4622	51
10	0.68412	0.72937	0.93797	1.0661	1.3710	1.4617	50
11	.68433	.72917	.93851	.0655	.3714	.4613	49
12	.68455	.72897	.93906	.0649	.3718	.4608	48
13	.68476	.72877	.93961	.0643	.3722	.4604	47
14	.68497	.72857	.94016	.0636	.3725	.4599	46
15	0.68518	0.72837	0.94071	1.0630	1.3729	1.4595	45
16	.68539	.72817	.94125	.0624	.3733	.4590	44
17	.68561	.72797	.94180	.0618	.3737	.4586	43
18	.68582	.72777	.94235	.0612	.3740	.4581	42
19	.68603	.72757	.94290	.0605	.3744	.4577	41
20	0.68624	0.72737	0.94345	1.0599	1.3748	1.4572	40
21	.68645	.72717	.94400	.0593	.3752	.4568	39
22	.68666	.72697	.94455	.0587	.3756	.4563	38
23	.68688	.72677	.94510	.0581	.3759	.4559	37
24	.68709	.72657	.94565	.0575	.3763	.4554	36
25	0.68730	0.72637	0.94620	1.0568	1.3767	1.4550	35
26	.68751	.72617	.94675	.0562	.3771	.4545	34
27	.68772	.72597	.94731	.0556	.3774	.4541	33
28	.68793	.72577	.94786	.0550	.3778	.4536	32
29	.68814	.72557	.94841	.0544	.3782	.4532	31
30	0.68835	0.72537	0.94896	1.0538	1.3786	1.4527	30
31	.68856	.72517	.94952	.0532	.3790	.4523	29
32	.68878	.72497	.95007	.0525	.3794	.4518	28
33	.68899	.72477	.95062	.0519	.3797	.4514	27
34	.68920	.72457	.95118	.0513	.3801	.4510	26
35	0.68941	0.72437	0.95173	1.0507	1.3805	1.4505	25
36	.68962	.72417	.95229	.0501	.3809	.4501	24
37	.68983	.72397	.95284	.0495	.3813	.4496	23
38	.69004	.72377	.95340	.0489	.3816	.4492	22
39	.69025	.72357	.95395	.0483	.3820	.4487	21
40	0.69046	0.72337	0.95451	1.0476	1.3824	1.4483	20
41	.69067	.72317	.95506	.0470	.3828	.4479	19
42	.69088	.72297	.95562	.0464	.3832	.4474	18
43	.69109	.72277	.95618	.0458	.3836	.4470	17
44	.69130	.72256	.95673	.0452	.3839	.4465	16
45	0.69151	0.72236	0.95729	1.0446	1.3843	1.4461	15
46	.69172	.72216	.95785	.0440	.3847	.4457	14
47	.69193	.72196	.95841	.0434	.3851	.4452	13
48	.69214	.72176	.95896	.0428	.3855	.4448	12
49	.69235	.72156	.95952	.0422	.3859	.4443	11
50	0.69256	0.72136	0.96008	1.0416	1.3863	1.4439	10
51	.69277	.72115	.96064	.0410	.3867	.4435	9
52	.69298	.72095	.96120	.0404	.3870	.4430	8
53	.69319	.72075	.96176	.0397	.3874	.4426	7
54	.69340	.72055	.96232	.0391	.3878	.4422	6
55	0.69361	0.72035	0.96288	1.0385	1.3882	1.4417	5
56	.69382	.72015	.96344	.0379	.3886	.4413	4
57	.69403	.71994	.96400	.0373	.3890	.4408	3
58	.69424	.71974	.96456	.0367	.3894	.4404	2
59	.69445	.71954	.96513	.0361	.3898	.4400	1
60	0.69466	0.71934	0.96569	1.0355	1.3902	1.4395	0
M	Cosine	Sine	Cotan.	Tan.	Cosec.	Secant	M
133°							46°

44°							135°
M	Sine	Cosine	Tan.	Cotan.	Secant	Cosec.	M
0	0.69466	0.71934	0.96569	1.0355	1.3902	1.4395	60
1	.69487	.71914	.96625	.0349	.3905	.4391	59
2	.69508	.71893	.96681	.0343	.3909	.4387	58
3	.69528	.71873	.96738	.0337	.3913	.4382	57
4	.69549	.71853	.96794	.0331	.3917	.4378	56
5	0.69570	0.71833	0.96850	1.0325	1.3921	1.4374	55
6	.69591	.71813	.96907	.0319	.3925	.4370	54
7	.69612	.71792	.96963	.0313	.3929	.4365	53
8	.69633	.71772	.97020	.0307	.3933	.4361	52
9	.69654	.71752	.97076	.0301	.3937	.4357	51
10	0.69675	0.71732	0.97133	1.0295	1.3941	1.4352	50
11	.69696	.71711	.97189	.0289	.3945	.4348	49
12	.69716	.71691	.97246	.0283	.3949	.4344	48
13	.69737	.71671	.97302	.0277	.3953	.4339	47
14	.69758	.71650	.97359	.0271	.3957	.4335	46
15	0.69779	0.71630	0.97416	1.0265	1.3960	1.4331	45
16	.69800	.71610	.97472	.0259	.3964	.4327	44
17	.69821	.71589	.97529	.0253	.3968	.4322	43
18	.69841	.71569	.97586	.0247	.3972	.4318	42
19	.69862	.71549	.97643	.0241	.3976	.4314	41
20	0.69883	0.71529	0.97700	1.0235	1.3980	1.4310	40
21	.69904	.71508	.97756	.0229	.3984	.4305	39
22	.69925	.71488	.97813	.0223	.3988	.4301	38
23	.69945	.71468	.97870	.0218	.3992	.4297	37
24	.69966	.71447	.97927	.0212	.3996	.4292	36
25	0.69987	0.71427	0.97984	1.0206	1.4000	1.4288	35
26	.70008	.71406	.98041	.0200	.4004	.4284	34
27	.70029	.71386	.98098	.0194	.4008	.4280	33
28	.70049	.71366	.98155	.0188	.4012	.4276	32
29	.70070	.71345	.98212	.0182	.4016	.4271	31
30	0.70091	0.71325	0.98270	1.0176	1.4020	1.4267	30
31	.70112	.71305	.98327	.0170	.4024	.4263	29
32	.70132	.71284	.98384	.0164	.4028	.4259	28
33	.70153	.71264	.98441	.0158	.4032	.4254	27
34	.70174	.71243	.98499	.0152	.4036	.4250	26
35	0.70194	0.71223	0.98556	1.0146	1.4040	1.4246	25
36	.70215	.71203	.98613	.0141	.4044	.4242	24
37	.70236	.71182	.98671	.0135	.4048	.4238	23
38	.70257	.71162	.98728	.0129	.4052	.4233	22
39	.70277	.71141	.98786	.0123	.4056	.4229	21
40	0.70298	0.71121	0.98843	1.0117	1.4060	1.4225	20
41	.70319	.71100	.98901	.0111	.4065	.4221	19
42	.70339	.71080	.98958	.0105	.4069	.4217	18
43	.70360	.71059	.99016	.0099	.4073	.4212	17
44	.70381	.71039	.99073	.0093	.4077	.4208	16
45	0.70401	0.71018	0.99131	1.0088	1.4081	1.4204	15
46	.70422	.70998	.99189	.0082	.4085	.4200	14
47	.70443	.70977	.99246	.0076	.4089	.4196	13
48	.70463	.70957	.99304	.0070	.4093	.4192	12
49	.70484	.70936	.99362	.0064	.4097	.4188	11
50	0.70505	0.70916	0.99420	1.0058	1.4101	1.4183	10
51	.70525	.70895	.99478	.0052	.4105	.4179	9
52	.70546	.70875	.99536	.0047	.4109	.4175	8
53	.70566	.70854	.99593	.0041	.4113	.4171	7
54	.70587	.70834	.99651	.0035	.4117	.4167	6
55	0.70608	0.70813	0.99709	1.0029	1.4122	1.4163	5
56	.70628	.70793	.99767	.0023	.4126	.4159	4
57	.70649	.70772	.99826	.0017	.4130	.4154	3
58	.70669	.70752	.99884	.0012	.4134	.4150	2
59	.70690	.70731	.99942	.0006	.4138	.4146	1
60	0.70711	0.70711	1.00000	1.0000	1.4142	1.4142	0
M	Cosine	Sine	Cotan.	Tan.	Cosec.	Secant	M
134°							45°

For angles from 0° to 45°, read degrees and function at top and minutes at left; for 45° to 90°, read degrees and function at bottom and minutes at right; for 90° to 135°, read degrees and function at bottom and minutes at left; and for 135° to 180°, read degrees and function at top and minutes at right.

(*Text continued from page 89.*)

$$\text{Tangent} = \frac{\text{Side Opposite}}{\text{Side Adjacent}} = \text{Tangent } A = \frac{a}{b};$$

$$\text{Tangent } B = \frac{b}{a}$$

$$\text{Cotangent} = \frac{\text{Side Adjacent}}{\text{Side Opposite}} = \text{Cotangent } A = \frac{b}{a};$$

$$\text{Cotangent } B = \frac{a}{b}$$

$$\text{Secant} = \frac{\text{Hypotenuse}}{\text{Side Adjacent}} = \text{Secant } A = \frac{c}{b};$$

$$\text{Secant } B = \frac{c}{a}$$

$$\text{Cosecant} = \frac{\text{Hypotenuse}}{\text{Side Opposite}} = \text{Cosecant } A = \frac{c}{a};$$

$$\text{Cosecant } B = \frac{c}{b}$$

The functions of the Sine and Cosine are always less than unity. Their respective reciprocals, the Cosecant and the Secant, are always greater than unity. Functions of the Tangent and Cotangent may be less than, or greater than, unity. See Tables 3-5 and 3-6.

Table 3-6. Condensed Trigonometry Tables *

Values of the Trigonometric Functions							Logarithms of the Trigonometric Functions*						
Angle, degree	Sin	Cos	Tan	Angle, degree	Sin	Cos	Tan	Angle, degree	L Sin	L Cos	L Tan	Angle, degree	L Sin	L Cos	L Tan
1	.0175	.9998	.0175	46	.7193	.6947	1.0355	1	8.2419	9.9999	8.2419	46	9.8569	9.8418	10.0152
2	.0349	.9994	.0349	47	.7314	.6820	1.0724	2	8.5428	9.9997	8.5431	47	9.8641	9.8338	10.0303
3	.0523	.9986	.0524	48	.7431	.6691	1.1106	3	8.7188	9.9994	8.7194	48	9.8711	9.8255	10.0456
4	.0698	.9976	.0699	49	.7547	.6561	1.1504	4	8.8436	9.9989	8.8446	49	9.8778	9.8169	10.0608
5	.0872	.9962	.0875	50	.7660	.6428	1.1918	5	8.9403	9.9983	8.9420	50	9.8843	9.8081	10.0762
6	.1045	.9945	.1051	51	.7771	.6293	1.2349	6	9.0192	9.9976	9.0216	51	9.8905	9.7989	10.0916
7	.1219	.9925	.1228	52	.7880	.6157	1.2799	7	9.0859	9.9968	9.0891	52	9.8965	9.7893	10.1072
8	.1392	.9903	.1405	53	.7986	.6018	1.3270	8	9.1436	9.9958	9.1478	53	9.9023	9.7795	10.1229
9	.1564	.9877	.1584	54	.8090	.5878	1.3764	9	9.1943	9.9946	9.1997	54	9.9080	9.7692	10.1387
10	.1736	.9848	.1763	55	.8192	.5736	1.4281	10	9.2397	9.9934	9.2463	55	9.9134	9.7586	10.1548
11	.1908	.9816	.1944	56	.8290	.5592	1.4826	11	9.2806	9.9919	9.2887	56	9.9186	9.7476	10.1710
12	.2079	.9781	.2126	57	.8387	.5446	1.5399	12	9.3179	9.9904	9.3275	57	9.9236	9.7361	10.1875
13	.2250	.9744	.2309	58	.8480	.5299	1.6003	13	9.3521	9.9887	9.3634	58	9.9284	9.7242	10.2042
14	.2419	.9703	.2493	59	.8572	.5150	1.6643	14	9.3837	9.9869	9.3968	59	9.9331	9.7118	10.2212
15	.2588	.9659	.2679	60	.8660	.5000	1.7321	15	9.4130	9.9849	9.4281	60	9.9375	9.6990	10.2386
16	.2756	.9613	.2867	61	.8746	.4848	1.8040	16	9.4403	9.9828	9.4575	61	9.9418	9.6856	10.2562
17	.2924	.9563	.3057	62	.8829	.4695	1.8807	17	9.4659	9.9806	9.4853	62	9.9459	9.6716	10.2743
18	.3090	.9511	.3249	63	.8910	.4540	1.9626	18	9.4900	9.9782	9.5118	63	9.9499	9.6650	10.2928
19	.3256	.9455	.3443	64	.8988	.4384	2.0503	19	9.5126	9.9757	9.5370	64	9.9537	9.6418	10.3118
20	.3420	.9397	.3640	65	.9063	.4226	2.1445	20	9.5341	9.9730	9.5611	65	9.9573	9.6259	10.3313
21	.3584	.9336	.3839	66	.9135	.4067	2.2460	21	9.5543	9.9702	9.5842	66	9.9607	9.6093	10.3514
22	.3746	.9272	.4040	67	.9205	.3907	2.3559	22	9.5736	9.9672	9.6064	67	9.9640	9.5919	10.3721
23	.3907	.9205	.4245	68	.9272	.3746	2.4751	23	9.5919	9.9640	9.6279	68	9.9672	9.5736	10.3936
24	.4067	.9135	.4452	69	.9336	.3584	2.6051	24	9.6093	9.9607	9.6486	69	9.9702	9.5543	10.4158
25	.4226	.9063	.4663	70	.9397	.3420	2.7475	25	9.6259	9.9573	9.6687	70	9.9730	9.5341	10.4389
26	.4384	.8988	.4877	71	.9455	.3256	2.9042	26	9.6418	9.9537	9.6882	71	9.9757	9.5126	10.4630
27	.4540	.8910	.5095	72	.9511	.3090	3.0777	27	9.6570	9.9499	9.7072	72	9.9782	9.4900	10.4882
28	.4695	.8829	.5317	73	.9563	.2924	3.2709	28	9.6716	9.9459	9.7257	73	9.9806	9.4659	10.5147
29	.4848	.8746	.5543	74	.9613	.2756	3.4874	29	9.6856	9.9418	9.7438	74	9.9828	9.4403	10.5425
30	.5000	.8660	.5774	75	.9659	.2588	3.7321	30	9.6990	9.9375	9.7614	75	9.9849	9.4130	10.5719
31	.5150	.8572	.6009	76	.9703	.2419	4.0108	31	9.7118	9.9331	9.7788	76	9.9869	9.3837	10.6032
32	.5299	.8480	.6249	77	.9744	.2250	4.3315	32	9.7242	9.9284	9.7958	77	9.9887	9.3521	10.6366
33	.5446	.8387	.6494	78	.9781	.2079	4.7046	33	9.7361	9.9236	9.8125	78	9.9904	9.3179	10.6725
34	.5592	.8290	.6745	79	.9816	.1908	5.1446	34	9.7476	9.9186	9.8290	79	9.9919	9.2806	10.7113
35	.5736	.8192	.7002	80	.9848	.1736	5.6713	35	9.7586	9.9134	9.8452	80	9.9934	9.2397	10.7537
36	.5578	.8090	.7265	81	.9877	.1564	6.3138	36	9.7692	9.9080	9.8613	81	9.9946	9.1943	10.8003
37	.6018	.7986	.7536	82	.9903	.1392	7.1154	37	9.7795	9.9023	9.8771	82	9.9958	9.1436	10.8522
38	.6157	.7880	.7813	83	.9925	.1219	8.1443	38	9.7893	9.8965	9.8928	83	9.9968	9.0859	10.9109
39	.6293	.7771	.8098	84	.9945	.1045	9.5144	39	9.7989	9.8905	9.9084	84	9.9976	9.0192	10.9784
40	.6428	.7660	.8391	85	.9962	.0872	11.4301	40	9.8081	9.8843	9.9238	85	9.9983	8.9403	11.0580
41	.6561	.7547	.8693	86	.9976	.0698	14.3007	41	9.8169	9.8778	9.9392	86	9.9989	8.8436	11.1554
42	.6691	.7431	.9004	87	.9986	.0523	19.0811	42	9.8255	9.8711	9.9544	87	9.9994	8.7188	11.2806
43	.6820	.7314	.9325	88	.9994	.0349	28.6363	43	9.8338	9.8641	9.9697	88	9.9997	8.5428	11.4569
44	.6947	.7193	.9657	89	.9998	.0175	57.2900	44	9.8418	9.8569	9.9848	89	9.9999	8.2419	11.7581
45	.7071	.7071	1.0000	90	1.0000	.0000	...	45	9.8495	9.8495	10.0000	90	10.0000	...	...

* This table gives the logarithm increased by 10. Hence, in each case 10 should be subtracted.

It is more convenient to multiply rather than to divide with three- or four-place decimal functions. Therefore, when an angle and side of a right triangle are known, formulas from Table 3-7 are useful.

Table 3-7. Computing Dimensions of a Right Triangle when One Angle and One Side are Known

To find length of:	Multiply
Side Opposite	Hypotenuse by Sine
	Adjacent by Tangent
Side Adjacent	Hypotenuse by Cosine
	Opposite by Cotangent
Hypotenuse	Opposite by Cosecant
	Adjacent by Secant

Table 3-8 is a condensed list of trig functions used most often for standard bevels of 30°, 45°, and 60°.

Table 3-8. Functions of Commonly Used Angles

Angle	Sine	Cosine	Tangent	Cotangent	Secant	Co-secant
7½°	.1305	.991	.1317	7.596	1.009	7.661
15°	.259	.966	.268	3.732	1.035	3.864
22½°	.383	.924	.414	2.414	1.082	2.613
30°	.500	.866	.577	1.732	1.155	2.000
45°	.707	.707	1.000	1.000	1.414	1.414
60°	.866	.500	1.732	.577	2.000	1.155

Computing Parts of a Right Triangle

If two parts of a right triangle are known, any of the other sides or angles may be computed. A rough drawing of a problem is generally helpful so that the proper formula can be selected.

Example:

In a 30° right triangle, the length of one of the legs is 41½". Find the length of the hypotenuse and the other leg.

Table 3-9. Products of Numbers

NO.	.268	.414	.577	1.155	1.414	1.732	NO.	.268	.414	.577	1.155	1.414	1.732
1/8	1/32	1/16	1/16	1/8	3/16	1/4	23	6 1/8	9 1/2	13 1/4	26 1/2	32 1/2	39 7/8
1/4	1/16	1/8	1/8	1/4	3/8	3/8	24	6 3/8	9 7/8	13 7/8	27 3/4	33 7/8	41 5/8
3/8	1/8	1/8	1/4	3/8	1/2	5/8	25	6 3/4	9 7/8	14 3/8	28 7/8	35 3/8	43 1/4
1/2	1/8	1/4	1/4	5/8	3/4	7/8	26	7	10 3/4	15	30	36 3/8	45
5/8	3/16	1/4	3/8	3/4	7/8	1 1/8	27	7 1/4	11 1/8	15 5/8	31 1/8	38 1/8	46 3/4
3/4	1/4	5/16	3/8	7/8	1 1/16	1 1/4	28	7 1/2	11 5/8	16 1/8	32 3/8	39 5/8	48 1/2
7/8	1/4	3/8	1/2	1	1 1/4	1 1/2	29	7 3/4	12	16 3/4	33 1/2	41	50 1/4
1	1/4	3/8	5/8	1 1/8	1 3/8	1 3/4	30	8	12 3/8	17 1/4	34 5/8	42 3/8	52
2	1/2	7/8	1 1/8	2 1/4	2 7/8	3 1/2	31	8 1/4	12 7/8	17 7/8	35 3/4	43 7/8	53 3/4
3	3/4	1 1/4	1 3/4	3 1/2	4 1/4	5 1/4	32	8 5/8	13 1/4	18 1/2	37	45 1/4	55 3/8
4	1 1/8	1 5/8	2 1/4	4 5/8	5 5/8	6 7/8	33	8 7/8	12 5/8	19	38 1/8	46 5/8	57 1/8
5	1 3/8	2 1/8	2 7/8	5 3/4	7 1/8	8 5/8	34	9 1/8	14 1/8	19 5/8	39 1/4	48 1/8	58 7/8
6	1 5/8	2 1/2	3 1/2	6 7/8	8 1/2	10 3/8	35	9 3/8	14 1/2	20 1/4	40 3/8	49 1/2	60 5/8
7	1 7/8	2 7/8	4	8 1/8	9 7/8	12 1/8	36	9 5/8	14 7/8	20 3/4	41 5/8	50 7/8	62 3/8
8	2 1/8	3 3/8	4 5/8	9 1/4	11 3/8	13 7/8	37	9 7/8	15 3/8	21 3/8	42 3/4	52 3/8	64 1/8
9	2 3/8	3 3/4	5 1/4	10 3/8	12 3/4	15 5/8	38	10 1/8	15 3/4	21 7/8	43 7/8	53 3/4	65 7/8
10	2 5/8	4 1/8	5 3/4	11 1/2	14 1/8	17 3/8	39	10 1/2	16 1/8	22 1/2	45	55 1/8	67 1/2
11	3	4 1/2	6 3/8	12 3/4	15 1/2	19	40	10 3/4	16 1/2	23 1/8	46 1/4	56 1/2	69 1/4
12	3 1/4	5	6 7/8	13 7/8	17	20 3/4	41	11	17	23 5/8	47 3/8	58	71
13	3 1/2	5 3/8	7 1/2	15	18 3/8	22 1/2	42	11 1/4	17 3/8	24 1/4	48 1/2	59 3/8	72 3/4
14	3 3/4	5 3/4	8 1/8	16 1/8	19 3/4	24 1/4	43	11 1/2	17 3/4	24 3/4	49 5/8	60 3/4	74 1/2
15	4	6 1/8	8 5/8	17 3/8	21 1/4	26	44	11 3/4	18 1/4	25 3/8	50 7/8	62 1/4	76 1/4
16	4 1/4	6 5/8	9 1/4	18 1/2	22 5/8	27 3/4	45	12 1/8	18 5/8	26	52	63 5/8	78
17	4 1/2	7	9 3/4	19 5/8	24	29 1/2	46	12 3/8	19	26 1/2	53 1/8	65	79 5/8
18	4 7/8	7 1/2	10 3/8	20 3/4	25 1/2	31 1/8	47	12 5/8	19 1/2	27 1/8	54 1/4	66 1/2	81 3/8
19	5 1/8	7 7/8	11	22	26 7/8	32 7/8	48	12 7/8	19 7/8	27 3/4	55 1/2	67 7/8	83 1/8
20	5 3/8	8 1/4	11 1/2	23 1/8	28 1/4	34 5/8	49	13 1/8	20 1/4	28 1/4	56 5/8	69 1/4	84 7/8
21	5 5/8	8 3/4	12 1/8	24 1/4	29 3/4	36 3/8	50	13 3/8	20 3/4	28 7/8	57 3/4	70 3/4	86 5/8
22	5 7/8	9 1/8	12 3/4	25 3/8	31 1/8	38 1/8	51	13 5/8	21 1/8	29 1/8	58 7/8	72 1/8	88 3/8

* Examples to Find Products Not Directly Read: 50⅝ × .414 = (50 + ⅝) .414 = 20¾ + ¼ = 21   96 × 1.414 = (50 + 46) 1.414 = 70¾ + 65 = 135¾

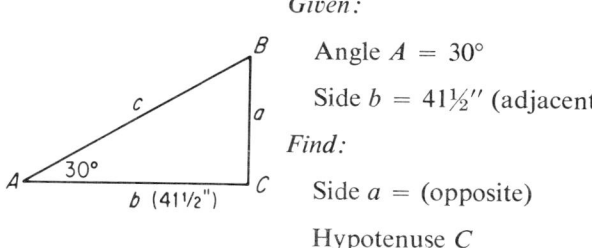

Given:

Angle $A = 30°$

Side $b = 41\frac{1}{2}''$ (adjacent)

Find:

Side $a = $ (opposite)

Hypotenuse C

Side opposite = Side adjacent × Tangent angle A

$$a = 41.5'' \times .577$$
$$= 24'' \quad Ans$$

Hypotenuse = Side opposite × Cosecant angle A

$$c = 24'' \times 2.0$$
$$= 48'' \quad Ans$$

Table 3-9 provides the products of numbers for computations of duct offsets using standard bevels of 30°, 45°, or 60°: tan 15°(.268), tan 22½° (.414), tan 30° and cot 60°(.577), sec. 30° and cosec 60°(1.155), cosec 45°(1.414), cot 30° and tan 60°(1.732).

Table 3-10 indicates the hypotenuse increase, or gain, of right triangles with leg lengths across and down from "B." For example, if the length of a large transition fitting with a 12" slope was 44" instead of 46", the 47⅛" cut size (44", plus 1½" slip end, plus 1⅝" **gain**) is convenient for a 48" standard sheet width.

USE OF TRIGONOMETRY IN COMPUTATIONS FOR OFFSETS

Legend Symbols

Angle a = Angle of set

Angle b = One-half angle a

S = Set, straight side to straight side

W = Large cheek width

w = Small cheek width

RL = Running length

TL = True length

(See Figs. 3-1, 3-3, 3-6, and 3-7 for additional legend symbols.)

The "Ogee Set"

When a rectangular duct-run changes height, and a raise or drop fitting is required, the draftsman should calculate the set fitting dimensions rather than resort to a sectional view. Such calculations will indicate whether the fitting will have a reasonable throat radius and also determine if it can be fabricated economically. In cases where ducts are greater than standard sheet widths, by reducing the set length an inch or two, transverse seams in wrappers can be avoided.

The throat radius of a rectangular-duct offset is a function of the cheek width and the distance in which the set occurs. (See Fig. 3-1.)

Example:

A high-pressure cold-air duct is 64" x 28" in plan view and requires a drop of 1'-10" in 3'-4" to avoid a plumbing pipe. Connections are matched angle frames. Compute the throat radius and wrapper length to the nearest quarter inch.

$$S = 1'\text{-}10'' = 22'' \qquad Find: r \text{ and } WL$$
$$L = 3'\text{-}4'' = 40''$$
$$W = 28''$$

Tangent angle $a = \dfrac{S}{L} = \dfrac{22}{40} = .55 = 28°\text{-}50'$ or $28.8°$

Cosecant angle $a = 2.0735 \qquad \dfrac{S}{4} = \dfrac{22}{4} = 5.5$

$$\dfrac{W}{2} = \dfrac{28}{2} = 14$$

$$R = (\text{cosecant angle } a)^2 \times \dfrac{S}{4}$$

$$R = (2.0735)^2 \times 5.5$$
$$= (4.3)5.5$$

$$R = 23\tfrac{3}{4}''$$

$$r = 23\tfrac{3}{4}'' - 14 = 9\tfrac{3}{4}'' \quad Ans$$

$WL = 4R(.01745)$ angle a (See Fig. 3-1.)

$$= 4(23.75)(.01745)28.8$$

$$WL = 47\tfrac{3}{4}'' \quad Ans$$

The drop fitting is quite abrupt as indicated by the ratio of throat radius to cheek width, 9.75 to

(Text continued on page 113.)

Table 3-10. Gain Chart

B	1	2	3	4	5	6	7	8	9	10	11	12
1	7/16	1/4	3/16	1/8	1/8	1/16	1/16	1/16	1/16	1/16	1/16	1/16
2	1/4	13/16	5/8	1/2	3/8	5/16	5/16	1/4	3/16	3/16	3/16	3/16
3	3/16	5/8	1 1/4	1	13/16	11/16	5/8	1/2	1/2	7/16	3/8	3/8
4	1/8	1/2	1	1 11/16	1 7/16	1 3/16	1 1/16	15/16	13/16	3/4	11/16	5/8
5	1/8	3/8	13/16	1 7/16	2 1/16	1 13/16	1 5/8	1 7/16	1 5/16	1 3/16	1 1/16	1
6	1/16	5/16	11/16	1 3/16	1 13/16	2 1/2	2 1/4	2	1 13/16	1 11/16	1 1/2	1 7/16
7	1/16	5/16	5/8	1 1/16	1 5/8	2 1/4	2 15/16	2 5/8	2 3/8	2 3/16	2 1/16	1 7/8
8	1/16	1/4	1/2	15/16	1 7/16	2	2 5/8	3 5/16	3 1/16	2 13/16	2 5/8	2 7/16
9	1/16	3/16	1/2	13/16	1 5/16	1 13/16	2 3/8	3 1/16	3 3/4	3 7/16	3 1/4	3
10	1/16	3/16	7/16	3/4	1 3/16	1 11/16	2 3/16	2 13/16	3 7/16	4 1/8	3 7/8	3 5/8
11	1/16	3/16	3/8	11/16	1 1/16	1 1/2	2 1/16	2 5/8	3 1/4	3 7/8	4 3/16	4 1/4
12	1/16	3/16	3/8	5/8	1	1 7/16	1 7/8	2 7/16	3	3 5/8	4 1/4	5
13	1/16	3/16	3/8	5/8	15/16	1 5/16	1 3/4	2 1/4	2 13/16	3 3/8	4 1/16	4 11/16
14	1/16	3/16	5/16	9/16	7/8	1 1/4	1 5/8	2 1/8	2 3/8	3 3/16	3 13/16	4 7/16
15	1/16	1/8	5/16	1/2	13/16	1 1/8	1 9/16	2	2 1/2	3	3 5/8	4 3/16
16	1/16	1/8	5/16	1/2	3/4	1 1/16	1 7/16	1 7/8	2 3/8	2 7/8	3 7/16	4
17	1/16	1/8	1/4	1/2	11/16	1	1 3/8	1 3/4	2 1/4	2 3/4	3 1/4	3 13/16
18	1/16	1/8	1/4	7/16	11/16	1	1 5/16	1 11/16	2 1/8	2 5/8	3 1/8	3 5/8
19	1/16	1/8	1/4	7/16	11/16	15/16	1 1/4	1 5/8	2 1/16	2 1/2	3	3 1/2
20	1/16	1/8	1/4	7/16	5/8	7/8	1 3/16	1 9/16	1 15/16	2 3/8	2 7/8	3 3/8
21	0	1/8	1/4	3/8	9/16	7/8	1 1/8	1 1/2	1 7/8	2 1/4	2 3/4	3 3/16
22	0	1/8	1/4	3/8	9/16	13/16	1 1/16	1 7/16	1 3/4	2 1/8	2 5/8	3 1/16
23	0	1/8	3/16	3/8	9/16	13/16	1	1 3/8	1 11/16	2 1/16	2 1/2	2 15/16
24	0	1/16	3/16	5/16	1/2	3/4	1	1 5/16	1 5/8	2	2 7/16	2 7/8
25	0	1/16	3/16	5/16	1/2	3/4	15/16	1 1/4	1 9/16	1 15/16	2 5/16	2 3/4
26	0	1/16	3/16	5/16	1/2	11/16	15/16	1 3/16	1 1/2	1 7/8	2 1/4	2 5/8
27	0	1/16	3/16	5/16	7/16	11/16	7/8	1 3/16	1 1/2	1 13/16	2 3/16	2 9/16
28	0	1/16	1/8	1/4	7/16	5/8	7/8	1 1/8	1 7/16	1 3/4	2 1/8	2 1/2
29	0	1/16	1/8	1/4	7/16	5/8	7/8	1 1/16	1 3/8	1 11/16	2	2 3/8
30	0	1/16	1/8	1/4	7/16	5/8	13/16	1 1/16	1 5/16	1 5/8	1 15/16	2 5/16
31	0	1/16	1/8	1/4	3/8	9/16	13/16	1	1 5/16	1 9/16	1 7/8	2 1/4
32	0	1/16	1/8	1/4	3/8	9/16	3/4	1	1 1/4	1 1/2	1 7/8	2 3/16
33	0	1/16	1/8	1/4	3/8	1/2	3/4	15/16	1 1/4	1 1/2	1 13/16	2 1/8
34	0	1/16	1/8	1/4	3/8	1/2	3/4	15/16	1 3/16	1 7/16	1 3/4	2 1/16
35	0	1/16	1/8	1/4	3/8	1/2	11/16	15/16	1 1/8	1 3/8	1 11/16	2
36	0	1/16	1/8	3/16	3/8	1/2	11/16	15/16	1 1/8	1 3/8	1 5/8	1 15/16
37	0	1/16	1/16	3/8	7/16	5/8	7/8	1 1/16	1 5/16	1 5/8	1 7/8	
38	0	1/16	1/16	3/8	7/16	5/8	7/8	1 1/16	1 1/4	1 9/16	1 7/8	
39	0	1/16	1/16	3/8	7/16	5/8	7/8	1	1 1/4	1 1/2	1 13/16	
40	0	1/16	1/16	3/16	3/8	7/16	9/16	13/16	1	1 1/4	1 1/2	1 3/4
41	0	1/16	1/16	3/16	3/8	3/8	9/16	3/4	1	1 3/16	1 7/16	1 3/4
42	0	1/16	1/16	3/16	5/16	3/8	9/16	3/4	1	1 3/16	1 7/16	1 11/16
43	0	1/16	1/16	3/16	1/4	3/8	9/16	3/4	15/16	1 1/8	1 3/8	1 5/8
44	0	1/16	1/16	3/16	1/4	3/8	1/2	3/4	7/8	1 1/8	1 3/8	1 5/8
45	0	1/16	1/16	3/16	1/4	3/8	1/2	11/16	7/8	1 1/8	1 5/16	1 9/16
46	0	1/16	1/16	3/16	1/4	3/8	1/2	11/16	7/8	1 1/8	1 5/16	1 9/16
47	0	1/16	1/16	3/16	1/4	3/8	1/2	11/16	7/8	1 1/16	1 1/4	1 1/2
48	0	1/16	1/16	3/16	1/4	3/8	1/2	11/16	7/8	1 1/16	1 1/4	1 1/2

Table 3-10 (Continued). Gain Chart

B	13	14	15	16	17	18	19	20	21	22	23	24
1	1/16	1/16	1/16	1/16	1/16	1/16	1/16	1/16	0	0	0	0
2	3/16	3/16	1/8	1/8	1/8	1/8	1/8	1/8	1/8	1/8	1/8	1/8
3	3/8	5/16	5/16	5/16	1/4	1/4	1/4	1/4	1/4	1/4	3/16	3/16
4	5/8	9/16	1/2	1/2	1/2	7/16	7/16	7/16	3/8	3/8	3/8	5/16
5	15/16	7/8	13/16	3/4	11/16	11/16	11/16	5/8	9/16	9/16	9/16	1/2
6	1 5/16	1 1/4	1 1/8	1 1/16	1	1	15/16	7/8	7/8	13/16	13/16	3/4
7	1 3/4	1 5/8	1 9/16	1 7/16	1 3/8	1 5/16	1 1/4	1 3/16	1 1/8	1 1/16	1	1
8	2 1/4	2 1/8	2	1 7/8	1 3/4	1 11/16	1 5/8	1 9/16	1 1/2	1 7/16	1 3/8	1 5/16
9	2 13/16	2 5/8	2 1/2	2 3/8	2 1/4	2 1/8	2 1/16	1 15/16	1 7/8	1 3/4	1 11/16	1 5/8
10	3 3/8	3 3/16	3	2 7/8	2 3/4	2 5/8	2 1/2	2 3/8	2 1/4	2 1/8	2 1/16	2
11	4 1/16	3 13/16	3 5/8	3 7/16	3 1/4	3 1/8	3	2 7/8	2 3/4	2 5/8	2 1/2	2 7/16
12	4 11/16	4 7/16	4 3/16	4	3 13/16	3 5/8	3 1/2	3 3/8	3 3/16	3 1/16	2 15/16	2 7/8
13	5 3/8	5 1/8	4 7/8	4 5/8	4 7/16	4 1/4	4 1/16	3 7/8	3 3/4	3 9/16	3 7/16	3 5/16
14	5 1/8	5 13/16	5 7/16	5 1/4	5	4 13/16	4 5/8	4 7/16	4 1/4	4 1/8	3 15/16	3 13/16
15	4 7/8	5 7/16	6 1/4	5 15/16	5 11/16	5 7/16	5 1/4	5 1/16	4 13/16	4 5/8	4 1/2	4 5/16
16	4 5/8	5 1/4	5 15/16	6 5/8	6 3/8	6 1/8	5 7/8	5 5/8	5 7/16	5 1/4	5 1/16	4 7/8
17	4 7/16	5	5 11/16	6 3/8	7 1/16	6 3/4	6 1/2	6 1/4	6 1/16	5 13/16	5 5/8	5 7/16
18	4 1/4	4 13/16	5 7/16	6 1/8	6 3/4	7 7/16	7 3/16	6 15/16	6 11/16	6 7/16	6 1/4	6
19	4 1/16	4 5/8	5 1/4	5 7/8	6 1/2	7 3/16	7 7/8	7 5/8	7 3/8	7 1/8	6 7/8	6 5/8
20	3 7/8	4 7/16	5 1/16	5 5/8	6 1/4	6 15/16	7 5/8	8 5/16	8	7 3/4	7 1/2	7 1/4
21	3 3/4	4 1/4	4 13/16	5 7/16	6 1/16	6 11/16	7 5/16	8	8 3/4	8 7/16	8 3/16	7 7/8
22	3 9/16	4 1/8	4 5/8	5 1/4	5 13/16	6 7/16	7 1/16	7 3/4	8 7/16	9 1/8	8 7/8	8 9/16
23	3 7/16	3 15/16	4 1/2	5 1/16	5 5/8	6 1/4	6 7/8	7 1/2	8 3/16	8 7/8	9 9/16	9 1/4
24	3 5/16	3 13/16	4 5/16	4 7/8	5 7/16	6	6 5/8	7 1/4	7 15/16	8 9/16	9 1/4	9 15/16
25	3 3/16	3 11/16	4 3/16	4 11/16	5 1/4	5 13/16	6 7/16	7 1/16	7 11/16	8 5/16	9	9 11/16
26	3 1/16	3 9/16	4 1/16	4 9/16	5 1/16	5 5/8	6 3/16	6 13/16	7 7/16	8 1/16	8 3/4	9 3/8
27	3	3 7/16	3 15/16	4 3/8	4 15/16	5 1/2	6 1/16	6 5/8	7 1/4	7 7/8	8 1/2	9 1/8
28	2 7/8	3 5/16	3 3/4	4 1/4	4 3/4	5 5/16	5 7/8	6 7/16	7	7 5/8	8 1/4	8 7/8
29	2 13/16	3 3/16	3 11/16	4 1/8	4 5/8	5 1/8	5 11/16	6 1/4	6 13/16	7 7/16	8	8 5/8
30	2 3/4	3 1/8	3 9/16	4	4 1/2	5	5 1/2	6 1/16	6 5/8	7 1/4	7 7/8	8 3/8
31	2 5/8	3	3 7/16	3 15/16	4 3/8	4 7/8	5 3/8	5 15/16	6 1/2	7	7 5/8	8 1/4
32	2 9/16	2 15/16	3 3/8	3 3/4	4 1/4	4 3/4	5 1/4	5 3/4	6 5/16	6 7/8	7 7/16	8
33	2 1/2	2 7/8	3 1/4	3 11/16	4 1/8	4 5/8	5 1/8	5 5/8	6 1/8	6 11/16	7 1/4	7 13/16
34	2 7/16	2 13/16	3 3/16	3 5/8	4	4 1/2	5	5 7/16	6	6 1/2	7 1/16	7 5/8
35	2 3/8	2 3/4	3 1/8	3 1/2	3 15/16	4 3/8	4 7/8	5 5/16	5 13/16	6 3/8	6 7/8	7 7/16
36	2 1/4	2 5/8	3	3 3/8	3 13/16	4 1/4	4 11/16	5 3/16	5 11/16	6 3/16	6 3/4	7 1/4
37	2 1/4	2 9/16	2 15/16	3 5/16	3 3/4	4 1/8	4 9/16	5 1/16	5 9/16	6 1/16	6 9/16	7 1/8
38	2 3/16	2 1/2	2 7/8	3 1/4	3 5/8	4 1/16	4 1/2	4 15/16	5 7/16	5 15/16	6 7/16	6 15/16
39	2 1/8	2 7/16	2 13/16	3 3/16	3 9/16	3 15/16	4 3/8	4 7/8	5 5/16	5 13/16	6 5/16	6 13/16
40	2 1/16	2 3/8	2 3/4	3 1/8	3 1/2	3 7/8	4 5/16	4 3/4	5 3/16	5 11/16	6 3/16	6 11/16
41	2 1/16	2 5/16	2 11/16	3	3 3/8	3 3/4	4 3/16	4 5/8	5 1/16	5 9/16	6	6 1/2
42	2	2 1/4	2 5/8	2 15/16	3 5/16	3 11/16	4 1/8	4 1/2	4 15/16	5 7/16	5 15/16	6 3/8
43	1 15/16	2 1/4	2 9/16	2 7/8	3 1/4	3 5/8	4	4 7/16	4 7/8	5 5/16	5 3/4	6 1/4
44	1 7/8	2 3/16	2 1/2	2 13/16	3 3/16	3 9/16	3 15/16	4 3/8	4 3/4	5 3/16	5 11/16	6 1/8
45	1 7/8	2 1/8	2 7/16	2 3/4	3 1/8	3 1/2	3 7/8	4 1/4	4 11/16	5 1/8	5 9/16	6
46	1 13/16	2 1/8	2 3/8	2 3/4	3 1/16	3 3/8	3 3/4	4 3/16	4 9/16	5	5 7/16	5 7/8
47	1 3/4	2 1/8	2 3/8	2 5/8	3	3 5/16	3 11/16	4 1/16	4 1/2	4 7/8	5 5/16	5 3/4
48	1 3/4	2	2 5/16	2 5/8	2 15/16	3 1/4	3 5/8	4	4 3/8	4 13/16	5 1/4	5 11/16

Table 3-10 (Continued). Gain Chart

B	1	2	3	4	5	6	7	8	9	10	11	12
49	0	0	1/16	3/16	1/4	5/16	7/16	11/16	13/16	1	1 3/16	1 1/2
50	0	0	1/16	3/16	1/4	5/16	7/16	11/16	13/16	1	1 3/16	1 7/16
51	0	0	1/16	3/16	1/4	5/16	7/16	5/8	13/16	1	1 3/16	1 7/16
52	0	0	1/16	3/16	3/16	5/16	7/16	5/8	3/4	1	1 1/8	1 3/8
53	0	0	1/16	3/16	3/16	5/16	7/16	5/8	3/4	15/16	1 1/8	1 3/8
54	0	0	1/16	3/16	3/16	5/16	7/16	5/8	3/4	15/16	1 1/8	1 5/16
55	0	0	1/16	3/16	3/16	5/16	7/16	5/8	3/4	15/16	1 1/16	1 5/16
56	0	0	1/16	3/16	3/16	5/16	7/16	5/8	11/16	7/8	1	1 1/4
57	0	0	1/16	3/16	3/16	5/16	7/16	5/8	11/16	7/8	1	1 1/4
58	0	0	1/16	3/16	3/16	5/16	7/16	5/8	11/16	7/8	1	1 1/4
59	0	0	1/16	3/16	3/16	5/16	7/16	9/16	11/16	7/8	1	1 3/16
60	0	0	1/16	3/16	3/16	5/16	7/16	9/16	11/16	7/8	1	1 3/16
61	0	0	1/16	3/16	3/16	1/4	3/8	9/16	11/16	13/16	15/16	1 3/16
62	0	0	1/16	3/16	3/16	1/4	3/8	9/16	5/8	13/16	15/16	1 3/16
63	0	0	1/16	1/8	3/16	1/4	3/8	9/16	5/8	13/16	15/16	1 1/8
64	0	0	1/16	1/8	3/16	1/4	3/8	1/2	5/8	13/16	15/16	1 1/8
65	0	0	1/16	1/8	3/16	1/4	3/8	1/2	5/8	3/4	15/16	1 1/8
66	0	0	1/16	1/8	3/16	1/4	3/8	1/2	5/8	3/4	7/8	1 1/16
67	0	0	1/16	1/8	3/16	1/4	3/8	1/2	5/8	3/4	7/8	1 1/16
68	0	0	1/16	1/8	3/16	1/4	3/8	1/2	5/8	3/4	7/8	1 1/16
69	0	0	1/16	1/8	3/16	1/4	3/8	1/2	5/8	3/4	7/8	1 1/16
70	0	0	1/16	1/8	3/16	1/4	3/8	1/2	5/8	3/4	7/8	1
71	0	0	1/16	1/8	3/16	1/4	5/16	1/2	9/16	5/8	3/4	1
72	0	0	1/16	1/8	3/16	1/4	5/16	1/2	9/16	5/8	3/4	1
73	0	0	0	1/8	3/16	1/4	5/16	3/8	9/16	5/8	3/4	1
74	0	0	0	1/8	3/16	1/4	5/16	3/8	9/16	5/8	3/4	1
75	0	0	0	1/8	3/16	1/4	5/16	3/8	9/16	5/8	3/4	1
76	0	0	0	1/8	3/16	1/4	5/16	3/8	9/16	5/8	3/4	15/16
77	0	0	0	1/8	3/16	1/4	5/16	3/8	9/16	5/8	3/4	15/16
78	0	0	0	1/8	3/16	1/4	5/16	3/8	9/16	5/8	3/4	15/16
79	0	0	0	1/8	3/16	1/4	5/16	3/8	9/16	5/8	3/4	15/16
80	0	0	0	1/8	3/16	1/4	5/16	3/8	9/16	5/8	3/4	7/8
81	0	0	0	0	1/8	3/16	1/4	5/16	9/16	5/8	11/16	7/8
82	0	0	0	0	1/8	3/16	1/4	5/16	9/16	5/8	11/16	7/8
83	0	0	0	0	1/8	3/16	1/4	5/16	9/16	5/8	11/16	7/8
84	0	0	0	0	1/8	3/16	1/4	5/16	9/16	5/8	11/16	7/8
85	0	0	0	0	1/8	3/16	1/4	5/16	9/16	5/8	11/16	13/16
86	0	0	0	0	1/8	3/16	1/4	5/16	9/16	5/8	11/16	13/16
87	0	0	0	0	1/8	3/16	1/4	5/16	9/16	5/8	11/16	13/16
88	0	0	0	0	1/8	3/16	1/4	5/16	9/16	5/8	11/16	13/16
89	0	0	0	0	1/8	3/16	1/4	5/16	9/16	5/8	11/16	13/16
90	0	0	0	0	1/8	3/16	1/4	5/16	9/16	5/8	11/16	13/16
91	0	0	0	0	1/8	3/16	1/4	5/16	9/16	5/8	11/16	3/4
92	0	0	0	0	1/8	3/16	1/4	5/16	9/16	5/8	11/16	3/4
93	0	0	0	0	1/8	3/16	1/4	5/16	9/16	5/8	11/16	3/4
94	0	0	0	0	1/8	3/16	1/4	5/16	9/16	5/8	11/16	3/4
95	0	0	0	0	1/8	3/16	1/4	5/16	9/16	5/8	11/16	3/4
96	0	0	0	0	1/8	3/16	1/4	5/16	9/16	5/8	11/16	3/4

Table 3-10 (*Concluded*). Gain Chart

B	13	14	15	16	17	18	19	20	21	22	23	24
49	1 11/16	1 15/16	2 1/4	2 9/16	2 7/8	3 1/4	3 9/16	3 15/16	4 5/16	4 11/16	5 1/8	5 9/16
50	1 11/16	1 15/16	2 3/16	2 1/2	2 13/16	3 1/8	3 1/2	3 7/8	4 1/4	4 5/8	5 1/16	5 7/16
51	1 5/8	1 7/8	2 3/16	2 7/16	2 3/4	3 1/8	3 7/16	3 13/16	4 3/16	4 9/16	4 15/16	5 3/8
52	1 5/8	1 7/8	2 1/8	2 3/8	2 11/16	3 1/16	3 3/8	3 3/4	4 1/8	4 7/16	4 13/16	5 5/16
53	1 9/16	1 7/8	2 1/8	2 3/8	2 11/16	3	3 5/16	3 11/16	4	4 3/8	4 3/4	5 1/4
54	1 9/16	1 13/16	2 1/16	2 5/16	2 5/8	2 15/16	3 1/4	3 5/8	3 15/16	4 5/16	4 3/4	5 1/8
55	1 1/2	1 3/4	2	2 5/16	2 9/16	2 7/8	3 3/16	3 9/16	3 7/8	4 1/4	4 5/8	5
56	1 1/2	1 3/4	2	2 1/4	2 9/16	2 7/8	3 1/8	3 1/2	3 13/16	4 3/16	4 9/16	4 15/16
57	1 1/2	1 11/16	1 15/16	2 3/16	2 1/2	2 13/16	3 1/16	3 7/16	3 3/4	4 1/8	4 1/2	4 7/8
58	1 7/16	1 11/16	1 15/16	2 3/16	2 1/2	2 3/4	3 1/16	3 3/8	3 11/16	4 1/16	4 7/16	4 3/4
59	1 7/16	1 5/8	1 7/8	2 1/8	2 7/16	2 11/16	3	3 5/16	3 5/8	4	4 3/8	4 11/16
60	1 7/16	1 5/8	1 7/8	2 1/8	2 3/8	2 5/8	2 15/16	3 1/4	3 9/16	3 15/16	4 1/4	4 5/8
61	1 3/8	1 5/8	1 13/16	2 1/16	2 5/16	2 5/8	2 15/16	3 3/16	3 1/2	3 7/8	4 3/16	4 9/16
62	1 3/8	1 9/16	1 13/16	2	2 5/16	2 9/16	2 7/8	3 3/16	3 7/16	3 13/16	4 1/8	4 1/2
63	1 3/8	1 9/16	1 3/4	2	2 1/4	2 1/2	2 11/16	3 1/8	3 7/16	3 11/16	4 1/16	4 7/16
64	1 5/16	1 1/2	1 3/4	2	2 1/4	2 1/2	2 3/4	3 1/16	3 3/8	3 11/16	4	4 3/8
65	1 5/16	1 1/2	1 3/4	1 15/16	2 3/16	2 7/16	2 3/4	3	3 5/16	3 5/8	4	4 3/16
66	1 1/4	1 7/16	1 11/16	1 15/16	2 3/16	2 7/16	2 11/16	3	3 1/4	3 9/16	3 7/8	4 1/4
67	1 1/4	1 7/16	1 11/16	1 7/8	2 1/8	2 3/8	2 11/16	2 15/16	3 3/16	3 1/2	3 7/8	4 3/16
68	1 1/4	1 7/16	1 5/8	1 7/8	2 1/8	2 5/16	2 5/8	2 7/8	3 3/16	3 1/2	3 13/16	4 1/8
69	1 1/4	1 7/16	1 5/8	1 13/16	2 1/8	2 5/16	2 9/16	2 7/8	3 1/8	3 3/8	3 3/4	4 1/16
70	1 3/16	1 3/8	1 5/8	1 13/16	2 1/16	2 1/4	2 9/16	2 13/16	3 1/8	3 5/16	3 3/4	4
71	1 3/16	1 3/8	1 9/16	1 13/16	2 1/16	2 3/16	2 1/2	2 3/4	3 1/16	3 5/16	3 11/16	3 15/16
72	1 3/16	1 3/8	1 9/16	1 3/4	2	2 3/16	2 1/2	2 3/4	3	3 5/16	3 5/8	3 7/8
73	1 3/16	1 5/16	1 9/16	1 3/4	2	2 3/16	2 7/16	2 3/4	3	3 1/4	3 9/16	3 13/16
74	1 3/16	1 5/16	1 1/2	1 3/4	2	2 3/16	2 7/16	2 11/16	2 15/16	3 1/4	3 1/2	3 13/16
75	1 1/8	1 5/16	1 1/2	1 3/4	2	2 1/8	2 3/8	2 11/16	2 15/16	3 3/16	3 1/2	3 3/4
76	1 1/8	1 5/16	1 1/2	1 11/16	1 15/16	2 1/8	2 3/8	2 5/8	2 7/8	3 1/8	3 7/16	3 11/16
77	1 1/8	1 1/4	1 7/16	1 5/8	1 15/16	2 1/8	2 5/16	2 9/16	2 7/8	3 1/8	3 3/8	3 11/16
78	1 1/8	1 1/4	1 7/16	1 5/8	1 7/8	2 1/16	2 5/16	2 9/16	2 13/16	3 1/16	3 3/8	3 5/8
79	1 1/16	1 1/4	1 7/16	1 5/8	1 7/8	2 1/16	2 5/16	2 1/2	2 3/4	3	3 5/16	3 5/8
80	1 1/16	1 1/4	1 7/16	1 5/8	1 13/16	2	2 1/4	2 1/2	2 3/4	3	3 1/4	3 9/16
81	1 1/16	1 1/4	1 3/8	1 9/16	1 13/16	2	2 1/4	2 7/16	2 11/16	3	3 1/4	3 1/2
82	1 1/16	1 3/16	1 3/8	1 9/16	1 3/4	1 15/16	2 1/4	2 7/16	2 11/16	2 15/16	3 3/16	3 7/16
83	1 1/16	1 3/16	1 3/8	1 9/16	1 3/4	1 15/16	2 3/16	2 3/8	2 5/8	2 7/8	3 3/16	3 7/16
84	1 1/16	1 3/16	1 3/8	1 9/16	1 3/4	1 7/8	2 3/16	2 3/8	2 5/8	2 7/8	3 1/8	3 3/8
85	1	1 3/16	1 5/16	1 1/2	1 3/4	1 7/8	2 1/8	2 5/16	2 9/16	2 13/16	3 1/8	3 5/16
86	1	1 3/16	1 5/16	1 1/2	1 11/16	1 7/8	2 1/8	2 5/16	2 9/16	2 13/16	3 1/8	3 5/16
87	1	1 1/8	1 5/16	1 1/2	1 11/16	1 7/8	2 1/8	2 5/16	2 9/16	2 3/4	3 1/16	3 1/4
88	1	1 1/16	1 1/4	1 1/2	1 5/8	1 7/8	2 1/16	2 1/4	2 1/2	2 3/4	3	3 1/4
89	1	1 1/16	1 1/4	1 7/16	1 5/8	1 13/16	2 1/16	2 1/4	2 1/2	2 11/16	3	3 3/16
90	1	1 1/16	1 1/4	1 7/16	1 5/8	1 13/16	2 1/16	2 3/16	2 7/16	2 11/16	2 15/16	3 3/16
91	15/16	1 1/16	1 1/4	1 3/8	1 5/8	1 3/4	2	2 3/16	2 7/16	2 5/8	2 15/16	3 1/8
92	15/16	1 1/16	1 1/4	1 3/8	1 9/16	1 3/4	2	2 3/16	2 3/8	2 5/8	2 7/8	3 1/8
93	15/16	1 1/16	1 3/16	1 3/8	1 9/16	1 3/4	1 15/16	2 1/8	2 3/8	2 9/16	2 7/8	3 1/16
94	15/16	1 1/16	1 3/16	1 3/8	1 9/16	1 11/16	1 15/16	2 1/8	2 5/16	2 9/16	2 13/16	3 1/16
95	15/16	1 1/16	1 3/16	1 5/16	1 1/2	1 11/16	1 15/16	2 1/8	2 5/16	2 9/16	2 13/16	3
96	15/16	1 1/16	1 3/16	1 5/16	1 1/2	1 11/16	1 7/8	2 1/16	2 5/16	2 1/2	2 3/4	3

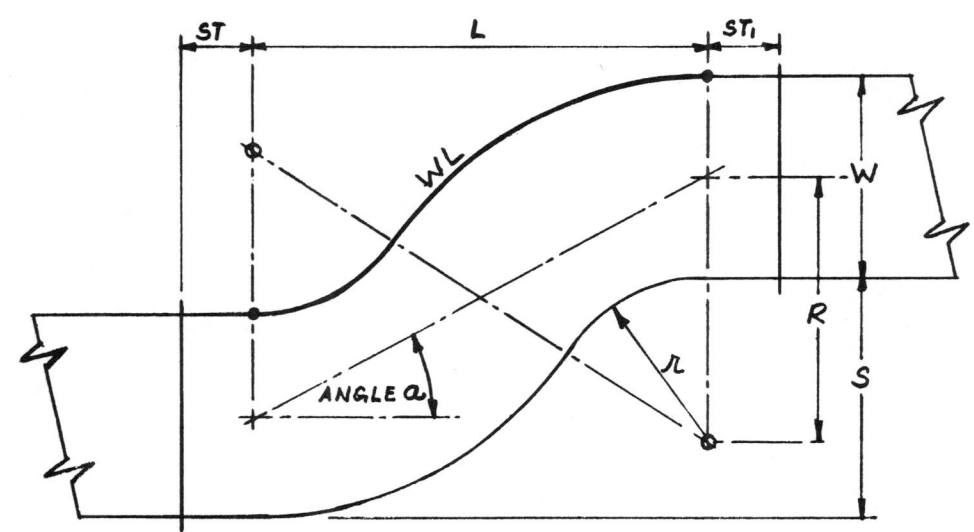

Fig. 3-1. Equal cheek ogee set.

(*Text continued from page 107.*)

28. From a fabrication standpoint, the wrapper length (47¾″) is advantageous because it can be cut from a 48″ sheet.

Centerline radius R, can also be calculated by simple arithmetic: $R = \dfrac{L^2 + S^2}{4S}$

For example:

Find R (to the nearest quarter inch), when $L = 40''$, and $S = 22''$

$$R = \frac{L^2 + S^2}{4S} = \frac{40^2 + 22^2}{4 \times 22} = 23.68$$

$$23.68 = 23\tfrac{3}{4}'' \quad Ans$$

Figures 3-2 to 3-5 are used for trigonometric calculations of rectangular duct offsets.

Figure 3-6 is for round offset calculations.

ROLLING BEVELS OR LATERALS

When a 90° elbow or tee is rolled to satisfy a difference in elevation and change in direction for round duct, a level plane of the duct centerline is resumed by using a bevel fitting of a degree equal to the amount of roll. (See Fig. 1-2.) For instance, a tee rolled up 30° would use a 30° recovery elbow to level off the duct.

If the centerline location of a duct run changes in plan or elevation view and the run remains parallel to the original, an offset, raise, drop, or rolling set would satisfy the condition. (See Figs. 3-6 and 3-7.)

However, when a take-off fitting at an angle of less than 90° to the duct run changes elevation and direction, the draftsman must compute the recovery angle of the fitting. A typical application for rolling laterals is shown in Fig. 3-8. Although plan view centerline, AC, identifies a horizontal duct, principles and uses of the formulas are for ducts in any position. Computations are in relationship to the take-off and roll angles, and dimensions of offset and change in elevation. (See Figs. 3-8 and 3-9.)

Legend Symbols

DAC = angle a = angle of lateral or bevel used

DAB = angle b = angle of rotation, or roll

E = change of elevation, centerline of main to centerline of branch

S = offset, centerline of main to centerline of branch

L = centerline distance parallel to main run

TL = True Length of AB along centerline of slope

angle x = true recovery angle of bevel elbow

Formulas

Tangent angle $b = \dfrac{E}{S}$

Sine angle $b \;\;\; = \dfrac{E}{R}$

$R = E$ (cosecant angle b)

$L = R$ (cotangent angle a)

$S = E$ (cotangent angle b)

$TL = E$ (cosecant angle b)(cotangent angle a) (secant angle a)

Cosine angle x = (sine angle a)(cosine angle b)

TRIGONOMETRY FUNCTIONS FOR STANDARD LATERALS

Angle a	Sine	Secant	Cotangent
30°	0.500	1.155	1.732
45°	0.707	1.414	1.000
60°	0.866	2.000	0.577

Figure 3-9 is an application of the previous formulas for standard laterals.

Example:

An 8″-diameter 45° lateral will be used as a branch take-off from a 16″-diameter high-pressure cold duct. In order to avoid the adjacent hot duct, the lateral will be rotated upward 14″ within a plan distance of 29″. All measurements are from centerline of branch to centerline of main. Calculate dimensions required for fabrication and installation.

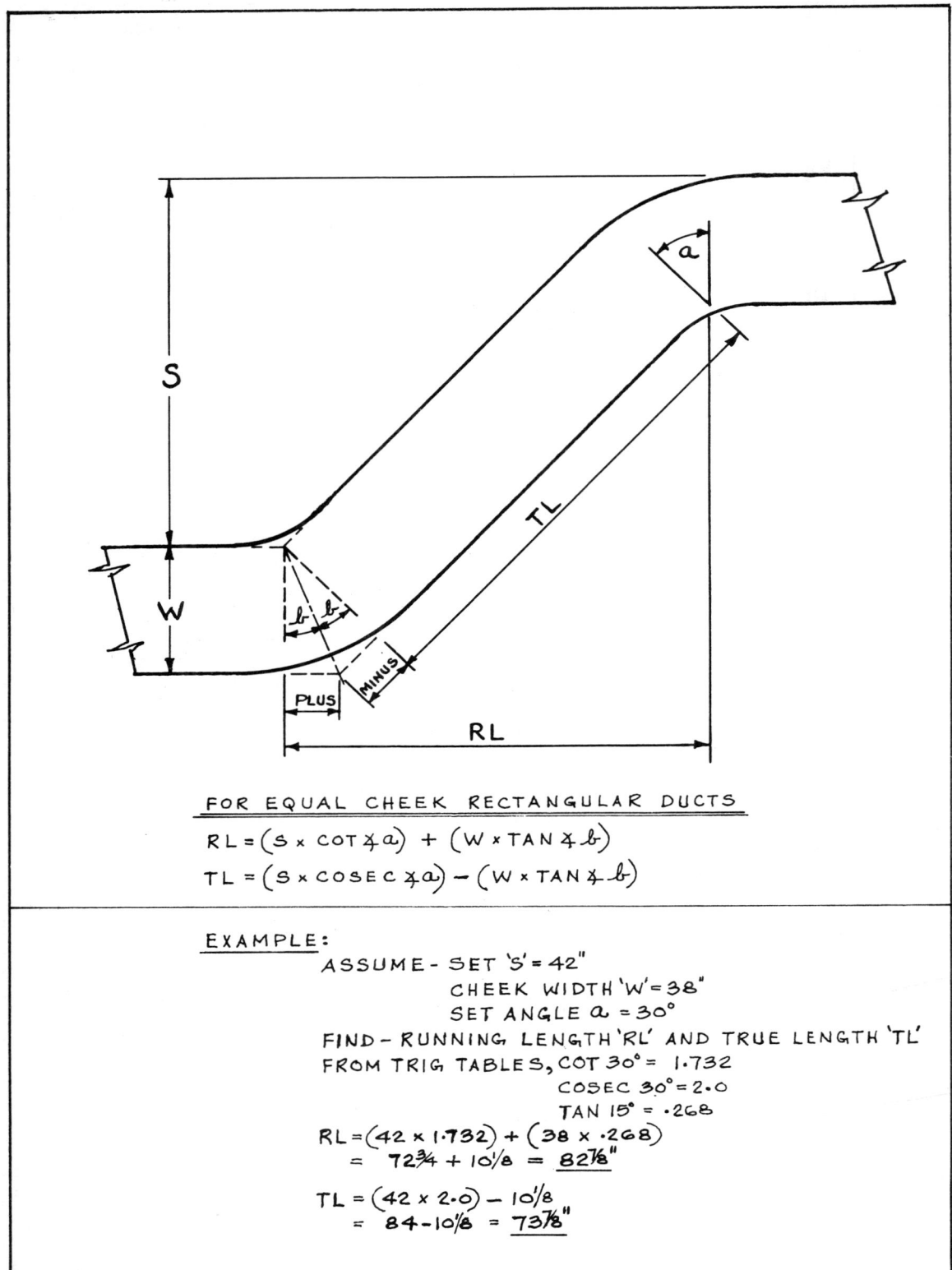

Fig. 3-2. For equal cheek rectangular ducts.

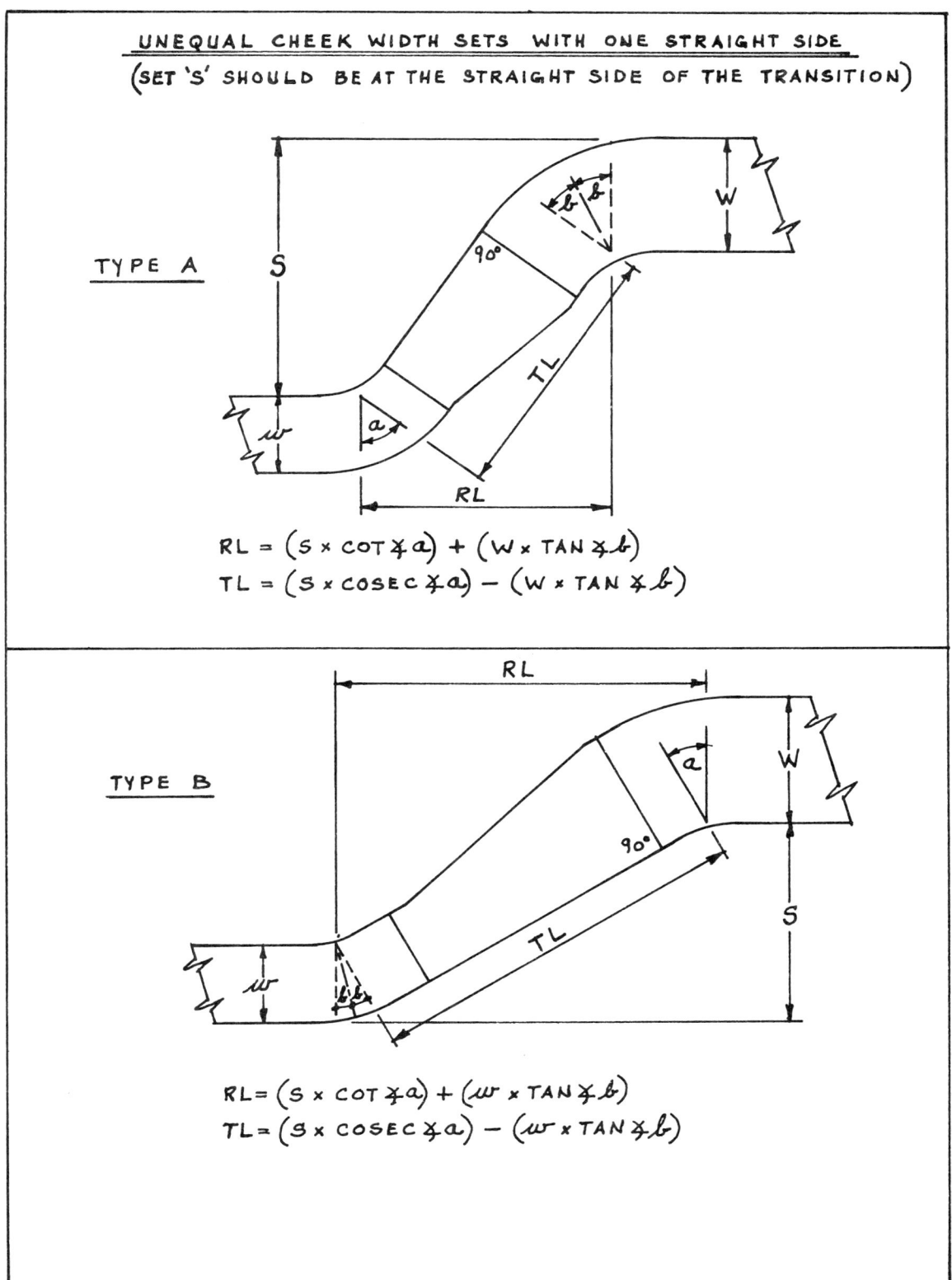

Fig. 3-3. Unequal cheek width sets with one straight side.

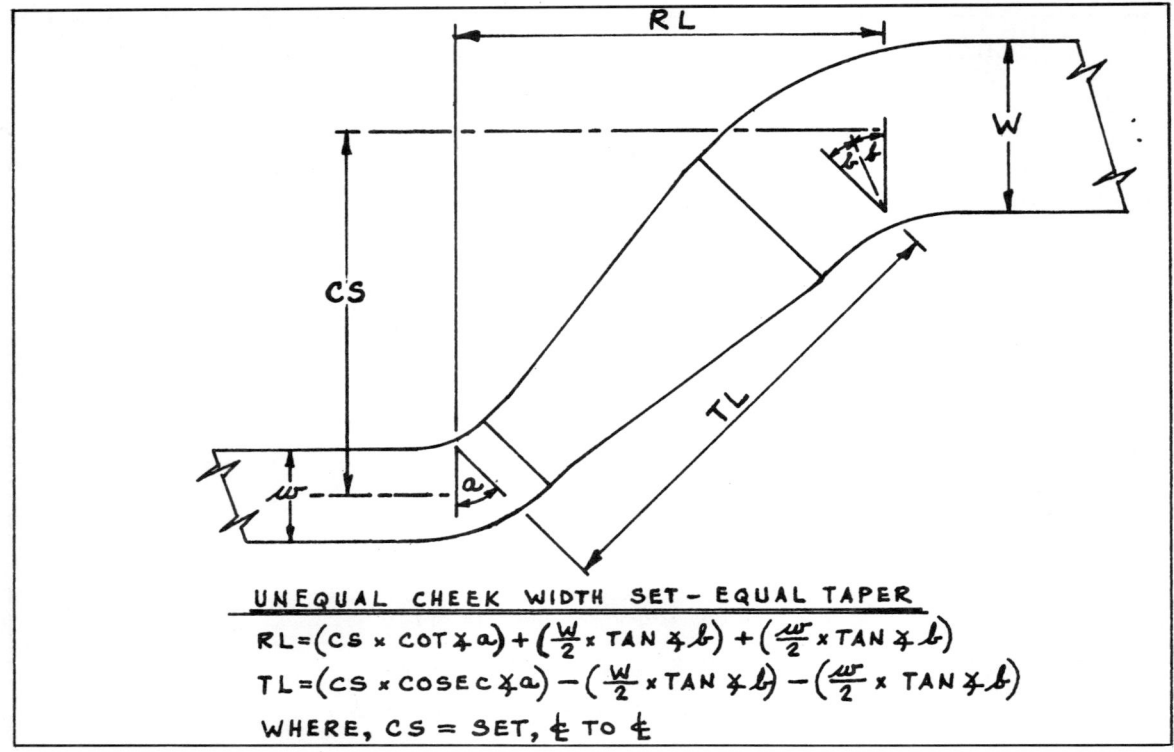

Fig. 3-4 (Top). Unequal cheek width set — equal taper.

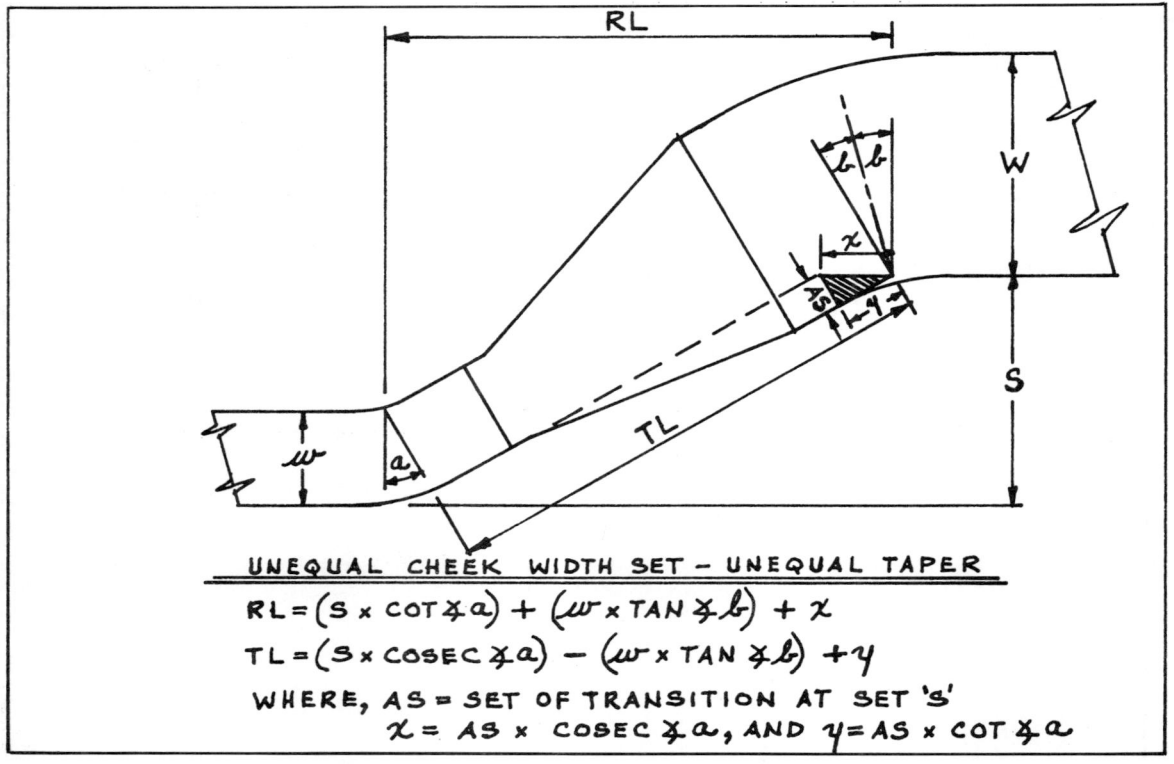

Fig. 3-5 (Bottom). Unequal cheek width set — unequal taper.

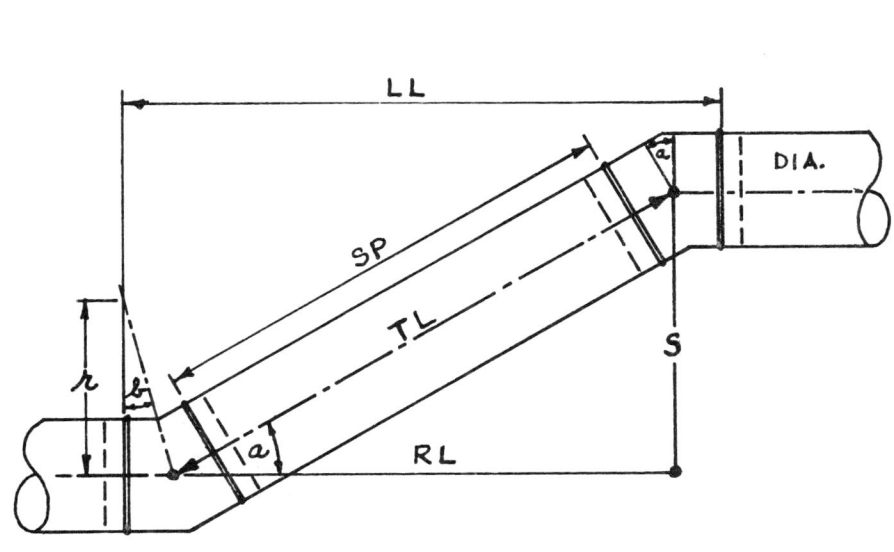

ROUND DUCT OFFSET

RL = S × COT ½∠a
TL = S × COSEC ½∠a
LL = RL + 2(TAN ½∠b × ₰)
SP = TL − 2(TAN ½∠b × ₰)

∠a = ANGLE OF SET
∠b = ONE HALF ∠a
₰ = CENTERLINE RADIUS OF BEVEL FITTING
S = SET
RL = RUNNING LENGTH
TL = TRUE LENGTH
LL = INSTALLED LENGTH
SP = INTERMEDIATE STRAIGHT PIPE LENGTH

EXAMPLE —

USING STANDARD FITTINGS AND SPIRAL CONDUIT, FIND LL AND SP WHEN, DUCT = 14" DIA.

∠a = 45°
₰ = 21"
S = 5'-2

2(TAN ½∠b × ₰) = 2(.414 × 21) = 17½"
RL = S × COT ½∠a = 62 × 1.0 = 62"
LL = RL + 2(TAN ½∠b × ₰) = 62 + 17½ = 79½" = 6'-7½

TL = S × COSEC ½∠a = 62 × 1.414 = 87¾"
SP = TL − 2(TAN ½∠b × ₰) = 87¾ − 17½ = 70¼ = 5'-10¼

LL = 6'-7½
SP = 5'-10¼

Fig. 3-6. Round duct offset.

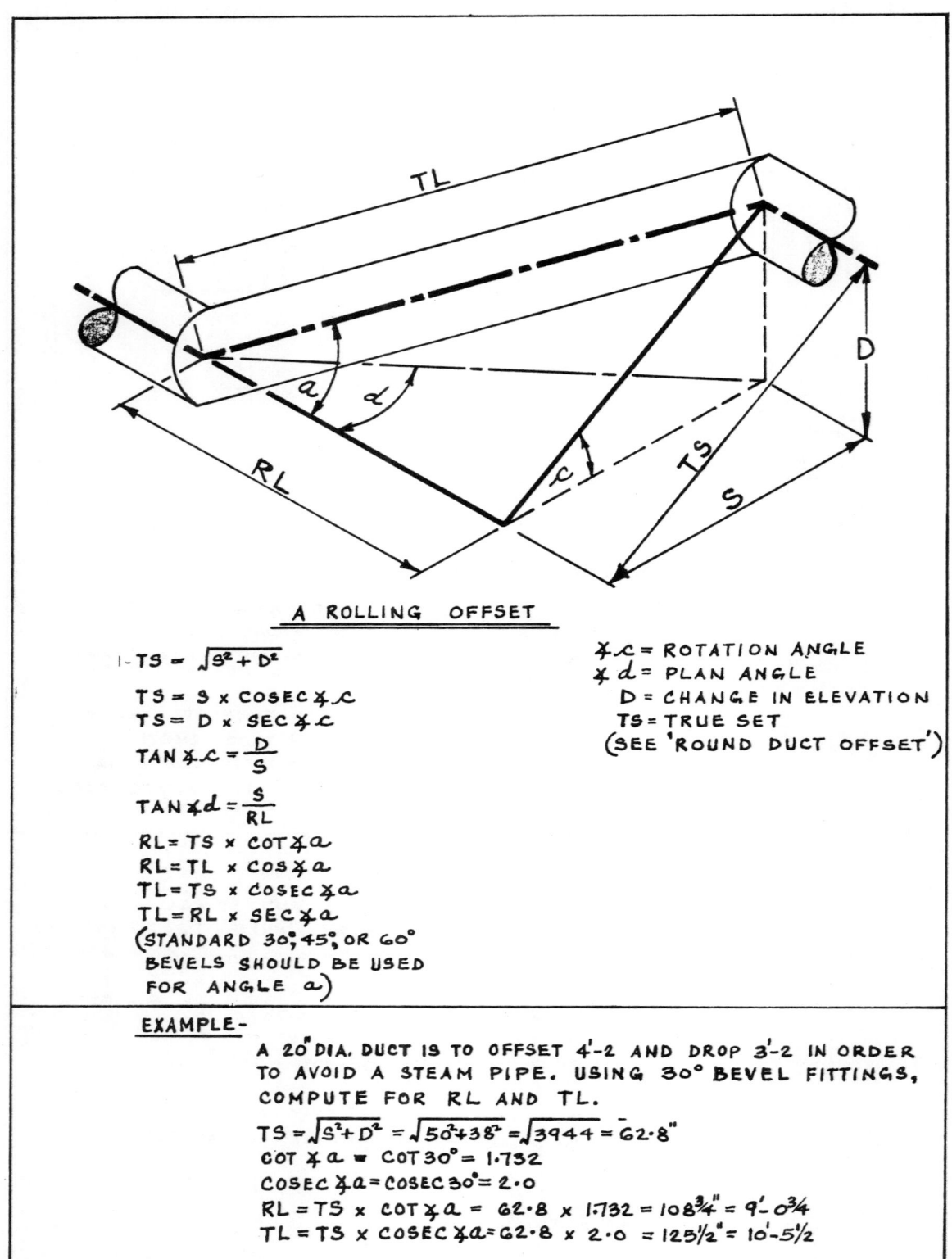

Fig. 3-7. Computing a rolling offset.

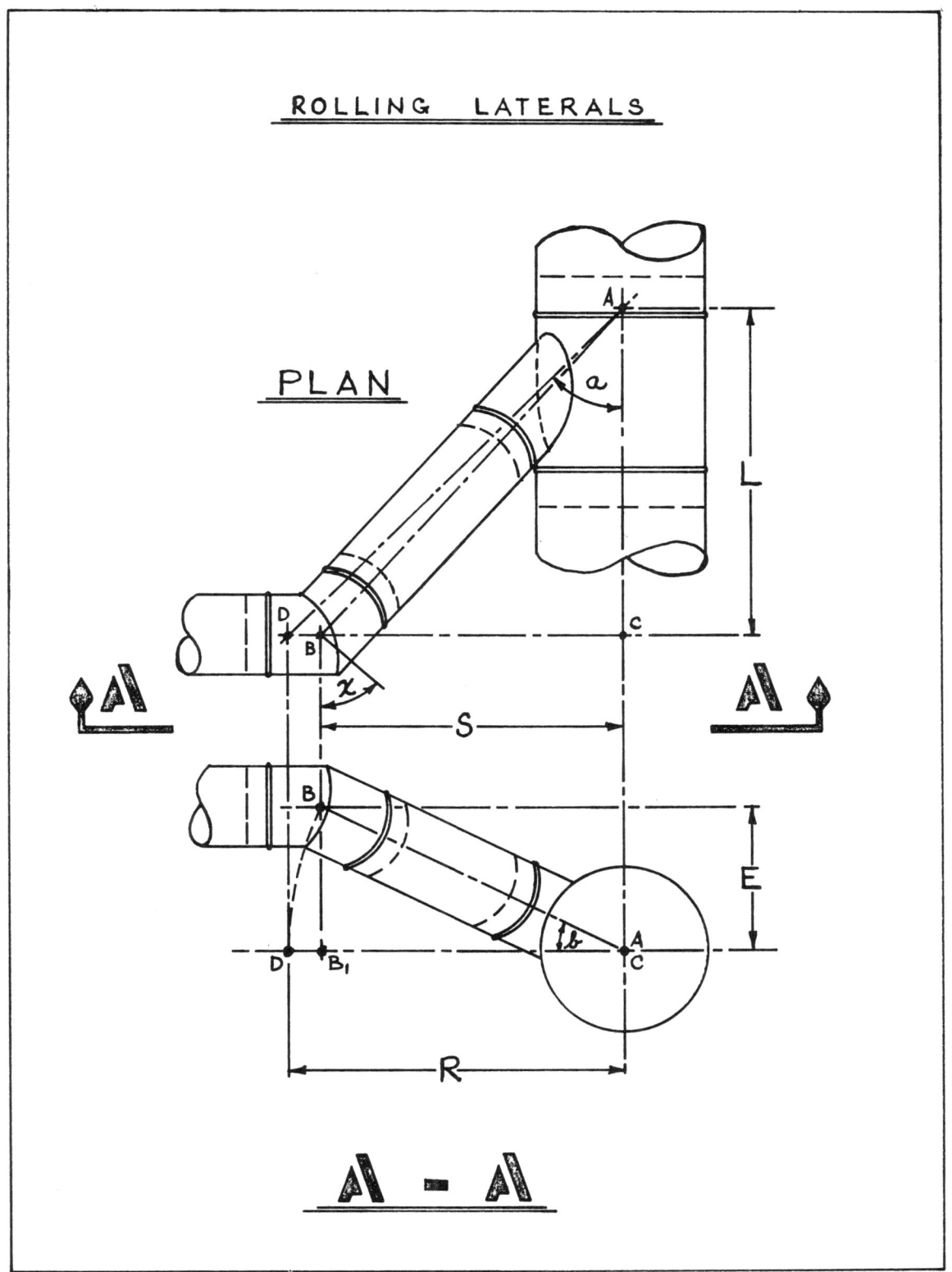

Fig. 3-8. Plan and elevation of a rolling lateral.

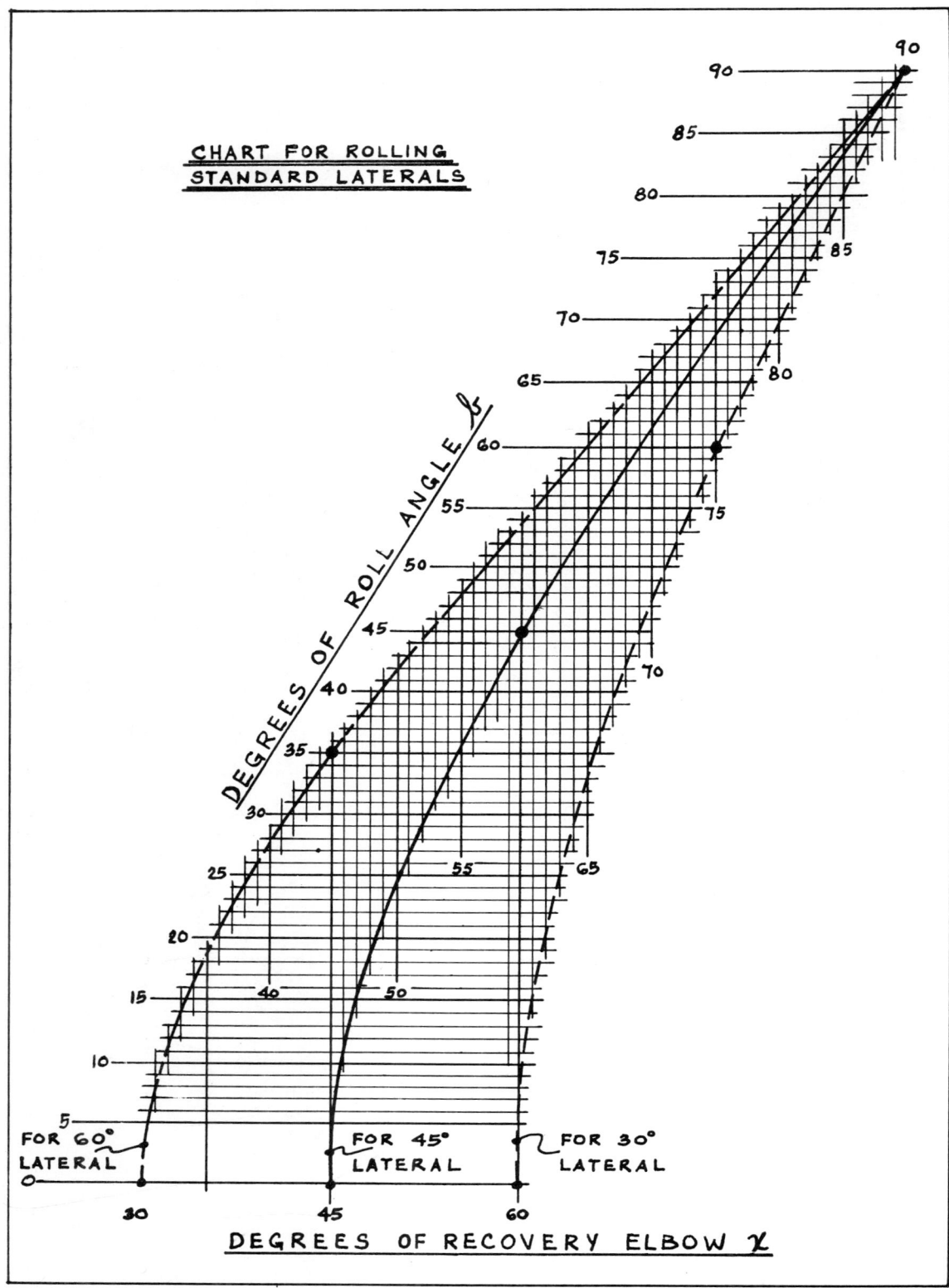

Fig. 3-9. Chart for rolling standard laterals.

Dimensions to be calculated: *TL*, *L*, and angle *x*

For angle a: = 45°

Cotangent a = 1.00

Secant a = 1.414

Sine a = 0.707

$E = 14''$ and $S = 29''$

For angle b:

Tangent $= \dfrac{E}{S} = \dfrac{14}{29} = .4828 = 25°\text{-}46'$

Cosecant = 2.30

Cosine = 0.90

$R = E(\text{cosecant angle } b)$
$= 14(2.3) = 32.2$

$L = R(\text{cotangent angle } a)$
$= 32.2(1.0) = 32.2$

$TL = E(\text{cosecant angle } b)(\text{cotangent angle } a)$
$\qquad (\text{secant angle } a)$
$= 14(2.3 \times 1.0 \times 1.414)$

$TL = 32.2 \times 1.414 = 45.53 = 45\tfrac{1}{2}''$

Cosine angle x = (sine angle a)(cosine angle b)
$= (0.707)(0.9) = 0.6363$

angle $x = 50°\text{-}30'$

See Graph in Fig. 3-9 for Check of Calculations

Angle $a = 45°$

Angle $b = 25°\text{-}46'$

Recovery angle $x = 50\tfrac{1}{2}°$

Figure 3-9 indicates convenient combinations of rolling laterals or bevels such as: a 60° take-off rotated 35° uses a standard 45° recovery elbow; a 45° take-off rotated 45° uses a 60° recovery elbow; and a 30° take-off rotated 60° uses a 75° recovery elbow.

ANGLE OF A HIP

When a duct run is required to follow the slope of a hip corner in a structure, the draftsman must calculate the true interior angle of the hip so that clearances can be maintained for the desired duct location. Figure 3-10 shows a plan view of a typical 90° corner layout.

By Geometry

Draw plan view *ACDFEB* as in Fig. 3-10.

Connect points *A-F*; *AF* is plan of hip miter line.

At point *F*, erect a perpendicular to line *AF* and extend the line to intersect line *AB* at point *G*, and line *AC* at point *H*.

From point *F*, set off distance *h* as *FJ*.

Connect *AJ*.

AFJ is true section through hip miter line, as *M-M*.

From point *F*, erect perpendicular to line *AJ*, as *FK*.

From point *F* as center, set compass to *FK* and swing an arc to intersect line *AF* at point *L*.

True hip angle is *GLH*.

By Trigonometry

Tangent angle $a = \dfrac{w}{W}$

Tangent angle $b = \dfrac{AF}{h} = \dfrac{w(\text{cosecant angle } a)}{h}$

Tangent angle x = (secant angle a)(sine angle a)
$\qquad (\text{secant angle } b)$

Tangent angle y = (cotangent angle a)(secant angle b)

True hip angle GLH = angle x + angle y

Example:

Using Fig. 3-10, find the true interior angle of a hip by trigonometry, when $w = 18''$, $W = 28''$, and $h = 30''$.

For angle a:

Tangent $= \dfrac{w}{W} = \dfrac{18}{28} = .6428 = 32°\text{-}44'$

Cosecant = 1.849

Secant = 1.888

Sine = 0.54073

Cotangent = 1.5557

For angle b:

Tangent $= \dfrac{w(\text{cosecant angle } a)}{h} = \dfrac{18 \times 1.849}{30}$

Tangent = 1.109 = 47°-58′

Secant = 1.4935

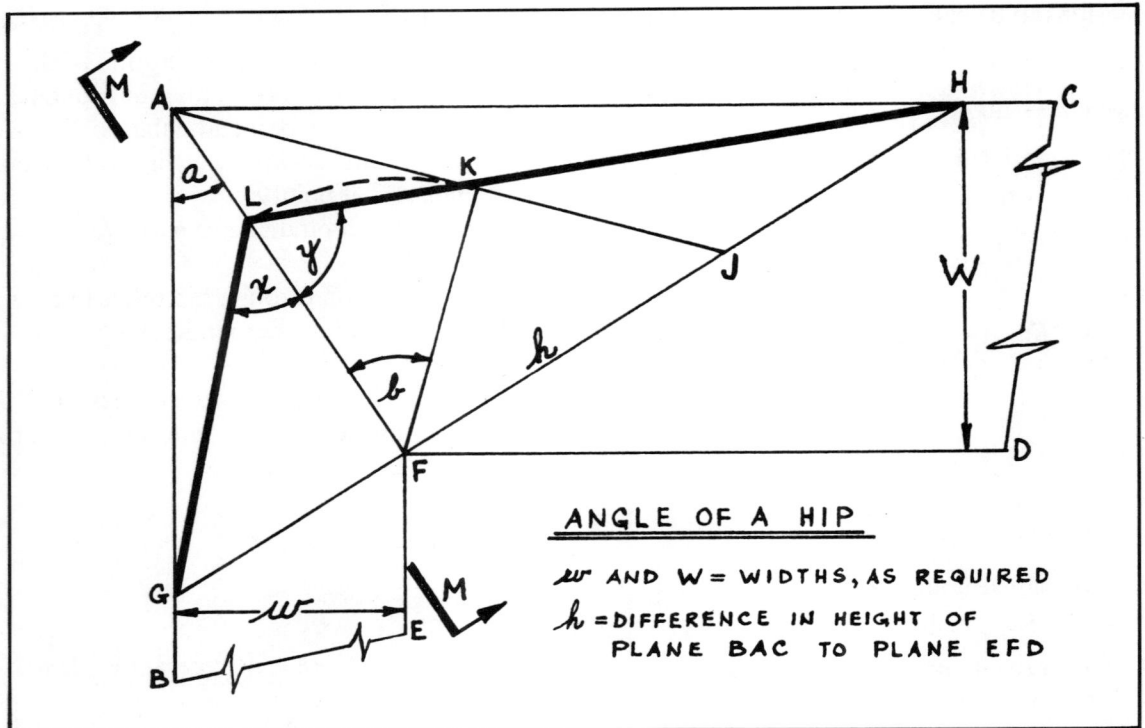

Fig. 3-10 (Top). To develop the true angle of a hip.

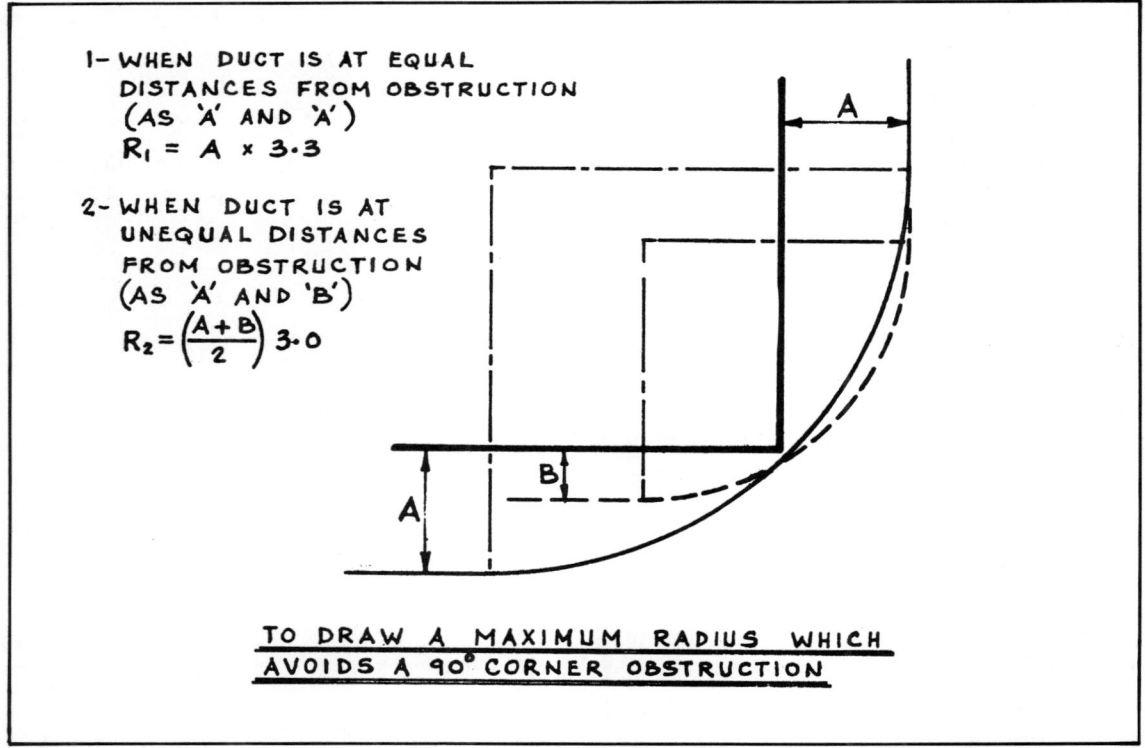

Fig. 3-11 (Bottom). Radius to clear a corner.

Tangent angle x = (secant angle a)(sine angle a)
        (secant angle b)
    = (1.1888)(0.54073)(1.4935)
    = 0.9575

    angle x = 43°-45′

Tangent angle y = (cotangent angle a)(secant angle b)
    = (1.5557)(1.4935)
    = 2.3244

    angle y = 66°-43′

True angle GLH = (angle x) + (angle y)
    = 43°-45′ + 66°-43′
    = 110°-28′

    angle GLH = 110°   *Ans*

When the throat radius of an elbow must clear the corner of a column or beam, one of two methods (as shown in Fig. 3-11) can be used.

CHAPTER 4

Duct Design

Although design drawings prepared by a consulting engineer are used for reference by sheet metal draftsmen, it is often necessary to resize or reroute ducts, calculate branch-neck tap sizes, or make other changes in the air-handling systems. Knowledge of the principles of duct-system design and their application will enable draftsmen to efficiently accomplish these changes during preparation of shop drawings.

The amount of air to be circulated in a duct system for heating, for air-conditioning, and for ventilation must be calculated separately. Consideration must be given to the volume of the space, limits of perceptibility of air movement, occupancy or space usage, noise factors, ventilation requirements, and heating and cooling loads. Comprehensive explanations for determining heat and moisture gains or losses may be found in various published texts.

The British Thermal Unit (Btu) is the measuring standard for heat transfer. One Btu is the amount of heat needed to raise the temperature of one pound of water one degree Fahrenheit at atmospheric pressure.

Heating and cooling loads are measured in Btuh, the letter h indicating a time factor of one hour.

Air Quantity for Heating

The heating load is the sum of the heat-transmission losses through the building enclosure, plus the heat required for ventilation and infiltration, minus any significant internal-heat gains.

To satisfy this load, the supply-air volume Q, in cubic feet per minute (cfm) can be expressed approximately as:

$$Q = \frac{H}{1.085 \Delta T}$$

where:

H = heat load in space, Btuh
ΔT = temperature difference between supply air and the temperature to be maintained in the space.

Example 1:

Find the amount of air required to heat a house which has a heat loss of 120,000 Btuh, when the house temperature is to be maintained at 70F, and the supply-air temperature is 110F.

where:

$H = 120,000$
$\Delta T = 110 - 70 = 40$

$$Q = \frac{H}{(1.085)\Delta T}$$

$$Q = \frac{120,000}{(1.085)40}$$

$Q = 2,764$ cfm   *Ans*

Example 2:

How much heat is transferred from a hot water coil to a volume of air of 2,000 cfm which shall raise the air temperature from 40F to 120F?

where:

$Q = 2,000$ cfm
$\Delta T = 120 - 40 = 80$

$$Q = \frac{H}{(1.085)\Delta T}$$

$$H = Q(1.085)\Delta T$$
$$= 2,000 \times 1.085 \times 80$$

$H = 174,000$ Btuh   *Ans*

Air Quantity for Cooling

The cooling load is the sum of the external and internal heat and moisture gains.

DUCT DESIGN

To find the supply-air quantity through a bank of cooling coils:

$$Q = \frac{H}{4.5\,(iwb-fwb)}$$

where:
- iwb = initial wet-bulb temperature, total heat content
- fwb = final wet-bulb temperature, total heat content

Table 4-1 shows the heat removed per pound of dry air Btu, at various wet-bulb temperatures.

Example:

Calculate the amount of supply air required to satisfy an office space which has a total heat gain

Table 4-1. Heat Removed per Pound of Dry Air*

Final Wet Bulb Temp., F	Initial Wet Bulb Temperature, F															
	60	61	62	63	64	65	66	67	68	69	70	71	72	73	74	75
	Heat Removed per Lb. Dry Air, Btu															
40	11.20	11.89	12.59	13.31	14.04	14.79	15.56	16.34	17.14	17.96	18.80	19.66	20.54	21.43	22.36	23.30
41	10.74	11.43	12.13	12.85	13.58	14.33	15.10	15.88	16.68	17.50	18.34	19.20	20.08	20.97	21.90	22.84
42	10.26	10.95	11.65	12.37	13.10	13.85	14.62	15.40	16.20	17.02	17.86	18.72	19.60	20.49	21.42	22.36
43	9.78	10.47	11.17	11.89	12.62	13.37	14.14	14.92	15.72	16.54	17.38	18.24	19.12	20.01	20.94	21.88
44	9.29	9.98	10.68	11.40	12.13	12.88	13.65	14.43	15.23	16.05	16.89	17.75	18.63	19.52	20.45	21.39
45	8.79	9.48	10.18	10.90	11.63	12.38	13.15	13.93	14.73	15.55	16.39	17.25	18.13	19.02	19.95	20.89
46	8.28	8.97	9.67	10.39	11.12	11.87	12.64	13.42	14.22	15.04	15.88	16.74	17.62	18.51	19.44	20.38
47	7.76	8.45	9.15	9.87	10.60	11.35	12.12	12.90	13.70	14.52	15.36	16.22	17.10	17.99	18.92	19.86
48	7.23	7.92	8.62	9.34	10.07	10.82	11.59	12.37	13.17	13.99	14.83	15.69	16.57	17.46	18.39	19.33
49	6.69	7.38	8.08	8.80	9.53	10.28	11.05	11.83	12.63	13.45	14.29	15.15	16.03	16.92	17.85	18.79
50	6.14	6.83	7.53	8.25	8.98	9.73	10.50	11.28	12.08	12.90	13.74	14.60	15.48	16.37	17.30	18.24
51	5.58	6.27	6.97	7.69	8.42	9.17	9.94	10.72	11.52	12.34	13.18	14.04	14.92	15.81	16.74	17.68
52	5.01	5.70	6.40	7.12	7.85	8.60	9.37	10.15	10.95	11.77	12.61	13.47	14.35	15.24	16.17	17.11
53	4.43	5.12	5.82	6.54	7.27	8.02	8.79	9.57	10.37	11.19	12.03	12.89	13.77	14.66	15.59	16.53
54	3.83	4.52	5.22	5.94	6.67	7.42	8.19	8.97	9.77	10.59	11.43	12.29	13.17	14.06	14.99	15.93
55	3.23	3.92	4.62	5.34	6.07	6.82	7.59	8.37	9.17	9.99	10.83	11.69	12.57	13.46	14.39	15.33
56	2.61	3.30	4.00	4.72	5.45	6.20	6.97	7.75	8.55	9.37	10.21	11.07	11.95	12.84	13.77	14.71
57	1.98	2.67	3.37	4.09	4.82	5.57	6.34	7.12	7.92	8.74	9.58	10.44	11.32	12.21	13.14	14.08
58	1.33	2.02	2.72	3.44	4.17	4.92	5.69	6.47	7.27	8.09	8.93	9.79	10.67	11.56	12.49	13.43
59	.67	1.36	2.06	2.78	3.51	4.26	5.03	5.81	6.61	7.43	8.27	9.13	10.01	10.90	11.83	12.77
60		.69	1.39	2.11	2.84	3.59	4.36	5.14	5.94	6.76	7.60	8.46	9.34	10.23	11.16	12.10
61			.70	1.42	2.15	2.90	3.67	4.45	5.25	6.07	6.91	7.77	8.65	9.54	10.47	11.41
62				.72	1.45	2.20	2.97	3.75	4.55	5.37	6.21	7.07	7.95	8.84	9.97	10.71
63					.73	1.48	2.25	3.03	3.83	4.65	5.49	6.35	7.23	8.12	9.05	9.99
64						.75	1.52	2.30	3.10	3.92	4.76	5.62	6.50	7.39	8.32	9.26
65							.77	1.55	2.35	3.17	4.01	4.87	5.75	6.64	7.57	8.51
66								.78	1.58	2.40	3.24	4.10	4.98	5.87	6.80	7.74
67									.80	1.62	2.46	3.32	4.20	5.09	6.02	6.96
68										.82	1.66	2.52	3.40	4.29	5.22	6.16
69											.84	1.70	2.58	3.47	4.40	5.34
70												.86	1.74	2.63	3.56	4.50
71													.88	1.77	2.70	3.64
72														.89	1.82	2.76
73															.93	1.87
74																.94

* Reproduced from *Handbook of Air Conditioning, Heating and Ventilating*. Edited by Clifford Strock and Richard L. Koral. New York: Industrial Press Inc., 1965.

Table 4-1 (Continued). Heat Removed per Pound of Dry Air

Final Wet Bulb Temp., F.	Initial Wet Bulb Temperature, F.														
	76	77	78	79	80	81	82	83	84	85	86	87	88	89	90
	Heat Removed per Lb. Dry Air, Btu														
40	24.26	25.25	26.26	27.30	28.36	29.44	30.56	31.70	32.87	34.07	35.30	36.56	37.85	39.18	40.54
41	23.80	24.79	25.80	26.84	27.90	28.98	30.10	31.24	32.41	33.61	34.84	36.10	37.39	38.72	40.08
42	23.32	24.31	25.32	26.36	27.42	28.50	29.62	30.76	31.93	33.13	34.36	35.62	36.91	38.24	39.60
43	22.84	23.83	24.84	25.88	26.94	28.02	29.14	30.28	31.45	32.65	33.88	35.14	36.43	37.76	39.12
44	22.35	23.34	24.35	25.39	26.45	27.53	28.65	29.79	30.96	32.16	33.39	34.65	35.94	37.27	38.63
45	21.85	22.84	23.85	24.89	25.95	27.03	28.15	29.29	30.46	31.66	32.89	34.15	35.44	36.77	38.13
46	21.34	22.33	23.34	24.38	25.44	26.52	27.64	28.78	29.95	31.15	32.38	33.64	34.93	36.26	37.62
47	20.82	21.81	22.82	23.86	24.92	26.00	27.12	28.26	29.43	30.63	31.86	33.12	34.41	35.74	37.10
48	20.29	21.28	22.29	23.33	24.39	25.47	26.59	27.73	28.90	30.10	31.33	32.59	33.88	35.21	36.57
49	19.75	20.74	21.75	22.79	23.85	24.93	26.05	27.19	28.36	29.56	30.79	32.05	33.34	34.67	36.03
50	19.20	20.19	21.20	22.24	23.30	24.38	25.50	26.64	27.81	29.01	30.24	31.50	32.79	34.12	35.48
51	18.64	19.63	20.64	21.68	22.74	23.82	24.94	26.08	27.25	28.45	29.68	30.94	32.23	33.56	34.92
52	18.07	19.06	20.07	21.11	22.17	23.25	24.37	25.51	26.68	27.88	29.11	30.37	31.66	32.99	34.35
53	17.49	18.48	19.49	20.53	21.59	22.67	23.79	24.93	26.10	27.30	28.53	29.79	31.08	32.41	33.77
54	16.89	17.88	18.89	19.93	20.99	22.07	23.19	24.33	25.50	26.70	27.93	29.19	30.48	31.81	33.17
55	16.29	17.28	18.29	19.33	20.39	21.47	22.59	23.73	24.90	26.10	27.33	28.59	29.88	31.21	32.57
56	15.67	16.66	17.67	18.71	19.77	20.85	21.97	23.11	24.28	25.48	26.71	27.97	29.26	30.59	31.95
57	15.04	16.03	17.04	18.08	19.14	20.22	21.34	22.48	23.65	24.85	26.08	27.34	28.63	29.96	31.32
58	14.39	15.38	16.39	17.43	18.49	19.57	20.69	21.83	23.00	24.20	25.43	26.69	27.98	29.31	30.67
59	13.73	14.72	15.73	16.77	17.83	18.91	20.03	21.17	22.34	23.54	24.77	26.03	27.32	28.65	30.01
60	13.06	14.05	15.06	16.10	17.16	18.24	19.36	20.50	21.67	22.87	24.10	25.36	26.65	27.98	29.34
61	12.37	13.36	14.37	15.41	16.47	17.55	18.67	19.81	20.98	22.18	23.41	24.67	25.96	27.29	28.65
62	11.67	12.66	13.67	14.71	15.77	16.85	17.97	19.11	20.28	21.48	22.71	23.97	25.26	26.59	27.95
63	10.95	11.94	12.95	13.99	15.05	16.13	17.25	18.39	19.56	20.76	21.99	23.25	24.54	25.87	27.23
64	10.22	11.21	12.22	13.26	14.32	15.40	16.52	17.66	18.83	20.03	21.26	22.52	23.81	25.14	26.50
65	9.47	10.46	11.47	12.51	13.57	14.65	15.77	16.91	18.08	19.28	20.51	21.77	23.06	24.39	25.75
66	8.70	9.69	10.70	11.74	12.80	13.88	15.00	16.14	17.31	18.51	19.74	21.00	22.29	23.62	24.98
67	7.92	8.91	9.92	10.96	12.02	13.10	14.22	15.36	16.53	17.73	18.96	20.22	21.51	22.84	24.20
68	7.12	8.11	9.12	10.16	11.22	12.30	13.42	14.56	15.73	16.93	18.16	19.42	20.71	22.04	23.40
69	6.30	7.29	8.30	9.34	10.40	11.48	12.60	13.74	14.91	16.11	17.34	18.60	19.89	21.22	22.58
70	5.46	6.45	7.46	8.50	9.56	10.64	11.76	12.90	14.07	15.27	16.50	17.76	19.05	20.38	21.74
71	4.60	5.59	6.60	7.64	8.70	9.78	10.90	12.04	13.21	14.41	15.64	16.90	18.19	19.52	20.88
72	3.72	4.71	5.72	6.76	7.82	8.90	10.02	11.16	12.33	13.53	14.76	16.02	17.31	18.64	20.00
73	2.83	3.82	4.83	5.87	6.93	8.01	9.13	10.27	11.44	12.64	13.87	15.13	16.42	17.75	19.11
74	1.90	2.89	3.90	4.94	6.00	7.08	8.20	9.34	10.51	11.71	12.94	14.20	15.49	16.82	18.18
75	.96	1.95	2.96	4.00	5.06	6.14	7.26	8.40	9.57	10.77	12.00	13.26	14.55	15.88	17.24
76		.99	2.00	3.04	4.10	5.18	6.30	7.44	8.61	9.81	11.04	12.30	13.59	14.92	16.28
77			1.01	2.05	3.11	4.19	5.31	6.45	7.62	8.82	10.05	11.31	12.60	13.93	15.29
78				1.04	2.10	3.18	4.30	5.44	6.61	7.81	9.04	10.30	11.59	12.92	14.28
79					1.06	2.14	3.26	4.40	5.57	6.77	8.00	9.26	10.55	11.88	13.24
80						1.08	2.20	3.34	4.51	5.71	6.94	8.20	9.49	10.82	12.18

DUCT DESIGN

of 72,000 Btuh. The initial wet-bulb temperature entering the cooling coils is 62F, and the air leaving the bank of coils is 55F wet bulb.

From Table 4-1: Difference of total heat content, iwb of 62 and fwb of 55 = 4.62.

$$\text{cfm} = \frac{72,000}{4.5 \, (\text{iwb-fwb})}$$
$$= \frac{72,000}{4.5 \times 4.62}$$
$$\text{cfm} = 3,450 \quad Ans$$

Air Quantity Required for Ventilation

The air-change method is often used to determine ventilation-system capacities. In this method, the volume of the space is computed and divided by a rate-of-air-change factor. The result will be the cubic feet per minute (cfm) of ventilation air.

$$\text{Air quantity} = \frac{\text{Volume of space}}{\text{Number of minutes of air change}}$$

Table 4-2. Air Changes in Various Spaces

Minutes per Change of Air	Type of Space
1	Boiler Room, Projection Booth
2	Engine Room, Kitchen
3	Bakeries, Banquet Hall, Laundry, Packing House, Toilet
4	Printing Plant
5	Auditorium, Bowling Alley, Cafeteria, Cleaning and Dyeing Room, Clubroom, Church, Dance Hall, Dining Room, Foundry, Laboratory, Meeting Room, Mill, Office, Recreation Room, Rest Room, Store
6	Classroom, Creamery, Garage, Machine Shop, Factory
10	Ship's Hold, Tunnel

Example:

How much ventilation air must be circulated in a garage which is 60 ft. long, 35 ft. wide, and has a 10-ft ceiling?

From Table 4-2: air change is 6 minutes

$$\text{Air quantity} = \frac{\text{Volume of space}}{\text{Number of minutes of air change}}$$
$$= \frac{60 \times 35 \times 10}{6}$$
$$= 3,500 \text{ cfm} \quad Ans$$

Air Pressures in Buildings

New buildings are being designed with a minimum of windows or with windows that have a fixed sash and must rely on mechanical ventilation for occupancy comfort. Because of this relatively airtight construction feature, allowances must be considered for air pressures in the space. When outside air is introduced, a positive pressure buildup is prevented by a spill air-duct system or relief vents. Conversely, a building with an extensive exhaust system develops a negative pressure unless air is recirculated through an air reclamation system or if make-up air is provided.

Excess negative pressure in a building may create adverse infiltration at windows and doors, poor efficiency of the exhaust system, back drafting of flues and combustion equipment, and other problems. Make-up air of sufficient quantity to equalize this pressure, should be supplied through the heating or air-conditioning apparatus so that comfort conditions are maintained in the space. It may be economically advantageous to minimize the volume of outside makeup air by recirculating a predetermined quantity of air from the exhaust system. This reduces the load on the heat-transfer apparatus.

Ventilation air for occupancy comfort may be outside air treated through the conditioning apparatus or recirculated air which as been reactivated and decontaminated by a charcoal filter system. The amount of air required per person varies from a minimum of 5 cfm for no-smoking areas, to a maximum of 30 cfm where heavy smoking may occur.

Industrial Exhaust Systems

The use of air cleaning, moving, collecting, and reclaiming equipment for conveying dust, fumes, fibers, gases, metal particulates or other refuse, and design requirements of exhaust systems for industrial plants and laboratories, generally must comply with local and state health and labor codes. When shop drawings are prepared for jobs of this type, the draftsman must indicate any special fabrication and erection requirements that conform to the governing codes.

Industrial exhaust systems are basically high negative pressure, high-velocity ducts. Round-duct construction, with smooth interiors, and airtight seams, is commonly used. The velocity and volume required for each branch determines its diameter, and the main-duct areas at any point in a system is between 120 to 125 percent of the total area of all of the branch ducts serviced by the main duct to that point. (See Table 4-3.)

To compute a round duct with an area equivalent to a series of round branch-ducts, the square of each branch must be added and the square root extracted.

Example:

Find the diameter of a main duct equal in area to branch ducts of 6″, 9″, 10″, and 12″ diameters.

$$\text{Diameter of main} = \sqrt{6^2 + 9^2 + 10^2 + 12^2}$$

$$\text{Diameter} = 19 \text{ inches} \quad Ans$$

Table 4-3. Square Foot Areas of Round Ducts*

Dia.	Area	Dia.	Area	Dia.	Area
4	.0873	32	5.585	86	40.34
5	.1364	34	6.305	88	42.24
6	.1964	36	7.069	90	44.18
7	.2673	38	7.876	92	46.16
8	.3491	40	8.727	94	48.19
9	.4418	42	9.621	96	50.27
10	.5454	44	10.56	98	52.38
11	.6600	46	11.54	100	54.54
12	.7854	48	12.57	102	56.75
13	.9218	50	13.64	104	59.04
14	1.069	52	14.75	106	61.24
15	1.227	54	15.90	108	63.62
16	1.396	56	17.10	110	66.04
17	1.576	58	18.35	112	68.37
18	1.767	60	19.63	114	70.88
19	1.969	62	20.97	116	73.44
20	2.182	64	22.34	118	75.89
21	2.405	66	23.76	120	78.54
22	2.640	68	25.22	124	84.00
23	2.885	70	26.73	128	89.35
24	3.142	72	28.27	132	95.03
25	3.409	74	29.87	136	100.90
26	3.687	76	31.50	140	106.90
27	3.976	78	33.18	144	113.10
28	4.276	80	34.91	...	....
29	4.587	82	36.67	...	....
30	4.909	84	38.48	...	....

* Diameters are given in inches, and areas in square feet.

If a main was specified to be 125 percent for the total area, the areas of each branch must be added and multiplied by 1.25. The diameter is then found by application of the formula for the area of a circle ($A = .7854d^2$).

Pressure

Airflow in fan systems is always from a point of higher pressure to one of lower pressure. Total pressure is equal to the sum of velocity pressure and static pressure. Velocity pressure is the pressure corresponding to velocity of airflow. It is measured in the direction of flow and is always positive.

Static pressure is the expansive pressure in a supply duct, or the contractive pressure in a return or exhaust duct and will be positive or negative, respectively. It may exist in a duct or plenum chamber with no air movement. In the case of a fan blowing into a supply duct with a tightly shut damper, the total pressure would be that of the static pressure only. If the damper were open, the static pressure would decrease and velocity pressure would exist, the total pressure being their sum.

The relationship of velocity to velocity pressure for dry air at 70F and 29.92″ barometric pressure is:

$$V = 4005\sqrt{P_v}$$

or:

$$P_v = \left(\frac{V}{4005}\right)^2$$

where:

V = velocity /fpm
P_v = velocity pressure /in. wg

Table 4-4 is a solution of the above equations.

Example 1:

What is the velocity pressure in a duct with an airflow velocity of 1,500 fpm?

$$P_v = \left(\frac{1500}{4005}\right)^2$$

$$P_v = 0.14 \text{ in. wg} \quad Ans$$

Table 4-4. Velocity Equivalents of Velocity Pressures*

(For conversion of velocity pressure gage readings in inches of water to velocity in feet per minute when air is at 70 deg. and 29.92 inches barometric pressure.)

Pressure, P_v, Inches of Water	Velocity, V, Feet per Min.	Pressure, P_v, Inches of Water	Velocity, V, Feet per Min.	Pressure, P_v, Inches of Water	Velocity, V, Feet per Min.	Pressure, P_v, Inches of Water	Velocity, V, Feet per Min.
.01	400.5	.50	2,831.9	.99	3,985.0	1.48	4,872.3
.02	566.3	.51	2,859.0	1.00	4,005.0	1.49	4,888.7
.03	693.7	.52	2,888.4	1.01	4,021.0	1.50	4,905.1
.04	801.0	.53	2,915.6	1.02	4,044.8	1.51	4,921.4
.05	895.5	.54	2,942.9	1.03	4,064.6	1.52	4,937.7
.06	980.8	.55	2,970.1	1.04	4,084.3	1.53	4,953.9
.07	1,059.7	.56	2,996.9	1.05	4,103.9	1.54	4,970.1
.08	1,132.6	.57	3,023.8	1.06	4,123.4	1.55	4,986.2
.09	1,201.5	.58	3,050.2	1.07	4,142.8	1.56	5,002.2
.10	1,266.4	.59	3,076.2	1.08	4,162.1	1.57	5,018.3
.11	1,328.5	.60	3,102.3	1.09	4,181.3	1.58	5,034.2
.12	1,387.3	.61	3,127.9	1.10	4,200.5	1.59	5,050.1
.13	1,444.2	.62	3,153.5	1.11	4,219.5	1.60	5,066.0
.14	1,498.7	.63	3,178.8	1.12	4,238.5	1.61	5,081.8
.15	1,551.1	.64	3,204.0	1.13	4,257.4	1.62	5,097.5
.16	1,602.0	.65	3,228.8	1.14	4,276.2	1.63	5,113.2
.17	1,651.3	.66	3,253.7	1.15	4,294.9	1.64	5,128.9
.18	1,699.3	.67	3,254.1	1.16	4,313.5	1.65	5,144.5
.19	1,745.8	.68	3,302.5	1.17	4,332.1	1.66	5,160.0
.20	1,791.0	.69	3,327.0	1.18	4,350.6	1.67	5,175.5
.21	1,835.5	.70	3,351.0	1.19	4,368.9	1.68	5,191.1
.22	1,870.3	.71	3,374.6	1.20	4,387.3	1.69	5,206.5
.23	1,920.8	.72	3,398.2	1.21	4,405.5	1.70	5,221.9
.24	1,962.0	.73	3,421.9	1.22	4,423.7	1.71	5,237.2
.25	2,002.5	.74	3,445.1	1.23	4,441.7	1.72	5,252.5
.26	2,042.1	.75	3,468.3	1.24	4,459.8	1.73	5,267.7
.27	2,081.0	.76	3,491.6	1.25	4,477.7	1.74	5,283.0
.28	2,119.4	.77	3,514.4	1.26	4,495.6	1.75	5,298.1
.29	2,156.7	.78	3,537.2	1.27	4,513.4	1.76	5,313.2
.30	2,193.5	.79	3,559.6	1.28	4,531.1	1.77	5,328.3
.31	2,230.0	.80	3,582.1	1.29	4,548.8	1.78	5,343.4
.32	2,265.6	.81	3,604.5	1.30	4,566.4	1.79	5,358.3
.33	2,300.9	.82	3,626.5	1.31	4,583.9	1.80	5,373.3
.34	2,335.3	.83	3,648.6	1.32	4,601.4	1.81	5,388.2
.35	2,369.4	.84	3,670.6	1.33	4,618.8	1.82	5,403.0
.36	2,403.0	.85	3,692.6	1.34	4,636.1	1.83	5,417.8
.37	2,436.2	.86	3,714.2	1.35	4,653.4	1.84	5,432.7
.38	2,468.7	.87	3,735.5	1.36	4,670.6	1.85	5,447.4
.39	2,573.2	.88	3,757.1	1.37	4,687.7	1.86	5,462.1
.40	2,533.2	.89	3,778.3	1.38	4,704.8	1.87	5,476.8
.41	2,564.4	.90	3,799.5	1.39	4,721.8	1.88	5,491.4
.42	2,595.6	.91	3,820.4	1.40	4,738.8	1.89	5,506.0
.43	2,626.1	.92	3,841.6	1.41	4,755.7	1.90	5,520.5
.44	2,656.5	.93	3,862.4	1.42	4,772.5	1.91	5,535.0
.45	2,686.6	.94	3,882.8	1.43	4,789.3	1.92	5,549.5
.46	2,716.2	.95	3,903.7	1.44	4,806.0	1.93	5,563.9
.47	2,745.8	.96	3,924.1	1.45	4,822.7	1.94	5,578.3
.48	2,774.7	.97	3,944.5	1.46	4,839.2	1.95	5,592.7
.49	2,803.5	.98	3,964.5	1.47	4,855.8	1.96	5,607.0

* Reproduced from *Handbook of Air Conditioning, Heating and Ventilating*. Edited by Clifford Strock and Richard L. Koral. New York: Industrial Press Inc., 1965.

Table 4-4 (Continued). Velocity Equivalents of Velocity Pressures

Pressure, P_v, Inches of Water	Velocity, V, Feet per Min.	Pressure, P_v, Inches of Water	Velocity, V, Feet per Min.	Pressure, P_v, Inches of Water	Velocity, V, Feet per Min.	Pressure, P_v, Inches of Water	Velocity, V, Feet per Min.
1.97	5,621.3	2.32	6,100.2	2.67	6,544.2	3.02	6,959.9
1.98	5,635.5	2.33	6,113.4	2.68	6,556.5	3.03	6,971.5
1.99	5,649.7	2.34	6,126.5	2.69	6,568.7	3.04	6,983.0
2.00	5,663.9	2.35	6,139.5	2.70	6,580.9	3.05	6,994.4
2.01	5,678.0	2.36	6,152.6	2.71	6,593.1	3.06	7,005.9
2.02	5,692.2	2.37	6,165.6	2.72	6,605.2	3.07	7,017.3
2.03	5,706.2	2.38	6,178.6	2.73	6,617.3	3.08	7,028.7
2.04	5,720.3	2.39	6,191.6	2.74	6,629.4	3.09	7,040.1
2.05	5,734.3	2.40	6,204.5	2.75	6,641.5	3.10	7,051.5
2.06	5,748.3	2.41	6,217.4	2.76	6,653.6	3.11	7,062.9
2.07	5,762.2	2.42	6,230.3	2.77	6,665.6	3.12	7,074.2
2.08	5,776.1	2.43	6,243.2	2.78	6,677.7	3.13	7,085.6
2.09	5,789.9	2.44	6,256.0	2.79	6,689.7	3.14	7,096.9
2.10	5,803.8	2.45	6,268.9	2.80	6,701.6	3.15	7,108.2
2.11	5,817.6	2.46	6,281.6	2.81	6,713.6	3.16	7,119.4
2.12	5,831.4	2.47	6,294.3	2.82	6,725.6	3.17	7,130.7
2.13	5,845.1	2.48	6,307.0	2.83	6,737.5	3.18	7,142.0
2.14	5,858.8	2.49	6,319.8	2.84	6,749.3	3.19	7,153.2
2.15	5,872.5	2.50	6,332.5	2.85	6,761.2	3.20	7,164.3
2.16	5,886.1	2.51	6,345.1	2.86	6,773.1	3.21	7,175.6
2.17	5,899.7	2.52	6,357.7	2.87	6,784.9	3.22	7,186.7
2.18	5,913.3	2.53	6,370.4	2.88	6,796.7	3.23	7,197.9
2.19	5,926.8	2.54	6,382.9	2.89	6,808.5	3.24	7,209.0
2.20	5,940.4	2.55	6,395.5	2.90	6,820.3	3.25	7,220.1
2.21	5,953.9	2.56	6,408.0	2.91	6,832.0	3.26	7,231.2
2.22	5,967.3	2.57	6,420.5	2.92	6,843.7	3.27	7,242.3
2.23	5,980.7	2.58	6,433.0	2.93	6,855.4	3.28	7,253.4
2.24	5,994.1	2.59	6,445.4	2.94	6,867.1	3.29	7,264.4
2.25	6,007.5	2.60	6,457.9	2.95	6,878.8	3.30	7,275.4
2.26	6,020.8	2.61	6,470.3	2.96	6,890.5	3.31	7,286.5
2.27	6,034.1	2.62	6,482.7	2.97	6,902.1	3.32	7,297.5
2.28	6,047.4	2.63	6,495.0	2.98	6,913.7	3.33	7,308.4
2.29	6,060.6	2.64	6,507.4	2.99	6,925.3	3.34	7,319.4
2.30	6,073.9	2.65	6,519.7	3.00	6,936.9	3.35	7,330.4
2.31	6,087.0	2.66	6,532.0	3.01	6,948.4	3.36	7,341.3

Example 2:

Find the velocity in a duct which has a velocity pressure of .060 in. wg

$$V = 4005\sqrt{P_v}$$
$$= 4005\sqrt{0.60}$$

where:

$P_v = 0.60$

$V = 3,102$ fpm *Ans*

Purpose of the Fan in the Duct System

Because of the relatively low pressure involved in heating, ventilating, and air-conditioning systems, the unit of pressure measurement is the inch of water. A pressure of 1 in. of water gage (in. wg) is 5.196 lb per sq ft, or .5774 oz per sq in. Compare this with the normal average residential water-pressure of 20 to 30 lb per sq in.

The purpose of the fan is to supply enough pressure to the air system so that pressure differences will be sufficient for adequate distribution of air to the most remote space, and to the duct run with the greatest resistance to flow. Resistance to such flow is encountered at entrances such as louvers or goosenecks and at automatic dampers, filters, coils, moisture eliminators, fan connections, sound traps,

Ch. 4 DUCT DESIGN 131

branch-duct connections, splitter dampers, volume dampers, fire dampers, mixing units, and air outlets.

Pressure drops for many of these units are supplied by the manufacturer in units of inches of water or "equivalent feet of duct." Losses in duct elbows, plenums, transitions, offsets, and other fittings are converted to equivalent lengths of straight duct, then added, to obtain the total resistance. Popular charts and calculators, such as the Trane Ductulator (Fig. 4-1), and the Anemostat Air-

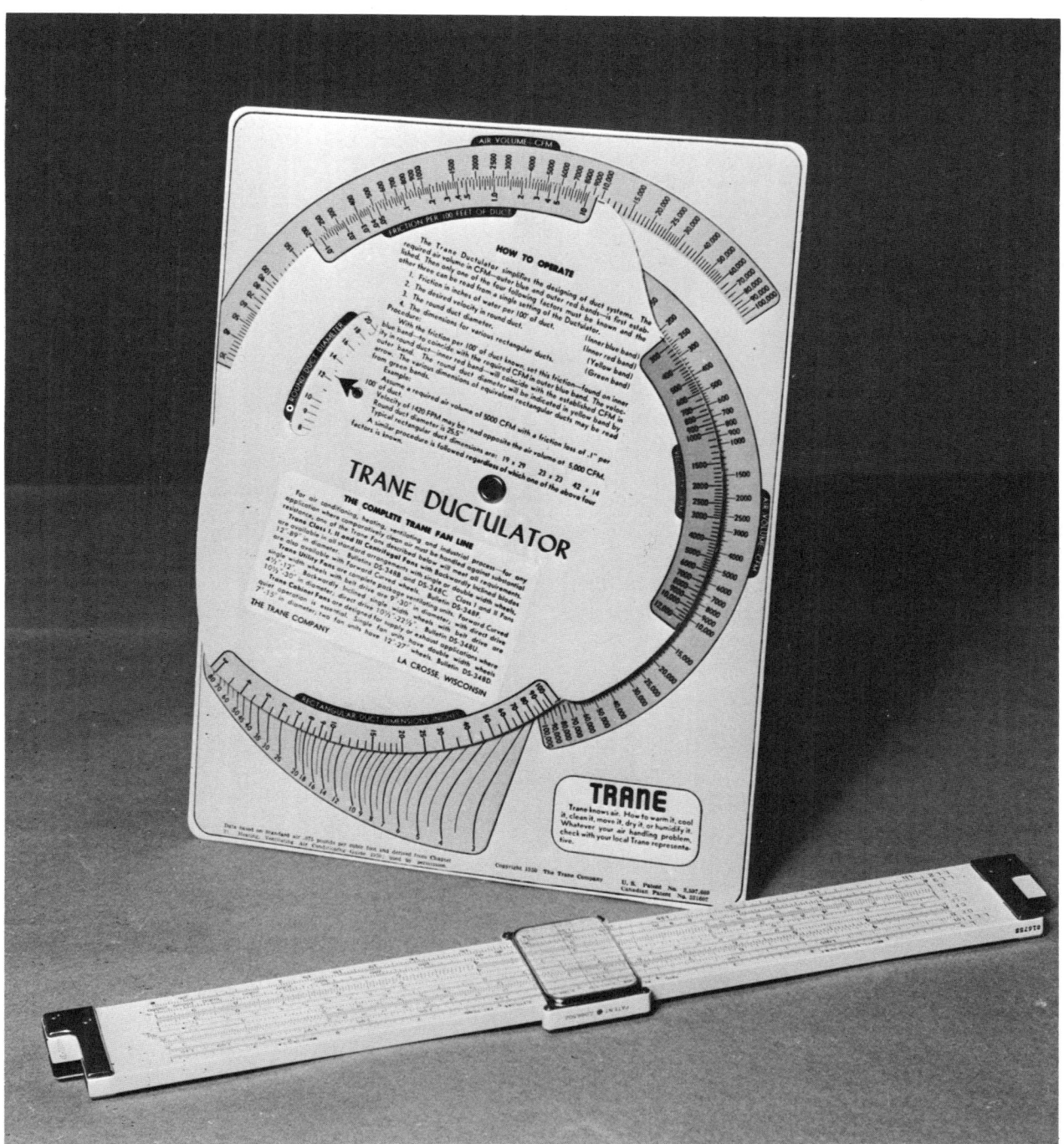

FIG. 4-1. Trane ductulator.

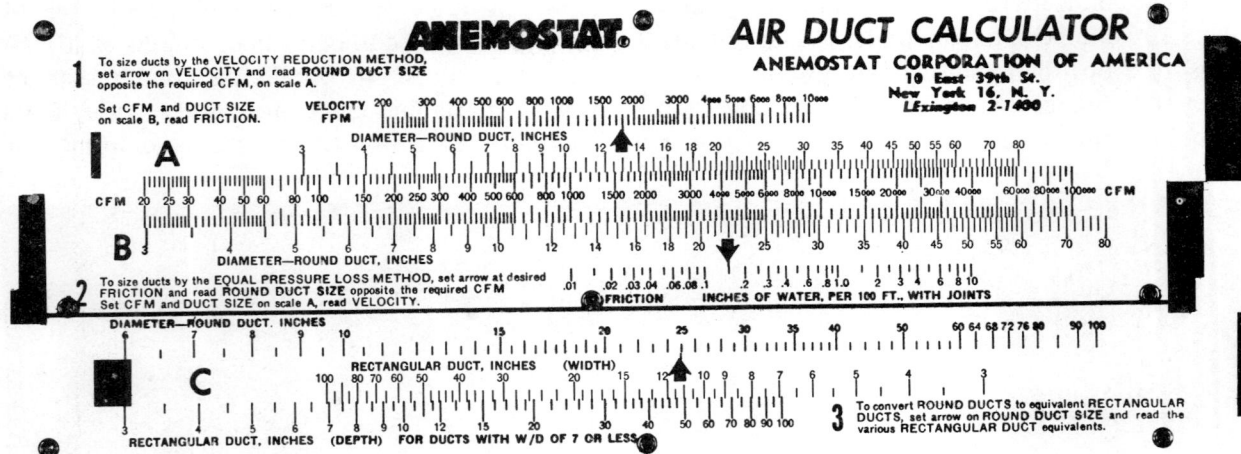

Fig. 4-2. Anemostat calculator.

Duct Calculator (Fig. 4-2), are convenient. They are based on the resistance of 100 lineal feet of *round duct* related to volume, velocity, and pressure. When it is necessary or desirable to change a rectangular-duct size shown on the design drawing, the draftsman must first determine a round duct of equivalent resistance per 100 feet and then convert to the duct size desired having the same friction.

Calculators give velocities in round ducts, therefore, to find a velocity in a rectangular duct it must be computed by the airflow formula:

$$Q = A \times V$$

where:

Q = volume /cfm
A = duct cross-sectional area, sq ft
V = velocity /lfm (lineal feet of air per minute)

also:

$$V = \frac{Q}{A}$$

$$A = \frac{Q}{V}$$

Example 1:

How much air will be delivered by a duct, 42″ x 18″, at a velocity of 1,500 fpm?

Converting inches to feet: 42″ = 3.5′; 18″ = 1.5′

$$Q = A \times V$$
$$= 3.5 \times 1.5 \times 1,500$$
$$= 7,850 \text{ cfm} \quad Ans$$

Example 2:

What size duct 16 inches deep is required to deliver 4,000 cfm at 1,200 fpm?

where:

W = required duct width, in inches

$$A = \frac{Q}{V}$$

$$A = \frac{4,000}{1,200}$$

$$A = \frac{40}{12} \text{ sq ft}$$

To convert A to sq in., multiply by 144 as:

$$\frac{40}{12} \times 144$$

The duct is 16″ deep, thus the sq in. area will be divided by 16, and the duct width is:

$$W = \frac{40}{12} \times 144 \div 16$$
$$= \frac{40}{12} \times \frac{144}{16}$$
$$W = 30″ \quad Ans$$

Ch. 4 DUCT DESIGN

Example 3:

What is the velocity of air in a 66″ x 27″ supply-duct carrying 26,000 cfm?

where:

66″ = 5.5′
27″ = 2.25′

$$Q = A \times V$$

$$V = \frac{Q}{A}$$

$$V = \frac{26,000}{(5.5)2.25}$$

$$V = 2,100 \text{ fpm} \quad Ans$$

It is suggested that the student draftsman convert the rectangular ducts of Examples 1, 2, and 3 to round ducts of equal friction, and check the changes in velocity.

Example 4:

Find the velocity of 22,000 cfm of air, flowing through two steam coils (one above the other in an apparatus casing), when each coil has a nominal tube-length of 8′-6, the coil-casing height being 29″. Free area through the coils is 87 percent of gross.

where free area of coils is:

$$2 \times 8.5 \times 2.417 \times .87 = 35.8 \text{ sq ft}$$

$$V = \frac{Q}{A}$$

$$V = \frac{22,000}{35.8}$$

$$V = 615 \text{ fpm} \quad Ans$$

An example for the student draftsman:

A central station, air-conditioning apparatus casing is connected to an outside-air intake-louver that is set into a masonry opening 12′-0 wide × 9′-6 high. The louver, together with bird-screen, has a 50 percent effective free area. Find the velocity through the louver when handling 35,000 cfm.

Duct Fitting, Friction Losses

The sheet metal draftsman, during preparation of the shop drawings, should analyze duct locations that are schematically indicated on the mechanical drawings. He must prevent these ducts from conflicting with other building components because certain conditions may require duct fittings with high resistance to airflow. When space permits, fittings should be selected with low friction-loss factors and economical fabrication and erection characteristics.

FRICTION LOSSES IN FITTINGS

Tables 4-5, 4-6, and 4-7 indicate losses of fittings most commonly used in heating, ventilating, and air-conditioning systems. The losses are in decimal fractions of the velocity pressure in the fitting. (See Table 4-4 for velocity equivalents of velocity pressures.)

Example 1:

Calculate the friction loss of a 24 × 12 straight clinch-tap to handle 4,000 cfm which connects to a supply plenum with an air velocity of 1,000 fpm

where:

Area of branch = 24 × 12 = 2 sq ft

Let, V_1 = branch-duct velocity

V_2 = plenum velocity = 1,000 fpm

$$V_1 = \frac{\text{cfm}}{A}$$

$$= \frac{4,000}{2}$$

$$V_1 = 2,000 \text{ fpm}$$

$$\frac{V_1}{V_2} = \frac{2,000}{1,000} = 2 \quad Ans$$

From Table 4-5, a sharp contraction which doubles in air velocity has a fraction of velocity pressure loss of 0.32.

Table 4-4, velocity pressure for 2,000 fpm = .25″ wg

$$\text{Fitting loss} = 0.32 \times 0.25$$
$$= .08 \text{ in. wg} \quad Ans$$

Example 2:

What is the friction loss of a 20-in. diameter, 5-piece 90° elbow with a 20″ throat radius and an

Table 4-5. Losses in Rectangular Ducts

Type	Application	Fraction of Velocity Pressure Loss	Condition
Sharp entry	Return-air inlet or outside-air intake-louver	0.50	—
Easy entry	Radius throat-clinch collar tap with boot or splitter damper	0.20	—
Sharp contraction	Straight clinch-tap from intake plenum	0.32	when air velocity doubles
		0.45	when air velocity increases four times
Increaser	Tapered fitting from fan discharge to supply duct, or fitting to coil	0.50	when included angle is approximately 20°
		0.60	when included angle is approximately 30°
Sharp increaser	Straight clinch tap to supply plenum	0.36	when air velocity decreases to 0.4 of initial air velocity
		0.64	when air velocity decreases to 0.2 of initial air velocity
Reducer	Fitting from coil to duct run, or for decrease in air volume	0.04	when included angle is approximately 45°
		0.07	when included angle is approximately 60°
Straight transition	Change of shape	0.15	when included angle is approximately 15°
90° branch take-off	Positive nest-branch connection	0.30	when area of branch is 0.4 to 0.6 that of main
		0.50	when area of branch is 0.15 to 0.3 that of main

Table 4-6. Losses in Round Ducts with Main Velocity Equal to Branch Velocity

Fitting	Fraction of Velocity-Pressure Loss
90° Tee	1.50
60° Lateral	0.75
45° Lateral	0.42
Long Cone	0.50
5-piece 90° Elbow*	0.24
3-piece 90° Elbow	0.33
3-piece 60° Elbow	0.18
3-piece 45° Elbow	0.12
2-piece 30° Elbow	0.09

* All elbows have centerline radius equal to 1.5 duct diameter.

air velocity of 3,100 fpm? From Table 4-6, the fraction of velocity pressure loss for an elbow of this type is 0.24.

Table 4-4, velocity pressure for 3,100 fpm = 0.60 in. wg

Fitting loss = 0.24 × 0.60
            = 0.144 inch   *Ans*

ELBOWS

Job specifications generally require rectangular elbows to have a throat radius equal to the cheek width. Where space conditions prevent installation of full-radius elbows, square throat, square

Table 4-7. Losses in 90° Rectangular-Duct Elbows

Type	Throat radius, r	Fraction of Velocity Pressure Loss, when:		
		$\frac{H}{W} = 0.75$ to 3.0	$\frac{H}{W} = 0.50$	$\frac{H}{W} = 0.25$
1	$0.75 - 1.0\ W$	0.13	0.17	0.27
2	$0.5\ W$	0.21	0.28	0.44
3	$0.25\ W$	0.36	0.44	0.54
4	Square throat and heel with double-thickness turning vanes	—	0.10	—

W = Cheek width
H = Height of elbow
$\frac{H}{W}$ = Aspect ratio

heel elbows with double-thickness turning vanes are usually used. The draftsman should draw radius elbows whenever possible as they are fabricated more economically than the vaned type. However, handling and field-erection conditions of large elbows should be studied to determine the most economical installation: one-piece radius elbows, multiple-section radius elbows, or square-throat vaned elbows.

Airflow resistance in elbows is influenced by the ratio of the throat radius to the cheek width and by aspect ratio, which is the relationship of the elbow height to the cheek width. The draftsman must also be aware that offsets are combinations of beveled elbows and similarly affect airflow resistance.

Using Type 1 elbow as a comparative standard, if its throat radius were reduced to equal one-half of its cheek width, the elbow resistance would be increased approximately 1.63 times, as in Type 2. If its throat radius were reduced to one-quarter the cheek width, the resistance would be 2.0 to 2.75 times greater, as in Type 3.

Example 1:

Find the resistance in a 90° elbow, 24″ x 24″, with a 22″ throat radius, and carrying air at 2,000 fpm velocity.

From Table 4-7: the elbow is Type 1 and has a fraction of velocity pressure loss of 0.13
From Table 4-4: an air velocity of 2,002 = 0.25 P_v
Elbow loss = 0.13 × 0.25 = 0.0325 in. wg   *Ans*

Example 2:

Find the resistance of a 90° elbow, 48″ wide by 12″ high, with a 12″ throat radius, and carrying 8,500 cfm of air.

$$V = \frac{Q}{A} = \frac{8,500}{8} = 2,125\ \text{fpm}$$

From Table 4-4: 2119.4 = 0.28 P_v
From Table 4-7: elbow is Type 3 with a fraction of P_v loss of 0.54
Elbow loss = 0.28 × 0.54 = 0.1512 in. wg   *Ans*

Example 3:

A 30″ x 20″ duct has an air velocity of 1,700 and a fitting 44″ long tapers equally to 54″ x 38″. Find the fitting pressure-drop.

To find the most severe angle of taper:

$$54'' - 30'' = 24\ \text{inches}$$

$$24 \div 2 = \text{taper of side} = 12\ \text{inches}$$

Use tangent formula for degree of slope:

$$\text{Tangent} = \frac{12}{44} = 15°$$

From Table 4-5: an increaser of this type has a fraction of P_v loss of 0.60
From Table 4-4: 1,700 = 0.18 P_v
Loss of fitting = 0.60 × 0.18 = .108 in. wg   *Ans*

The cross section of rectangular ducts should be as close to square as possible for airflow efficiency and material economy. For example — a 48″ x 12″ duct carrying 4,800 cfm at 1,200 fpm has a resistance of 0.103 in. wg per 100 feet of equivalent length, and a duct perimeter of 120 inches. A 24″ x 24″ duct with similar capacity and velocity has 0.084 in. wg friction and a perimeter of 96 inches. The 48″ x 12″ duct has 22.5 percent more friction per 100′ and requires 25 percent more material per running foot and also requires a heavier gage and greater stiffening. A 24″ x 22″ duct with similar capacity and friction would use 30 percent less metal per running foot. (See Figs. 4-1, and 4-2.)

For precise location of splitters in elbows, see Fig. 4-3.

Equal-Friction Method of Duct Design

To design a duct system by the *equal-friction method*, a friction factor per 100 ft of equivalent length is determined by selecting the maximum duct-velocity that would satisfy sound criteria for the space. (See Table 4-8.)

Example 1:

An air velocity of 1,440 fpm was selected for a public school ventilation system of 12,000 cfm. Find the duct width for a 20-in.-high main, and also the friction loss per 100 ft. of equivalent length.

Let W = required duct width

To find the sq ft area of a duct with a capacity of 12,000 cfm at 1,440 fpm:

$$A = \frac{Q}{V} = \frac{12,000}{1,440}$$

The duct area in sq ft = $\dfrac{W \times 20}{144}$

Substituting:

$$\frac{W \times 20}{144} = \frac{12,000}{1,440}$$

$$W = 60$$

Duct size = 60×20 *Ans*

From Table 4-9: a 60 x 20 duct is equivalent to one of 37 inch diameter

From Fig. 4-4: a 37 inch diameter duct at 12,000 cfm has a friction loss per 100 feet of 0.09 in. wg

The friction factor of 0.09 inch would be used for all duct sizes in an equal friction design.

Example 2:

Using a friction loss per 100 feet of 0.09 in. wg, calculate the velocity and size of a branch duct which is 14 inches high and has a capacity of 1,800 cfm.

From Fig. 4-4: a duct of 1,800 cfm at 0.09″ = 18″ diameter

From Table 4-9: an 18″ diameter = 19×14

where:

$$A = \frac{19 \times 14}{144} \quad \text{or} \quad 1.85 \text{ sq ft}$$

Table 4-8. Design Air Velocities for Conventional and High Velocity Systems*

Duct Element	Conventional						High Velocity	
	Residences		Public Bldgs.		Industrial Bldgs.		Commercial Bldgs.	
	Normal	Max.	Normal	Max.	Normal	Max.	Normal	Max.
	Design Velocities, fpm							
Main ducts	700	1200	1000	1600	1500	2200	2500	6000
Branch ducts	600	1000	800	1300	1000	1800	2000	4500
Fan outlets	1000	1700	1500	2200	2000	2800	2500	5000
Suction connections	700	900	800	1100	1000	1400	1700	3300
Outside air intakes	500	800	500	900	500	1200	600	1000
Filters	250	300	300	350	350	350	350	350
Heating coils	450	500	500	600	600	700	600	700
Air washers	500	...	500	...	...	...	500	...
Dehumidifiers	450	...	500	...	...	...	500	...

*Adapted from the data of 1963 ASHRAE Guide and Data Book, p. 177 by permission.

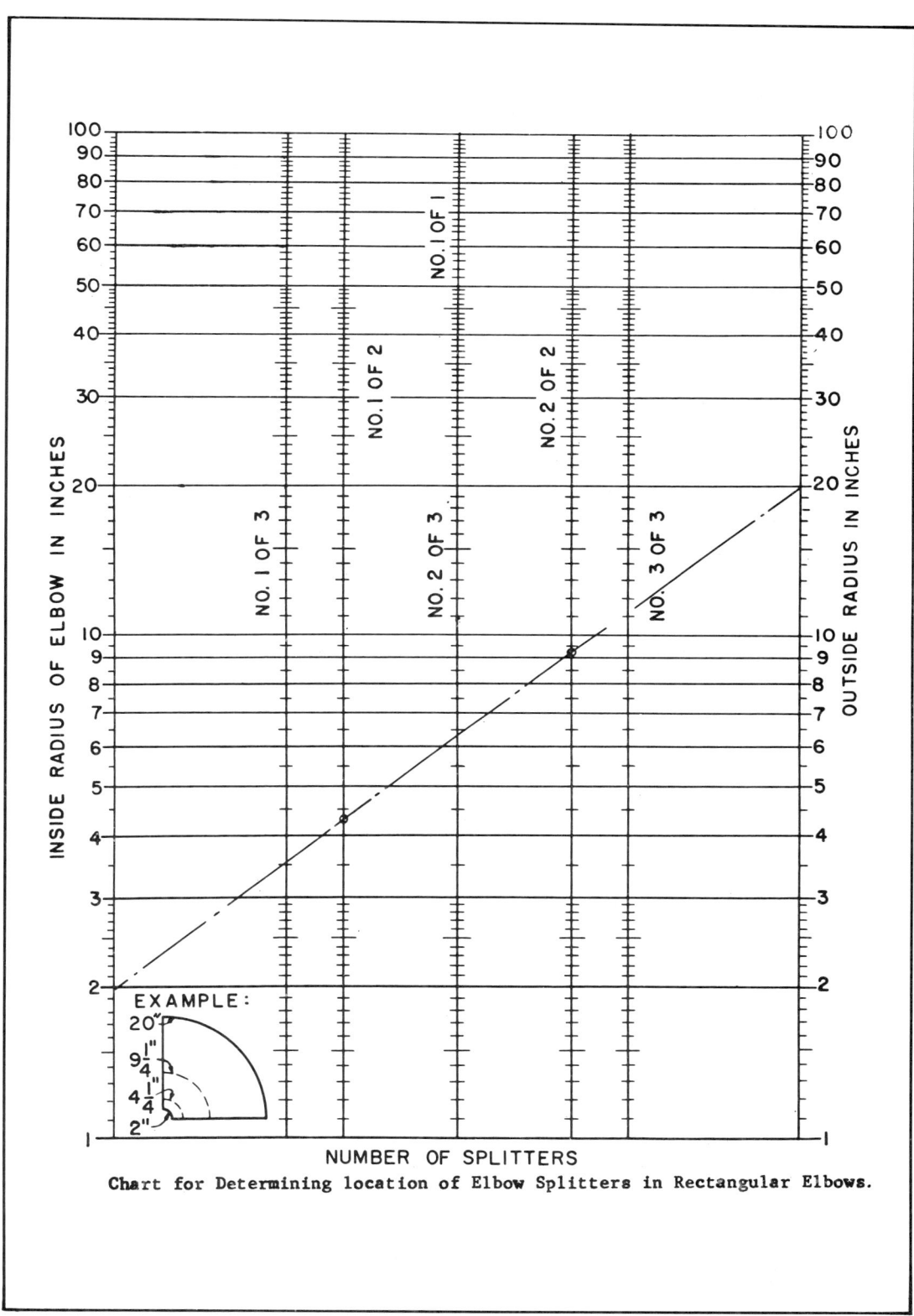

FIG. 4-3. Elbow splitter location chart. (Courtesy of the Buffalo Forge Co.)

$$V = \frac{Q}{A} = \frac{1,800}{1.85}$$

$V = 974$ fpm *Ans*

For low pressure systems of equal friction design the friction loss per 100 feet of equivalent length of 0.07–0.15 in. wg is usually selected. Total duct resistance is then determined by multiplying the equivalent length of the longest run by the factor selected and dividing by 100.

Example 1:

Find the friction loss of a duct run with an equivalent length of 350 feet and a friction loss of 0.09″ per 100 feet.

$$F.L. = \frac{EQ.L. \times F.F.}{100}$$

where:

$F.L.$ = Friction Loss
$EQ.L.$ = Equivalent Length
$F.F.$ = Friction Factor, loss per 100 feet

$$F.L. = \frac{350 \times .09}{100} = 0.315 \text{ in. wg} \text{ } Ans$$

When a decrease in velocity occurs in the duct at the fan discharge, a static pressure regain equal to 0.75 of the velocity pressure difference of the fan discharge and duct should be allowed:

Regain = $0.75(P_v$ fan $- P_v$ duct$)$

Example:

Find the static pressure gain when a fan outlet velocity of 2,400 fpm discharges into a duct increaser where the velocity decreases to 1,600 fpm.

Regain = $0.75(P_v$ fan $- P_v$ duct$)$

From Table 4-4: P_v 2,400 = 0.36 P_v 1,600 = 0.16

Regain = $0.75(0.36 - 0.16) = 0.75(0.20)$

Regain = 0.15 in. wg *Ans*

Table 4-9. Equivalent Round and Rectangular Ducts with Equal Friction*

One Side Rect'lar Duct, Inches	Other Side of Rectangular Duct, Inches												
	22	24	26	28	30	32	34	36	38	40	42	44	46	48
	Diameter of Equivalent Circular Duct, Inches												
22	24.2													
24	25.2	26.4												
26	26.3	27.5	28.6											
28	27.3	28.5	29.7	30.8										
30	28.2	29.5	30.7	31.9	33.0									
32	29.1	30.5	31.7	32.9	34.1	35.2								
34	30.0	31.3	32.7	33.9	35.1	36.3	37.4							
36	30.8	32.2	33.7	34.9	36.1	37.3	38.5	39.6						
38	31.5	33.1	34.6	35.9	37.1	38.4	39.5	40.7	41.8					
40	32.4	33.9	35.3	36.7	38.0	39.3	40.5	41.7	42.9	44.0				
42	33.0	34.5	36.0	37.6	39.0	40.3	41.5	42.7	44.0	45.1	46.2			
44	33.7	35.3	36.9	38.5	39.9	41.2	42.5	43.7	44.9	46.1	47.2	48.4		
46	34.6	36.2	37.8	39.3	40.8	42.2	43.5	44.8	46.0	47.2	48.4	49.5	50.6	
48	35.2	37.0	38.5	40.0	41.5	43.0	44.4	45.6	46.9	48.1	49.3	50.5	51.6	52.8
50	35.9	37.6	39.2	40.8	42.3	43.8	45.2	46.5	47.9	49.1	50.4	51.6	52.9	54.0
52	36.5	38.3	40.0	41.6	43.1	44.7	46.1	47.5	48.9	50.1	51.3	52.5	53.8	55.0
54	37.2	38.9	40.7	42.4	44.0	45.5	47.0	48.4	49.9	51.1	52.3	53.5	54.8	56.0
56	37.8	39.6	41.3	43.0	44.6	46.2	47.7	49.1	50.6	52.0	53.3	54.6	55.9	57.0
58	38.4	40.3	42.1	43.8	45.4	47.0	48.5	50.0	51.5	52.9	54.2	55.5	56.8	58.0
60	39.1	40.9	42.7	44.5	46.1	47.8	49.3	50.9	52.3	53.8	55.0	56.4	57.7	58.9
62	39.6	41.6	43.4	45.1	46.8	48.4	50.0	51.7	53.0	54.5	55.9	57.2	58.5	59.7
64	40.2	42.2	44.0	45.8	47.5	49.2	50.9	52.4	53.9	55.4	56.8	58.1	59.4	60.6
66	40.8	42.8	44.7	46.5	48.2	50.0	51.6	53.1	54.7	56.2	57.6	59.1	60.4	61.6

* Reproduced from *Handbook of Air Conditioning, Heating and Ventilating*. Edited by Clifford Strock and Richard L. Koral. New York: Industrial Press Inc., 1965.

Table 4.9 (Continued). Equivalent Round and Rectangular Ducts with Equal Friction

One Side Rect'lar Duct, Inches	Other Side of Rectangular Duct, Inches													
	3	4	5	6	7	8	9	10	11	12	14	16	18	20
	Diameter of Equivalent Circular Duct, Inches														
3½	—	—	—	—	—	5.7	6.0	6.3	6.6	6.9	7.4	7.8		
4	3.8	4.4	4.9	5.4	5.8	6.1	6.5	6.8	7.1	7.4	7.9	8.4		
4½	4.0	4.7	5.2	5.8	6.1	6.5	6.9	7.2	7.6	7.9	8.5	9.0		
5	4.3	4.9	5.5	6.3	6.5	6.9	7.3	7.7	8.0	8.3	8.9	9.5		
5½	4.4	5.2	5.9	6.4	6.8	7.3	7.7	8.1	8.5	8.8	9.5	10.1		
6	4.6	5.4	6.3	6.6	7.0	7.6	8.0	8.4	8.8	9.2	9.9	10.5		
7	5.0	5.8	6.5	7.0	7.7	8.2	8.7	9.2	9.6	10.0	10.8	11.4		
8	5.3	6.1	6.9	7.6	8.2	8.8	9.3	9.8	10.2	10.7	11.5	12.3		
9	5.5	6.5	7.3	8.0	8.7	9.3	9.9	10.4	10.9	11.4	12.3	13.1		
10	5.8	6.8	7.7	8.4	9.2	9.8	10.4	11.0	11.5	12.0	12.9	13.8		
11	6.1	7.1	8.0	8.8	9.6	10.2	10.9	11.5	12.1	12.6	13.6	14.5		
12	6.3	7.4	8.3	9.2	10.0	10.7	11.4	12.0	12.6	13.2	14.3	15.2		
13	6.4	7.6	8.7	9.6	10.4	11.1	11.8	12.5	13.1	13.7	14.8	15.8		
14	6.7	7.9	8.9	9.9	10.8	11.5	12.3	12.9	13.6	14.3	15.4	16.5		
15	7.0	8.2	9.2	10.2	11.1	11.9	12.7	13.4	14.1	14.7	16.0	17.1		
16	7.2	8.4	9.5	10.5	11.4	12.3	13.1	13.8	14.5	15.2	16.5	17.6		
17	7.3	8.6	9.8	10.8	11.8	12.6	13.5	14.2	15.0	15.7	17.0	18.2		
18	7.5	8.9	10.0	11.1	12.1	13.0	13.8	14.6	15.4	16.1	17.4	18.7	19.8	
19	7.7	9.1	10.3	11.4	12.4	13.3	14.2	15.0	15.8	16.5	17.9	19.2	20.4	
20	7.9	9.3	10.5	11.6	12.7	13.6	14.5	15.4	16.2	17.0	18.4	19.7	20.9	22.0
22	8.2	9.7	11.0	12.1	13.2	14.2	15.2	16.1	16.9	17.8	19.2	20.6	21.9	23.1
24	8.5	10.0	11.4	12.6	13.8	14.8	15.8	16.8	17.6	18.5	20.0	21.5	22.8	24.0
26	8.8	10.4	11.8	13.1	14.3	15.4	16.4	17.3	18.3	19.2	20.8	22.3	23.8	25.1
28	9.1	10.8	12.2	13.5	14.8	15.9	17.0	18.0	19.0	19.8	21.5	23.1	24.6	26.0
30	9.2	11.0	12.6	13.9	15.2	16.4	17.5	18.5	19.5	20.5	22.2	23.9	25.4	26.8
32	9.6	11.3	12.9	14.3	15.6	16.9	18.0	19.1	20.1	21.1	22.9	24.6	26.2	27.7
34	9.9	11.6	13.2	14.7	16.1	17.3	18.5	19.6	20.7	21.6	23.5	25.3	26.9	28.5
36	10.0	11.9	13.6	15.1	16.4	17.7	19.0	20.1	21.2	22.2	24.2	26.0	27.7	29.3
38	—	12.2	13.9	15.4	16.8	18.2	19.4	20.6	21.7	22.8	24.8	26.7	28.4	30.0
40	—	12.5	14.3	15.7	17.2	18.6	19.8	21.1	22.2	23.3	25.4	27.3	29.1	30.8
42	—	12.7	14.5	16.1	17.6	19.0	20.3	21.6	22.7	23.8	25.9	27.9	29.8	31.4
44	—	13.0	14.8	16.4	18.0	19.4	20.7	22.0	23.1	24.3	26.5	28.5	30.3	32.1
46	—	13.3	15.1	16.7	18.4	19.8	21.1	22.4	23.6	24.8	27.0	29.1	31.0	32.8
48	—	13.5	15.4	17.0	18.7	20.1	21.5	22.8	24.1	25.2	27.5	29.6	31.6	33.4
50	—	13.7	15.7	17.3	19.0	20.4	21.9	23.2	24.5	25.7	28.0	30.3	32.2	34.1
52	—	13.9	15.9	17.6	19.2	20.8	22.2	23.6	24.9	26.2	28.5	30.7	32.9	34.7
54	—	14.1	16.1	17.9	19.6	21.1	22.6	24.0	25.3	26.6	29.0	31.2	33.4	35.3
56	—	14.3	16.3	18.2	19.9	21.5	22.9	24.4	25.7	27.0	29.5	31.7	33.9	35.9
58	—	14.6	16.6	18.4	20.2	21.8	23.3	24.7	26.1	27.4	30.0	32.2	34.4	36.4
60	—	14.7	16.8	18.7	20.4	22.1	23.6	25.1	26.5	27.8	30.5	32.7	34.9	37.1
62	—	15.0	17.0	19.0	20.7	22.4	24.0	25.5	26.9	28.2	30.9	33.2	35.4	37.7
64	—	15.1	17.3	19.2	21.0	22.7	24.3	25.9	27.3	28.6	31.3	33.7	35.9	38.2
66	—	15.3	17.5	19.5	21.2	23.0	24.6	26.2	27.7	29.0	31.7	34.2	36.4	38.7

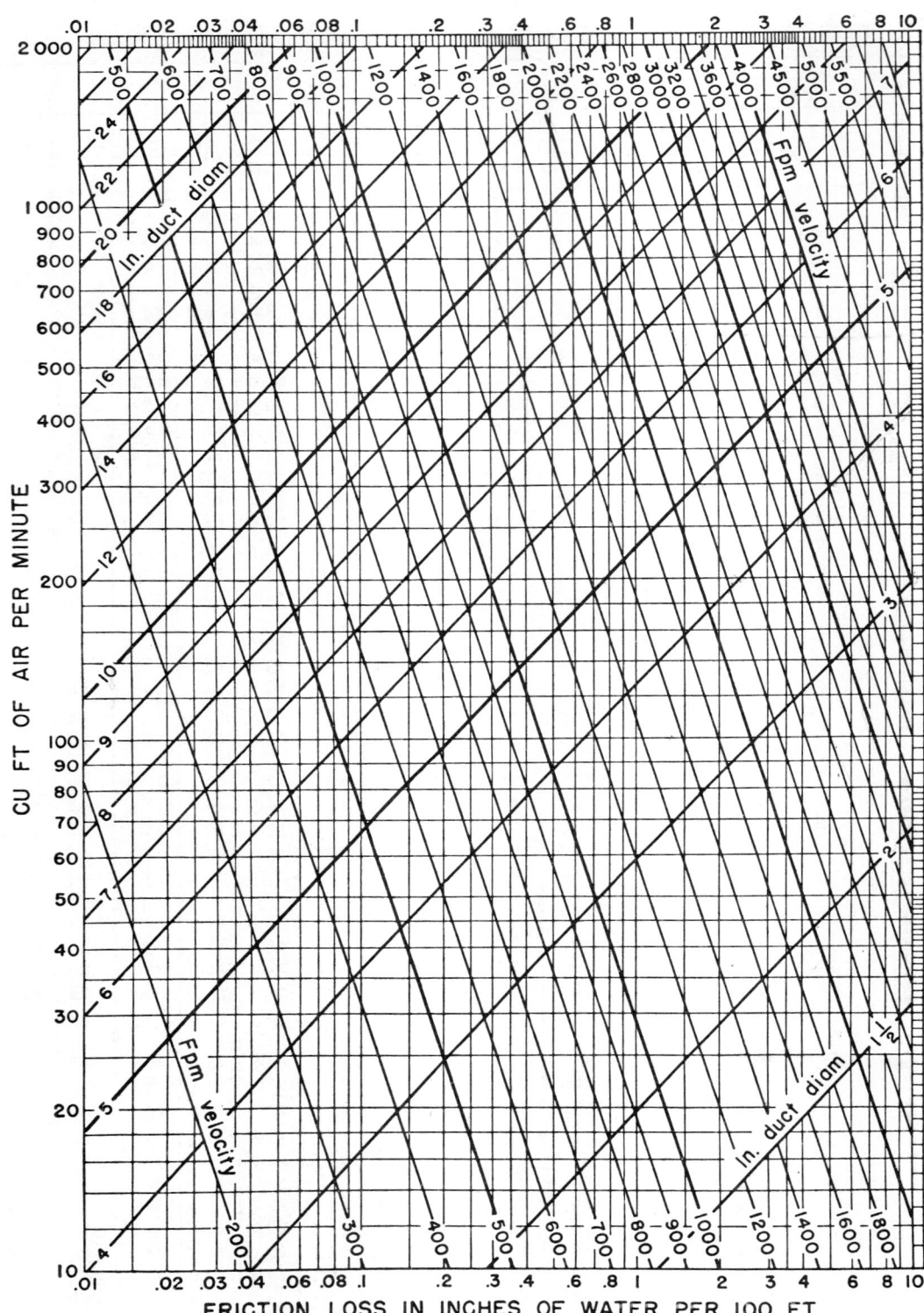

Fig. 4-4a. Friction of air in straight galvanized ducts for volumes of 10 to 2000 cfm. (Based on standard air of 0.075 lb per cu ft density flowing through average, clean, round, galvanized metal ducts having approximately 40 joints per 100 ft. Caution: Do not extrapolate below chart. Copyright by ASHRAE. Reprinted by permission of *ASHRAE Guide and Data Book,* 1969.)

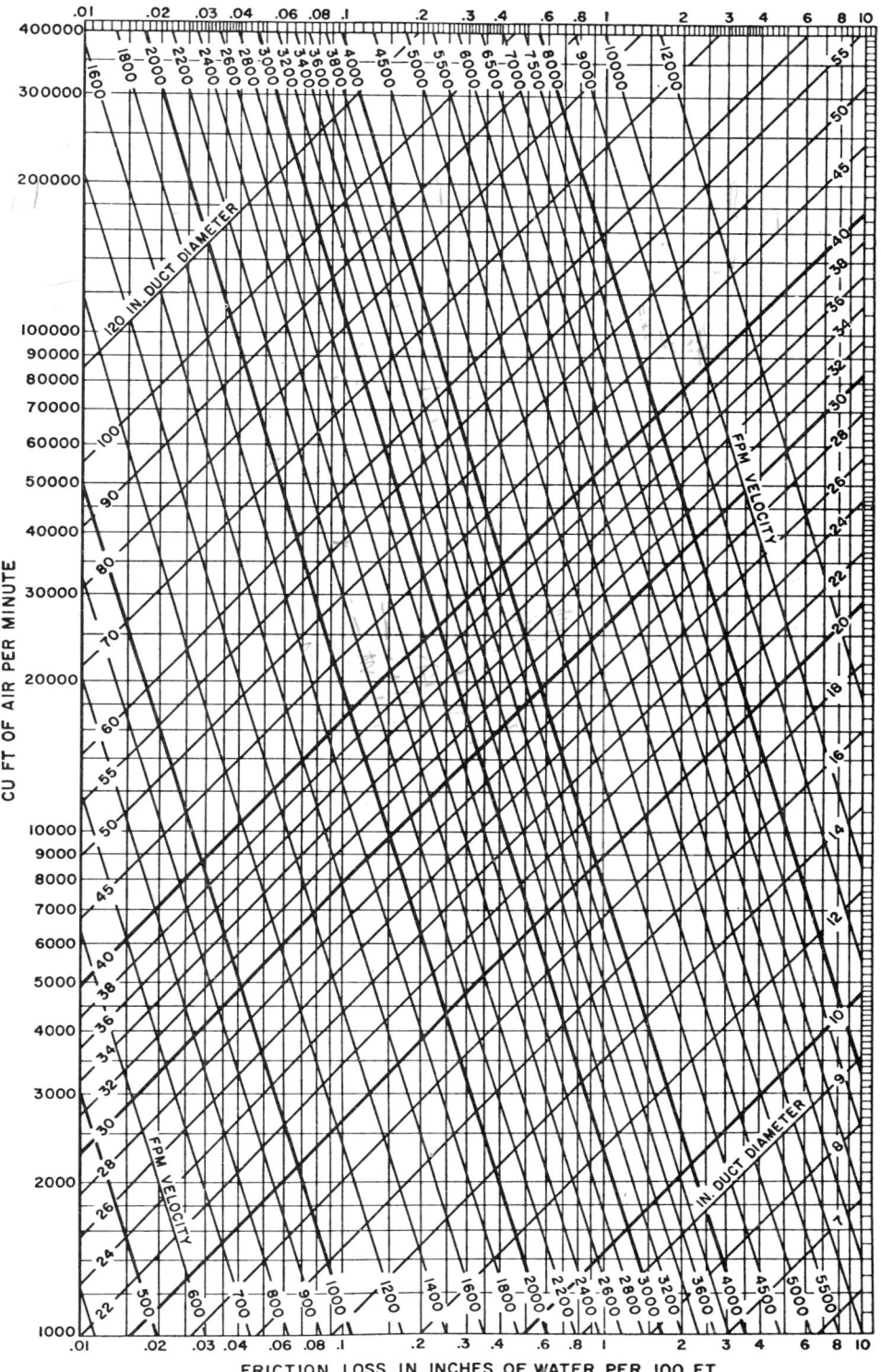

Fig. 4-4b. Friction of air in straight galvanized ducts for volumes of 1,000 to 400,000 cfm. (Same basis and source as Fig. 4-4a.)

Resistance in Low-Pressure Duct Systems

Total resistance in low-pressure duct systems is the sum of duct, equipment, entrance, and discharge losses, less the pressure gain at fan discharge.

Although the equal friction method is convenient because calculations are held to a minimum, branch ducts at close proximity to the fan have the greatest amount of pressure, require maximum dampering, and are not easily balanced for design air quantities. A refinement is to size these branch ducts at a greater friction-loss factor so the pressure available may be partially or fully utilized. However, velocities must neither be excessive nor generate objectionable noise in the space. (See Table 4-8.)

Example:

Calculate the size of a branch duct that will use pressure available at the take-off of a main in an air-conditioning system for a printing plant.

where:

Branch duct volume = 1,400 cfm
Branch duct depth = 10 inches
Equivalent length = 125'-0
Pressure loss at take-off = 0.035 in. wg
Pressure required at supply diffuser = 0.20 in. wg
Main duct pressure at take-off = 0.885 in. wg

From previous example:

$$F.L. = \frac{EQ.L. \times F.F.}{100}$$

converting,

$$F.F. = \frac{100 \, (F.L.)}{EQ.L.}$$

Loss at take-off = 0.035 inch
Pressure required at outlet = +0.200 inch
                              .235 inch

Friction loss of branch = 0.885 − 0.235
                        = 0.65 inch

$$F.F. = \frac{100(0.65'')}{125} = .52'' \text{ per 100 feet}$$

From Fig. 4-4:

1,400 at .52 inch = 12 inches diameter

From Table 4-9:

12 inch diameter = 12 inches × 10 inches *Ans*

To check velocity:

where:

Area of 12 × 10 = $\frac{120}{144}$, or .833 sq ft

$$V = \frac{Q}{A}$$

$$= \frac{1,400}{.833} = 1,680 \text{ fpm}$$

From Table 4-8: branch ducts in an industrial building may have velocities to 1,800 fpm, maximum.

Static-Regain Method

The following is a comparative analysis of two ducts for an air-conditioning supply system of 16,000 cfm:

Duct Size	Area, sq ft	Velocity, fpm	Velocity Pressure	Friction/ 100 ft
48" x 24"	8	2,000	.25 inch	.15 inch
24" x 24"	4	4,000	1.00 inch	.84 inch

In the 24 x 24 duct, the velocity pressure is 4 times greater and the friction factor is 5.6 times greater. To hold high-velocity duct losses at a minimum and to maintain a constant static pressure throughout the system, the *static regain method* is used, where the velocity pressure of each section in a duct run is reduced an amount equivalent to the friction loss of the previous section. The following formula is convenient to calculate duct sizes for a static regain which is assumed to have a 50 percent efficiency.

$$P_{vd} = P_{vu} - 2 \, F.L.$$

where:

P_{vd} = velocity pressure in downstream duct
P_{vu} = velocity pressure in upstream duct
$F.L.$ = Friction loss in preceding duct section

Ch. 4 DUCT DESIGN 143

Example:

A 22-inch-diameter cold duct has a volume of 12,500 cfm, a velocity of 5,000 fpm, and a 3,500 cfm branch take-off at a location 26 feet from a sound trap. The friction loss in this portion is 0.35 in. wg.

Find the required duct size after the branch tap.

where:

$P_{vu} = 5,000$ fpm $= 1.56$ inches
$F.L. = 0.35$ inch

cfm in downstream duct $= 12,500 - 3,500 = 9,000$

$$P_{vd} = P_{vu} - 2\ F.L.$$
$$= 1.56'' - 2(.35'')$$
$$P_{vd} = .86''$$
$$V = 3,714 \text{ fpm}$$

From Fig. 4-4: a duct for 9,000 cfm at 3,700 fpm is 22 inches in diameter.

The above example explains the plenum duct principle to achieve static regain. As each duct segment has a different friction loss, a separate tabulation should be used to summarize total system resistance and the most critical duct run (that having the most resistance), not necessarily the longest, is used for the total.

A branch duct to a terminal unit is generally made the size of the unit inlet. However, when a branch feeds more than one unit, the duct may be sized to use a portion of excessive pressure, such as a take-off at close proximity to the fan. Generally, a high-loss take-off fitting, such as a tee, is used to absorb some excess of pressure available for the total resistance requirement of the branch duct, fittings, terminal unit, and outlets.

Design range for high-velocity systems is 4,000 to 6,000 fpm for the initial main duct, at a friction loss of about 1.0 inch to 1.5 inches per 100 feet.

DUAL-DUCT SYSTEMS

In dual-duct systems, the cold duct is usually sized to carry the full design air quantity, while the hot duct may vary from 50 percent to 75 percent of this amount. The draftsman should always endeavor to locate parallel runs of hot and cold ducts at different elevations so that branch taps can cross mains easily.

Flexible duct is often used to connect rigid sheet metal to terminal units. Friction losses are greater in flexible round duct; and elbows, offsets, and other turns should be designed smoothly, with the total length kept to a minimum.

CASINGS

Built-up air conditioning systems generally use pull-through apparatus casings for single and double duct types, with the latter also utilizing a blow-through bank of coils so that hot and cold decks may be separated. Adequate bracing of casing panels is of prime importance, and consideration should also be given to a shear-type safety-panel section. A shear panel is connected to the main casing panels with bolts which would shear off at a predetermined design pressure. If a fan was started and return and outside-air dampers did not open, a buildup of excessive negative pressure would cause a relief blow off of the shear panel.

Casing cross-section is selected by maximum equipment size requirements, with safing for the perimeters of the smaller sections of equipment to prevent air bypass. Fans in casings should have a minimum clear distance at inlets of 0.75 wheel diameter between inlet and wall.

Sheet metal panels of large intake plenums, return-air fan casings, and apparatus casings are usually one of the following types:

A — Single-wall, which is either unlined, lined, or lined, but with perforated or expanded metal for maximum lining protection.

B — Double-wall, with thickness of panels required as insulation.

Single-wall panels are fabricated with perimeter flanges, angles, or standing seams, while double-wall panels are flush. As structural strength of the panels is primarily in the perimeter ribs, a selection should be made which will meet the design requirements and prevent panel pulsation or depression caused by system operating pressure.

Intermediate stiffening of double-wall panels is accomplished by fastening angles or Z-bars to the inside of the sheets before assembly. Cuts of standard sheet sizes and material bend allowances will determine finished panel dimensions. For example, a panel 19 inches wide, with ½" and 2" flanges on long sides will utilize 24 inches, or a cut of one-half a 48 inch sheet. (See Fig. 4-5.)

It is often necessary to connect a casing to a structural column encased in concrete or to a masonry wall. To compensate for possible construction misalignments, a covered angle or a bent channel is convenient for a mounting connection. This will provide an airtight seal and allow for adjustment of the single- or double-wall panels, respectively.

When equipment in a casing varies considerably in size and an advantage may be gained by reducing the casing cross-section, the reduction should be made by "easy" slope angles, of 30°, maximum. In mechanical-equipment room areas where the floor pitches to floor drains, it is desirable to mount the casing base-angle or channel on a level concrete pad or a perimeter curb approximately 4 inches high.

To prevent cooling-coil condensation carry-over, a bank of moisture eliminators is required on the air-leaving side of the coil. A drain pan should be mounted on insulation at a sufficient height to allow an adequate plumbing-trap seal for the fan system pressure. The drain-pan depth should be 3" high, minimum. Hot-dipped galvanized channels, equally spaced in the drain pan are used to support the coils in the pan. Man-size access doors should be provided for maintenance to all sides of equipment. Coil stands, fabricated from channels, angles, or both are also detailed on the shop drawing.

AIR FILTERS

Various types of air filters are used for air-conditioning applications. Manufacturers' detail cuts should be carefully checked for maintenance clearances. Mounting frames and accessories are usually furnished with throwaway, cleanable metal, replaceable dry media, and air-sleeve types. The automatic roll-curtain types are furnished in various stock sizes, depending on air-volume require-

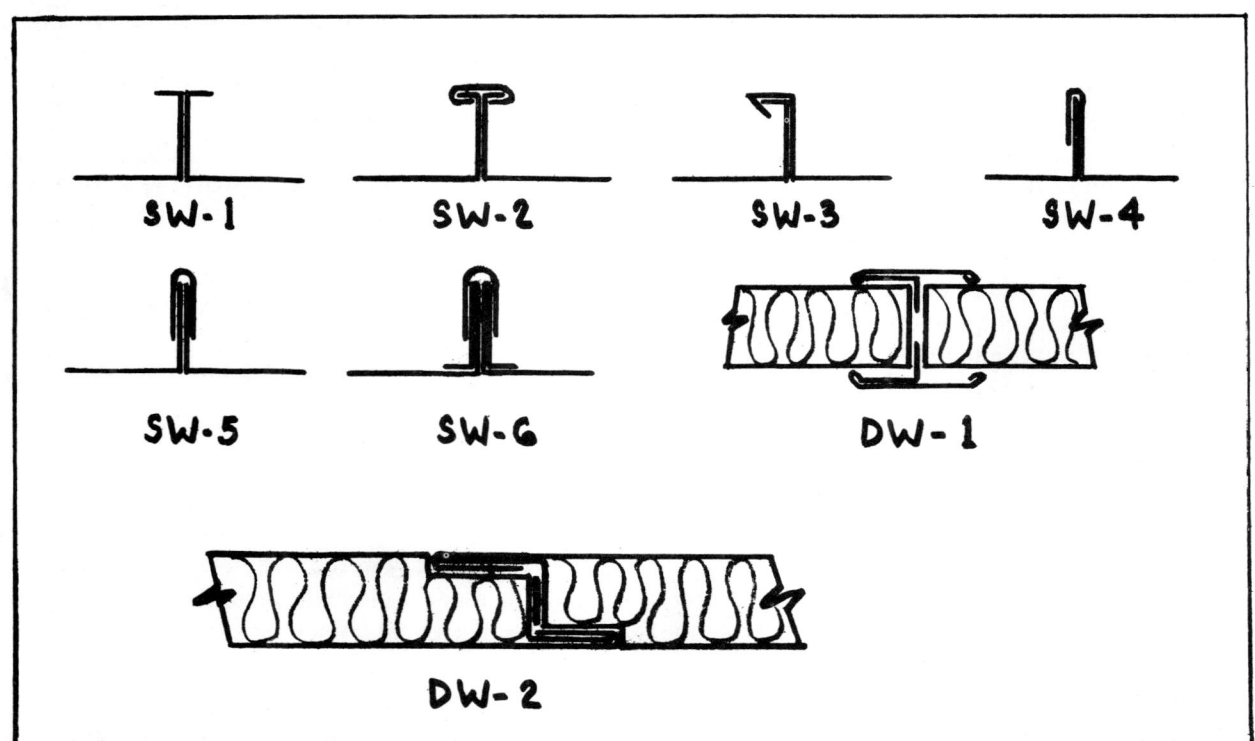

Fig. 4-5. Longitudinal casing details.

ments and the draftsman must take note of motor and drive locations on cuts. Electrostatic precipitator filters are arranged in cell banks, with some types having a self-washing feature.

HEATING COILS

Steam coils for pre-heat, heat, or re-heat must be mounted high enough to allow for pitch in condensate return piping. The draftsman should trace out the route of the return-pipe run so as to check minimum coil heights. When the drawing is coordinated with the piping contractor, all steam coil elevations should be verified. As this condition is primarily in the mechanical-equipment room, where space is at a premium, severe coordination changes to the sheet metal shop drawing can become an economic hardship. Coil banks, other than steam, need only be mounted on stands that provide enough room for the piping installation.

VIBRATION CONTROL

To prevent transmission of fan or duct vibration to the building structure, flexible fan connections, of canvas, asbestos, or neoprene as well as vibration isolation bases and hangers are used. Vibration-eliminator fan bases are generally placed under such equipment on concrete sub-bases, approximately 4 inches high. Inertia pad bases are primarily used for high-velocity fan equipment. These are large weight-mass concrete and steel bases with outrigger supports for spring-loaded mounts which may, or may not, be required to be fastened to the floor or sub-base.

LINING

While conventional, low-pressure duct systems can be designed to avoid excessive sound levels, noise generated by high-pressure fans and ducts requires special treatment. This problem is most acute for an installation having low background noise such as a recording studio or music rehearsal room. Glass fiber lining in plenums, casings, and ducts is commonly used for sound attenuation and serves both thermal as well as acoustical purposes. Design drawings most often indicate clear interior-duct sizes, so that actual duct dimensions must be increased to allow for lining thickness. Careful handling of lined ducts during fabrication and erection is necessary so that the insulation coating facing the air stream does not become damaged and erode under normal service. Raw edges of lining must be protected by metal or fabric nosing and should be so noted on the shop drawing, e.g., "stop lining with nosing."

Additional sound attenuation is accomplished by sound absorbers of various types which may have baffles, splitter plates, or cells and also by terminal units. When prefabricated traps are used, manufacturers' detail drawings must be checked for installation requirements, as connecting-angle frames may have to be furnished and this should be noted on the shop drawing. Lining is desirable in the low-pressure duct leaving terminal units. Unlined cold-air ducts routed through unconditioned spaces must be externally insulated.

ROOF-AIR INTAKES AND DISCHARGES

Intake or discharge air openings through the building structure (Refer to Figs. 5-6 and 5-7 in Chapter 5) may be accomplished by various weatherproof methods:

W-1. *Intake louver and bird screen* — A duct collar with an intake louver and bird screen is mounted to the masonry opening. The collar provides a mounting surface for a motor-operated damper.

W-2. *Exhaust louver* — A sheet-metal collar mounted to the louver frame encloses a gravity louver-damper and supports a propeller exhaust-fan.

W-3. *Wall-type exhaust fan* — A duct collar is fastened to the masonry opening and is then connected to the main duct.

W-4. *Gooseneck* — A rectangular-duct elbow with a ½ inch wire-mesh screened opening at 30° to the horizontal. A convenient application for an industrial building.

R-1. *Gooseneck and wire-mesh screened opening* for rectangular-duct roof opening and curb. Base flashing is by the roofer, cap flashing by HVAC contractor. The elbow is 135 to 150 with a small throat-radius. Screened-opening low point is generally a minimum height of approximately 3'-0 above roof line in snow localities. Although a small throat-radius is often used because of size limitations, the pressure drop may be decreased by expanding the cheek width to compensate for the area requirement of the wire-mesh screen.

R-2. *Roof exhaust fan* — The draftsman should indicate finished outside dimensions, including base flashing, for the roof curb opening.

R-3. *Roof ventilator* — Provides weather protection of discharge opening against wind-driven rain or snow.

R-4. *Discharge weather-cap with internal cone* — General application is in industrial exhaust systems.

R-5. *Louver house* — Has the advantage of a low profile.

A damper (motor operated or gravity-type) should be installed in the duct as close as possible to the building penetration, to prevent airflow through the system when the fan is not in use. Additional requirements for consideration are insect, bird, and burglar protection. The draftsman should be aware of types of materials, specified by the architect or mechanical design engineer, for metals exposed to the weather so that the shop drawing can be properly prepared.

Outside-air-intake velocities should be from 500 to 1,200 fpm, depending on the intake location and type of structure. Louvers and screens have a substantial free-area loss, as high as 50 percent of the framed opening area, while ½ inch wire-mesh, has a loss of approximately 20 percent.

FANS

Fans used in air-handling systems are of the following types: centrifugal, roof and wall types, cabinet units, axial, and propeller. They may be direct or belt driven, depending on the particular application. The centrifugal fan has an impeller or wheel with straight, forward curved, backward curved, or radial-tipped blades, mounted in a sheet or cast-metal scroll type of housing. This housing type is also used in many of the roof and wall exhaust fans, and cabinet unit arrangements. Axial fans are cylindrical and may have the motor or drive, in or out of the air stream. If the fan has guide vanes it is called "vaneaxial." Any of these fans is designed to move a wide range of air volumes at various pressures.

The propeller fan is primarily used to move large volumes of air at low pressure.

Inlet plenums or ducts to single-width, single-inlet centrifugal fans should have access doors for inspection and maintenance.

FAN SCROLL LAYOUT

A reasonably accurate fan-scroll profile with three radii is shown in Fig. 4-6. Each radius is similarly drawn by steps as in the following example:

To determine center and length of radius R1

A. With *C* as center, place off-distance *GC* on *BC*.

B. To *GC*, add ½ of the difference in length of *GC* and *BC*. This is the length of radius *R1*.

C. With *G* and *B* as centers, swing arcs equal to *R1* to intersect at *Point 1*, which is the center for radius *R1*.

Also shown is a spring-supported, vibration-eliminator, steel-frame fan base. The extreme outline in the plan is the concrete pad sub-base. If a 4 inch high pad was used, the fan base would be 9 inches above finished floor, as the spring-mount detail indicates an operating height of 5 inches. Note that plan tie-in dimensions are to centerlines of fan shaft and scroll housing.

ROOF EXHAUST FANS

The duct connection to a roof exhaust-fan is generally used as an intake plenum, and may have one or more branch tap-in connections. A 2 inch, or larger, flange turned out square on top, all around, is used for the plenum support at the top of the roof curb. Maximum duct-size should be

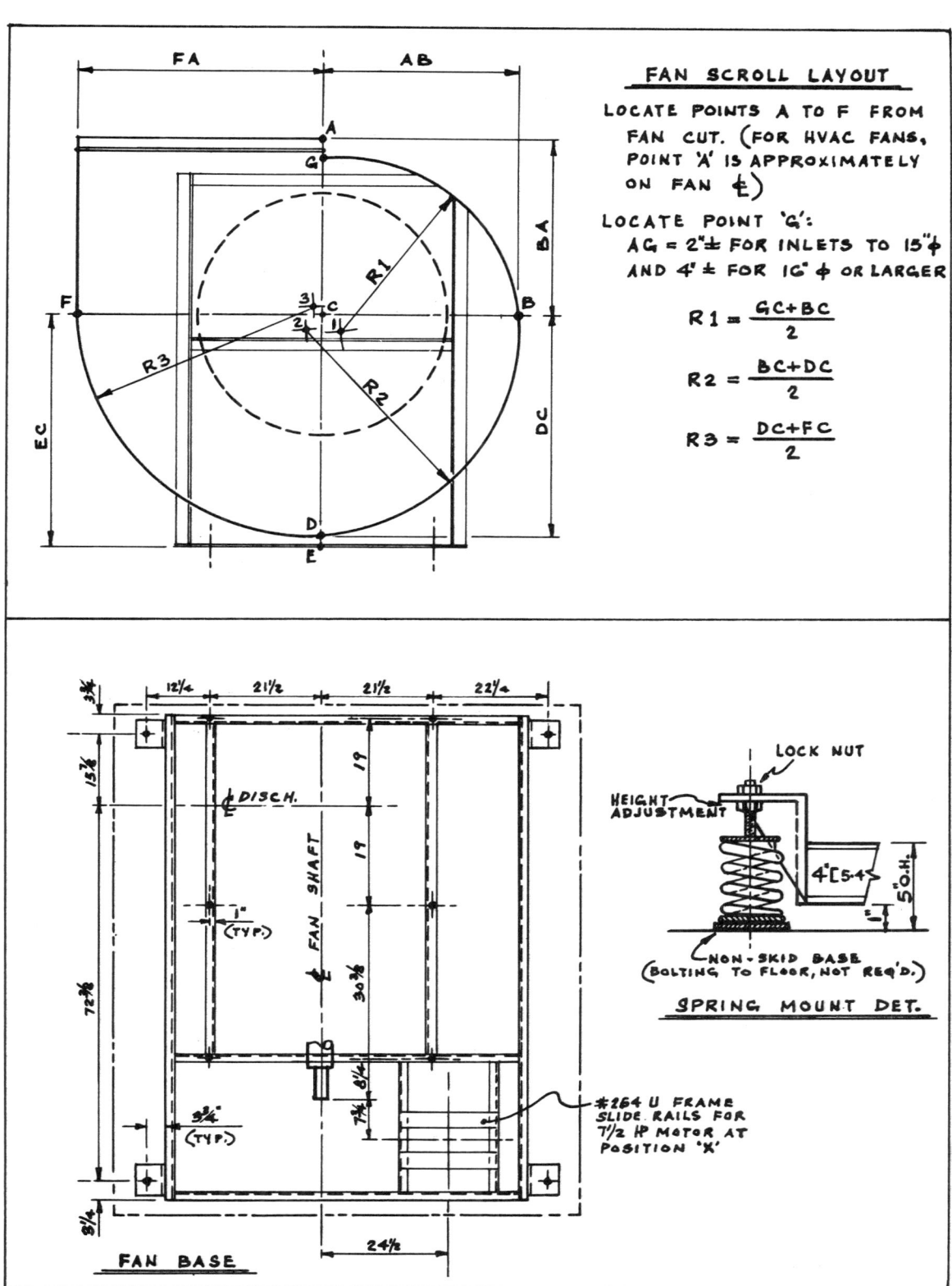

Fig. 4-6. Fan scroll layout and fan base.

½ inch less than the roof opening. The draftsman should check equipment cuts for electrical connection requirements.

CABINET FANS

Cabinet type fans are used for various air-handling applications. They may be suspended from the slab or floor mounted, and are available in horizontal, vertical, and combination arrangements. The manufacturers' detail drawings must be studied carefully to verify accessories furnished and their locations, and all dimensions necessary for the unit and duct-connection installation. Direct-fired warm-air units should have asbestos-cloth, flexible connections for inlet and discharge.

A sheet metal plenum-box is commonly used for an inlet to a cabinet unit. Depending on the job conditions, the plenum may be direct-connected or it may be isolated from the unit by a flexible connection. Each duct tap-in to the former must have a flexible connection. On draw-through units, it is desirable to converge the airstreams of separate outside- and return-air connections to the plenum so that entering air is blended and distributed evenly across the face of the coils for efficient heat transfer.

AXIAL FANS

Axial-flow fans may be arranged horizontally or vertically and either floor-mounted or suspended. Duct connections to and from this type of fan are "in-line" which is advantageous in areas of limited space conditions. These are often used in systems for fume exhaust with fans located on the roof-curb openings.

Propeller fans are generally mounted to a wall box which contains an automatic or gravity shutter to prevent airflow when the fan is not operating.

Fan classifications are determined by system operating total pressures as follows:

Class I   up to 3.75" TP   = Low Pressure
Class II 3.76" to 6.75" TP   = Medium Pressure
Class III 6.76" to 12.25" TP = High Pressure

Various air inlets and outlets may be selected for heating, ventilating, and air conditioning systems. An industrial building may use wire-mesh screen, perforated, or expanded metal for exhaust or return-air openings, and simple, plate-pan-type supply outlets. Air volume for each opening must be controlled in the duct itself by a splitter or volume damper.

To improve duct openings in function or appearance, manufactured devices called grilles are used. These may be of single-face bars, horizontal or vertical, stationary, or adjustable. Double-bar grilles, with horizontal and vertical bars of similar characteristics, are also available, as are floor type and ornamental, patterned grilles. Grilles do not have a self-contained control for air quantity. Registers are similar to grilles, but they have damper blades added behind the grille face so that air patterns and quantities may be controlled. Top registers with adjustable vertical and horizontal bars can be set manually, to desired airflow patterns.

Nominal duct sizes can usually accommodate registers or grilles. However, when a plaster-frame accessory is used for a register or grille, the nominal duct size is generally increased by ⅛ inch. Outlet sizes should be selected as per performance charts and data of the manufacturer.

Return-air register locations, while not as critical as supply selection, should consider efficiency in room air-circulation and a low draft-factor. Low sidewall or floor registers will pull warm air downward during the heating period and also induce cold floor-air into the system, while conditioned air for cooling will also have a more beneficial circulation.

As floor registers are natural dirt-catchers, provision must be made for maintenance. Mushroom-type inlets at the floor and below the seats of auditoriums or theaters are efficient for comfort conditions. Low velocities are used to transfer the air through the floor into plenums below the floor slab, and thence, into the return ducts.

High sidewall- or ceiling-returns have a tendency to short-circuit the flow of supply air before it has been distributed throughout the space. This can be particularly undesirable during the winter heating season as the cold air will lay along the floor level.

Return air is often circulated through a louvered door, an undercut door, a transfer grille through a wall, or a transfer duct from the room to the main return-air path. Acoustical treatment can also be accomplished with transfer ducts.

Many types and styles of ceiling diffusers are available — square, rectangular, round, half-round, and with blow patterns as required. The draftsman should check manufacturers' cuts for collar lengths, type of collar, inside or outside dimensions, ceiling opening required, and accessories supplied with diffuser. Blank-off baffles for installation in the diffuser neck should also be noted on the shop drawing. A diffuser collar tapped directly into a duct main should be avoided, especially if the noise level is critical and static pressure at the tap location is high. A branch duct to the diffuser is recommended as volume then can be controlled. A cushion head of approximately one neck diameter should be at the end of the duct branch. When short lengths are used for diffuser collars due to job conditions, distributing grids and vanes are recommended for uniform air-patterns.

Continuous, linear ceiling-diffusers and grilles require duct plenums with equally spaced, branch tap-ins. Ceiling construction and pattern layout must be carefully coordinated with the diffuser location. The draftsman must check manufacturers' cuts for type of diffuser, number of slots, type and size of plenum connection required, and installation tie-ins for the field.

Recessed lighting with distributing plenums for one or two sides of the light are convenient for ceiling outlets. Flexible duct should be used for connections to these units to compensate for slight installation irregularities, preferably in lengths as short as possible.

In cold climates, wall and glass downdrafts can be minimized by an installation of enclosures and grilles along the building periphery. The warm air supply at this surface will reduce discomfort to the occupants.

Perforated-pan ceiling panels are also used for conditioned-air-supply distribution. Manufacturers' data should always be checked for performance and installation requirements.

CHAPTER 5

Design Drawings and Specifications

Contract and Mechanical Drawings

When construction of a new building, or alteration in an existing building, is contemplated, the owner usually contracts for the professional services of architects and engineers. To satisfy the owner's purpose, his requirements, and his budget limitations for the work to be done, these architects and engineers will prepare design drawings and work specifications for the project. The drawings and specifications are of three major categories — mechanical, structural, and architectural. The drawings are generally drawn to scales of ⅛- or ¼-of-an-inch equal to one foot, and this scale is noted on the drawing. Estimates are prepared from these drawings by various contractors who will then submit bids for their phase of the installation. Bids are selected by the architect or engineer, and contracts signed on the basis of the design drawings which are then identified as the *contract* drawings. Subsequent revisions to the contract drawings may involve additional costs or credits to the owner. The sheet metal draftsman should check drawing number, scale, date, revision number (if any), orientation, and general notes of each design drawing used for reference while preparing *shop* drawings.

Basically, the *mechanical* drawings are schematic and intended to show general arrangements of equipment, sizes of ducts, piping, and wiring layouts. However, it is not uncommon for these drawings to locate ducts, pipes, or equipment by definite dimensions. If a draftsman is unable to achieve a location requirement, he should prominently note the reasons on his drawing.

HVAC Drawings

Most significant to the draftsman are the heating, ventilating, and air-conditioning drawings consisting of floor plans, mechanical-equipment room (MER) layouts, plans and sections, details, schedules, flow control and riser diagrams. Drawing numbers are generally prefixed: HVAC, AC, M, or H, corresponding to Heating Ventilating and Air Conditioning, Air Conditioning, Mechanical, or Heating, respectively. Although the draftsman is primarily interested in the duct systems, HVAC drawings also show piping for heating, cooling, refrigeration, and fuel oil. They may include pneumatic-tube systems, snow melting, automatic temperature-controls, and all mechanical equipment.

Double-line layouts are used most often for ductwork, although single-line drafting methods are not uncommon. The former distinguishes duct runs from piping layouts which are generally drawn single-line, so that duct arrangements and space requirements can be clearly analyzed. (See Figs. 5-1 to 5-7 for duct-system symbols and nomenclature.)

Floor Plans

Design drawing sheets are usually of adequate size to indicate entire floor-plan layouts. Buildings of very large area often require more than one drawing per floor. *Match lines* are used to connect points of adjacent drawings, along with notes indicating the continuation drawing number, e.g., (*Continued on Dwg. HVAC-6*). Also refer to Figs. 5-22 and 5-23, for match lines only.

HVAC floor plans indicate general arrangements of air inlets, outlets, and the related duct distribution systems. Using these drawings for reference, together with the corresponding structural and architectural design drawings, the sheet metal draftsman's responsibility is to locate air-system components on each shop drawing to conform to the intent of the design engineer.

(*Text continued on page 159.*)

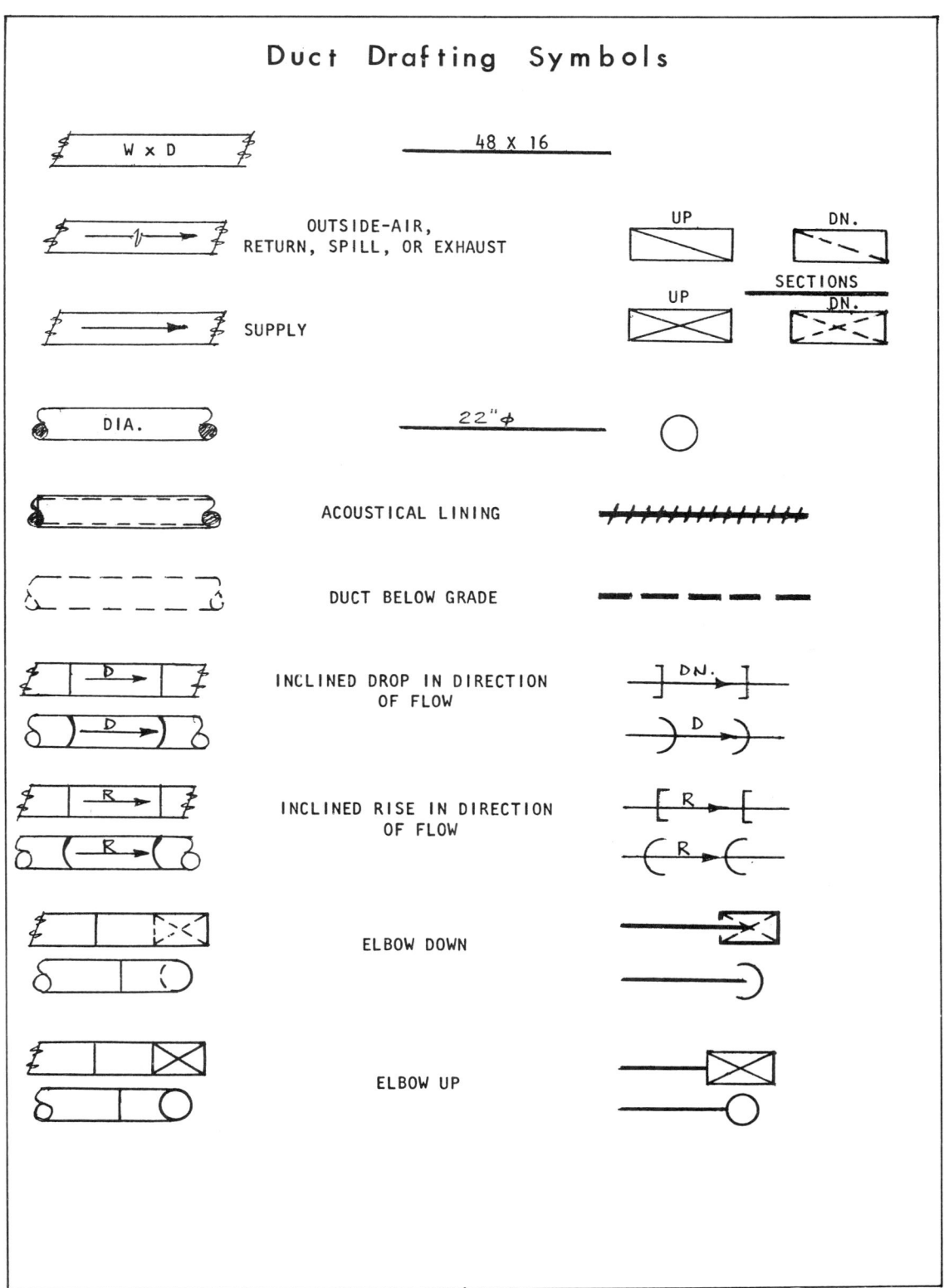

Fig. 5-1. Duct drafting symbols.

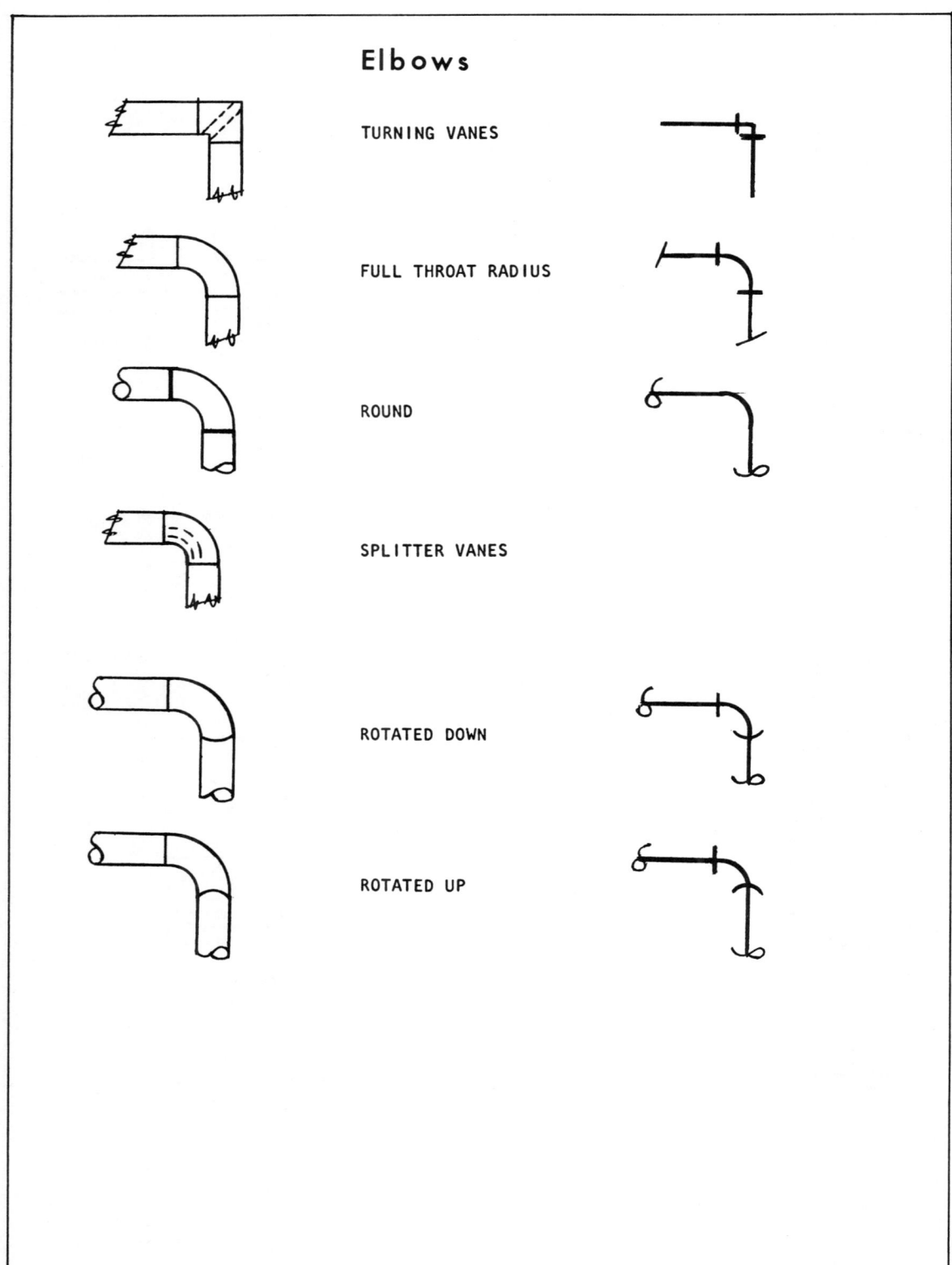

Fig. 5-2. Duct drafting symbols.

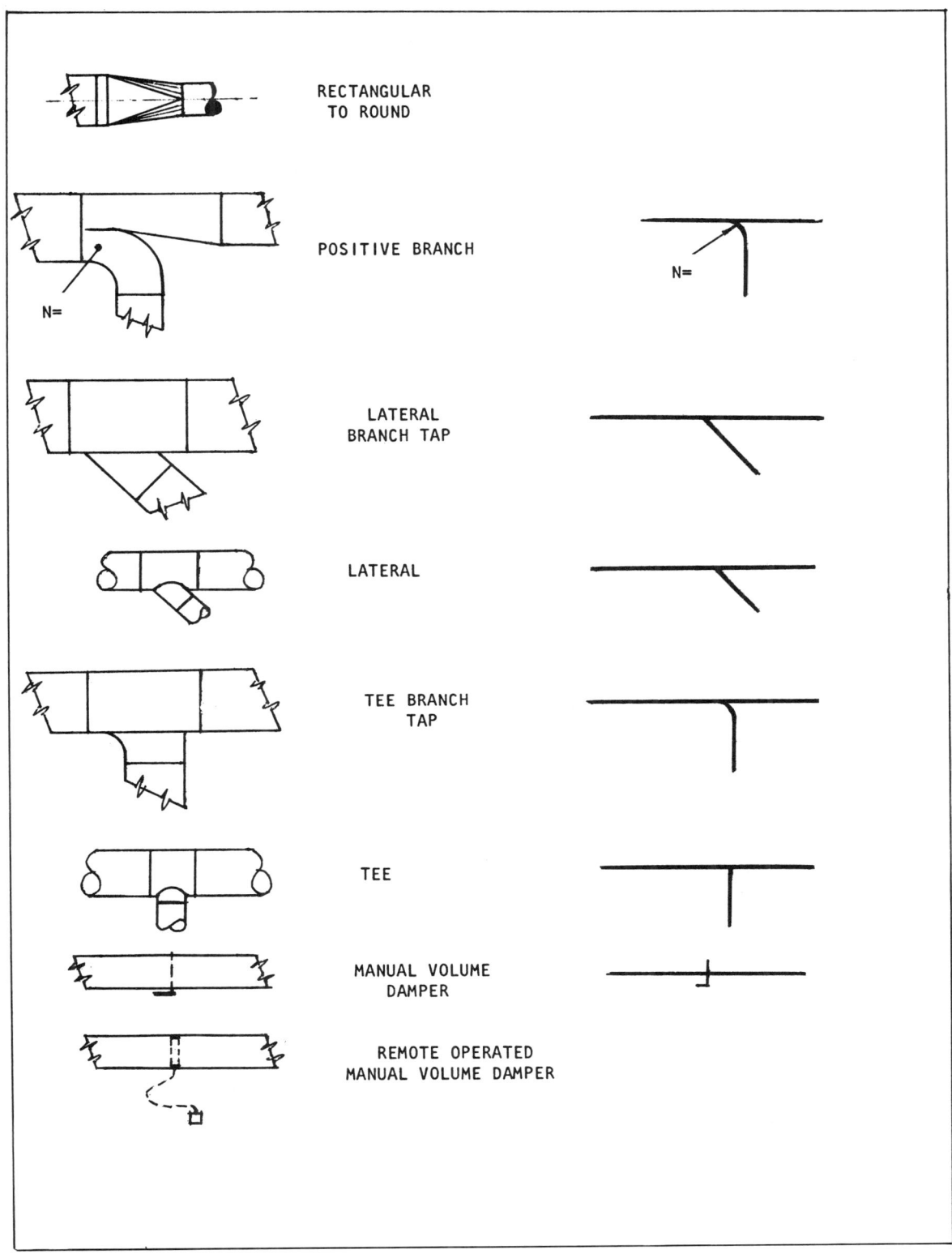

Fig. 5-3. Duct drafting symbols.

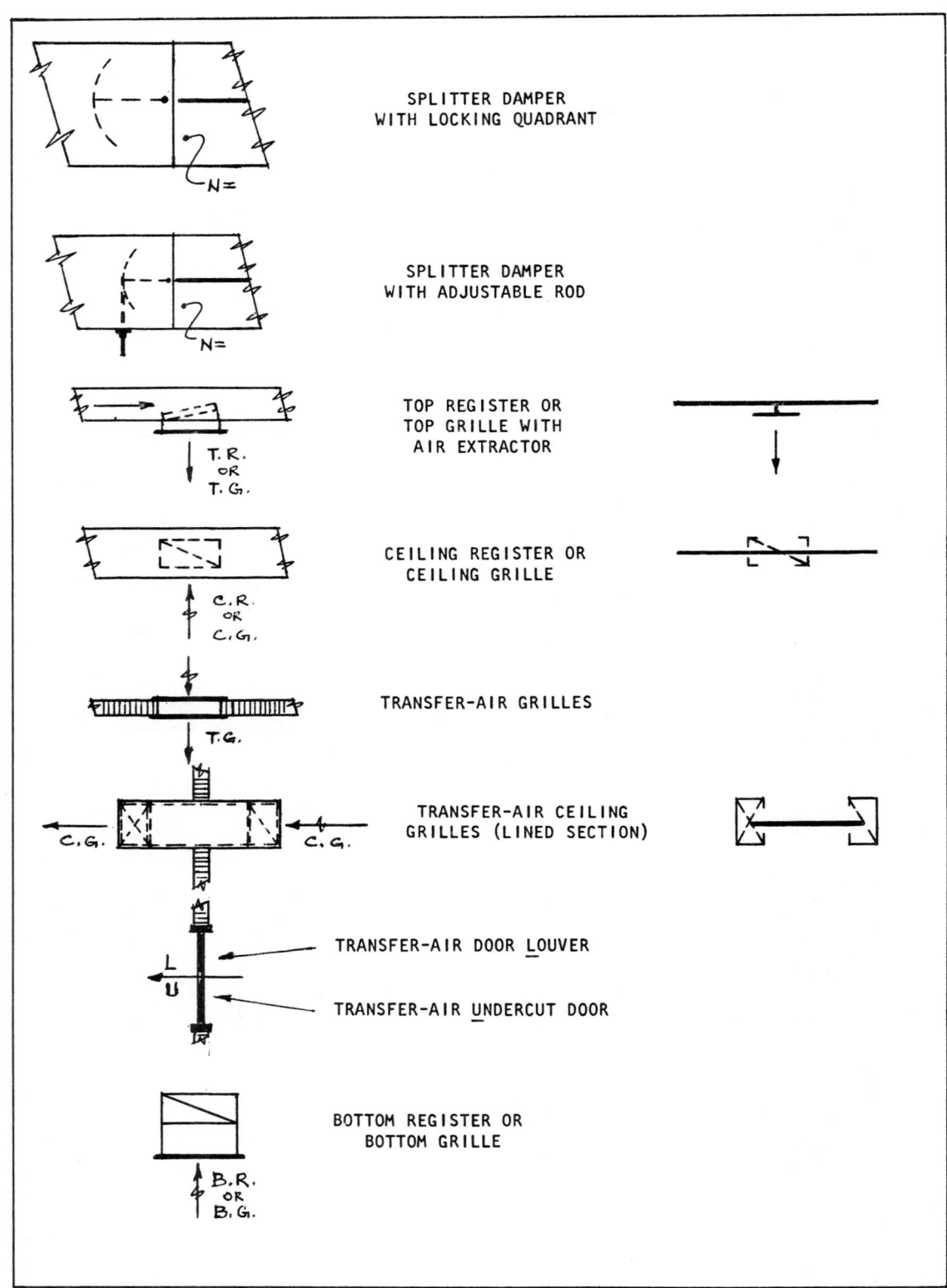

Fig. 5-4. Duct drafting symbols.

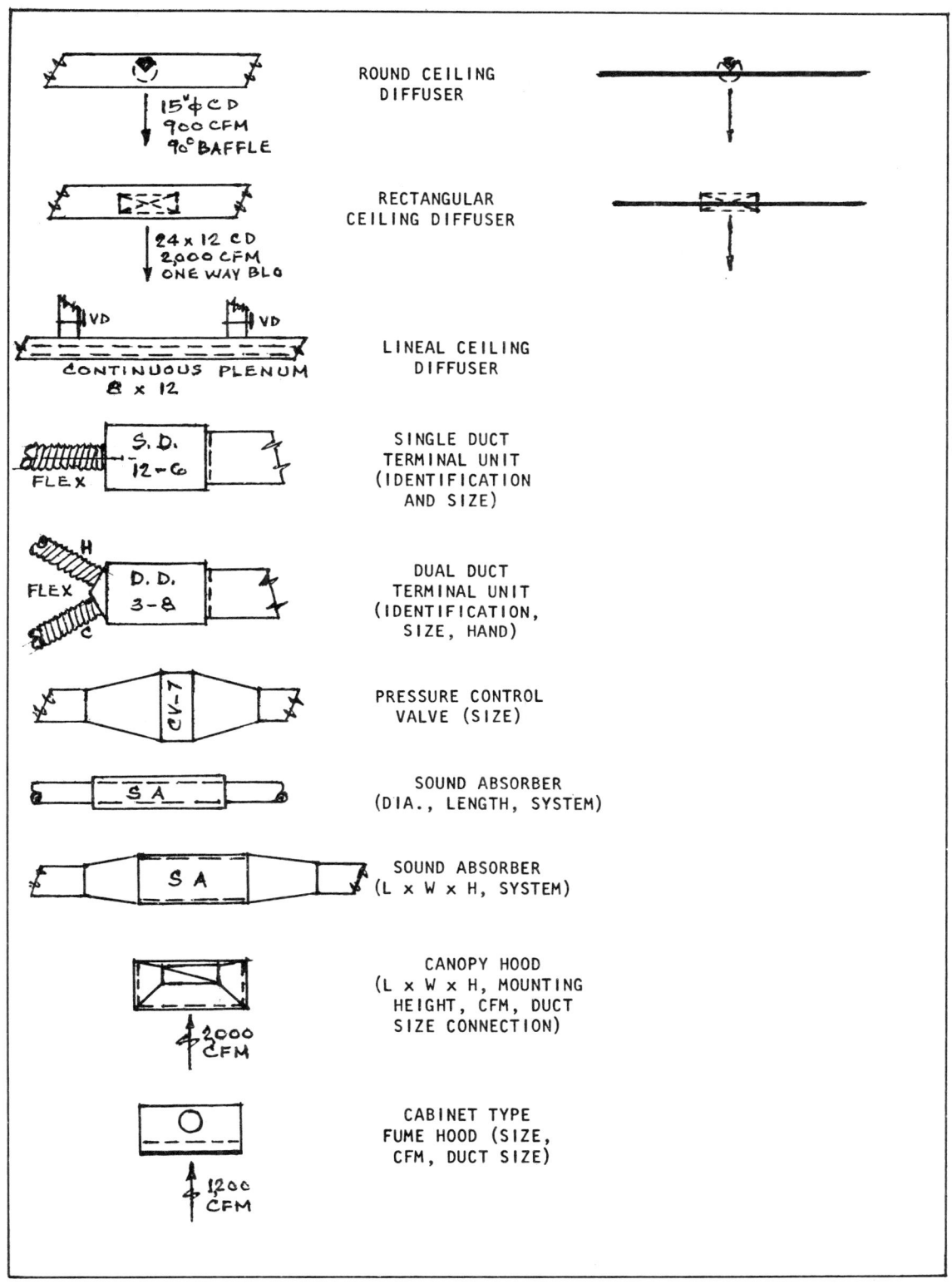

FIG. 5-5. Duct drafting symbols.

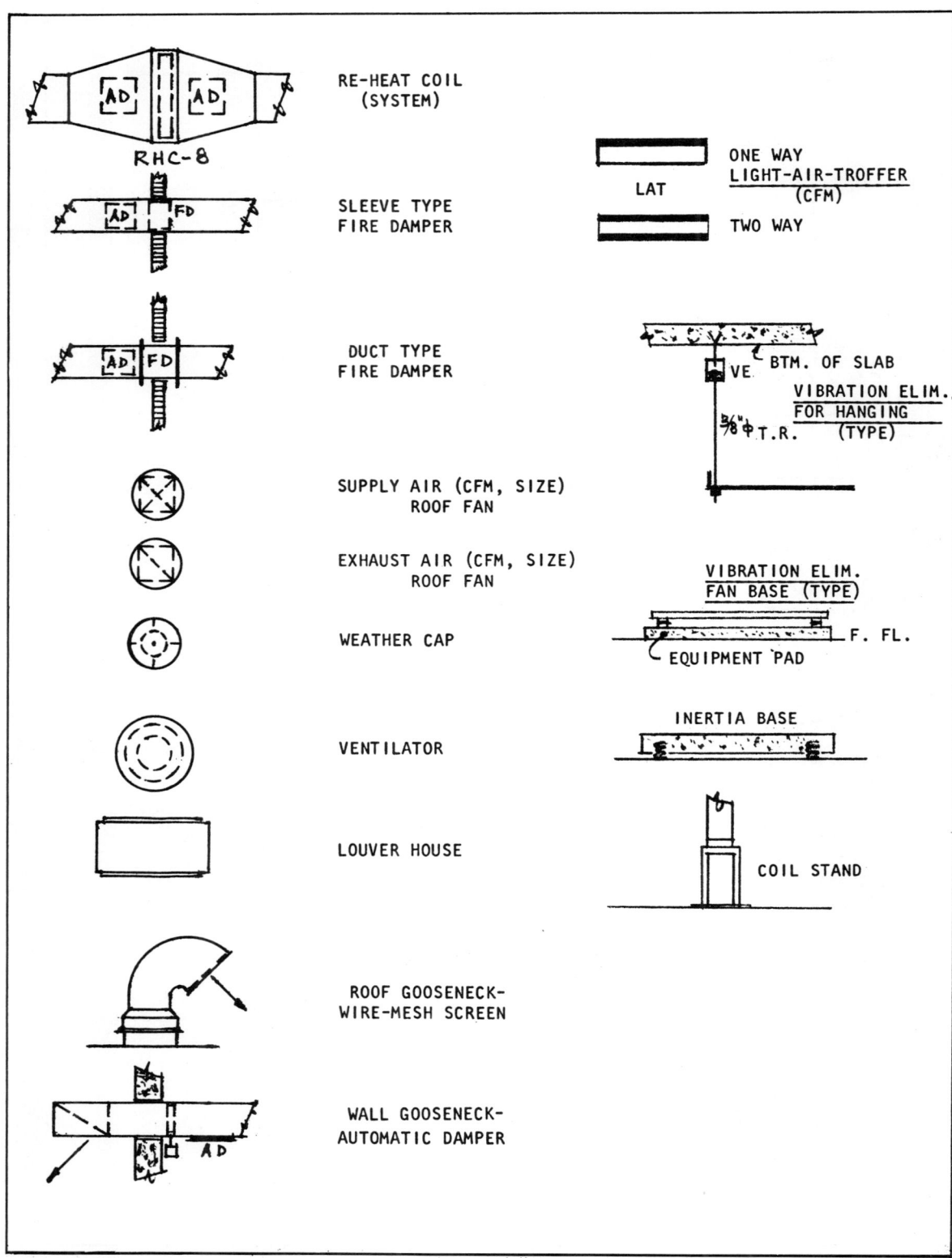

Fig. 5-6. Duct drafting symbols.

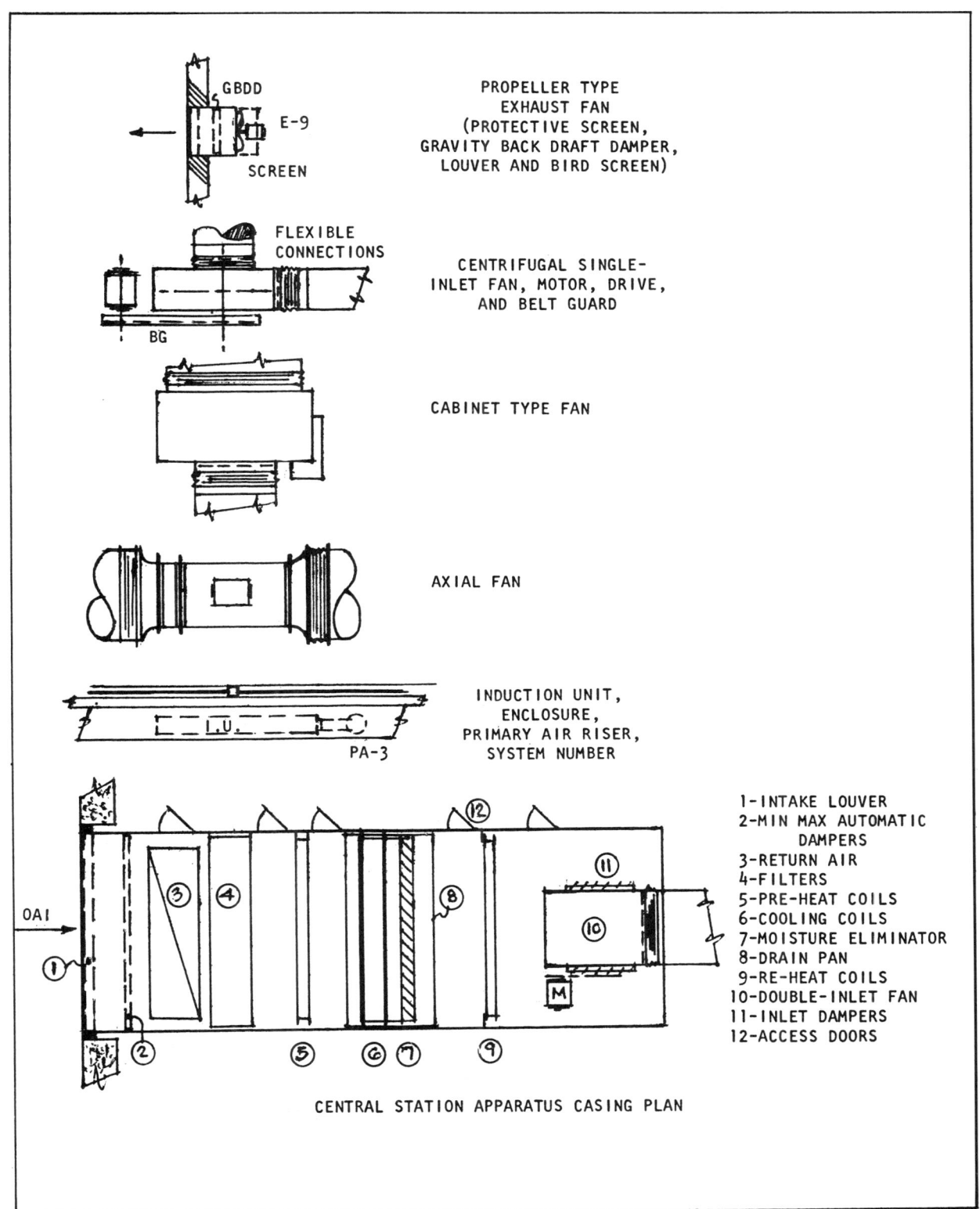

Fig. 5-7. Duct drafting symbols.

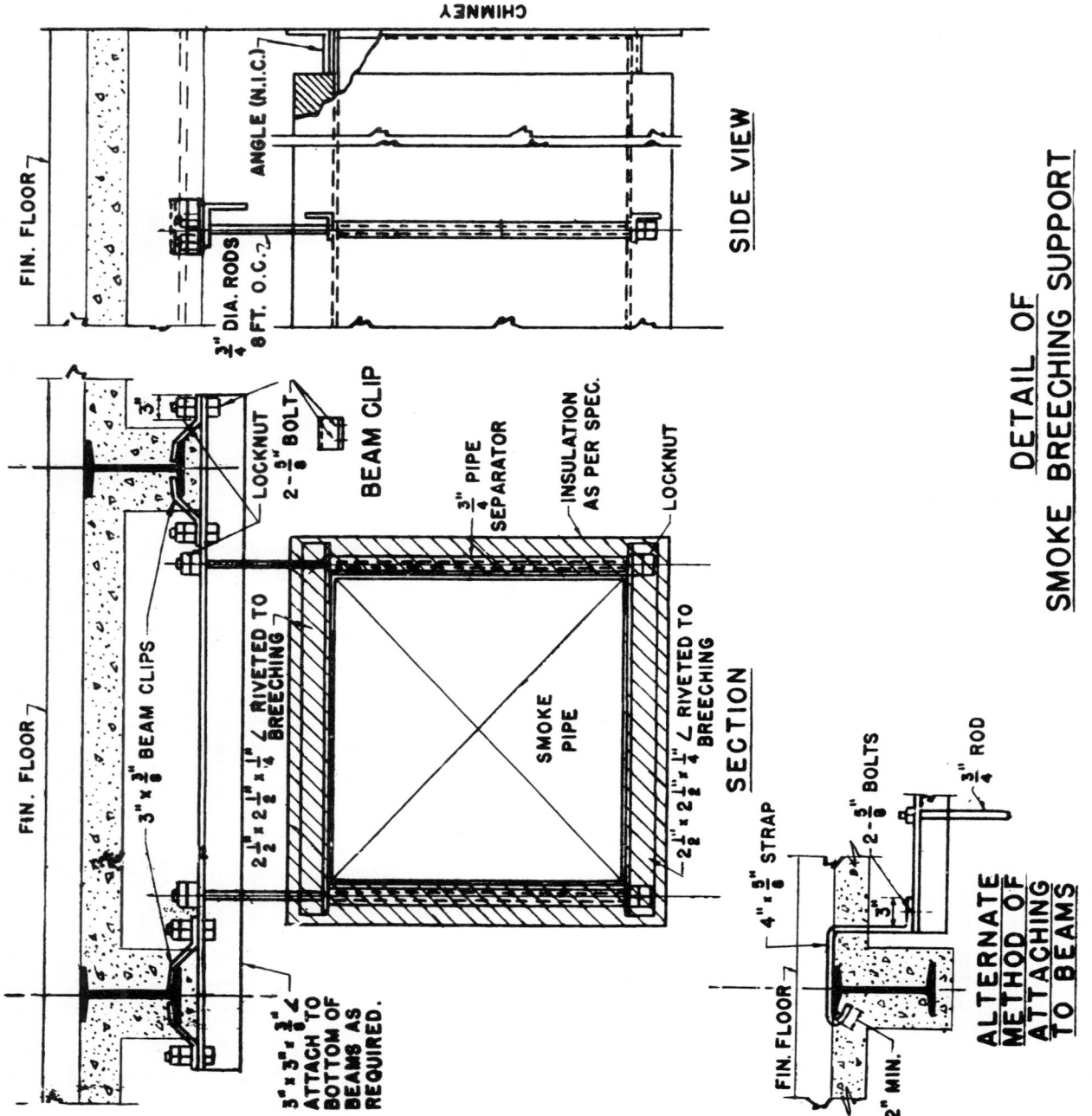

Fig. 5-8. Smoke breeching support.

(Text continued from page 150.)

Drafting procedures are affected by differences of duct-construction standards for various types of systems and methods of fabrication. The draftsman should make a visual analysis of the floor plans so that he may become aware of special material or construction requirements.

Duct materials for underground or below-grade installation are often specified as coated metal or rigid asbestos. Such materials must be ordered and the shop drawings prepared with sufficient lead-time so that ducts can be installed according to job-construction schedules and coordinated with work schedules of other mechanical and electrical trades.

Design drawings and specifications of duct systems used for boiler breechings, fume hoods, kitchen, dishwasher, and shower-room exhausts, as well as inlet and discharge ducts for interior cooling towers and evaporative condensers, must be checked for requirements of: types of materials, gages, watertight construction, expansion joints, and hanging details. (See Fig. 5-8 for smoke breeching, hanging details.) Concrete pour schedules at the job site will determine the sequence for making shop drawings. The drawings should indicate slab insert locations, slab, and concrete wall-openings for ducts. Although duct openings may be shown on architectural or structural drawings, they should be verified by the draftsman through field inspection, either personally or by a subordinate, during preparation of the drawing.

Plans must be checked against specifications of lined ducts to confirm net, clear interior-duct sizes, thickness, and extent of lining.

Single or double-duct supply systems to terminal units or pressure-reducing control valves are known as high-pressure ducts, and require special instructions from the draftsman. Ductwork downstream of mixing units or pressure-reducing boxes is low-pressure, and generally requires acoustical lining for a specified distance. Accessibility to these units for maintenance must be considered by the draftsman. (See Figs. 5-9 to 5-11.)

Supply- and return-air grilles, registers, and diffusers are available in many types and styles. Manufacturers' drawings indicate inside or outside dimensions for making sheet metal collars or other special requirements for connection to their equipment. The draftsman should note on the shop drawing where special grilles or air-turning devices are to be installed. (See Figs. 5-12 to 5-19.)

Air troffer units and linear diffusers, installed in coordination with lighting fixtures, are used extensively in many new buildings. Final connections to troffers are made by the use of flexible duct, with the sizes determined by manufacturers' dimensions. (See Figs. 5-20 and 5-21.)

Typical HVAC floor plans for the interior portion and also a section of an office building are shown in Figs. 5-22, 5-23, and 5-24.

Similarity in floor-plan layouts can offer drafting advantages and should always be exploited by the draftsman. Figure 5-25 is a partial plan of ducts near a riser shaft at the main level of an industrial plant. An identical, opposite-hand layout of a riser and branch ducts through a shaft, occurs at the other end of the building — near col. A-5, as shown by Fig. 5-26 in Sect. A-A.

Mechanical-Equipment Rooms

The shop drawings most difficult to prepare are those of the mechanical-equipment room (MER), more commonly called *fan rooms*. Equipment and ducts in equipment rooms are usually large, piping is abundant, and space is generally at a premium. Location of floor drains, concrete bases for equipment and casings, placement of equipment, and workable clearances for all trades are dependent on well-prepared shop drawings.

Partial duct layouts of fan-coil units and a centrifugal return-air fan are shown in Figs. 5-27 to 5-29. The double concentric circles are for enclosing shop drawing piece numbers of fittings which are acoustically lined.

Figures 5-30 and 5-31 are design drawings of a small fan room. Equipment, ducts, and piping are shown in general arrangement. Before a sheet metal shop drawing of such a room is started, the draftsman must have approved cuts of related equipment. The student draftsman should study Figs. 5-30 and 5-31, and prepare a list of equipment for which cuts would be required.

Built-up central-station apparatus casings are shown in Figs. 5-32, 5-33, and 5-34. Close coordination with the piping contractor is necessary

(Text continued on page 188.)

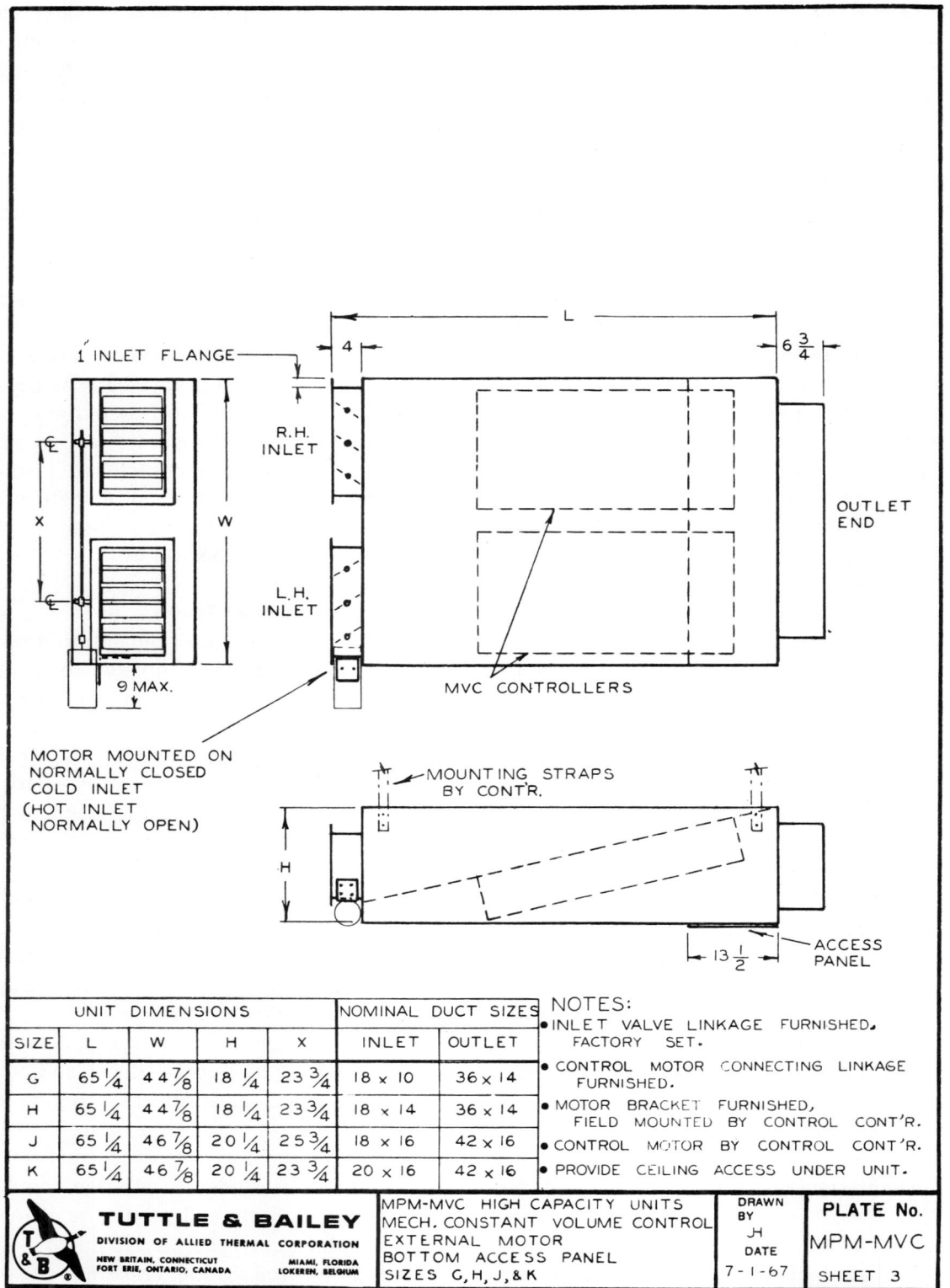

Fig. 5-9. Mixing unit cuts.

Ch. 5   DESIGN DRAWINGS AND SPECIFICATIONS   161

high velocity mixing box

SUBMITTAL SHEET   HV 4-8

	TYPE	VOLUME ADJUST	INLET
HV ___ ___ ___ ___	☐ E END DISCHARGE	☐ C MECHANICAL CONSTANT VOLUME	☐ D DUAL INLET
TYPE / SIZE / VOLUME ADJUST / INLET	☐ D DIFFUSER	☐ V VARIABLE CONSTANT VOLUME	☐ S SINGLE INLET WITH SHUT-OFF VALVE
	☐ O OCTOPUS	☐ D DIAL CONSTANT VOLUME	☐ O SINGLE OPEN INLET

HV	C ONLY 4	5	6	7	8
INLET O.D.	4	5	6	7	8
HVD OUTLET O.D.	5 or 6	6 or 8	8 or 10	10 or 12	12 or 15
HVO OUTLET O.D.	—	4	5	6	8
H	7 1/2	7 1/2	8 1/2	9 1/2	10 1/2
L	22	22	30	32	42
W	32 1/2	33 1/2	39	52	53
A	—	5 1/4	5 3/4	6	7 1/4
B	2 3/8	2 1/2	2 3/4	3 1/2	3 3/4
C	8 7/8	9 11/16	13	14 3/16	19 9/16
D	18	18	18	24	33 3/4
E	6 1/2	6 1/2	8	8	10 1/2
F	24	24	32	34	44
G	12	12	16	17	22
J	21	21	25 1/2	37 1/2	37 1/2
K	2 7/8	3	3 1/4	2 3/4	3 3/8
M	17 3/4	19 3/8	26	28 3/8	39 1/8
T	5 1/2	5 1/2	6	6	8
V	8	13 1/2	13 1/2	22	22

All dimensions in inches

Dual Inlet (D) shown. Hot Inlet normally open and may be L.H. or R.H. Single Inlet available with shut-off valve (S) normally open or normally closed and with open inlet (O).

Pneumatic Operators furnished by control contractor and internally mounted by Anemostat. With O PSI the pneumatic operator on the VCV regulator is set for minimum air volume.

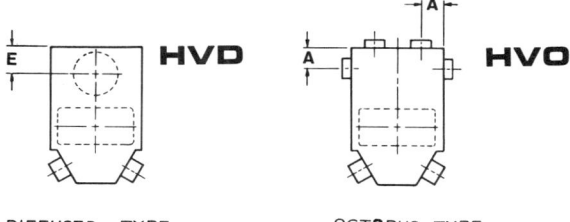

END DISCHARGE TYPE

DIFFUSER TYPE   OCTOPUS TYPE

Job Name & Location	Submitted by

ANEMOSTAT® registered trademark of **ANEMOSTAT PRODUCTS DIVISION, DYNAMICS CORPORATION OF AMERICA**, Scranton, Pennsylvania

Fig. 5-10. Mixing unit cuts.

162 DESIGN DRAWINGS AND SPECIFICATIONS Ch. 5

SUBMITTAL DATA							TYPE VB MIXING UNIT					
SIZE	NOMINAL CFM	A	B	C	D	DIA. E	F	G	H	J	K	S
4 VB	100-230	32	29	8	4¾	4	4½	13	9½	3	3½	22¾
5 VB	200-350	32	29¾	8	5½	5	5¾	14	9	3	3½	22¾
6 VB	300-450	36	31¾	9	6¼	6	6	15¾	10¼	4⅝	5½	24¼
7 VB	400-650	48	34½	10	6¾	7	6½	17¾	15¼	8¼	11½	25¼

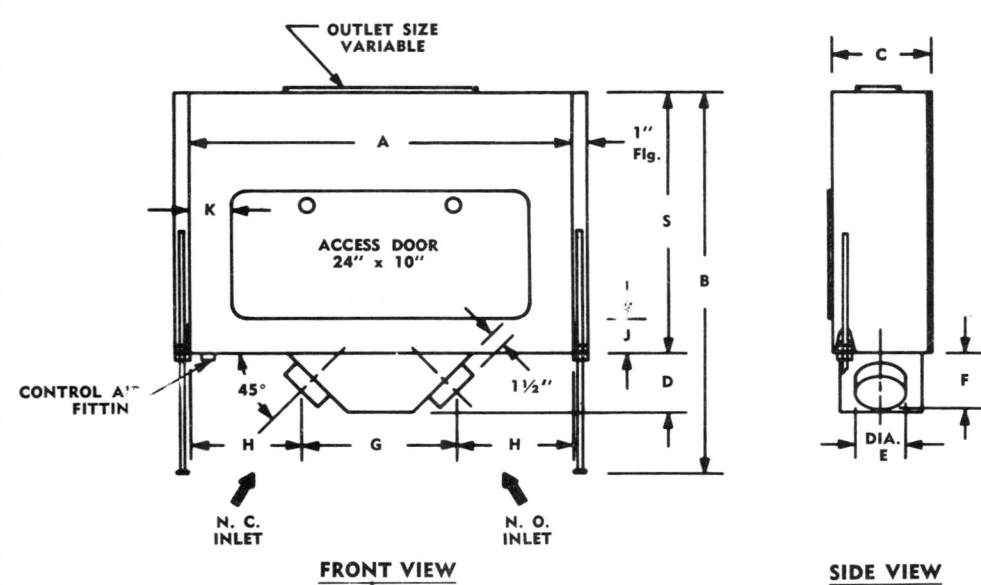

NOTES:

(1) Position of normally closed inlet, so marked when shipped, determines hand of unit. Unit shown is left hand. Units are furnished right or left hand as required.

(2) When controlled by direct acting thermostat, connect cold air to normally closed inlet.

(3) When controlled by reverse acting thermostat, connect warm air to normally closed inlet.

(4) Each unit is shipped complete with self-contained air volume regulator preset at the factory for specified air volume.

(5) Casings are aluminum.

(6) All units have built in sound attenuators and include thermal and acoustical lining.

(7) Control air fitting is ⅛".

(8) All dimensions are in inches and weights in pounds.

(9) Dimension "E" is connecting duct size.

(10) For minimum static pressure and sound levels see Bulletin No. DD-276-A.

(11) Dimensions also apply to units with Dual Volume (DV).

(12) Adjustable legs and brackets are furnished by Buensod for field mounting by others.

PROJECT: _____ BUENSOD AGENT: _____
_____ SUBMITTAL DATE: _____ BUENSOD JOB NO. _____

BUENSOD
DIVISION AERONCA, INC.
MANUFACTURED PRODUCTS

DRAWING NO. CS-571A
DATE 10-5-64 REV. 7-22-68

FIG. 5-11. Mixing unit cuts.

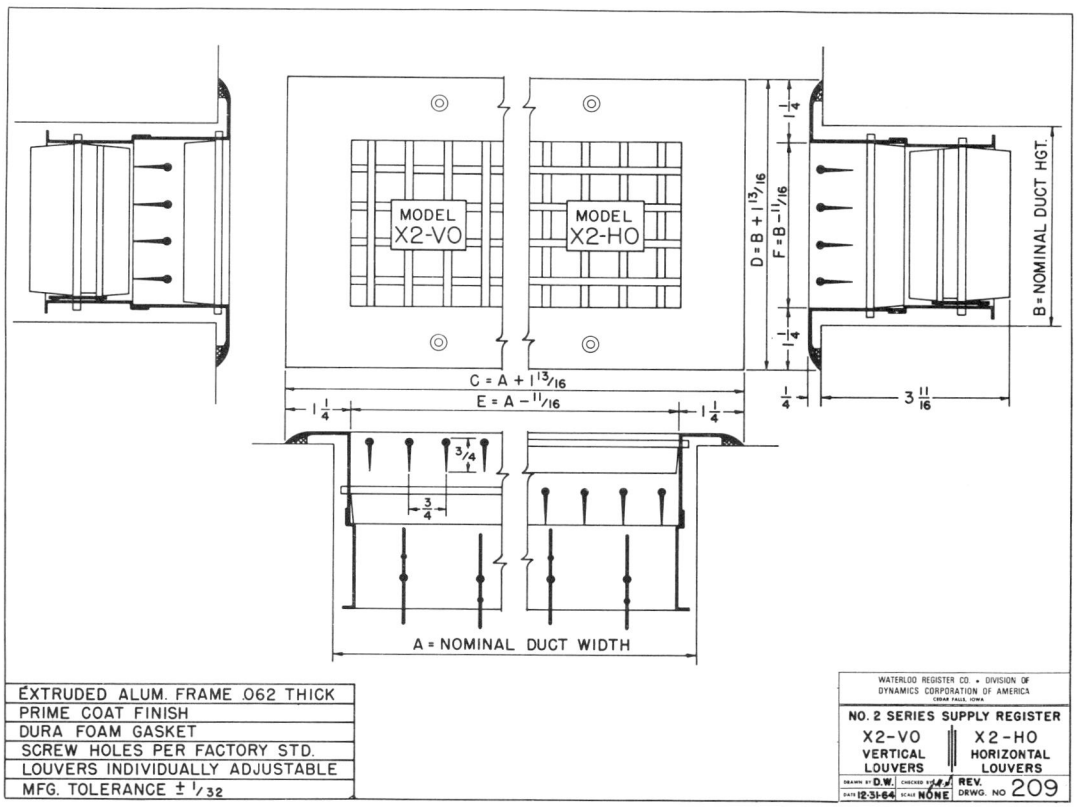

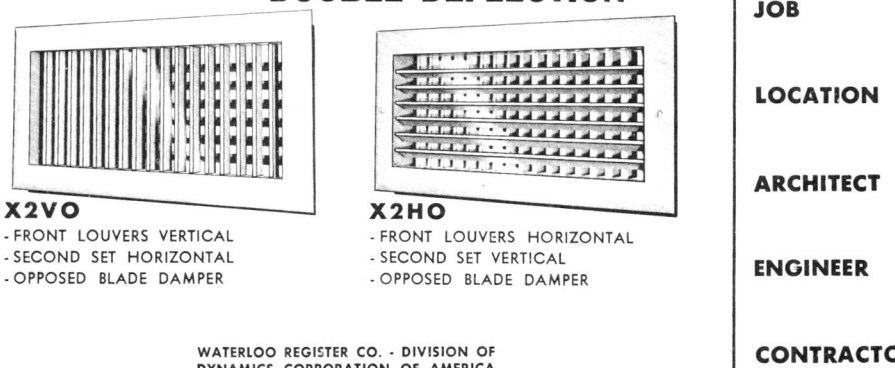

Fig. 5-12. Register, grilles and diffusers.

164   DESIGN DRAWINGS AND SPECIFICATIONS   Ch. 5

SUBMITTAL SHEET — CM-1

circular diffusers

flush-to-ceiling exposed ductwork

SUPPLY

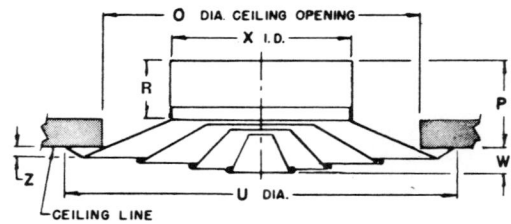

MATERIALS AND FINISH Steel construction with air-dried aluminum lacquer prime finish No. 25.

SIZE	O	P	U	W	X	Z	R
4	10½	2⅜	13	¾	4	⅜	1
5	10½	2⅛	13	¾	5	⅜	1
6	10½	2	13	¾	6	⅜	1
8	16	2½	18	1	8	½	1
10	21	3⅜	24	1⅛	10	½	1
12	21	2⅞	24	1⅛	12	½	1
15	31	4⅜	34	1⅜	15	½	1
18	31	3¾	34	1⅜	18	½	1
21	45	7	48	1⅞	21	¾	2
24	45	6¼	48	1⅞	24	⅝	2
30	54½	7½	60	2¼	30	¾	2
38	65	8¾	68	2⅝	37⅞	¾	2

All dimensions in inches

Job Name & Location	Submitted by

ANEMOSTAT® registered trademark of **ANEMOSTAT PRODUCTS DIVISION, DYNAMICS CORPORATION OF AMERICA,** Scranton, Pennsylvania

FIG. 5-13. Register, grilles and diffusers.

SUBMITTAL SHEET

accessories

CU-I

air flow equalization volume control

SUPPLY DAMPERS

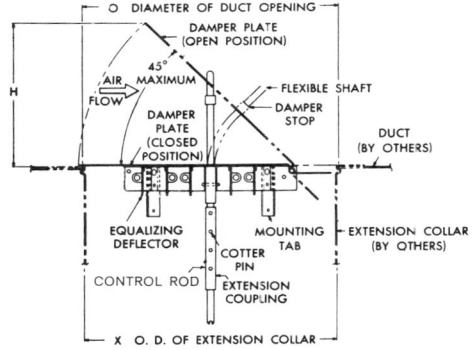

MATERIALS AND FINISH Steel construction with flat black lacquer finish No. 200.

ROD DIAMETERS
SIZES 4" THROUGH 18"—¼" DIA.
21" THROUGH 30"—11/16" DIA.

SIZE	H	O	X
4	2¼	4	3⅞
5	2⅞	5	4⅞
6	3½	6	5⅞
8	5¼	8	7⅞
10	6	10	9⅞
12	7½	12	11⅞
15	9⅞	15	14⅞
18	11⅞	18	17⅞
21	14	21	20⅞
24	15⅜	24	23⅞
30	19⅛	30	29⅞

All dimensions in inches

Job Name & Location	Submitted by

ANEMOSTAT® registered trademark of **ANEMOSTAT PRODUCTS DIVISION, DYNAMICS CORPORATION OF AMERICA,** Scranton, Pennsylvania

FIG. 5-14. Register, grilles and diffusers.

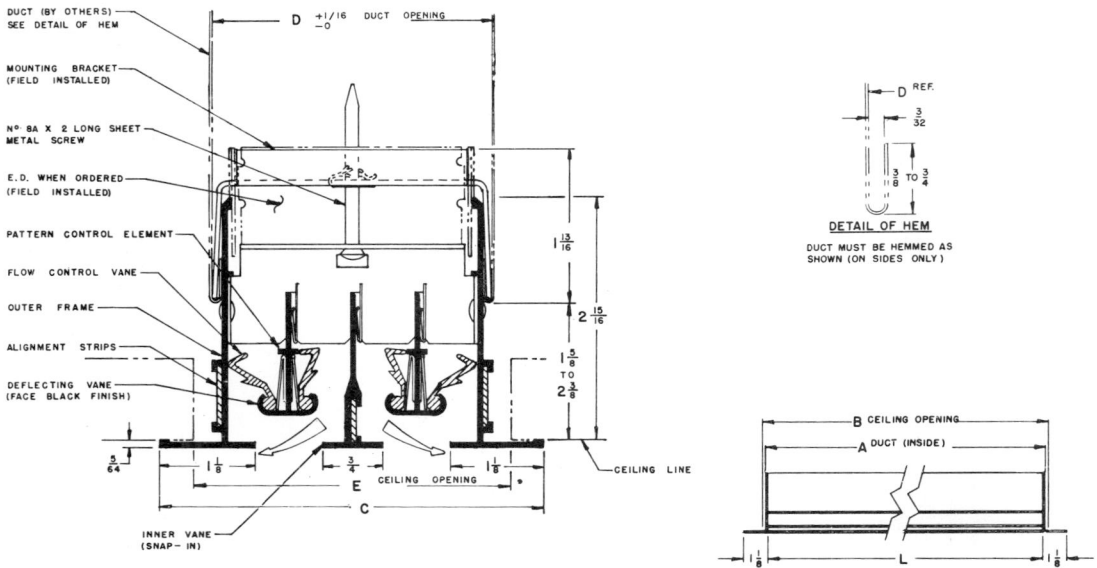

FIG. 5-15. Register, grilles and diffusers.

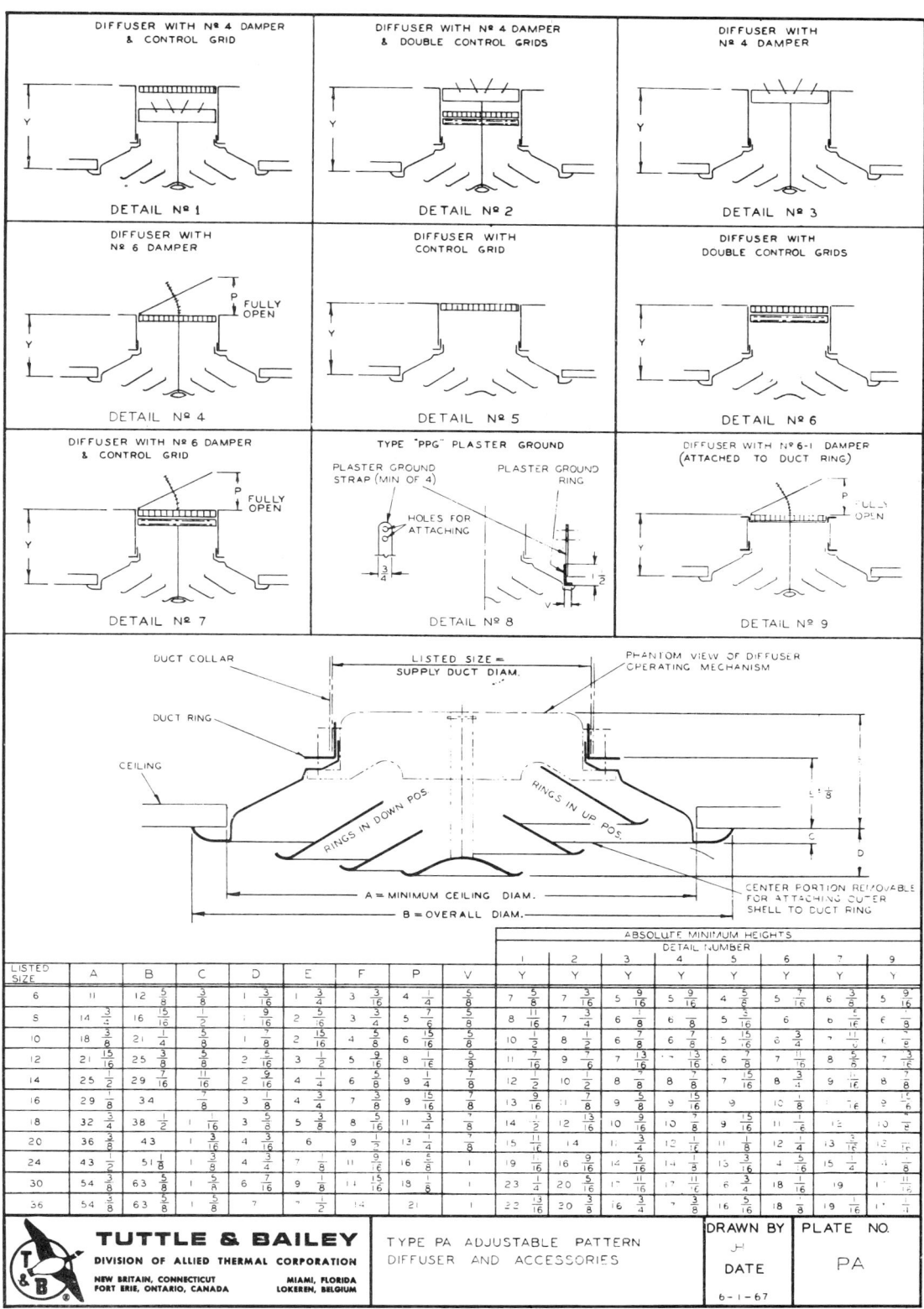

Fig. 5-16. Register, grilles and diffusers.

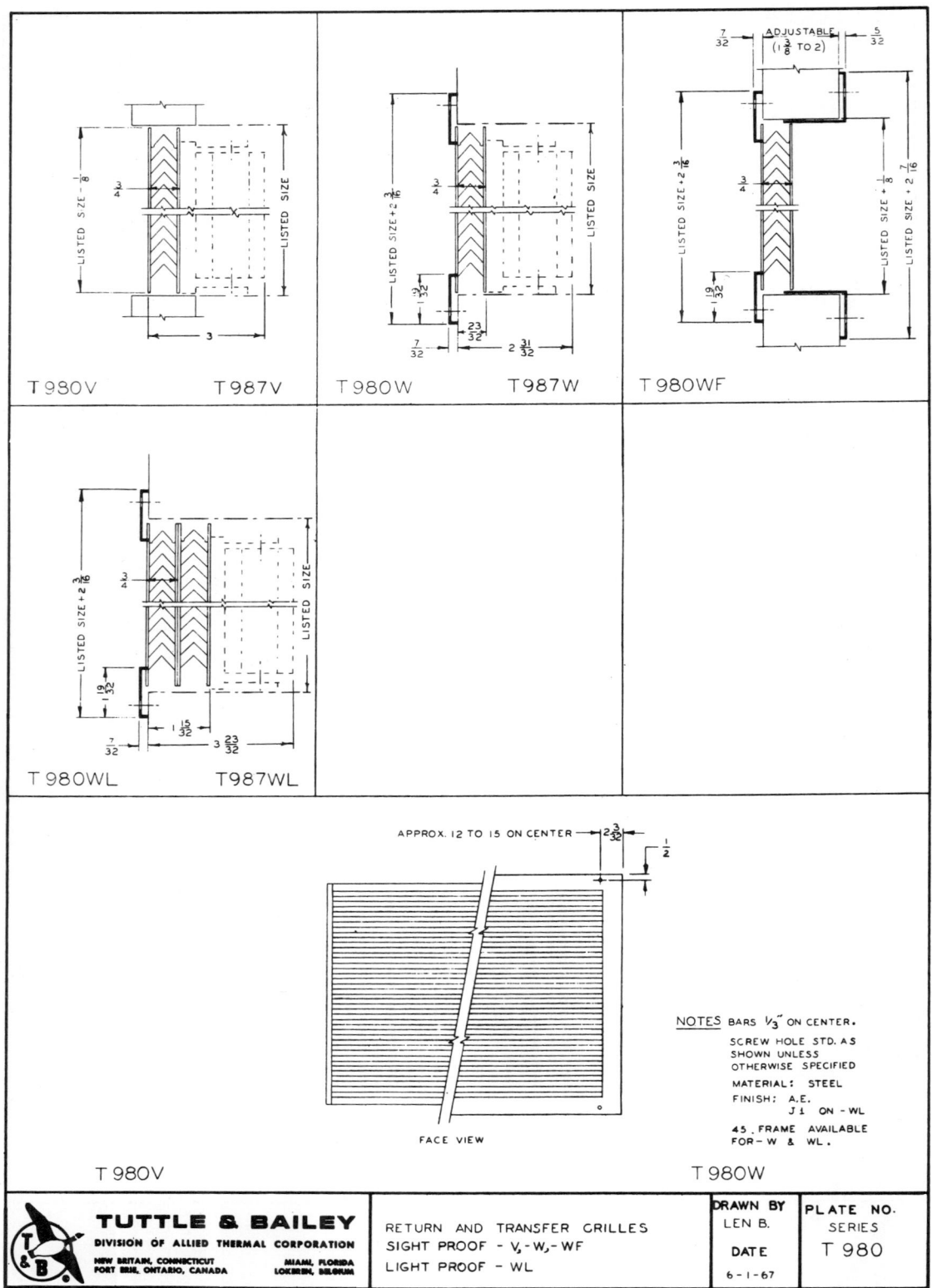

Fig. 5-17. Register, grilles and diffusers.

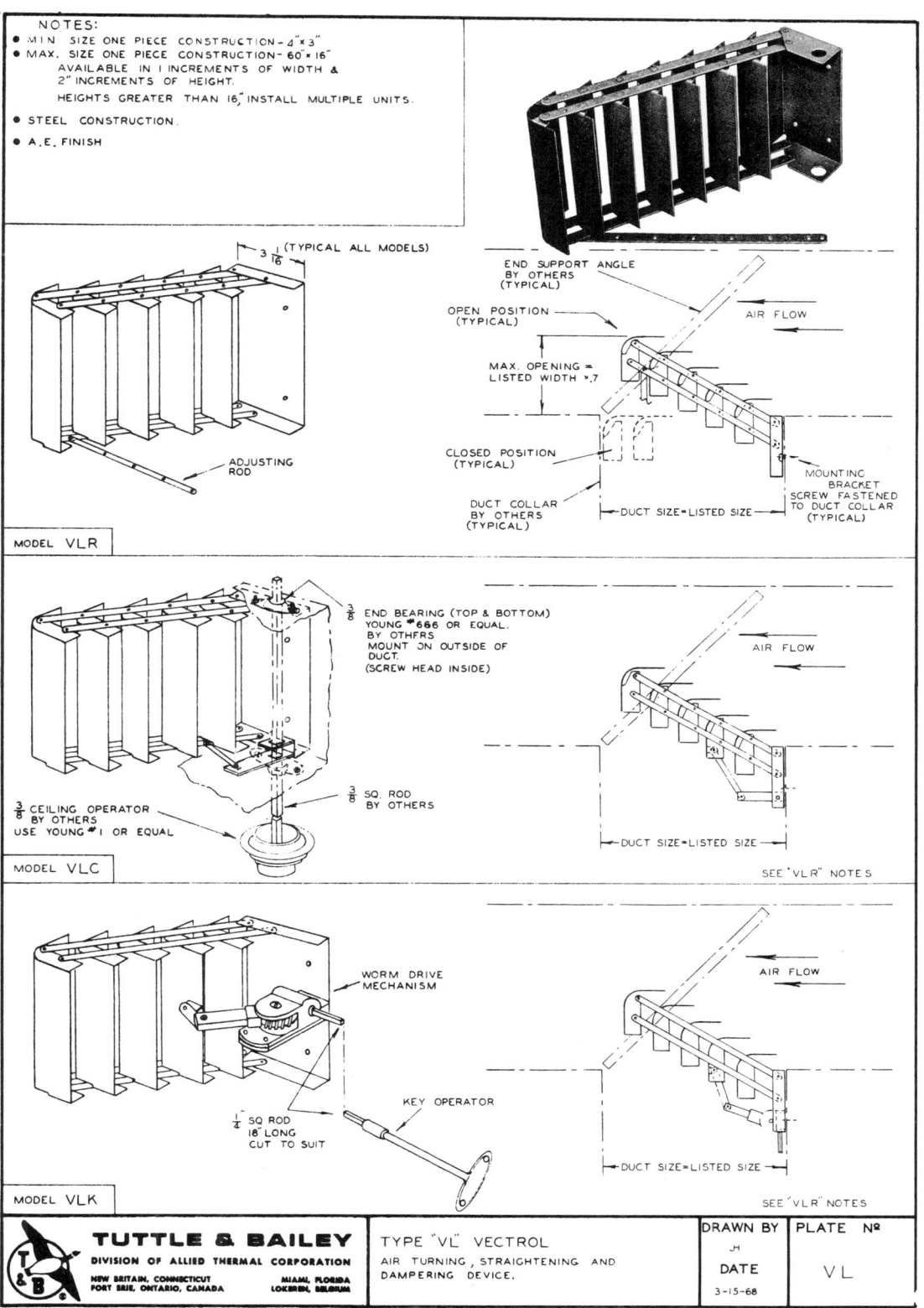

Fig. 5-18. Register, grilles and diffusers.

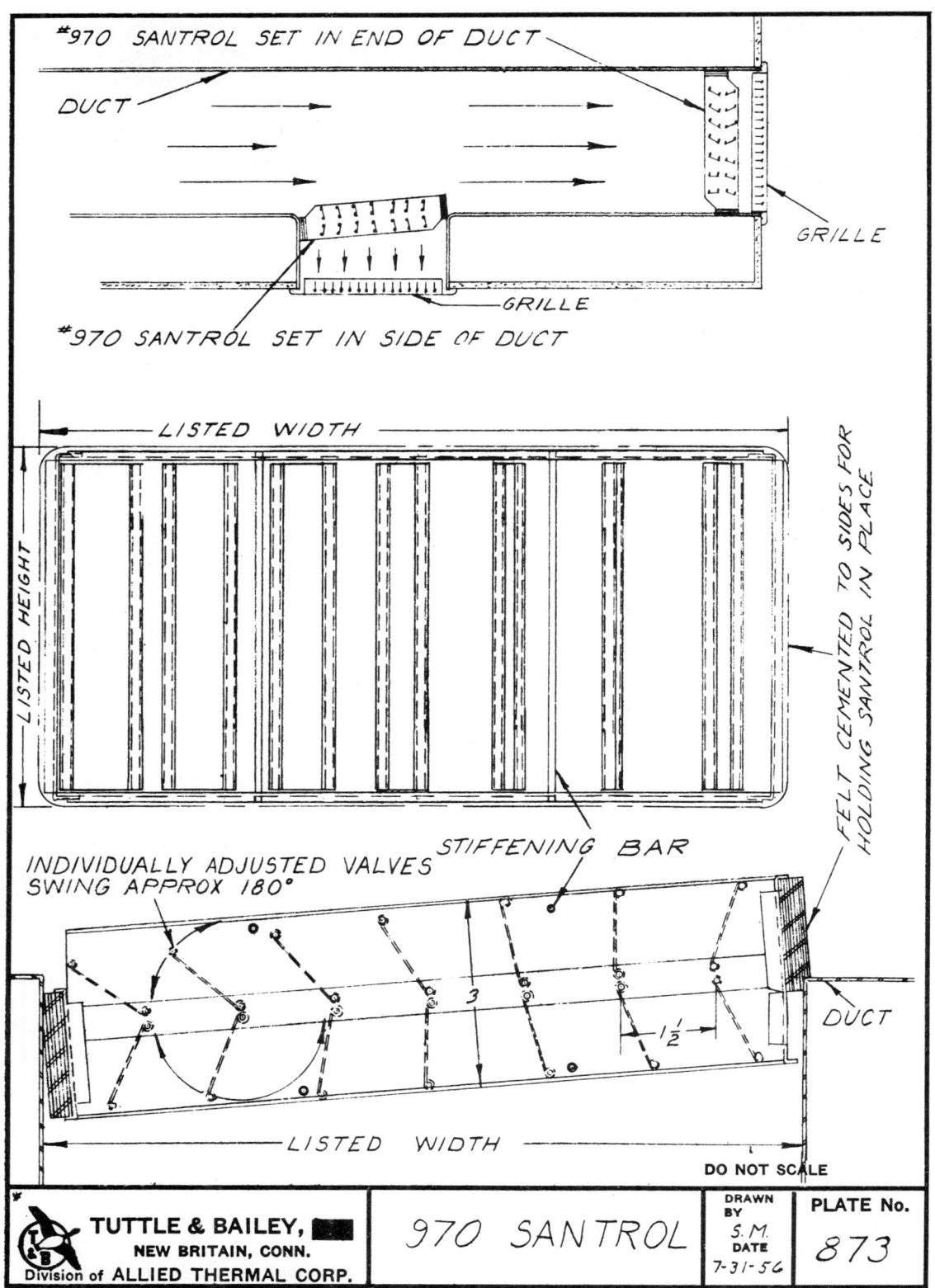

Fig. 5-19. Register, grilles and diffusers.

Dimensions and Performance

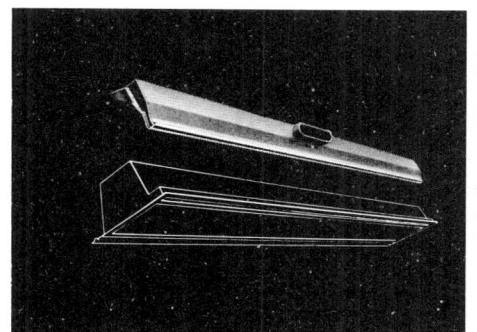

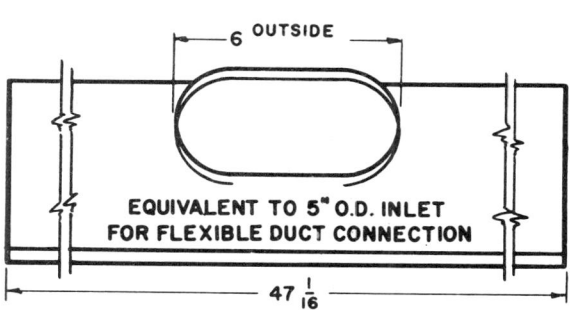

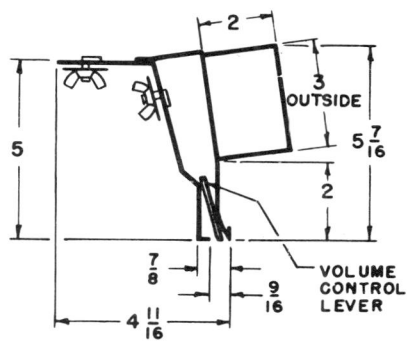

ANEMOSTAT TYPE CLD AIR DIFFUSER
Performance Table With One Unit Per Troffer

CAPACITY, cfm		60	70	80	90	100	110	120	130	140	150
STATIC PRESSURE, "w.g.		.07	.10	.13	.16	.20	.24	.29	.34	.40	.45
A-SOUND LEVEL, db		26	30	33	36	39	42	44	46	48	50
NC LEVEL, db		19	23	26	29	32	35	37	39	41	46
MIN.-MAX. DIFFUSION DISTANCE	FORWARD	3-10	4-11	5-12	6-12	7-13	8-14	9-14	10-15	11-16	12-16
	SPREAD	6-8	6-9	7-10	7-11	8-12	8-13	9-14	10-14	11-15	12-16

NOTE:
1. For two CLD Air Diffusers per troffer, increase A-sound level and NC level two decibels.
2. Static pressure and sound level are given with volume control in open position.
3. Static pressure measured in 5" diameter flexible duct 1' before CLD Air Diffuser.
4. Performance Data based on 9' ceiling height *and* 30° F maximum cooling differential.

No air conditioning system is better than its air distribution

FIG. 5-20. Troffer and lineal diffuser.

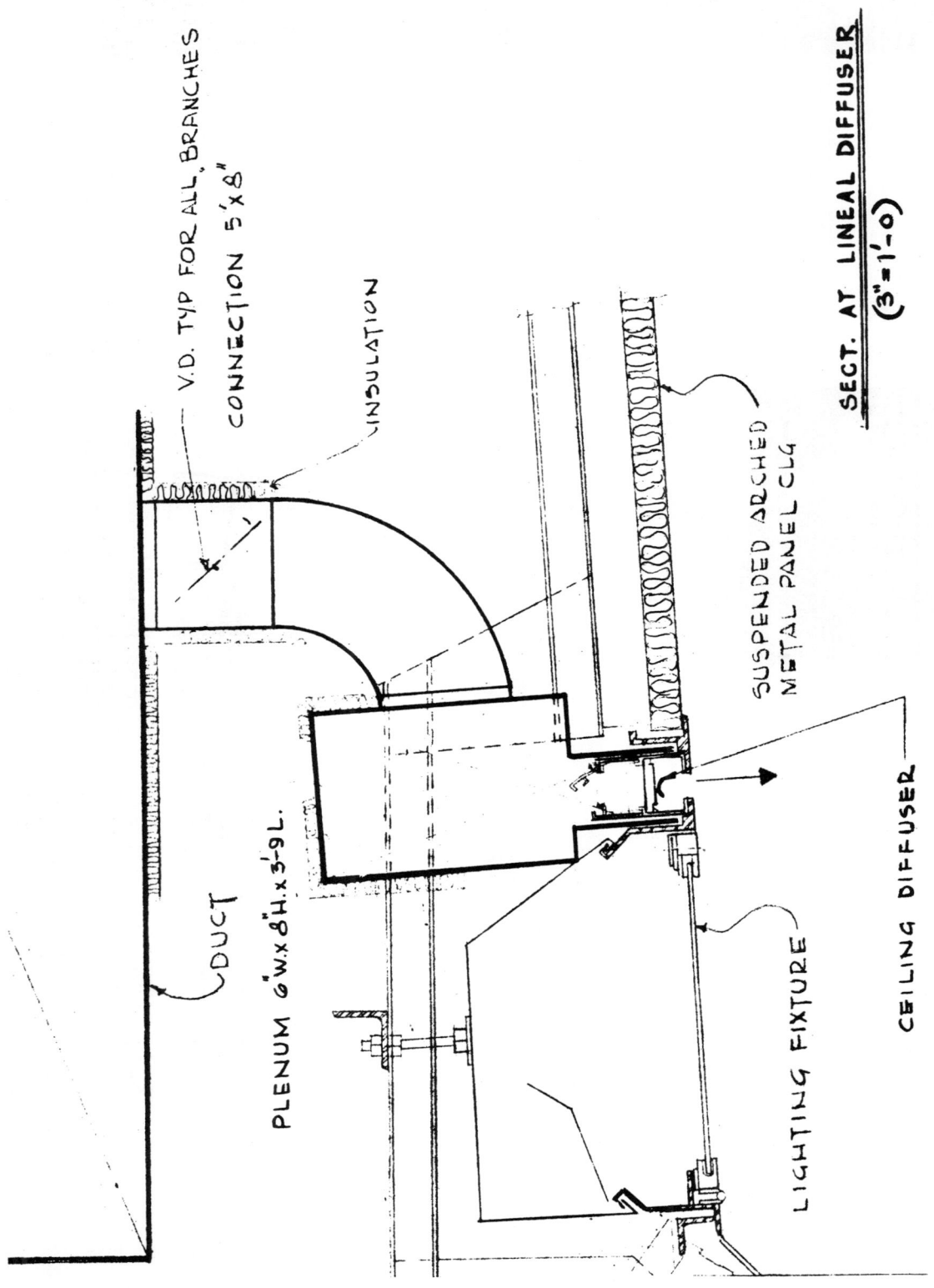

Fig. 5-21. Troffer and lineal diffuser.

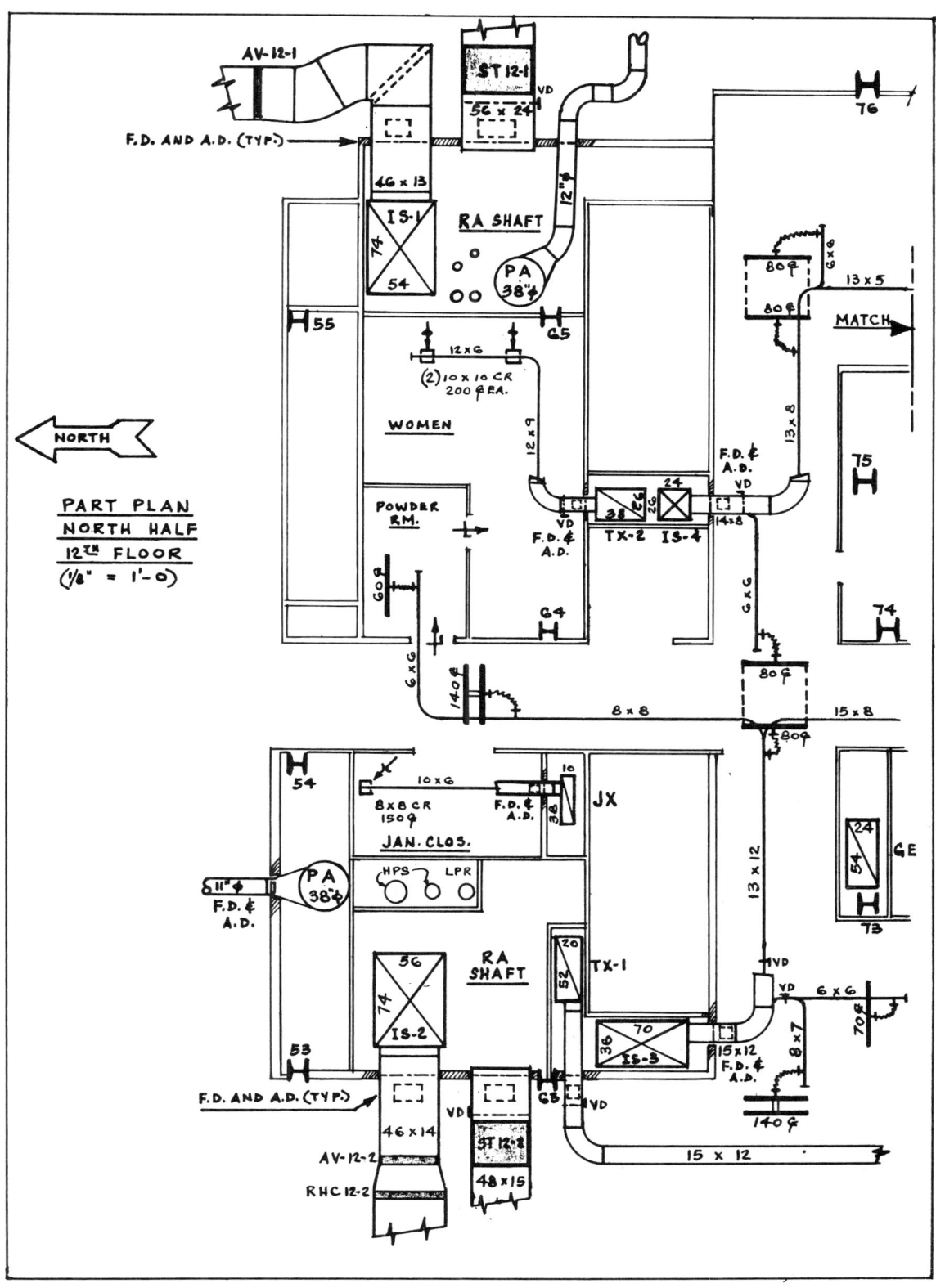

Fig. 5-22. Floor plans and section.

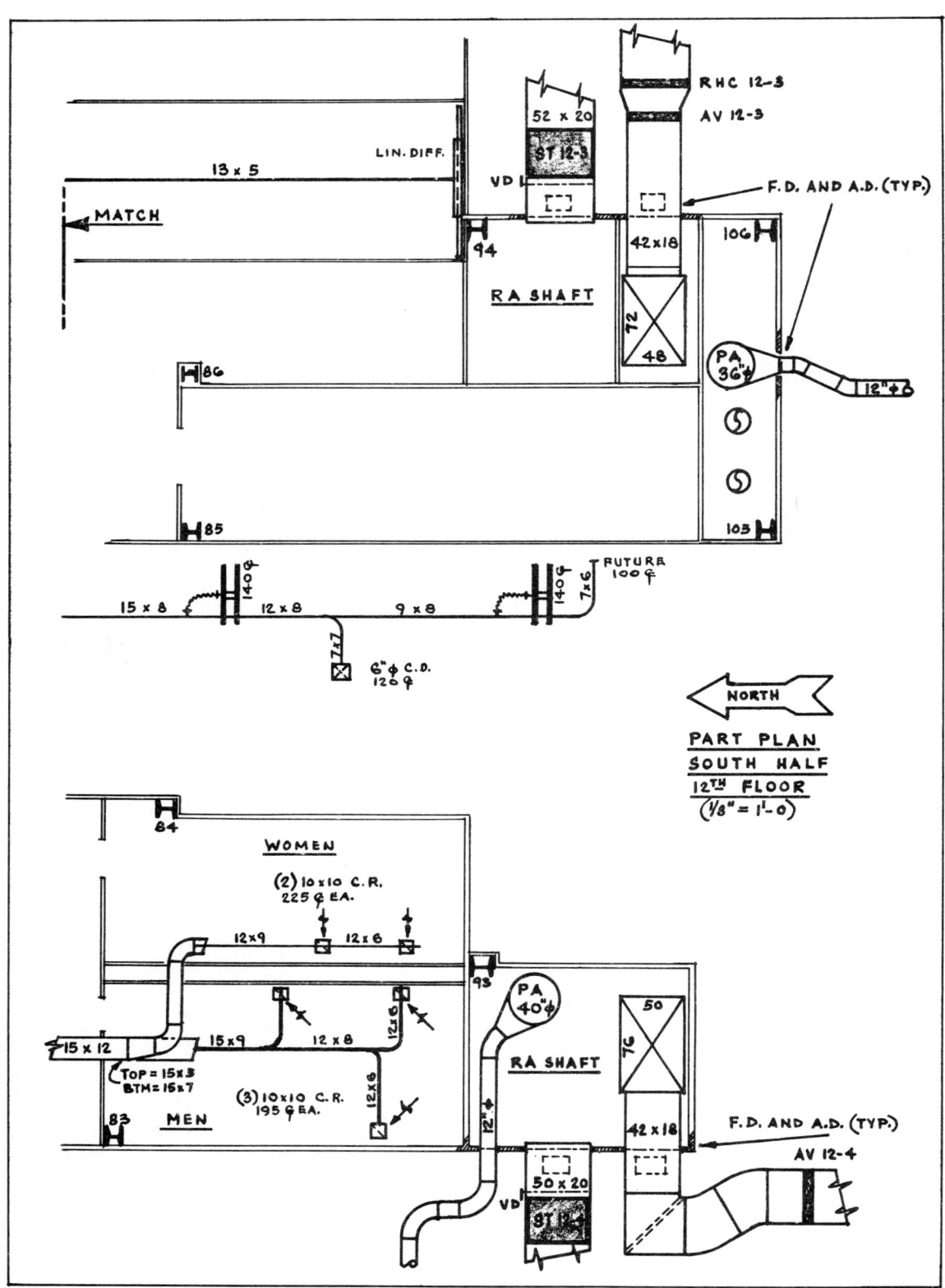

Fig. 5-23. Floor plans and section.

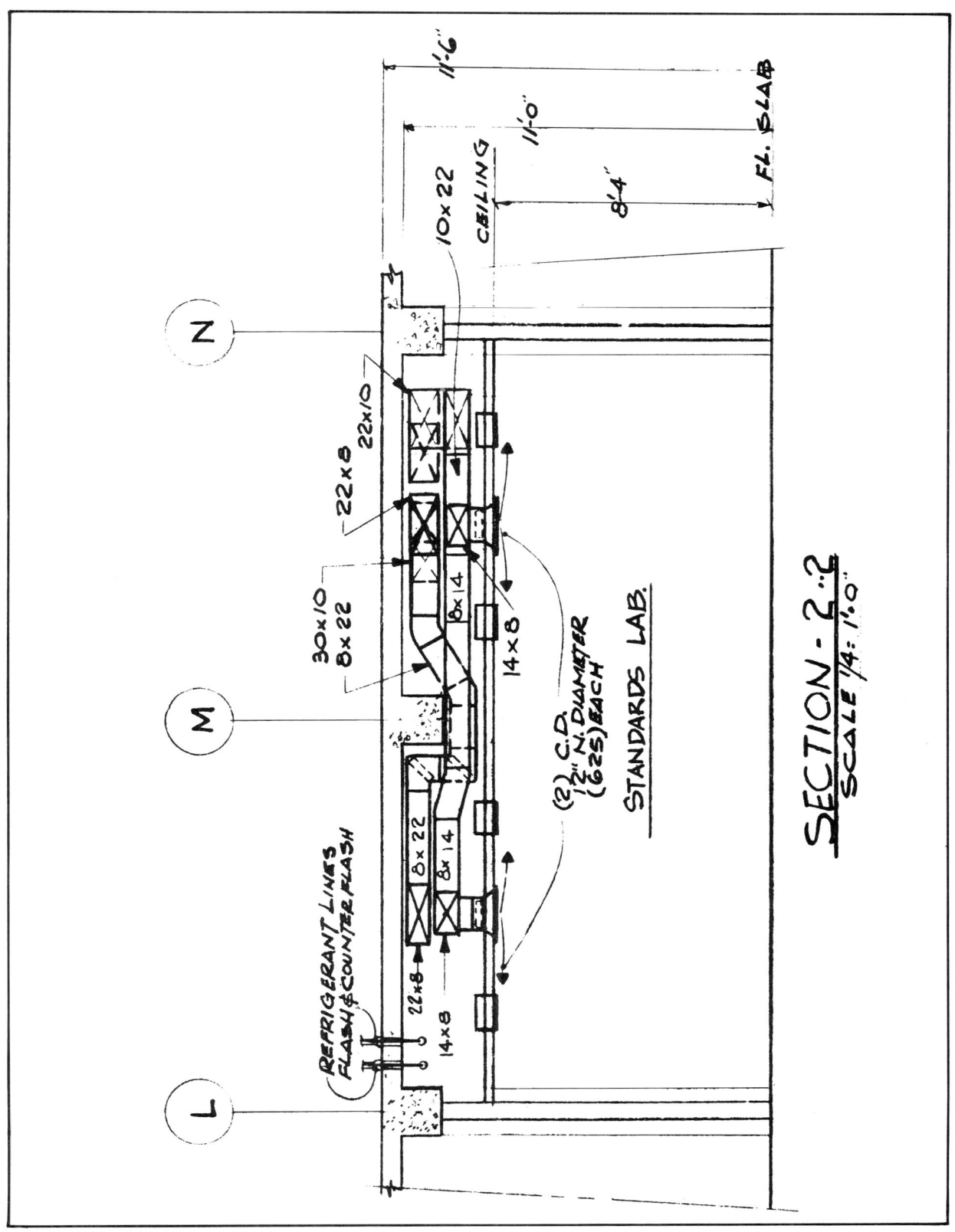

Fig. 5-24. Floor plans and section.

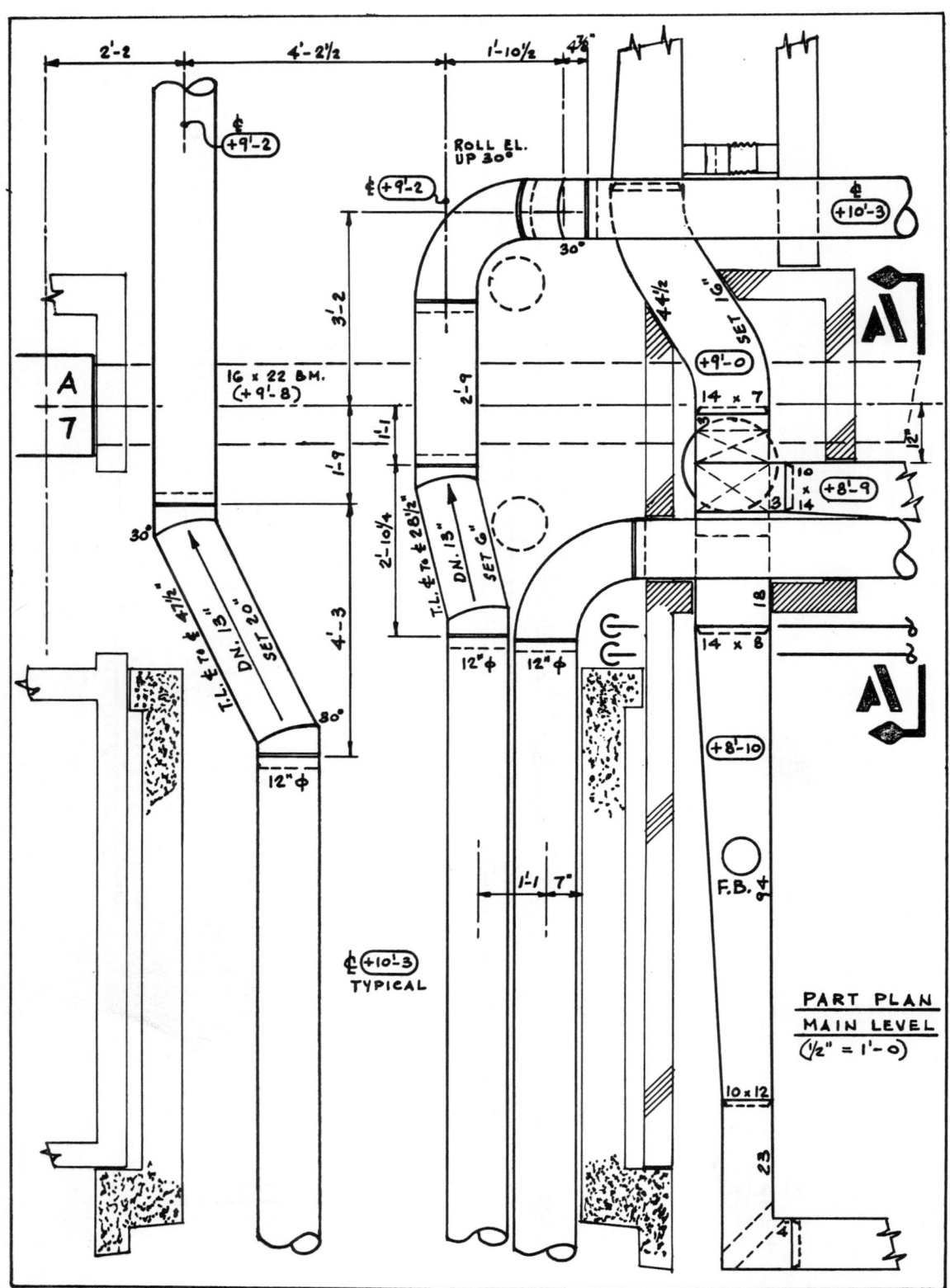

Fig. 5-25. Shop drawing, part plan.

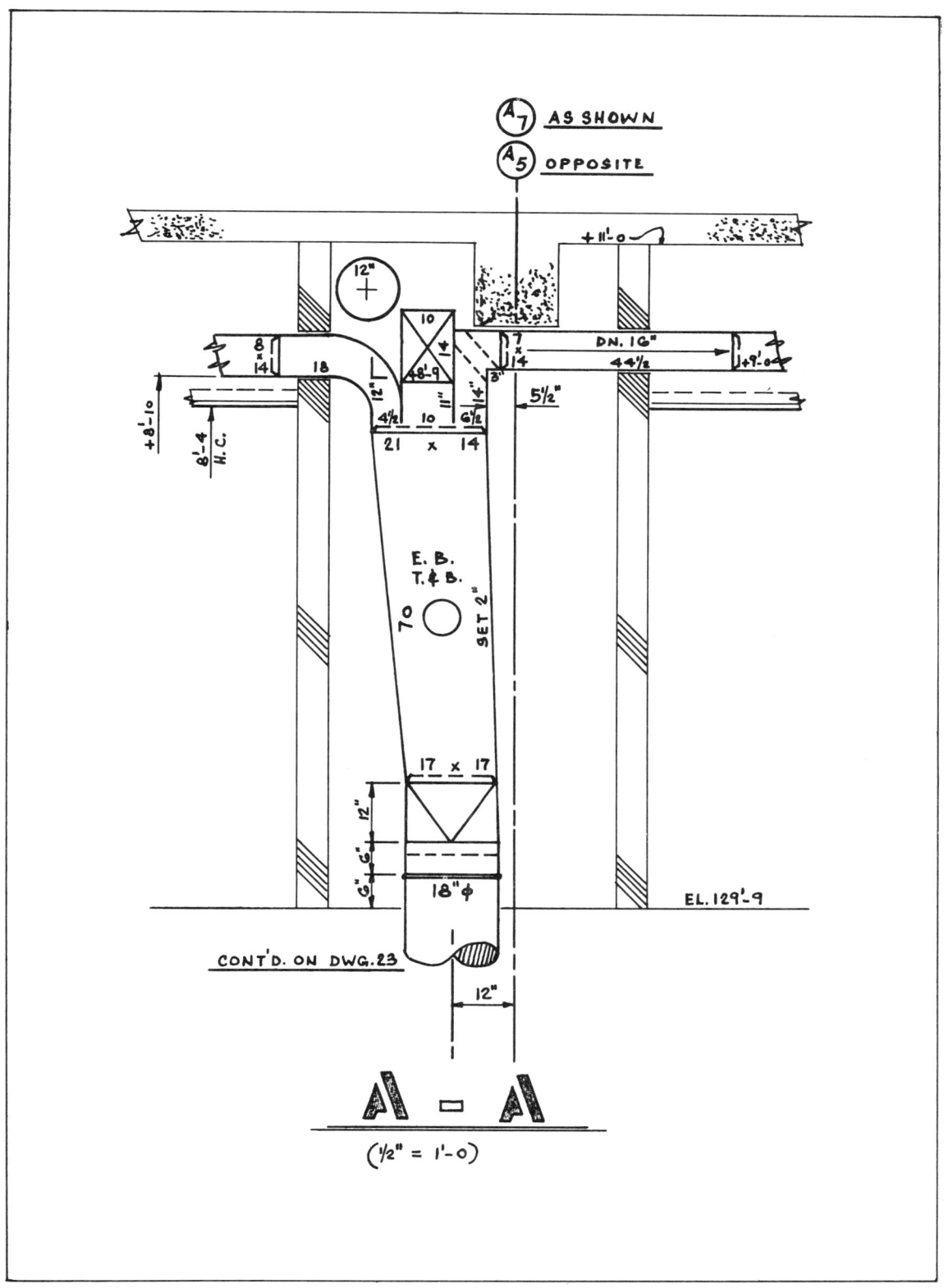

Fig. 5-26. Shop drawing, section.

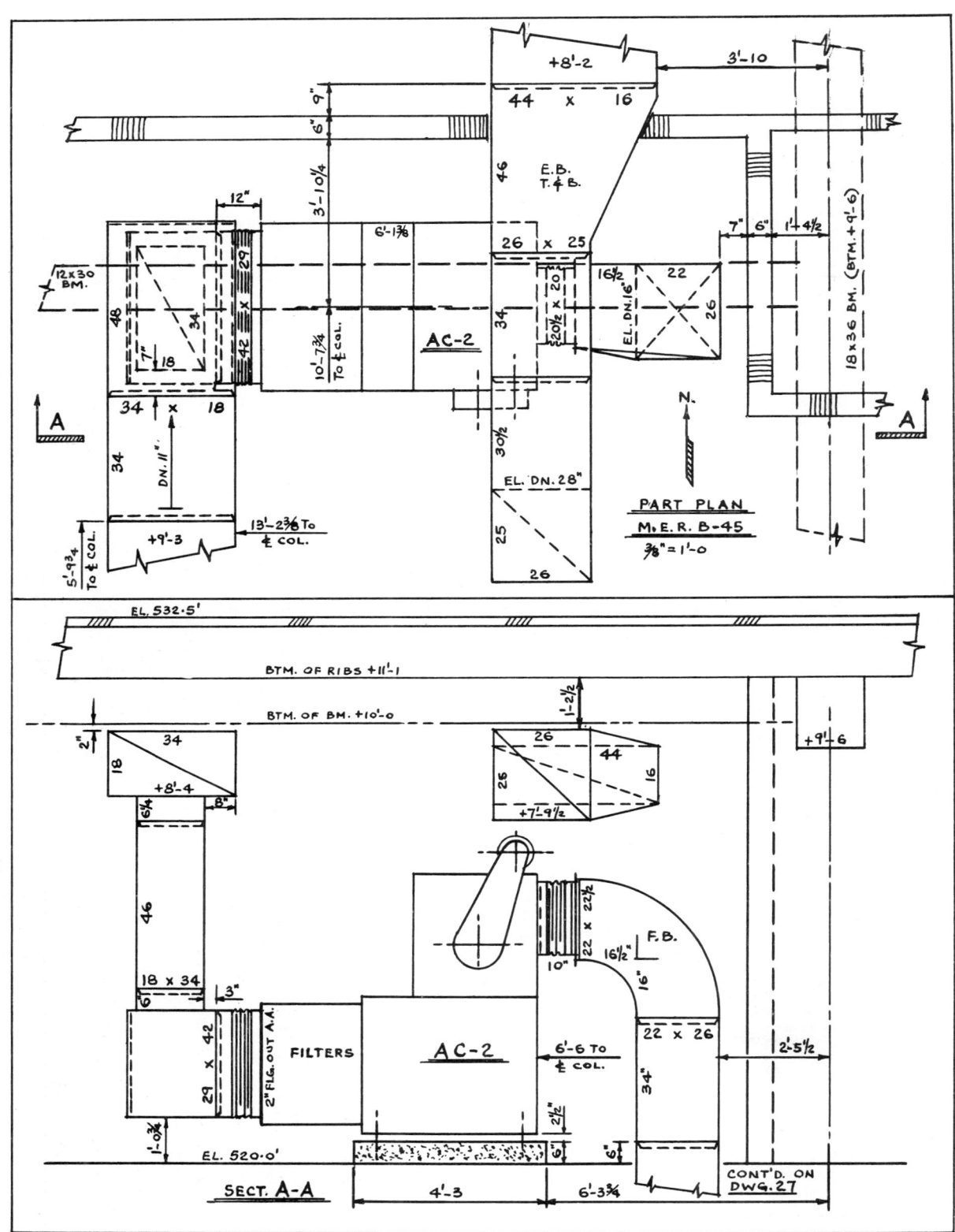

Fig. 5-27. Part, M.E.R. plans and section.

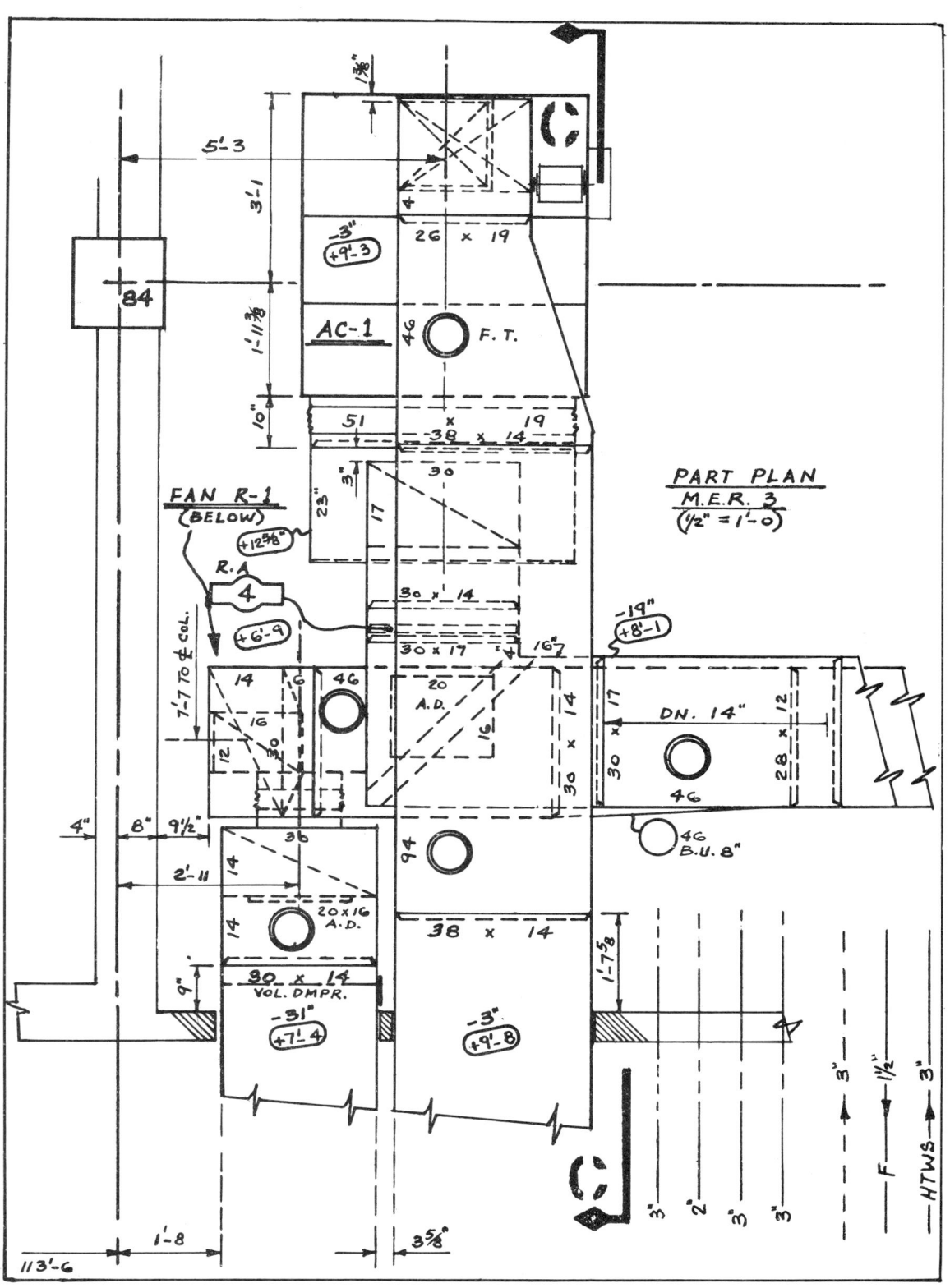

Fig. 5-28. Shop drawing, part plan.

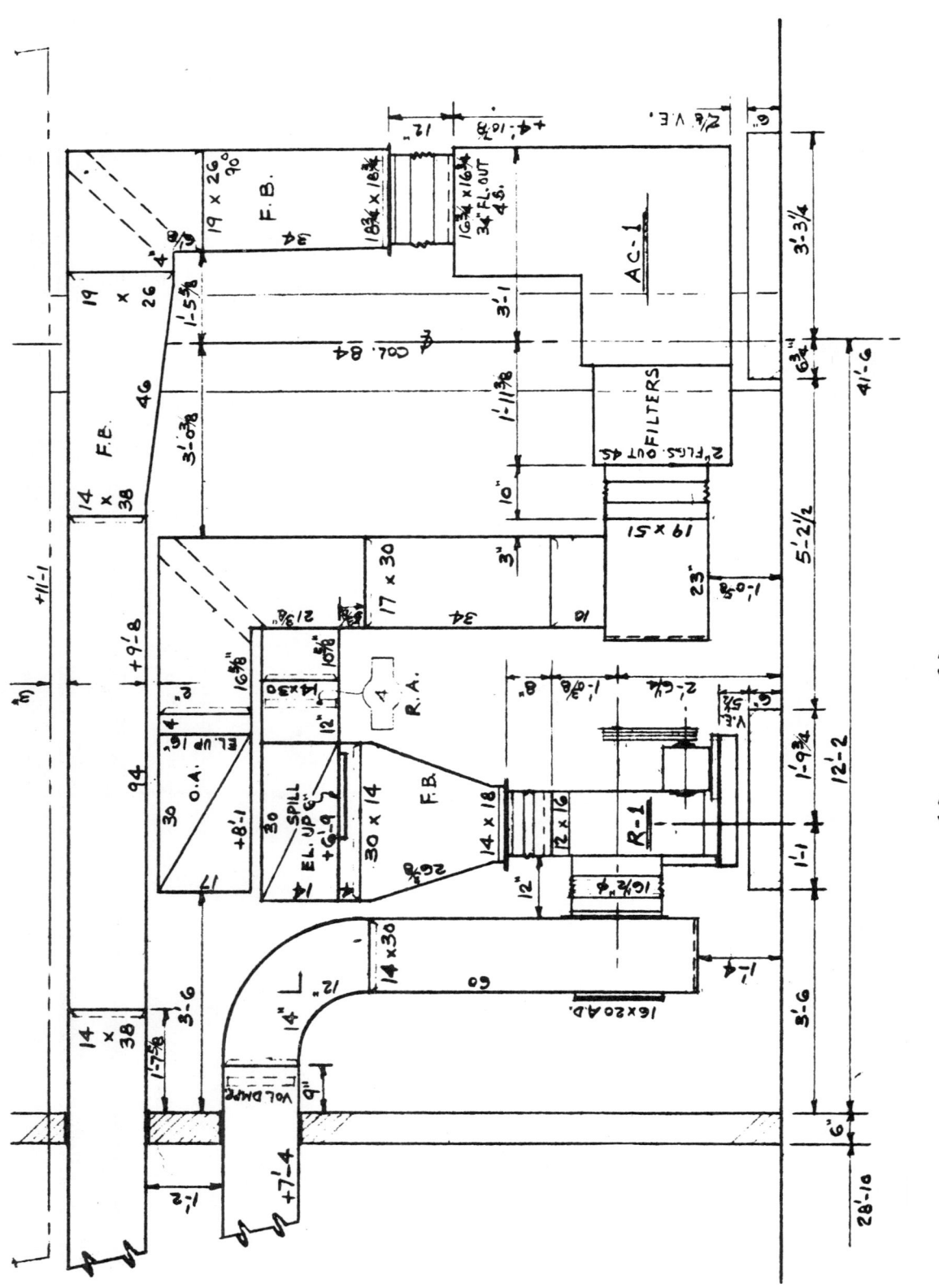

Fig. 5-29. Shop drawing, M.E.R. section.

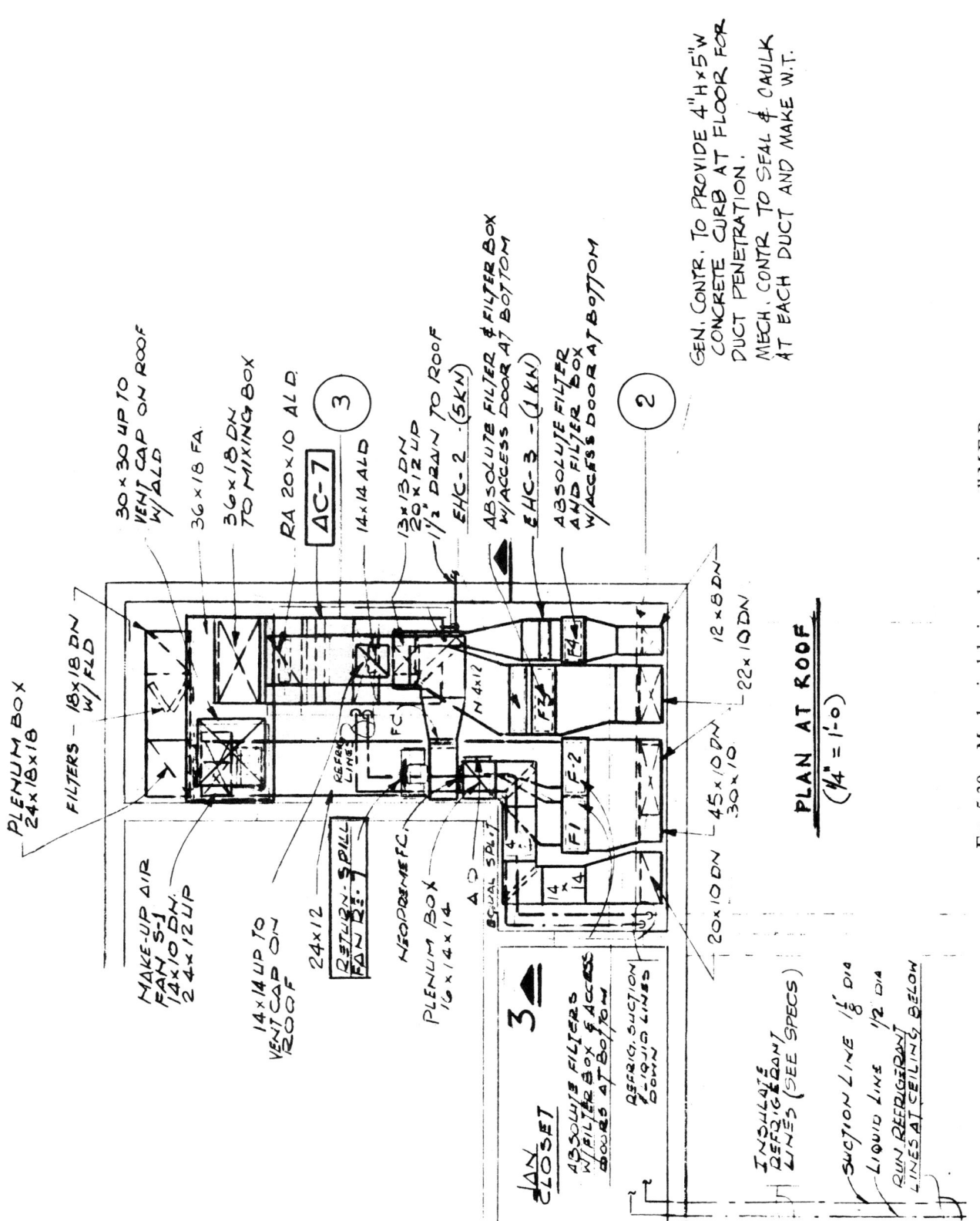

Fig. 5-30. Mechanical design drawings, small M.E.R.

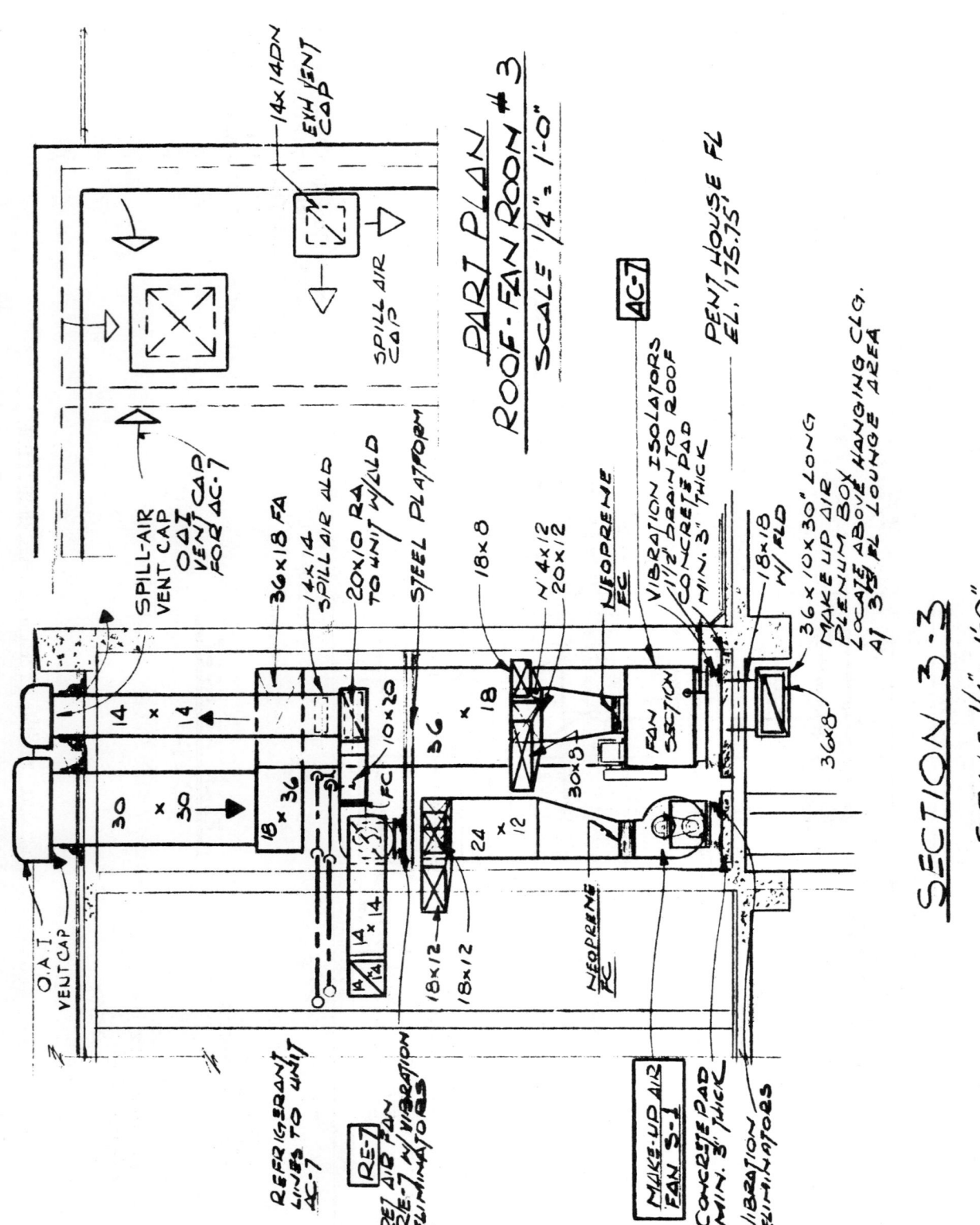

Fig. 5-31. Mechanical design drawings, small M.E.R.

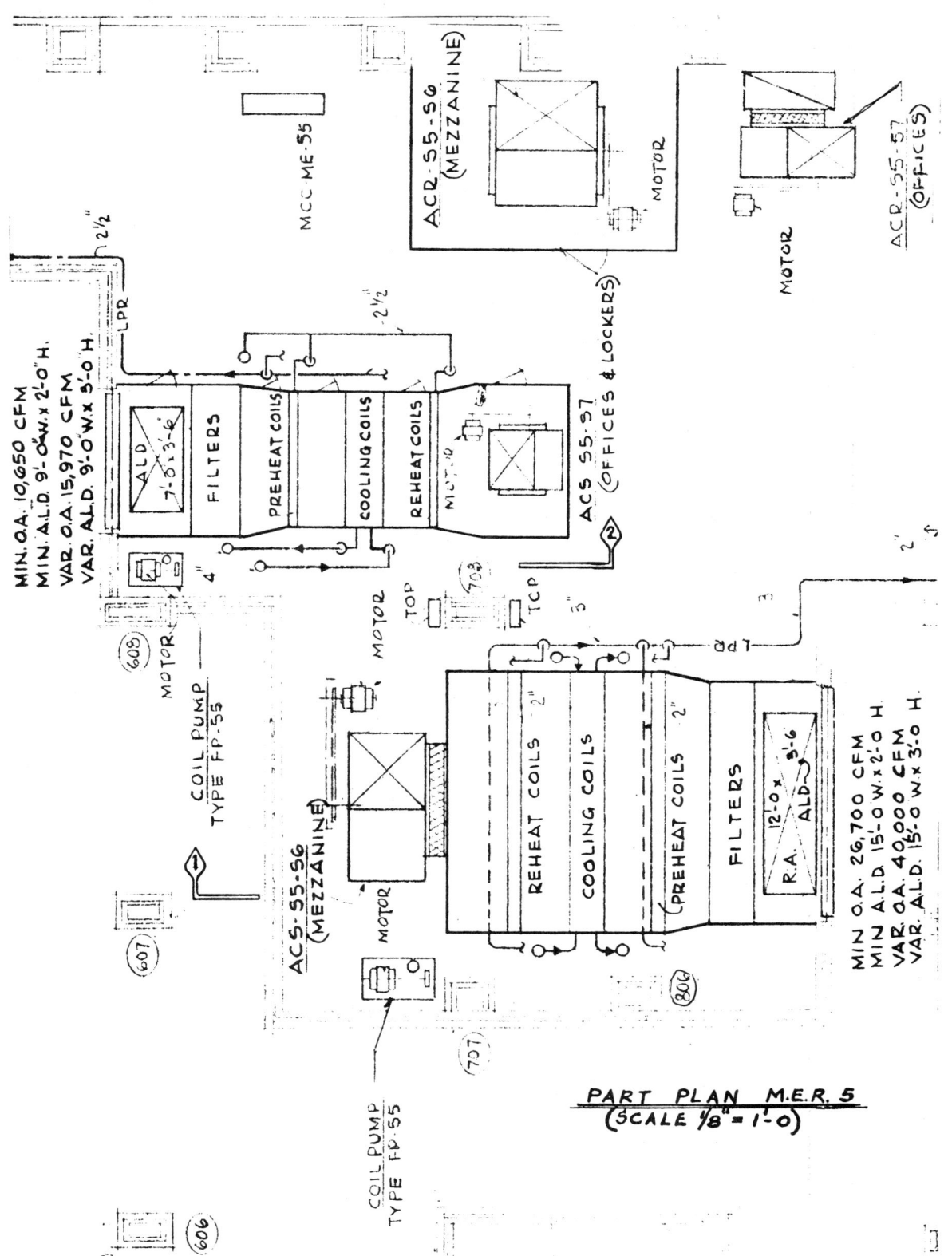

Fig. 5-32. Central sta. system M.E.R. and sections.

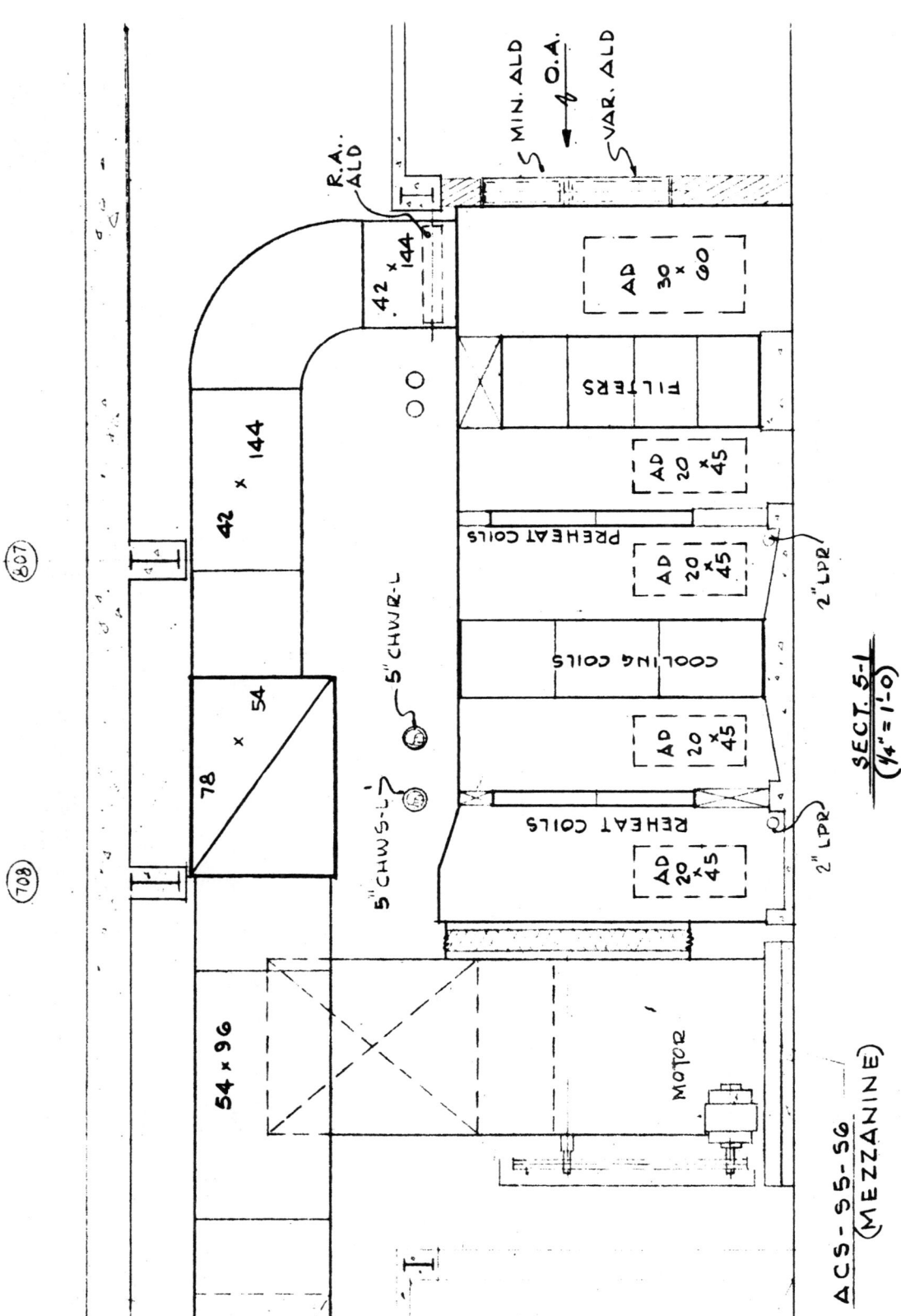

Fig. 5-33. Central sta. system M.E.R. and sections.

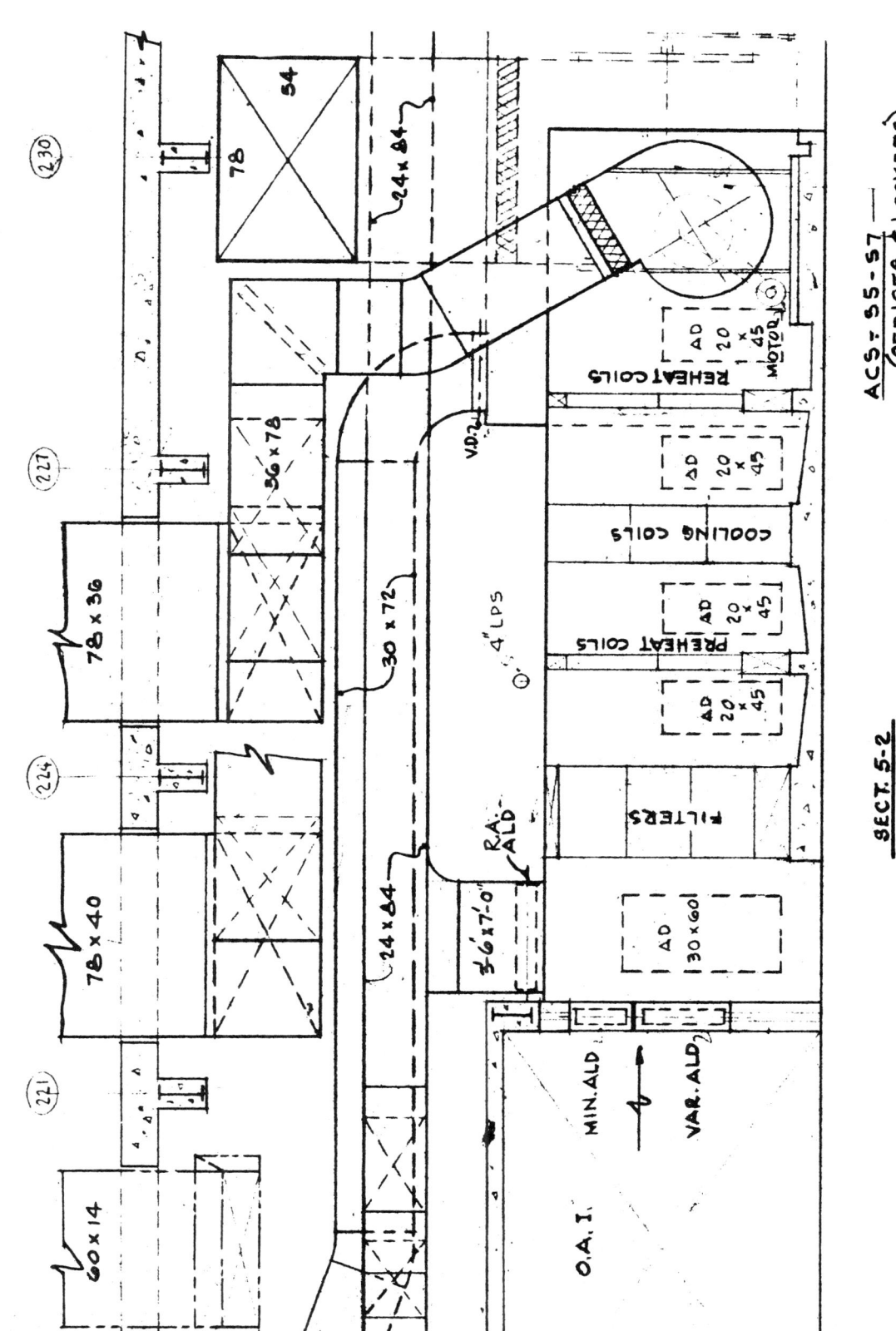

Fig. 5-34. Central sta. system M.E.R. and sections.

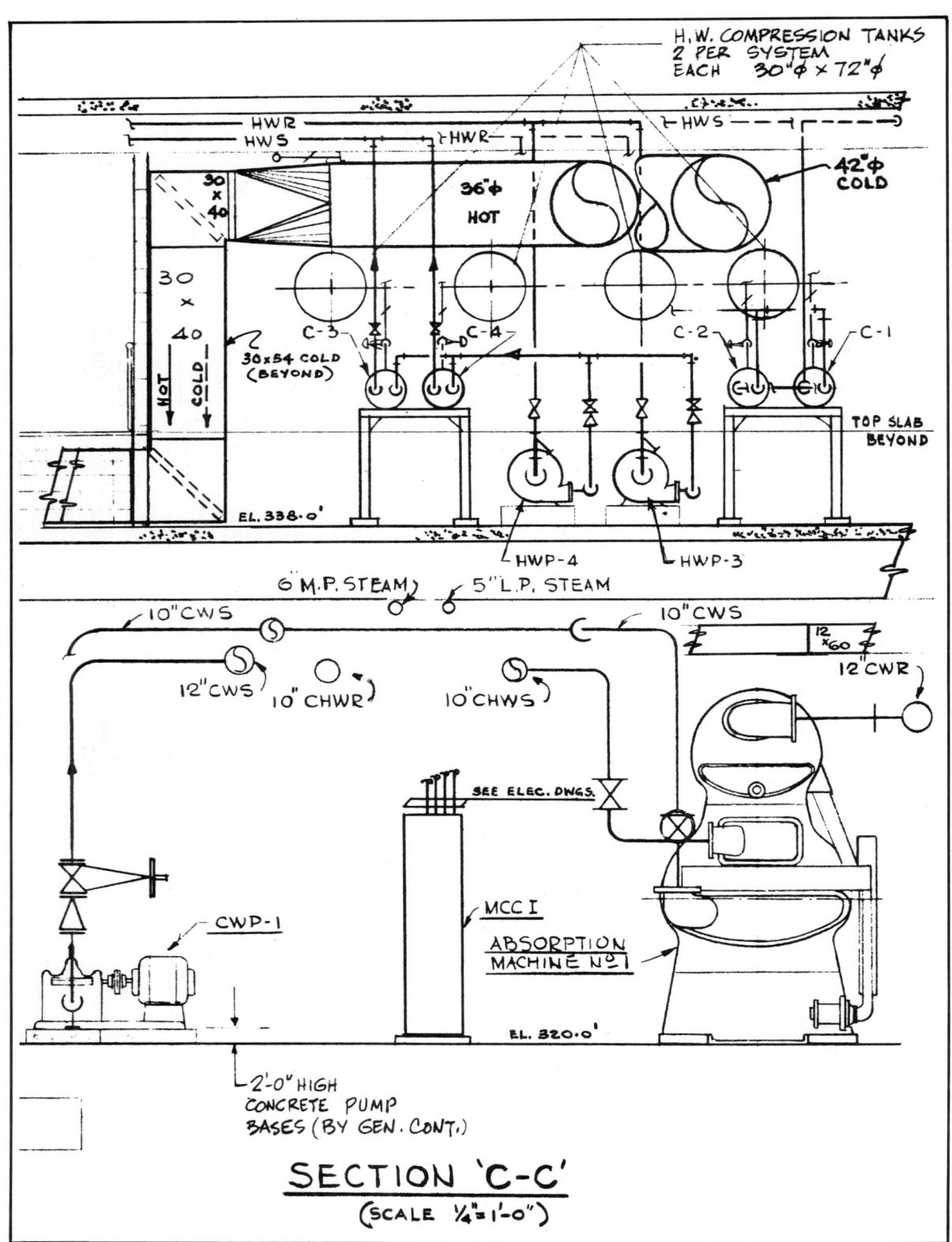

Fig. 5-35. Two-level M.E.R. section.

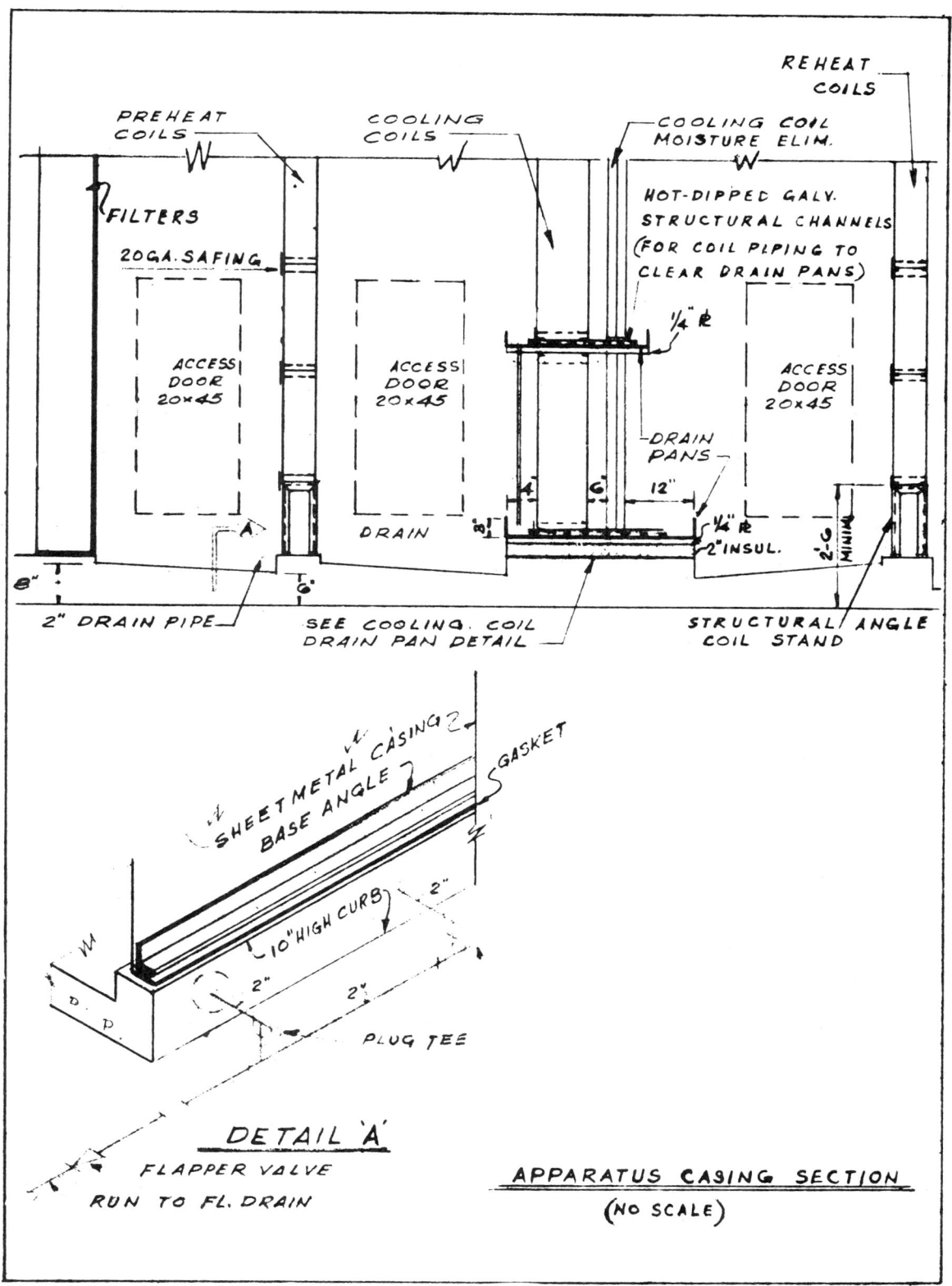

Fig. 5-36. Apparatus casing section.

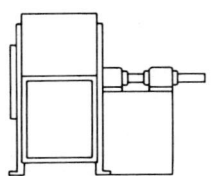
No. 1, SW, SI.
For belt drive or direct connection. Wheel overhung. Two bearings on base.

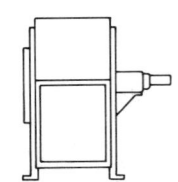
No. 2, SW, SI.
For belt drive or direct connection. Wheel overhung. Bearings in bracket supported by fan housing.

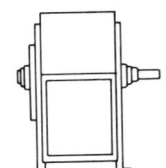
No. 3, SW, SI.
For belt drive or direct connection. One bearing on each side and supported by fan housing. Not recommended in sizes 27″ diam. and smaller.

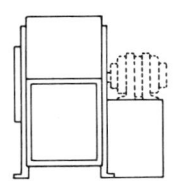
No. 4, SW, SI.
For direct drive. Wheel overhung on prime mover shaft. No bearings on fan. Base mounted or an integrally direct connected prime mover.

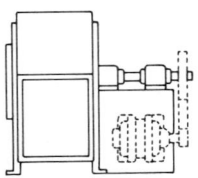
No. 9, SW, SI.
For belt drive. Arrangement No. 1 designed for mounting prime mover on side of base.

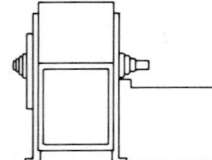
No. 7, SW, SI.
For belt drive on direct connection. Arrangement No. 3 plus base for prime mover. Not recommended in sizes 27″ diameter and smaller.

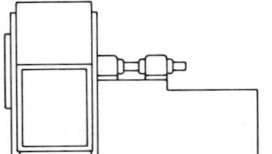
No. 8, SW, SI.
For belt drive or direct connection. Arrangement No. 1 plus base for prime mover.

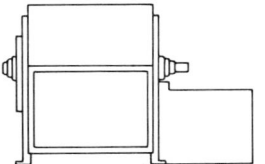
No. 7, DW, DI.
For belt drive or direct connection. Arrangement No. 3 plus base for prime mover.

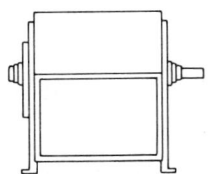
No. 3, DW, DI.
For belt drive or direct connection. One bearing on each side and supported by fan housing.

**SW indicates single width, DW double width
SI indicates single inlet, DI double inlet**

FIG. 5-37. Fan, drive and inlet box arrangements. (Reprinted by permission of Air Moving and Conditioning Association, Inc.)

(Text continued from page 159.)

for routing the low-pressure steam return line from the pre-heat coils. Figure 5-32 is a plan view from top of casings down to the floor line. Complicated layouts which may be difficult to draw and interpret clearly, can be clarified by a multiple-plan drafting method. When this technique is used, the draftsman should not indicate the same items in more than one plan. If coordination makes it necessary to do so, work shown in another plan should be shown with dashed lines.

Figure 5-35 is a partial section of a two-level mechanical room. A chilled-water pump (CWP-1), motor control center (MCC-1), and a large absorption refrigeration machine are located at the low level, and two hot-water pumps (HWP-3, HWP-4), tanks, and hot and cold high-pressure air-supply ducts are at the upper level.

Extensive coordination of HVAC equipment and ducts with work to be installed by other contractors, is necessary on mechanical equipment room shop drawings. To call attention to critical areas of this kind, the draftsman should indicate on shop drawings the location of such equipment, piping, and electrical work, wherever potential conflicts may occur with the sheet metal installation. Notes on the drawings are also helpful to point out specific problem areas.

Job specifications and details are always checked prior to preparation of shop drawings. For central-station apparatus casings, their dimensions are governed by size of equipment, such as heights of heating and cooling coils, moisture eliminators, and width of drain pans. Condensate return-coil connections, pitch, and length of pipe-run determine minimum heights of steam coils. Trap seals for cooling coil drain-pan piping are generally required to be 2 inches greater than the pressure of the air-handling system, therefore cooling-coil heights must allow for piping clearance. Typical design-drawing details of a draw-through apparatus casing are shown in Fig. 5-36. Standards for air moving

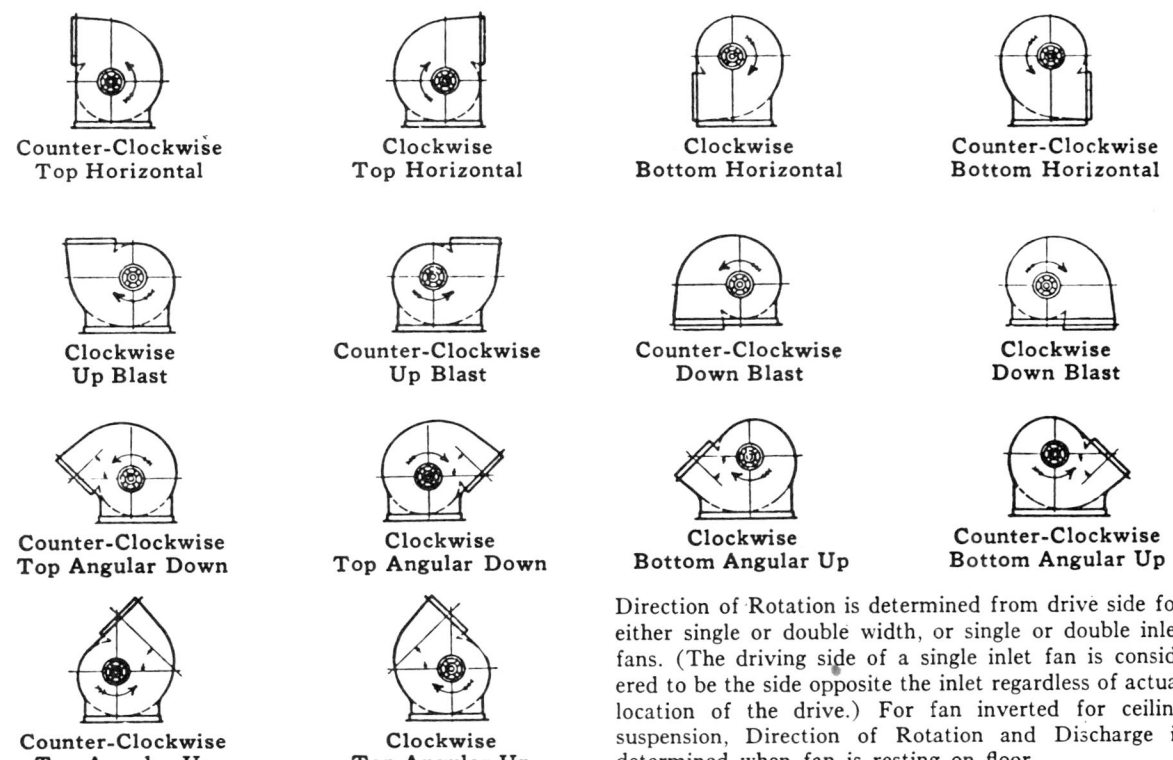

Fig. 5-38. Designation of direction of rotation and dicharge of fans. (Reprinted by permission of Air Moving and Conditioning Association, Inc.)

equipment used in HVAC and exhaust systems have been established by Air Moving and Conditioning Association, Inc. (AMCA). Some are listed below and are also shown in Figs. 5-37 to 5-41. Applications of these standards will greatly assist the draftsman in the precise interpretation of manufacturers' cuts.

PRODUCT DEFINITIONS

Air Moving Device (AMD)

Air Moving Device — A power-driven machine moving a continuous volume of air.

Central-Station Units

Central-Station Unit — A factory-assembled unit containing a fan, or fans, in a cabinet and which may contain cooling coils, heating coils, or other appurtenances to perform the function of circulating, cleaning, heating, cooling, humidifying, dehumidifying, mixing and/or distributing air, capable of operating at external static pressures greater than .25 inch of water gage (.25 in. wg). Specifically excluded are: self-contained air conditioning units which include the refrigeration machine as an integral component, fan-coil units specifically designed for room installations, either exposed or recessed, and steam and hot-water unit heaters.

Draw-Through Central-Station Unit — A central-station unit with the heat exchanger(s) upstream of the fan.

Blow-Through Central-Station Unit — A central-station unit with the heat exchanger(s) downstream of the fan.

Air-Conditioning Unit — A central-station unit designed primarily for cooling, including spray-coil units.

Heating and Ventilating Unit — A central-station unit with no provision for cooling, but which may contain a heating coil, or coils, using steam or hot water.

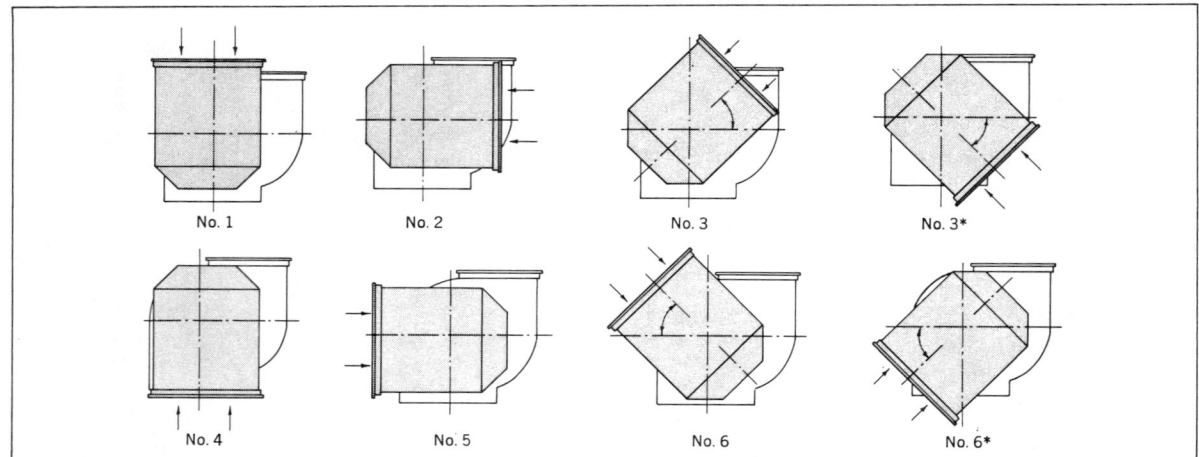

Definitions

Reference line is a horizontal line through center of fan shaft.

Air entry to inlet box is determined from drive side of fan.

On single inlet fans the drive side is always considered as the side opposite the fan inlet.

When drives are on both ends of fan shaft, the drive side is that side having the higher horsepower driving unit and is the same side from which the fan rotation is designated.

Air Entry Position Designation

1. Top Intake
2. Horizontal Right Intake
3. (Number of degrees) Above or below horizontal center line on right.
4. Bottom Intake
5. Horizontal Left Intake
6. (Number of degrees) Above or below horizontal center line on the left.

* It will be found in some cases that this arrangement interferes seriously with the framing of the floor structure by the amount of floor space required.

FIG. 5-39. Designation of position of inlet boxes. (Reprinted by permission of Air Moving and Conditioning Association, Inc.)

Industrial, Axial, and Propeller Fans

Axial Fan — AMD having airflow through the impeller which is predominantly parallel to the axis of rotation, contained in a cylindrical casing.

Tubeaxial Fan — An axial fan without guide vanes.

Vaneaxial Fan — An axial fan with either inlet or discharge vanes or both.

Propeller Fan — AMD with a propeller or disc wheel within a mounting ring or plate, through which the airflow is predominantly parallel to the axis of rotation.

Power Roof Ventilators

Power Roof Ventilator — AMD consisting of an impeller, which may be centrifugal, axial, or propeller type, and integral driver in a weather-resisting housing supported by a weather-resisting base designed to fit, usually by means of a curb, over a wall or roof opening. Specifically excluded are AMDs with integral air-tempering means.

Steam and Hot-Water Unit Heaters

Steam or Hot-Water Unit Heater — A factory-assembled device designed to heat and circulate air. Essential components are a heat-transfer element, using steam or hot water as the heating medium, and housing and fan(s) with driving motor(s). Normally designed for free delivery or recirculated air.

A unit heater may be provided with components for filtering, ventilating, and/or diffusing the discharge air.

Classification by Arrangement

Horizontal Propeller-Fan Type — A unit heater using a propeller fan and with a substantially horizontal air-discharge pattern, neglecting the effect of any diffusers or deflectors.

Vertical Propeller-Fan Type — A unit heater using a propeller fan and with a substantially vertical, downward air-discharge pattern, neglecting the effect of any diffusers or deflectors.

Cabinet — A unit designed for heating and recirculating air at free delivery to a maximum of .25 in. wg of external pressure with the essential elements, including, specifically, the fan motor with direct or belt-driven centrifugal fan or fans contained within the housing (cabinet).

When a sales engineer for a fan manufacturer prepares schedules for a customer, he selects fan sizes, arrangements, and performance ratings to agree with the original design drawing for approval by the architects and engineers. During the actual preparation for the drawing, the draftsman should use approved manufacturers' cuts and data sheets, such as those shown in Figs. 5-42 and 5-43.

Other HVAC design drawings used as references in producing the shop drawing, are air risers, automatic temperature-control flow diagrams, and general details. Duct riser-sizes, automatic-damper locations and their operation, and special duct-construction details are found on these drawings. See Figs. 5-44 and 5-45 for typical sleeve type fire damper details.

Miscellaneous Equipment

In a conventional refrigerant cycle, the liquid refrigerant, stored in a receiver, flows through a metered expansion-valve (pressure-reducing valve) to a cooling coil. In the coil tubes, the refrigerant absorbs heat from the air or water used for the cooling system. As the refrigerant absorbs heat it changes to a vapor and goes to a compressor. The compressor increases its pressure so that the condensing temperature of the vapor is higher than the air or water temperatures used in the condensing section. Heat is transferred from the vapor to the condensing agent, the vapor returns to a liquid and flows back to the receiver where it is stored again, to complete its cycle.

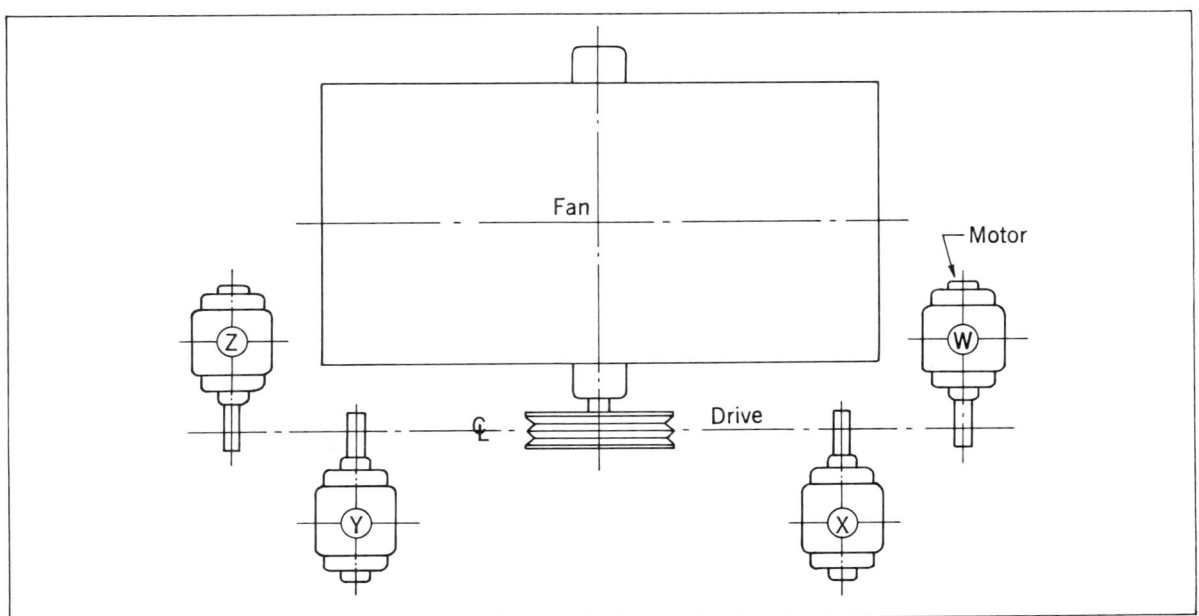

Location of motor is determined by facing the drive side of fan and designating the motor position by letters W, X, Y, or Z as the case may be.

FIG. 5-40. Motor position, belt or chain drive. (Reprinted by permission of Air Moving and Conditioning Association, Inc.)

AIR MOVING AND CONDITIONING ASSOCIATION, INC.

205 West Touhy Ave.
Park Ridge, Illinois 60068

SW - Single Width	DW - Double Width
SI - Single Inlet	DI - Double Inlet

Arrangements 1, 3, 7 and 8 are also available with bearings mounted on pedestals or base set independent of the fan housing.

For designation of rotation and discharge,	see AS 2406.
For motor position, belt or chain drive,	see AS 2407.
For designation of position of inlet boxes,	see AS 2405.

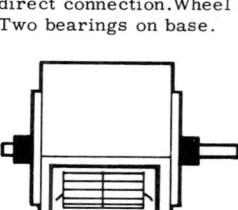

ARR. 1 SWSI For belt drive or direct connection. Wheel overhung. Two bearings on base.

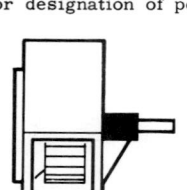

ARR. 2 SWSI For belt drive or direct connection. Wheel overhung. Bearings in bracket supported by fan housing.

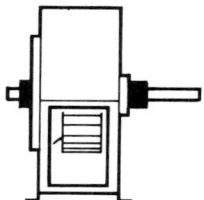

ARR. 3 SWSI For belt drive or direct connection. One bearing on each side and supported by fan housing. Not recommended in sizes 27-inch diameter wheel and smaller.

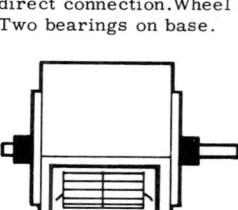

ARR. 3 DWDI For belt drive or direct connection. One bearing on each side and supported by fan housing.

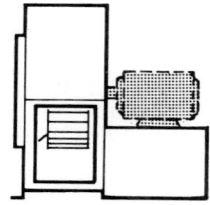

ARR. 4 SWSI For direct drive. Wheel overhung on prime mover shaft. No bearings on fan. Prime mover base mounted or integrally directly connected.

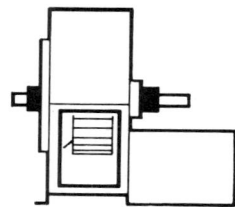

ARR. 7 SWSI For belt drive or direct connection. Arrangement 3 plus base for prime mover. Not recommended in sizes 27-inch diameter wheel and smaller.

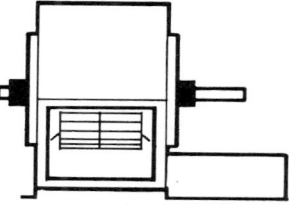

ARR. 7 DWDI For belt drive or direct connection. Arrangement 3 plus for prime mover.

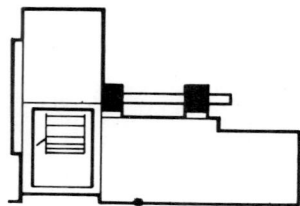

ARR. 8 SWSI For belt drive or direct connection. Arrangement 1 plus extended base for prime mover.

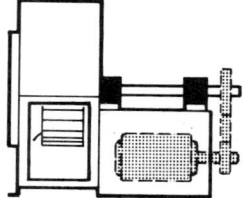

ARR. 9 SWSI For belt drive. Wheel overhung, two bearings, with prime mover outside base.

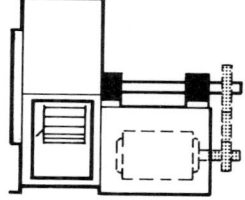

ARR. 10 SWSI For belt drive, wheel overhung, two bearings, with prime mover inside base.

DRIVE ARRANGEMENTS FOR CENTRIFUGAL FANS

Adopted 4-1-65
Revised 11-30-66

AMCA STANDARD
2404-66

FIG. 5-41. Drive arrangements for tubular centrifugal fans. (Reprinted by permission of Air Moving and Conditioning Association, Inc.)

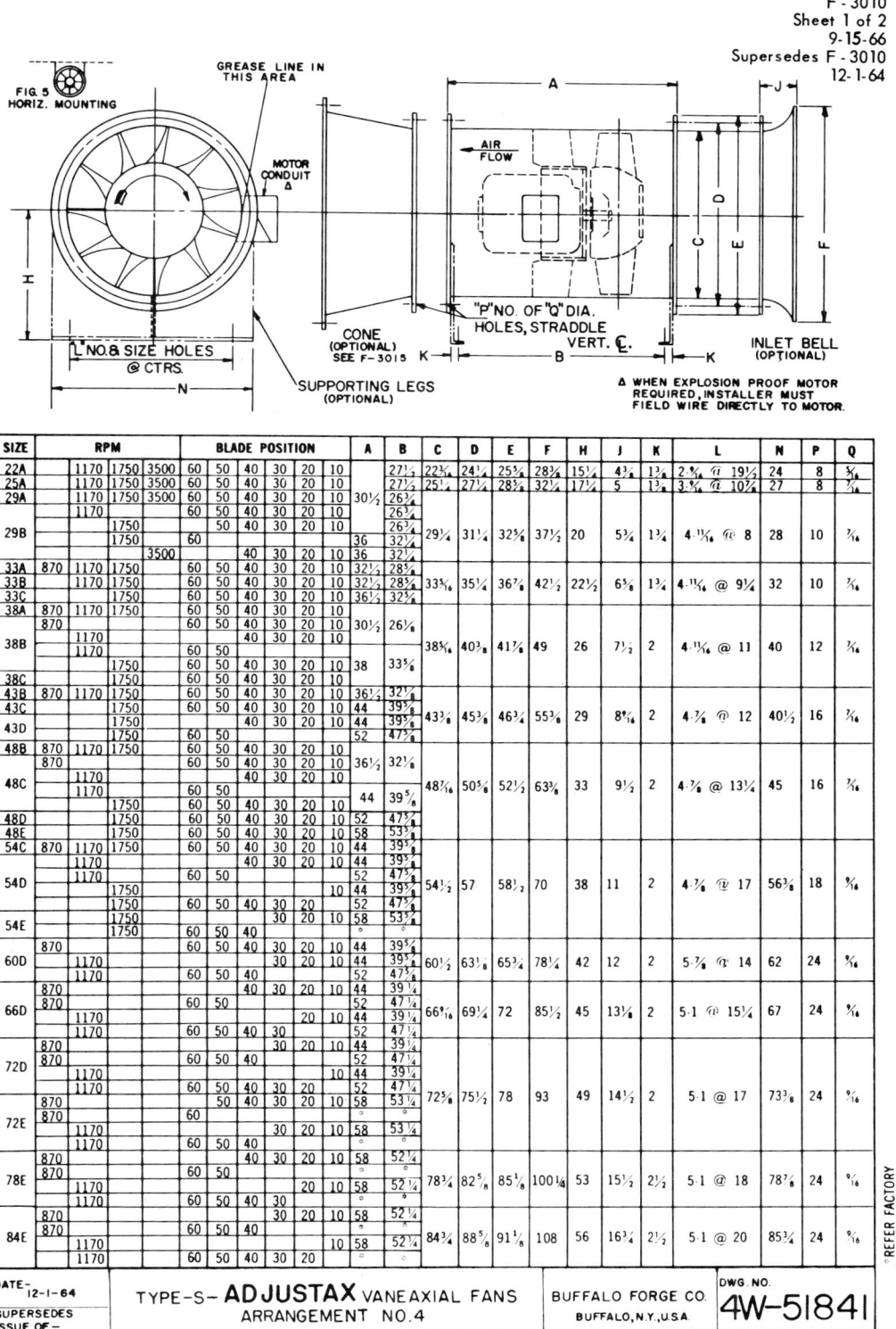

Fig. 5-42. Page 1, Type-S-ADJUSTAX vaneaxial fans. (Reprinted by permission of Buffalo Forge Co.)

F - 3010
Sheet 2 of 2
9-15-66
Supersedes F - 3010
12-1-64

CERTIFIED CORRECT ☐	FOR APPROVAL ☐	REVISED ☐
FOR CONSTRUCTION PURPOSES CH'K BY.......................... DATE............................	FABRICATION WILL NOT BEGIN UNTIL THIS PRINT IS RETURNED APPROVED, OR WITH CHANGES NOTED.	DESTROY ALL PRINTS SUBMITTED PREVIOUS TO: DATE............................

SOLD TO:

CUSTOMER'S ORDER NO.
BUFFALO FORGE CO. ORDER NO.
BRANCH OFFICE ORDER NO.

FURNISH

MARKS:

 — SIZE   TYPE S ADJUSTAX VANEAXIAL FAN ARR. NO. 4

EACH COMPLETE WITH:
 NEMA C FLANGE TEAO MOTOR FRAME
 RPM PH CYCLE VOLT
MODIFICATIONS:

RATING: CFM S.P. V.P. T.P.
 BLADE SETTING FULL LOAD AMPS

 MAXIMUM ALLOWABLE BLADE
 SETTING THIS MOTOR FULL LOAD AMPS

ACCESSORIES OR OPTIONAL EQUIPMENT:

 ☐ SUPPORTING LEGS
 ☐ SUSPENSION CLIPS
 ☐ INLET BELL
 ☐ INLET CONE ☐ WITH ACCESS DOOR
 ☐ OUTLET CONE ☐ WITH ACCESS DOOR
 ☐ INLET SCREEN
 ☐ OUTLET SCREEN
 ☐ VORTEX BREAKER
 ☐ INLET BOX ☐ WITH SUPPORT LEGS
 ☐ SOUND ATTENUATOR ☐ WITH SUPPORT LEGS
 ☐ VARIABLE INLET VANES
 ☐ SPECIAL COATINGS – PREPARATION PER CODE
  (SEE SPECIFICATION DO-100F-3 ATTACHED)

FIG. 5-43. Page 2, Type-S-ADJUSTAX vaneaxial fans. (Reprinted by permission of Buffalo Forge Co.)

When water is used as the condensing medium, a *cooling tower* sprays heat-laden water into the outdoor air. The sprayed water is thus cooled by evaporation and collects in a tank, for circulation in the refrigerant condenser.

An *evaporative condenser* contains a coil section through which refrigerant vapor passes, that is sprayed with water while air is forced over the tubes. The refrigerant vapor loses heat by this process and condenses to a liquid.

Cooling towers or evaporative condensers may be used within buildings by having outdoor and

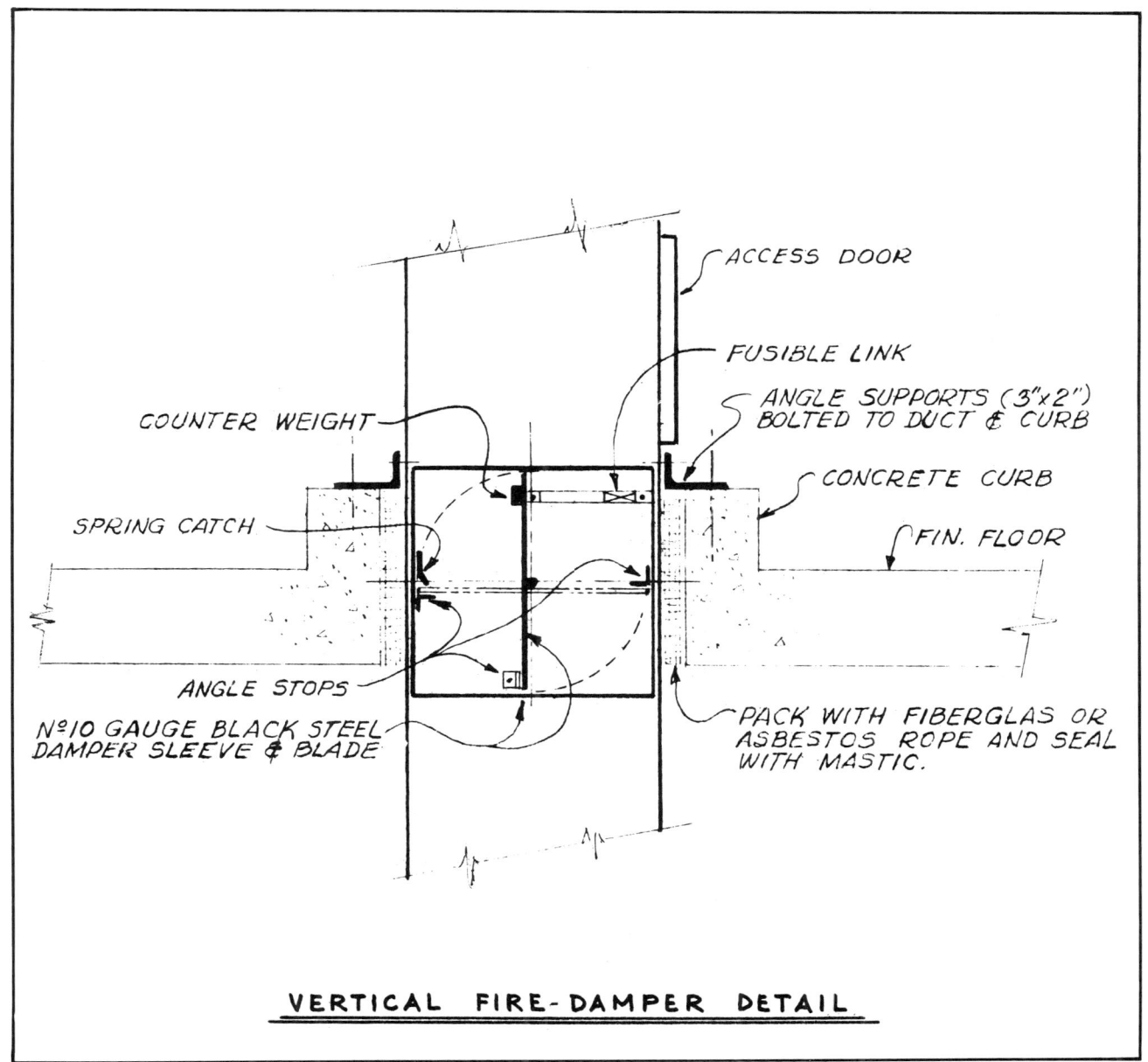

Fig. 5-44. Vertical fire damper, detail.

discharge-air duct connections. The draftsman should check specification requirements for these duct systems, however, as nonferrous metals and watertight duct construction are often used.

Illustrations, diagrams, dimensioned cuts, and installation data of multizone, air conditioning, and residential cabinet units, fans, heating coils, cooling coils, various types of vibration isolation equipment, and air filters are shown in Figs. 5-46 to 5-61. The design drawings and specifications indicate mechanical equipment by size, capacity, arrangement, type, horsepower, electrical characteristics, and other information that may be necessary for precise function of system components.

Vendors furnish equipment schedules with their method of product identification, and cross refer-

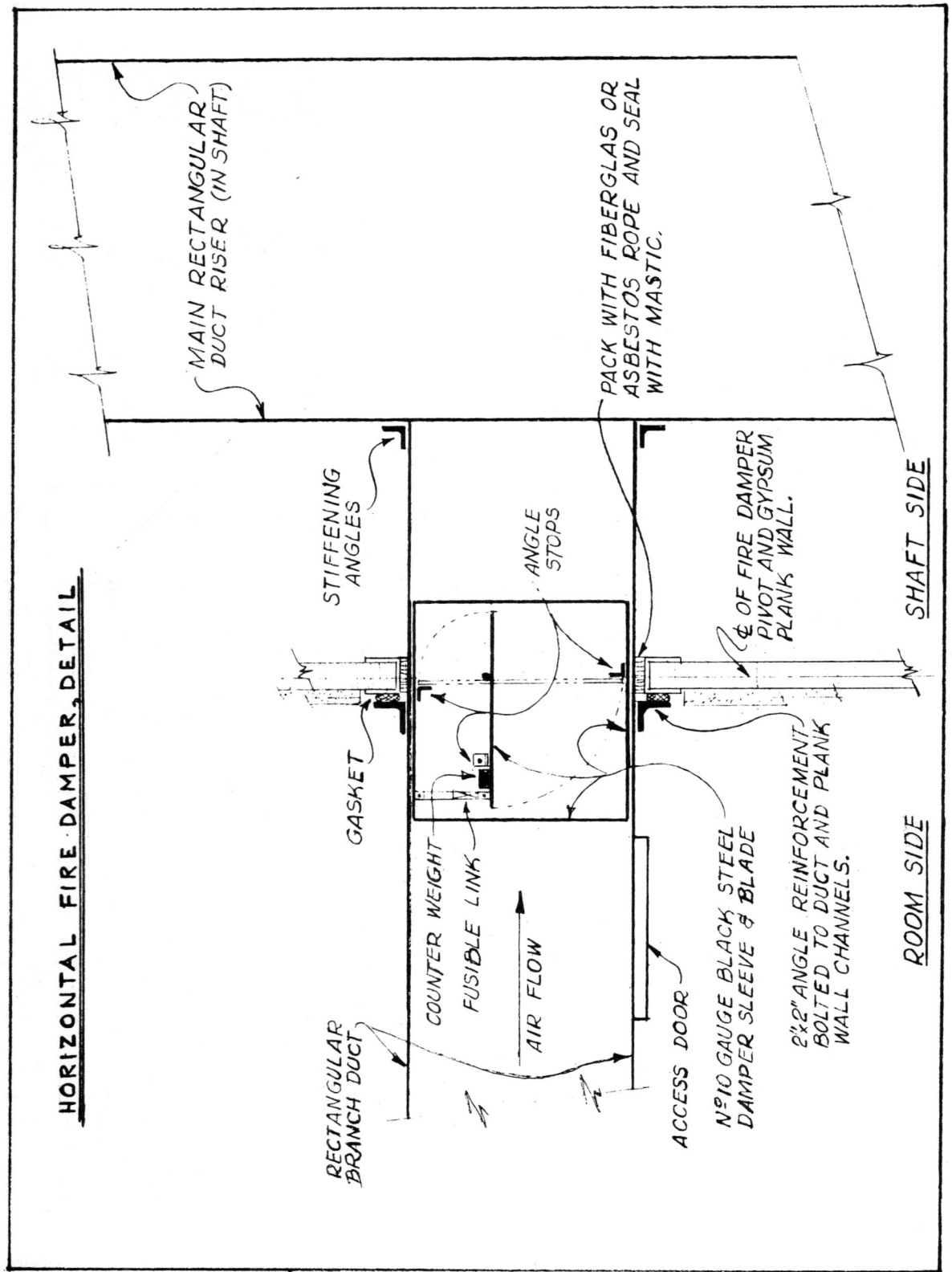

Fig. 5-45. Horizontal fire damper, detail.

Ch. 5   DESIGN DRAWINGS AND SPECIFICATIONS

ence these schedules with the items shown on the plans and in the specifications. The draftsman must use *approved* schedules and cuts for detailing layouts.

Interpretation of cuts is of prime importance and the student draftsman should study Figs. 5-46 to 5-61, in detail.

HVAC AND PLUMBING PIPING

The draftsman should be familiar with abbreviations, symbols, and drawing details of the various piping and electrical trades. While shop drawings are being prepared, potential conflicts can be held to a minimum when drawings of these trades are intelligently interpreted. Figures 5-62 to 5-83 show symbols, fitting dimensions, piping arrangements, pipe layouts and hangers.

Courtesy of The York Div., Borg-Warner Corp.

FIG. 5-46. Multi-zone unit.

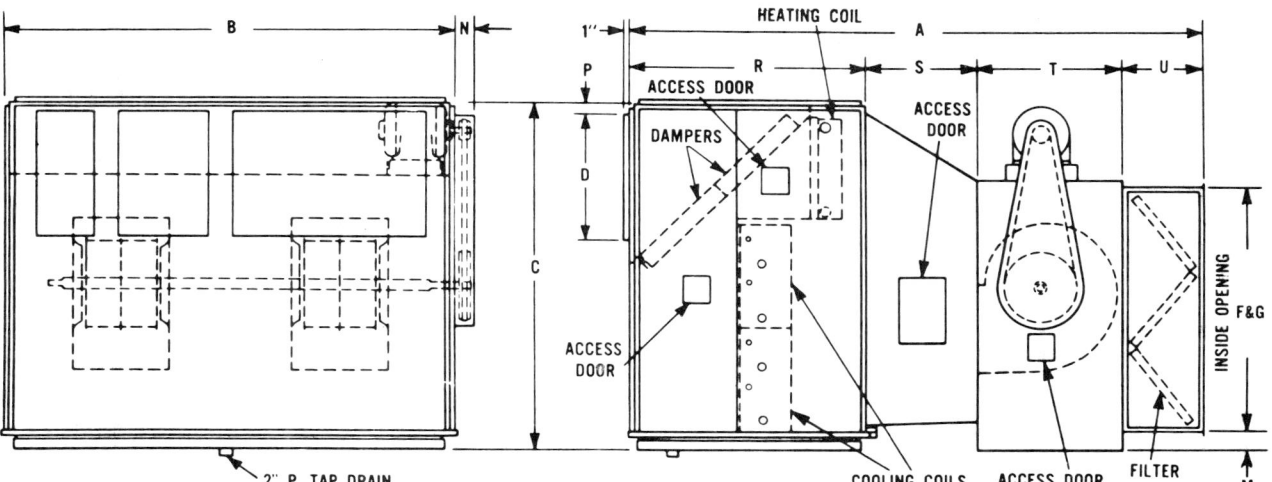

Casings are shipped in three sections: coil, fan, and plenum sections.

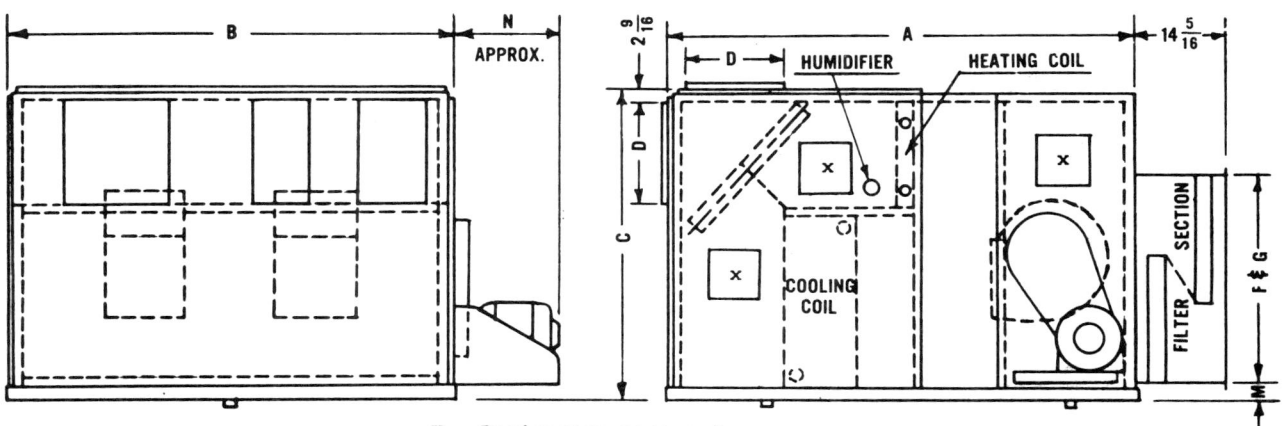

X — Designates access doors.

Courtesy of The Trane Company

FIG. 5-47. Dual-duct type unit.

Courtesy of The York Div., Borg-Warner Corp.

Fig. 5-48. Heating and air-conditioning unit.

Allowance for pitch in steam return, hot-water heating, sprinkler, vent, and drainage piping must be considered. While pressure pipes can be installed with traps, vent lines and gravity drainage pipes must maintain a constant pitch.

Clearance for insulation and hangers, access to valves, clean-outs, and traps, are other important considerations. Duct that must pass through pipe chase areas must be closely coordinated with piping contractor's plans.

Plumbing drain piping is often indicated by *invert elevations*. These are height designations to the inside bottom of the pipe from a fixed, base elevation. Elevations to pipe centerlines are also used.

SPRINKLER PIPING

A *branch line* is a pipe in which sprinklers are directly placed.

A *cross main* is a pipe directly supplying the branch lines.

A *riser* is a vertical pipe supplying the sprinkler systems.

A *feed main* is a bulk main which supplies a cross main or riser.

Horizontal piping should pitch ¼ inch per 10 feet to main drain valve. Auxiliary drains should be located at trapped low points of systems. Ducts should clear spray-patterns of sprinkler heads to conform to code requirements. General trade practice is to precut pipe to sketch; therefore, piping should be carefully coordinated with duct drawings.

See Figs. 5-28, and 5-30 through 5-35 for piping layouts on mechanical and shop drawings.

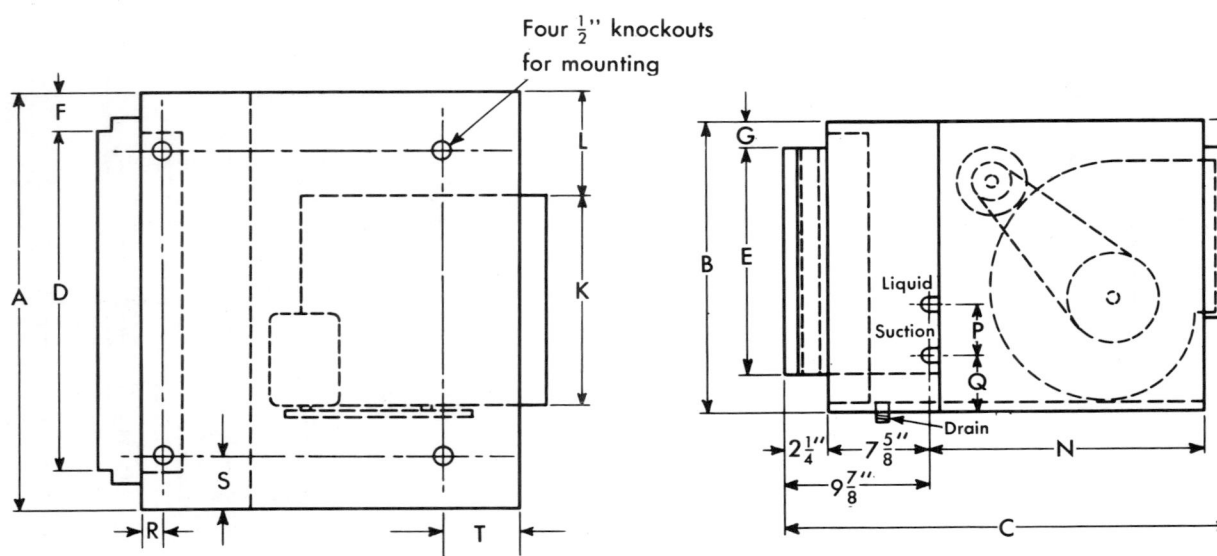

Courtesy of The Trane Company

Fig. 5-49. Direct expansion coil air-conditioning unit.

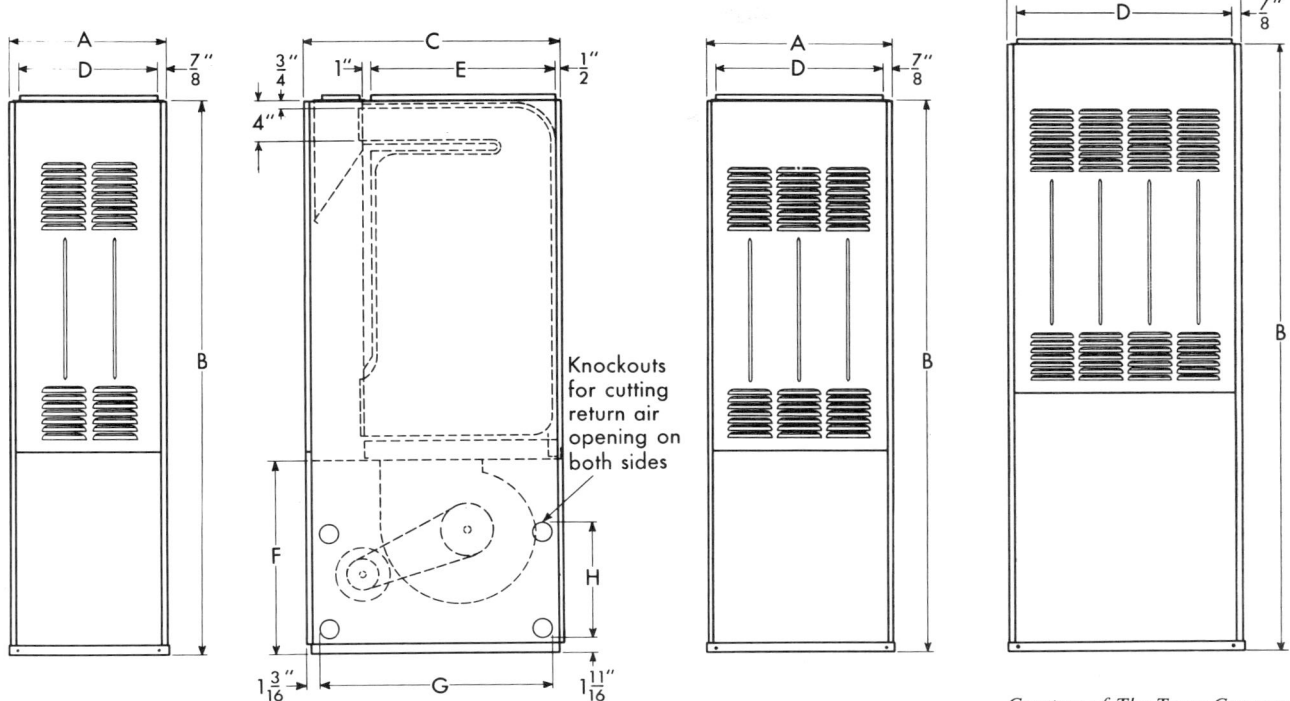

Fig. 5-50. Residential heating units.

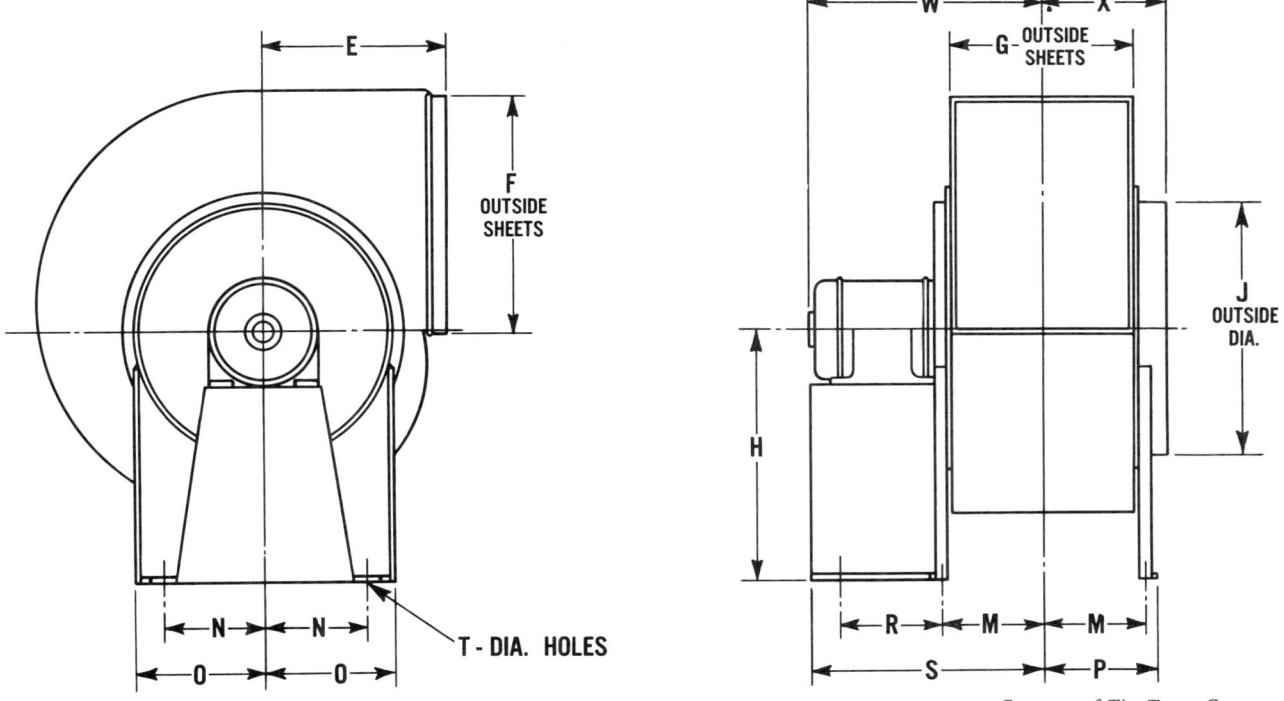

Fig. 5-51. Direct-drive centrifugal fan.

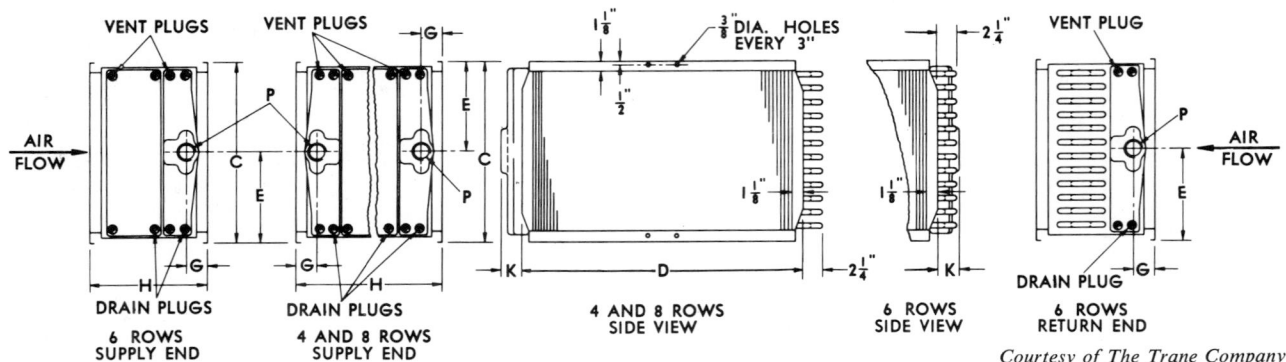

Fig. 5-52. Chilled-water cooling coils.

Courtesy of The Trane Company

Pneumatic Tube Systems

Installation of conveying tube systems is a special phase of the pipe-fitting trade. These systems use high-vacuum negative pressures developed by turbo blowers to convey carrier capsules in pipes to the desired station. Long-radius elbows are used for tube turns. Banks of conduit-type tubing to and from central receiving stations are not uncommon. Tube systems which cross or run at close proximity to HVAC ducts should be shown on the shop drawings so that coordination can be accomplished.

Electrical Systems

Electrical design drawings are often divided into two main categories: light and power. Lighting schedules are very important to the sheet metal draftsman as a key to the type of light shown on the design drawing. (See Fig. 5-84.) Ducts are often located close to hung ceilings and must avoid recessed lights. Spotlight height dimensions must be verified by the electrical contractor as they can easily be 18″ high, or more. Generally, if the bottom of ducts is held to a minimum of 8 inches above the finished ceiling line, clearance for standard, recessed fluorescent fixtures is maintained. Always check the design drawings for lighting types: fluorescent, incandescent, recessed, flush, mounted, pendant, spotlights, or cove lighting. Coordinate light patterns with reflected, ceiling architectural drawings, and secure cuts of light fixtures in critical areas. (See listing below, for electrical abbreviations and symbols.)

ELECTRICAL ABBREVIATIONS AND SYMBOLS

AFF	Above Finished Floor
ATS	Automatic Transfer Switch
₵	Centerline
CSB	Cable Support Box
DP	Distribution Panel
EC	Empty Conduit
EPC	Emergency Power Center
ERP	Emergency Receptacle Panel
ESB	Emergency Switch Board
Exp. Jt.	Expansion Joint
Exp.	Explosion Proof
F	Feeder
FBO	Furnished By Others
HC	Hung Ceiling
LP	Lighting Panel
LC	Load Center
MCC	Motor Control Center
NIC	Not In Contract

Courtesy of The York Div., Borg-Warner Corp.

Fig. 5-53. Chilled-water cooling coil.

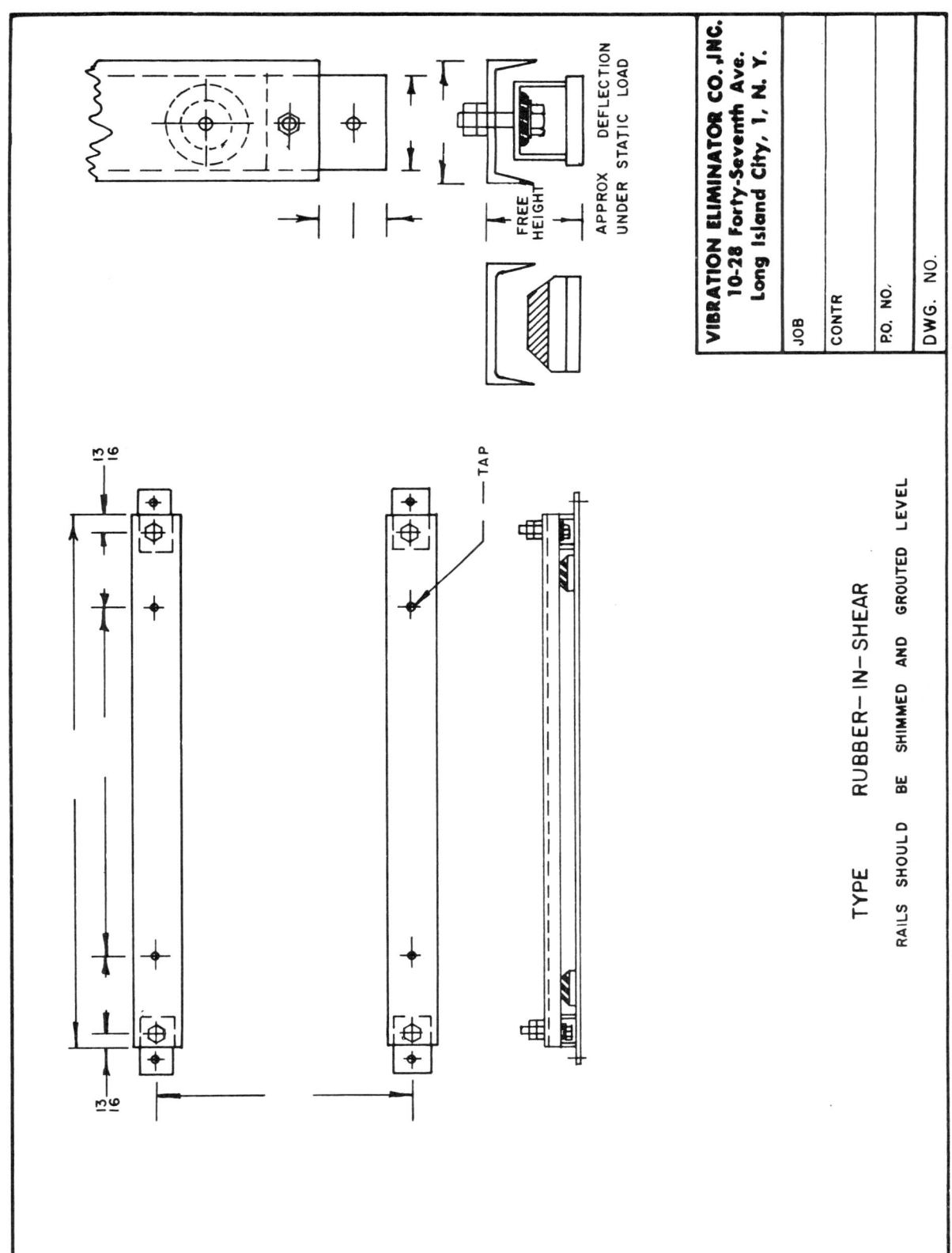

Fig. 5-54. Rubber-in-shear type vibration eliminator.

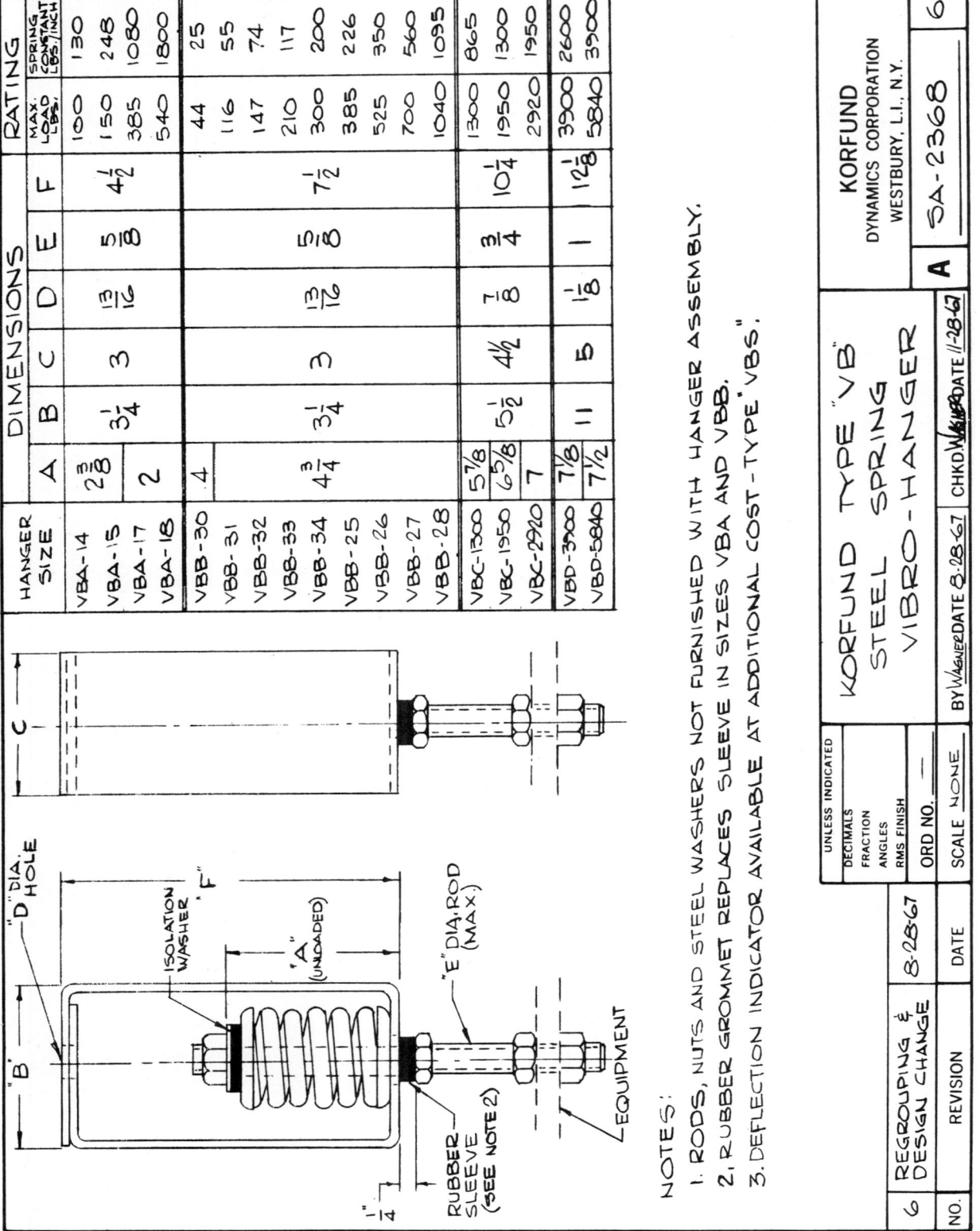

Fig. 5-55. Korfund type "VB" steel spring vibro-hanger.

Ch. 5   DESIGN DRAWINGS AND SPECIFICATIONS   203

PC	Power Center	VT	Vapor Tight
PP	Power Panel	WP	Weather Proof
PDC	Primary Distribution Center	▭	Recessed Fluorescent Fixture
PF	Primary Feeder	▭○	Surface or Pendant-Mounted Fluorescent Fixture
PB	Pull Box		
PC	Pull Chain	○	Recessed Incandescent Spot Light
RP	Receptacle Panel	PB	Panel Board
RPC	Refrigeration Power Center	MCC	Motor Control Center
S	Switchboard	———	Concealed Conduit
SS	Substation Switchboard	— — —	Conduit Concealed in Floor
T	Transformer	- - - -	Conduit Run Exposed
		—EC—	Empty Conduit
		——○	Conduit Up
		——•	Conduit Down

Duct drawings must be coordinated with design drawings of electrical power systems such as: motor control centers, main switchgear rooms, bus ducts, electrical closets, pull boxes, panel boards, conduit banks and risers, and main load-centers. See Fig. 5-85 for a typical electrical closet plan. Electrical motor sizes are shown in Fig. 5-86.

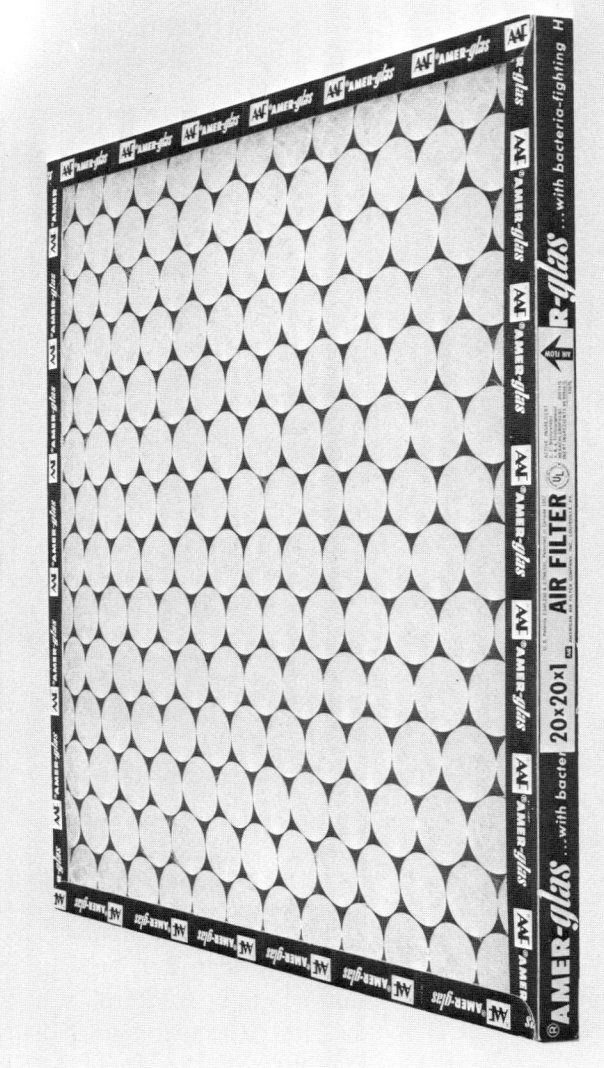

Courtesy of American Air Filter Company, Inc.

FIG. 5-56. Throw-away type filter.

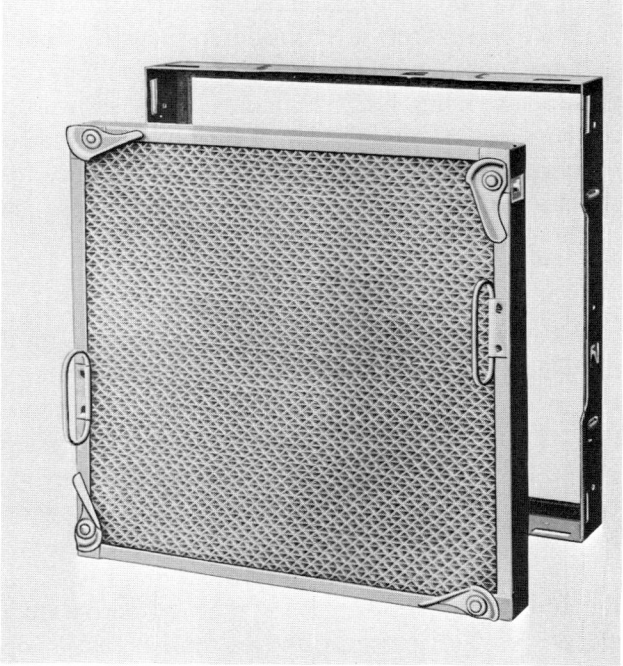

Courtesy of American Air Filter Company, Inc.

FIG. 5-57. Cleanable type filter and mounting frame.

Courtesy of American Air Filter Company, Inc.
FIG. 5-58. High-capacity air filter and frame.

Special Process Systems and Owners' Equipment

Hospitals and large manufacturing plants often purchase and let out separate contracts for X-ray equipment, high-vacuum sterilization chambers, paint spray-booths, and other large installations. As these contractors are considered specialty trades and may not be working for the general contractor, the ducts, piping, and mechanical work shown on the engineer's design drawings must be coordinated separately with the specialty trades. Reference installation drawings for jobs of this type must be made available to the sheet metal draftsman.

STRUCTURAL DRAWINGS

When a shop drawing is started, the structural drawings are used as reference for the draftsman's layout of the fixed building members. Horizontal duct-runs are often suspended from the underside of the floor above, therefore, the structural members of that level are indicated on the shop drawing for use in locating ducts, diffusers, and equipment.

The exception is for underground ducts wherein the foundation drawings are used to verify duct clearances with footings and column bases. Top of footing elevations may be indicated by decimal feet or in feet and inches. (See Fig. 5-S-1.)

Reinforced concrete beams and precast beams are both dimensioned in plan. The first dimension of such beams is the width. The given depth of reinforced concrete beams is from top of slab above, while depths of precast beams are separately noted on the design drawings.

Concrete columns can be round, square, or rectangular and should be checked in the column schedules.

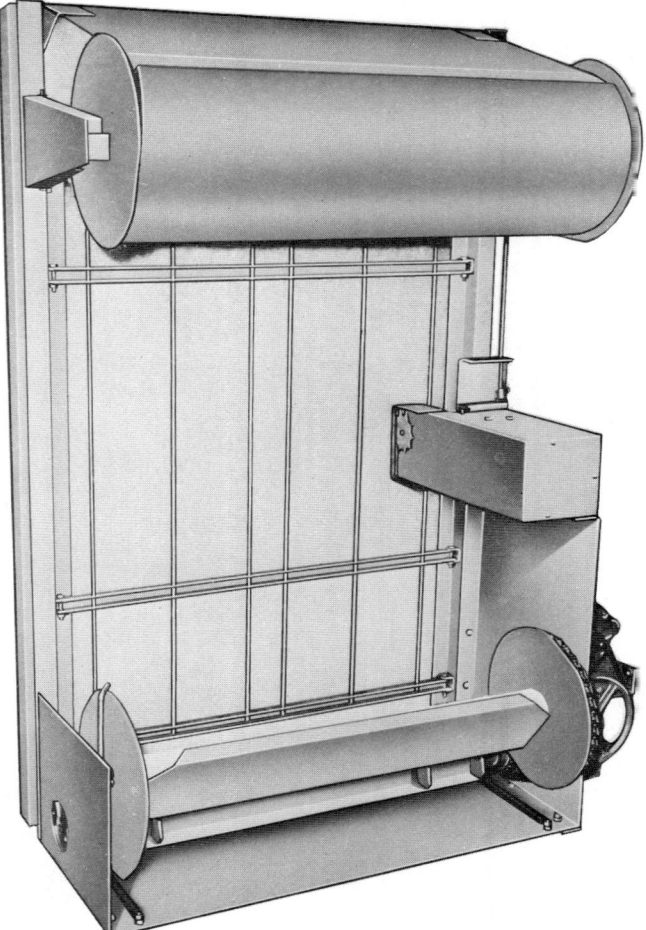

Courtesy of American Air Filter Company, Inc.
FIG. 5-59. Automatic roll-type filter.

When concrete columns are rectangular, the draftsman should verify the orientation of column dimensions. For example, the structural design drawing will show the first dimension as North-South or East-West. A note on the drawing will show which designation is used.

(Text continued on page 241.)

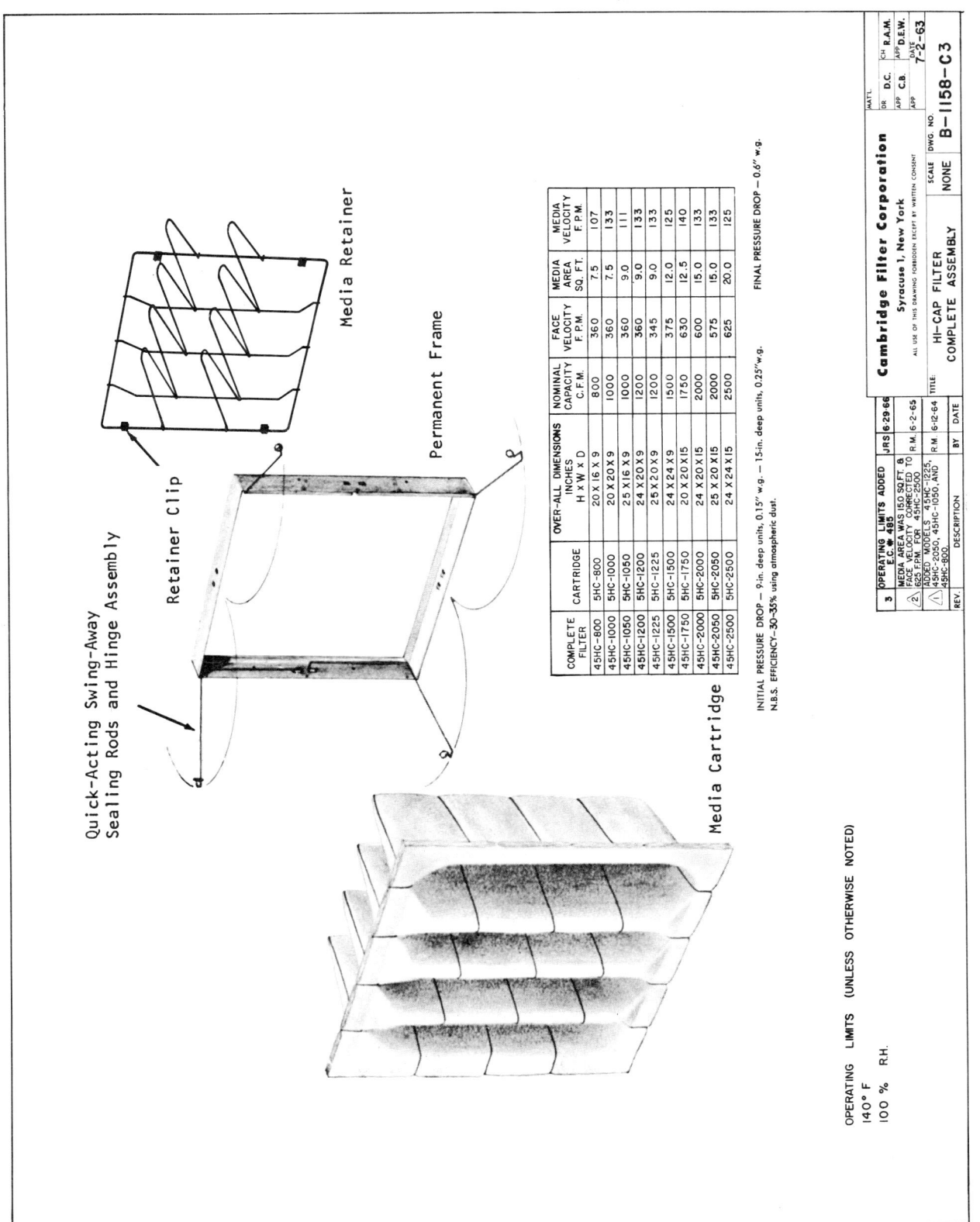

Fig. 5-60. High-capacity air filter, complete assembly.

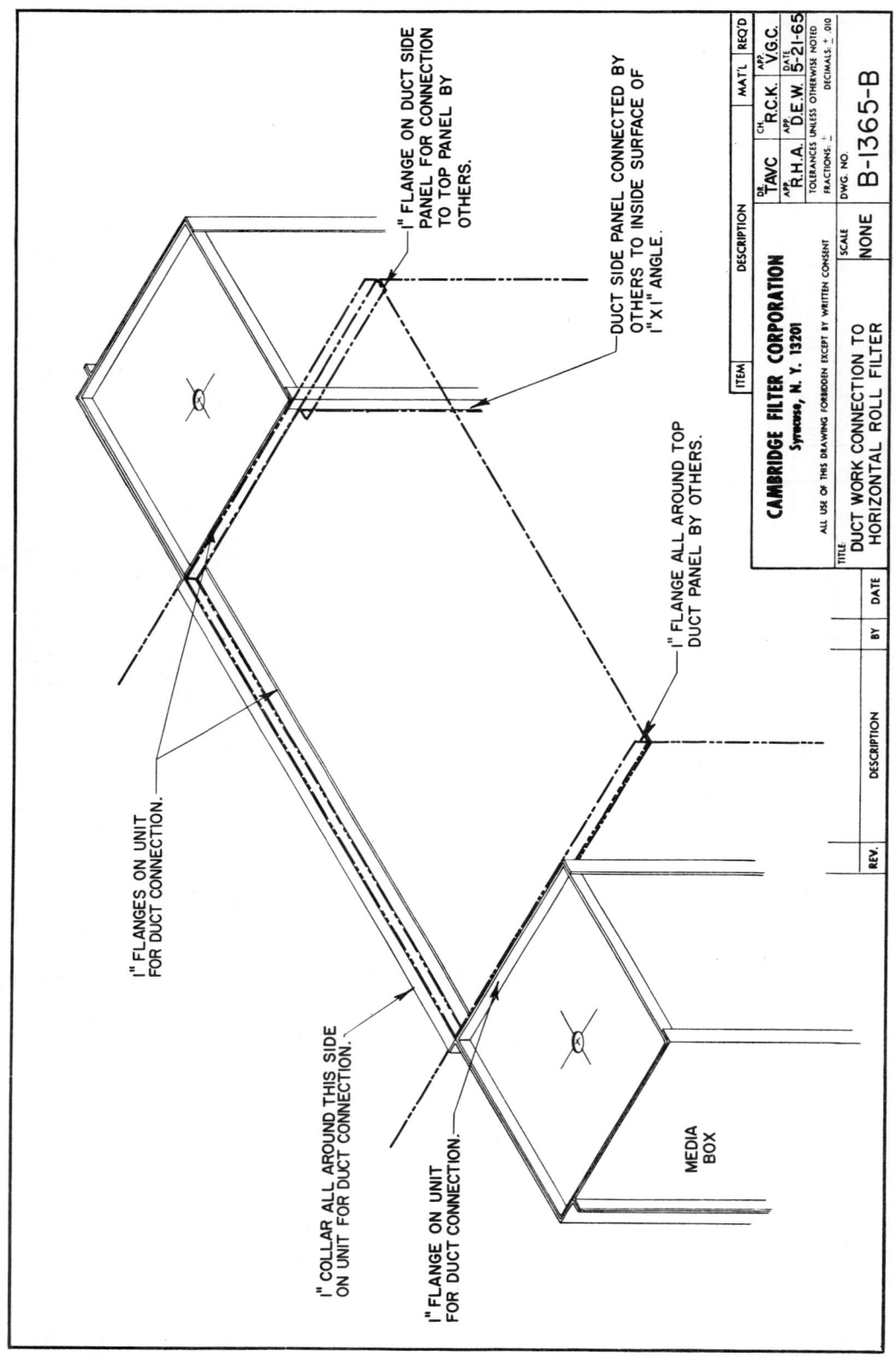

Fig. 5-61. Ductwork connection to horizontal-roll filter.

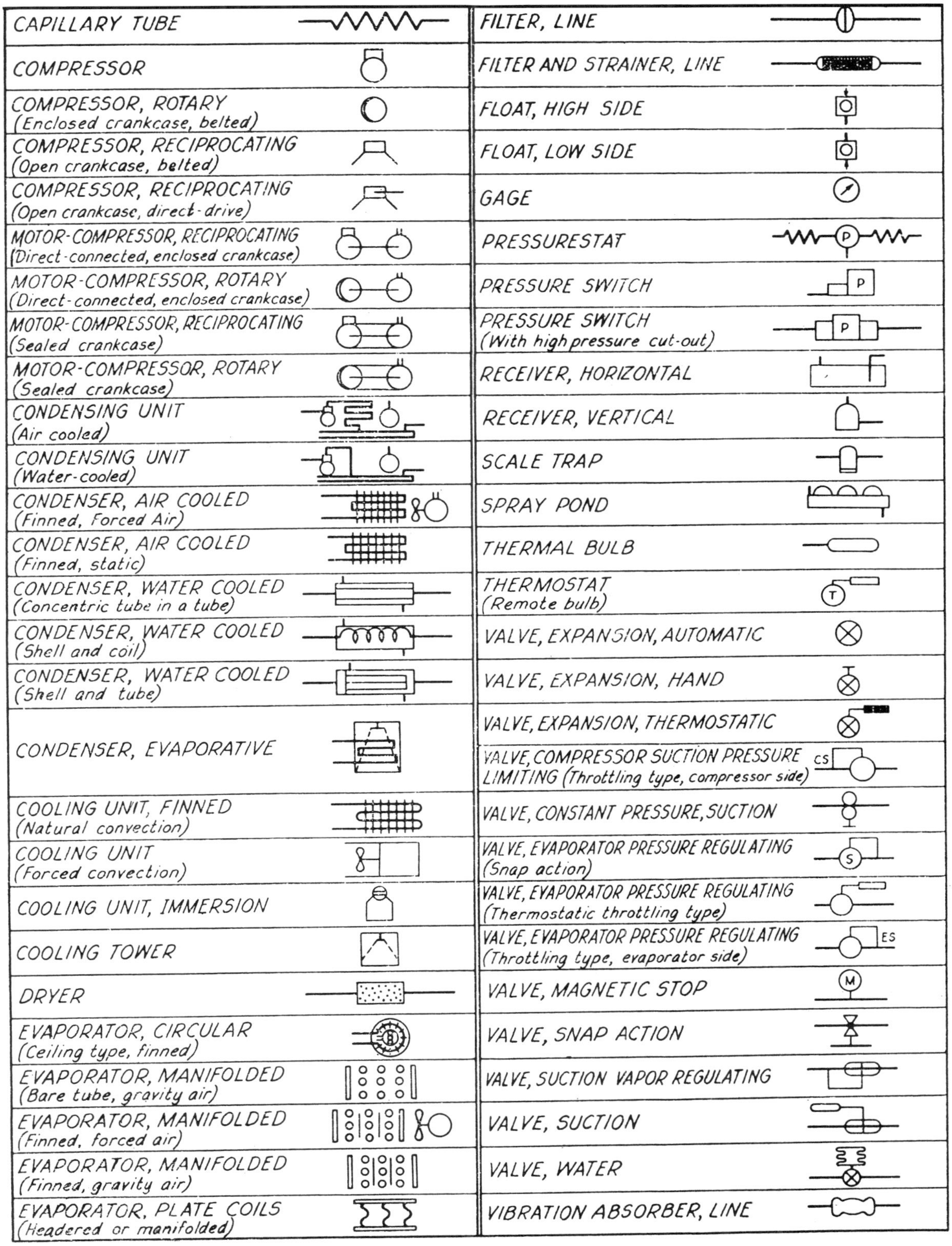

CAPILLARY TUBE		FILTER, LINE	
COMPRESSOR		FILTER AND STRAINER, LINE	
COMPRESSOR, ROTARY (Enclosed crankcase, belted)		FLOAT, HIGH SIDE	
COMPRESSOR, RECIPROCATING (Open crankcase, belted)		FLOAT, LOW SIDE	
COMPRESSOR, RECIPROCATING (Open crankcase, direct-drive)		GAGE	
MOTOR-COMPRESSOR, RECIPROCATING (Direct-connected, enclosed crankcase)		PRESSURESTAT	
MOTOR-COMPRESSOR, ROTARY (Direct-connected, enclosed crankcase)		PRESSURE SWITCH	
MOTOR-COMPRESSOR, RECIPROCATING (Sealed crankcase)		PRESSURE SWITCH (With high pressure cut-out)	
MOTOR-COMPRESSOR, ROTARY (Sealed crankcase)		RECEIVER, HORIZONTAL	
CONDENSING UNIT (Air cooled)		RECEIVER, VERTICAL	
CONDENSING UNIT (Water-cooled)		SCALE TRAP	
CONDENSER, AIR COOLED (Finned, Forced Air)		SPRAY POND	
CONDENSER, AIR COOLED (Finned, static)		THERMAL BULB	
CONDENSER, WATER COOLED (Concentric tube in a tube)		THERMOSTAT (Remote bulb)	
CONDENSER, WATER COOLED (Shell and coil)		VALVE, EXPANSION, AUTOMATIC	
CONDENSER, WATER COOLED (Shell and tube)		VALVE, EXPANSION, HAND	
CONDENSER, EVAPORATIVE		VALVE, EXPANSION, THERMOSTATIC	
		VALVE, COMPRESSOR SUCTION PRESSURE LIMITING (Throttling type, compressor side)	
COOLING UNIT, FINNED (Natural convection)		VALVE, CONSTANT PRESSURE, SUCTION	
COOLING UNIT (Forced convection)		VALVE, EVAPORATOR PRESSURE REGULATING (Snap action)	
COOLING UNIT, IMMERSION		VALVE, EVAPORATOR PRESSURE REGULATING (Thermostatic throttling type)	
COOLING TOWER		VALVE, EVAPORATOR PRESSURE REGULATING (Throttling type, evaporator side)	
DRYER		VALVE, MAGNETIC STOP	
EVAPORATOR, CIRCULAR (Ceiling type, finned)		VALVE, SNAP ACTION	
EVAPORATOR, MANIFOLDED (Bare tube, gravity air)		VALVE, SUCTION VAPOR REGULATING	
EVAPORATOR, MANIFOLDED (Finned, forced air)		VALVE, SUCTION	
EVAPORATOR, MANIFOLDED (Finned, gravity air)		VALVE, WATER	
EVAPORATOR, PLATE COILS (Headered or manifolded)		VIBRATION ABSORBER, LINE	

Fig. 5-62. Standard graphical symbols for air conditioning. (From American Standard Z32.2.4, sponsored by ASME and AIEE.)

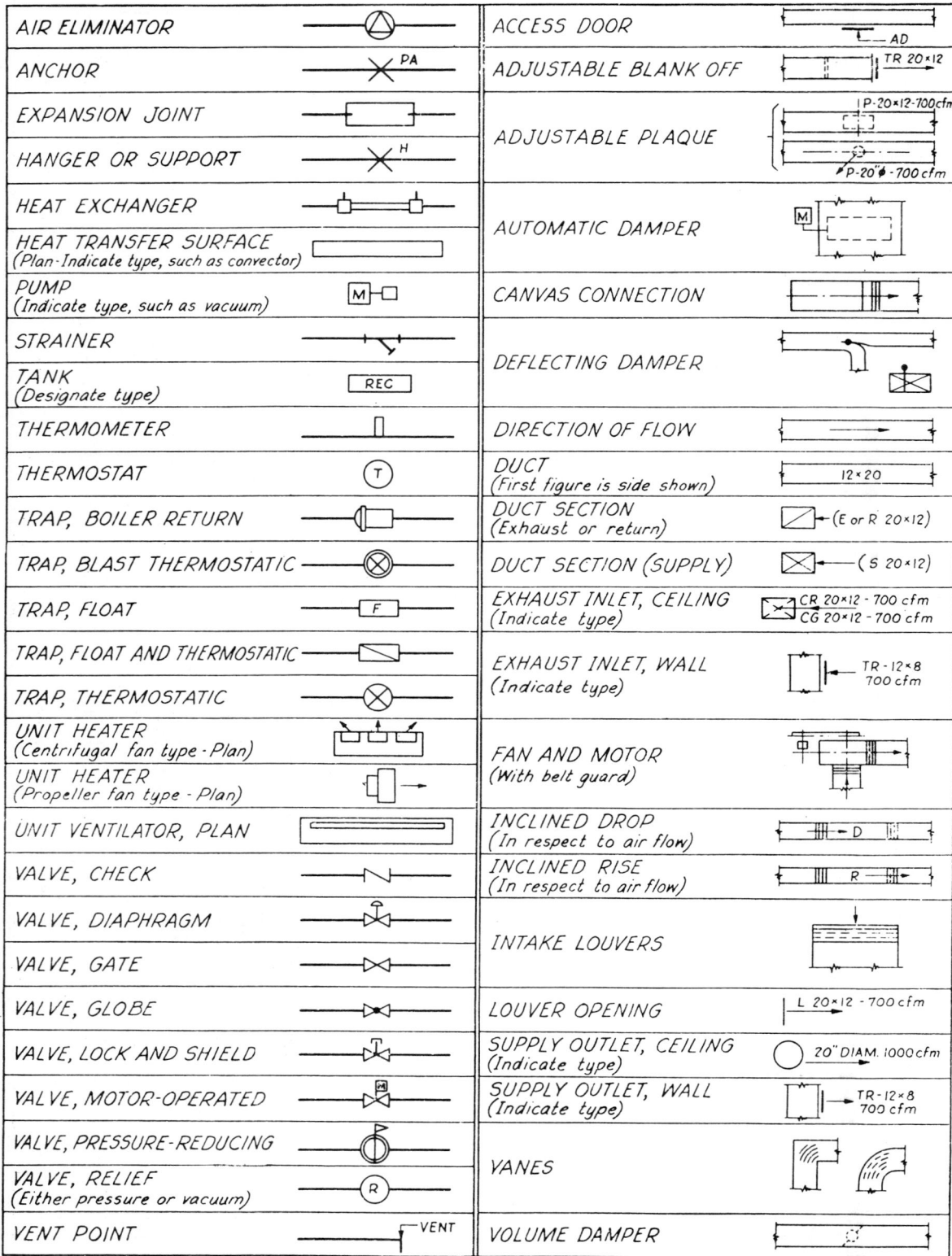

Fig. 5-63. Standard graphical symbols for heating and ventilating. (From American Standard Z32.2.4, sponsored by ASME and AIEE.)

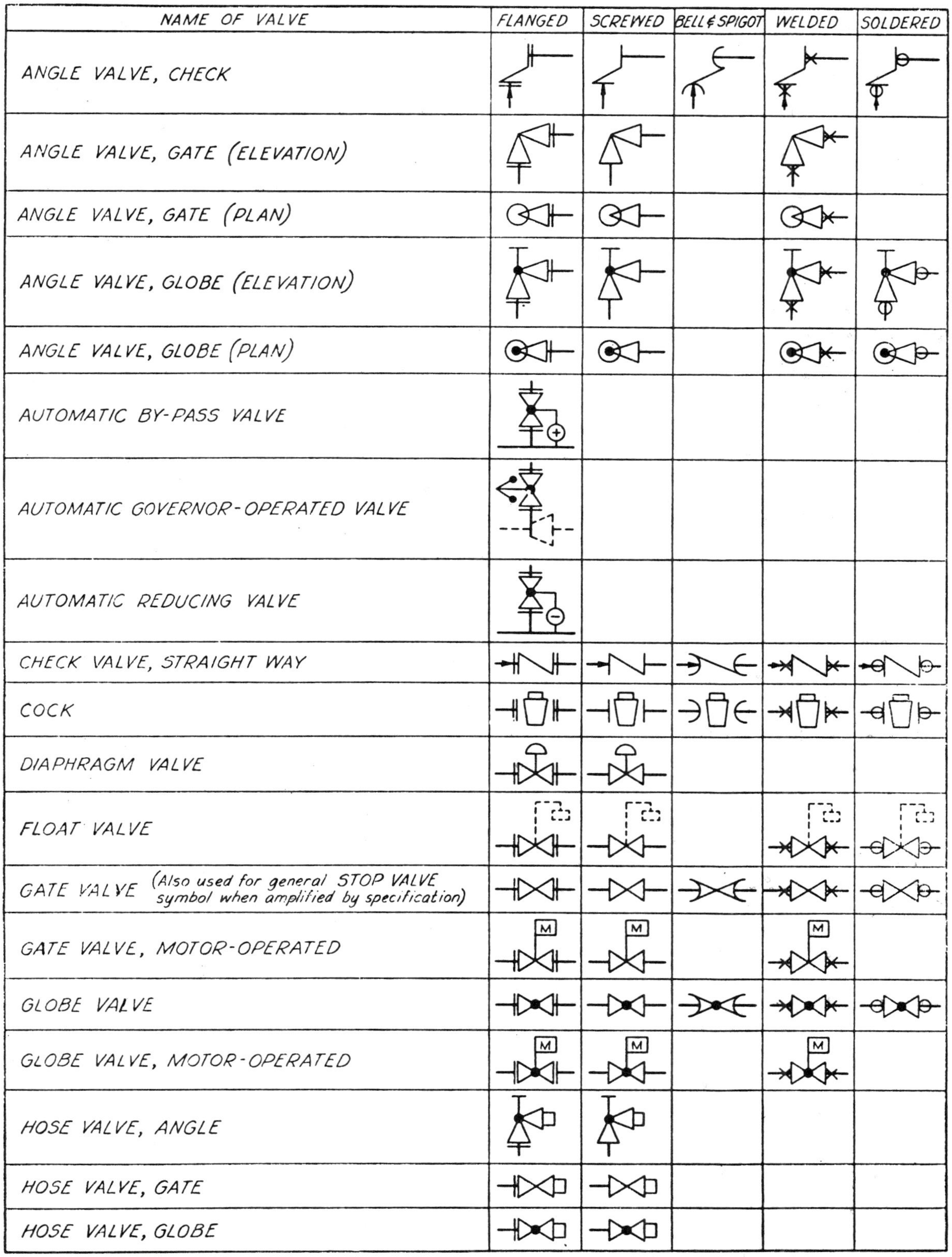

NAME OF VALVE	FLANGED	SCREWED	BELL & SPIGOT	WELDED	SOLDERED
ANGLE VALVE, CHECK	⌐	⌐	⌐	⌐	⌐
ANGLE VALVE, GATE (ELEVATION)	⌐	⌐		⌐	
ANGLE VALVE, GATE (PLAN)	⌐	⌐		⌐	
ANGLE VALVE, GLOBE (ELEVATION)	⌐	⌐		⌐	⌐
ANGLE VALVE, GLOBE (PLAN)	⌐	⌐		⌐	⌐
AUTOMATIC BY-PASS VALVE	⌐				
AUTOMATIC GOVERNOR-OPERATED VALVE	⌐				
AUTOMATIC REDUCING VALVE	⌐				
CHECK VALVE, STRAIGHT WAY	⌐	⌐	⌐	⌐	⌐
COCK	⌐	⌐	⌐	⌐	⌐
DIAPHRAGM VALVE	⌐	⌐			
FLOAT VALVE	⌐	⌐		⌐	⌐
GATE VALVE (Also used for general STOP VALVE symbol when amplified by specification)	⌐	⌐	⌐	⌐	⌐
GATE VALVE, MOTOR-OPERATED	⌐	⌐		⌐	
GLOBE VALVE	⌐	⌐	⌐	⌐	⌐
GLOBE VALVE, MOTOR-OPERATED	⌐	⌐		⌐	
HOSE VALVE, ANGLE	⌐	⌐			
HOSE VALVE, GATE	⌐	⌐			
HOSE VALVE, GLOBE	⌐	⌐			

FIG. 5-64. Standard graphical symbols for valves. (From American Standard Z32.2.4, sponsored by ASME and AIEE.)

NAME OF VALVE	FLANGED	SCREWED	BELL & SPIGOT	WELDED	SOLDERED
LOCKSHIELD VALVE	⊢⊠⊣	⊠			⊸⊠⊷
QUICK-OPENING VALVE	⊢⊠⊣	⊠		⊁⊠⊀	⊸⊠⊷
SAFETY VALVE	⊢⊠⊣	⊠	⊃⊠⊂	⊁⊠⊀	⊸⊠⊷

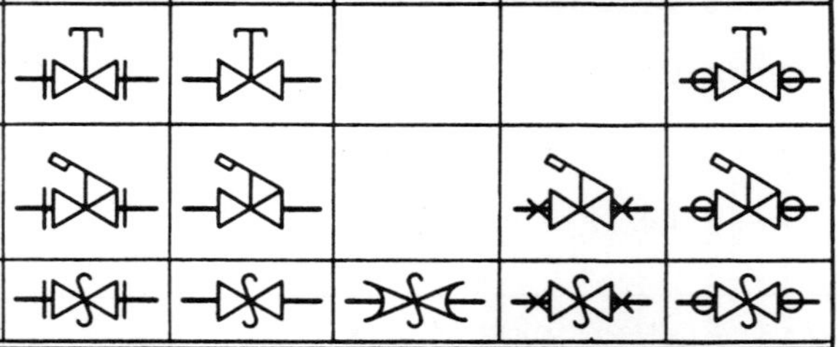

PIPING

AIR CONDITIONING			
BRINE RETURN	— — -BR- — —	LOW PRESSURE RETURN	— — — —
BRINE SUPPLY	——— B ———	LOW PRESSURE STEAM	———————
CHILLED OR HOT WATER FLOW (Circulating)	——— CH ———	MAKE-UP WATER	——·——·——
CHILLED OR HOT WATER RETURN (Circulating)	— — CHR — —	MEDIUM PRESSURE RETURN	⊹ — ⊹ —
CONDENSER WATER FLOW	——— C ———	MEDIUM PRESSURE STEAM	⊹ — ⊹
CONDENSER WATER RETURN	— — CR — —	**PLUMBING**	
DRAIN	——— D ———	ACID WASTE	——— ACID ———
HUMIDIFICATION LINE	—·— H —·—	COLD WATER	——·——·——
MAKE-UP WATER	——·——·——	COMPRESSED AIR	——— A ———
REFRIGERANT DISCHARGE	——— RD ———	DRINKING WATER FLOW	——·——·——
REFRIGERANT LIQUID	——— RL ———	DRINKING WATER RETURN	——··——··——
REFRIGERANT SUCTION	— — RS — —	FIRE LINE	——F———F——
HEATING		GAS	——G———G——
AIR RELIEF LINE	——·——·——	HOT WATER	——··——··——
BOILER BLOW-OFF	——— ———	HOT WATER RETURN	——···——···——
COMPRESSED AIR	——— A ———	SOIL, WASTE, OR LEADER (Above grade)	———————
CONDENSATE DISCHARGE	—o— —o— —o—	SOIL, WASTE, OR LEADER (Below grade)	— — — —
FEEDWATER PUMP DISCHARGE	—∞— —∞— —∞—	VACUUM CLEANING	——V———V——
FUEL-OIL FLOW	——— FOF ———	VENT	— — — — —
FUEL-OIL RETURN	— — FOR — —	**PNEUMATIC TUBES**	
FUEL-OIL TANK VENT	— — FOV — —	TUBE RUNS	═══════
HIGH PRESSURE RETURN	⊬ — ⊬ —	**SPRINKLERS**	
HIGH PRESSURE STEAM	⊬ — ⊬	BRANCH AND HEAD	——o———o——
HOT WATER HEATING RETURN	— — — —	DRAIN	— —S— — —S— —
HOT WATER HEATING SUPPLY	———————	MAIN SUPPLIES	——— S ———

FIG. 5-65. Standard graphical symbols for valves and piping. (From American Standard Z32.2.4, sponsored by ASME and AIEE.)

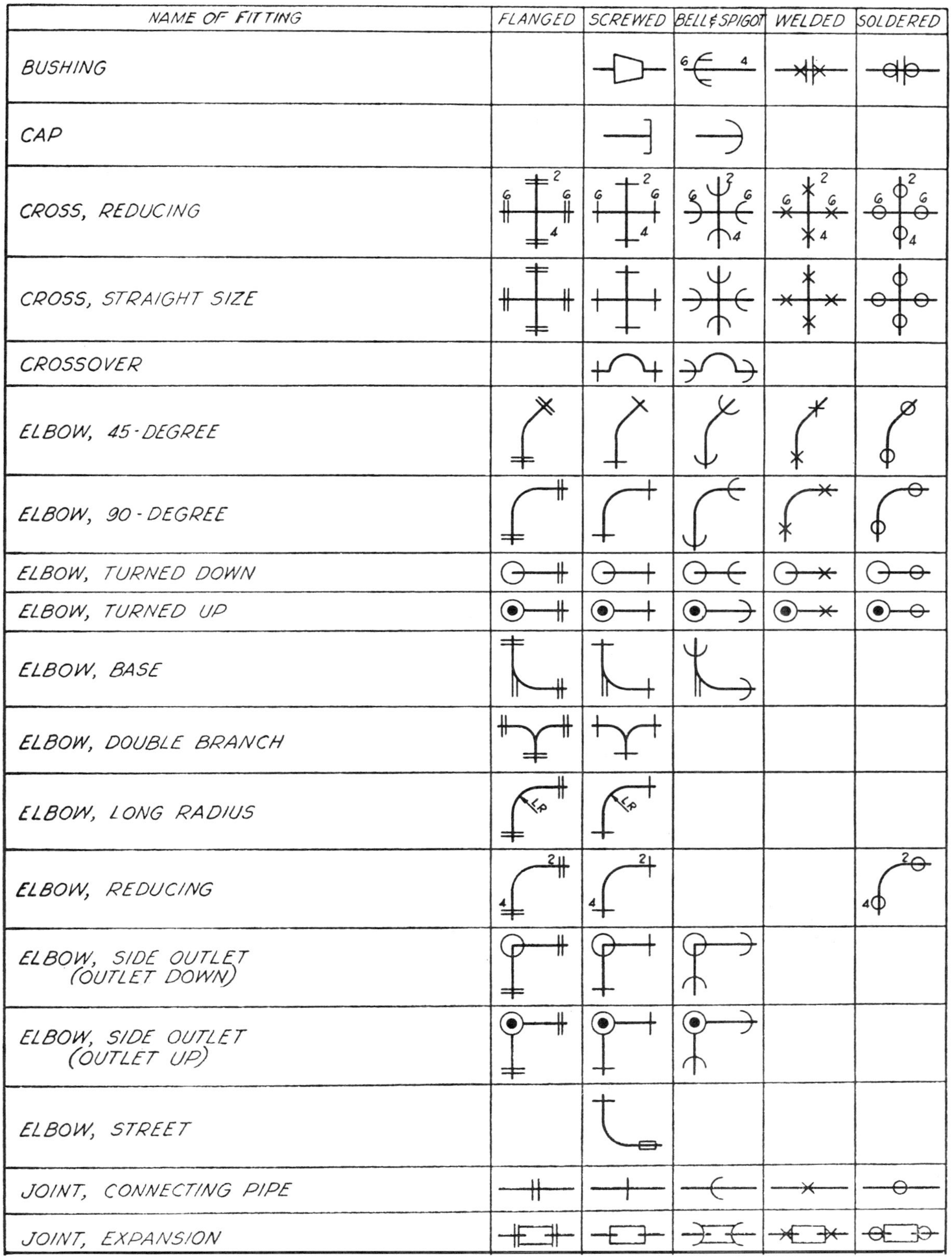

NAME OF FITTING	FLANGED	SCREWED	BELL & SPIGOT	WELDED	SOLDERED
BUSHING		⟶▷⟵	⁶⟶∈⟵⁴	⟶⋈⟵	⟶⊕⟵
CAP		⟶⊐	⟶⌒		
CROSS, REDUCING	⁶╫╬╫⁶ ²/⁴	⁶┼┼┼⁶ ²/⁴	⁶)╪(⁶ ²/⁴	⁶✕✳✕⁶ ²/⁴	⁶φ⊕φ⁶ ²/⁴
CROSS, STRAIGHT SIZE	╫╬╫	┼┼┼	)╪(	✕✳✕	φ⊕φ
CROSSOVER		┼⌒┼	)⌒⟶		
ELBOW, 45-DEGREE	⤢	⤢	⤢	⤢	⤢
ELBOW, 90-DEGREE	⌐	⌐	⌐	⌐	⌐
ELBOW, TURNED DOWN	⊖—╫	⊖—┼	⊖—∈	⊖—✕	⊖—⊖
ELBOW, TURNED UP	⊙—╫	⊙—┼	⊙—⟶	⊙—✕	⊙—⊖
ELBOW, BASE	⌐	⌐	⌐		
ELBOW, DOUBLE BRANCH	╫⋎╫	┼⋎┼			
ELBOW, LONG RADIUS	⌐LR	⌐LR			
ELBOW, REDUCING	²⌐╫ / ⁴	²⌐┼ / ⁴			²⌐⊖ / ⁴
ELBOW, SIDE OUTLET (OUTLET DOWN)	⊖⌐╫	⊖⌐┼	⊖⌐⟶		
ELBOW, SIDE OUTLET (OUTLET UP)	⊙⌐╫	⊙⌐┼	⊙⌐⟶		
ELBOW, STREET		⌐			
JOINT, CONNECTING PIPE	—╫—	—┼—	—∈	—✕—	—⊖—
JOINT, EXPANSION	╫⊏⊐╫	┼⊏⊐┼	)⊏⊐∈	✕⊏⊐✕	⊖⊏⊐⊖

FIG. 5-66. Standard graphical symbols for pipe fittings. (From American Standard Z32.2.4, sponsored by ASME and AIEE.)

Diameter, Inches			Wall Thickness, In.	Cross-Sectional Area, Sq. In.			Weight per Foot, Lb.		
Nominal	Actual Inside	Actual Outside		Outside	Inside	Metal	Of Pipe Alone	Of Water in Pipe	Of Pipe and Water
1/8	0.269	0.405	0.068	0.129	0.057	0.072	0.25	0.028	0.278
1/4	0.364	0.540	0.088	0.229	0.104	0.125	0.43	0.045	0.475
3/8	0.493	0.675	0.091	0.358	0.191	0.167	0.57	0.083	0.653
1/2	0.622	0.840	0.109	0.554	0.304	0.250	0.86	0.132	0.992
3/4	0.824	1.050	0.113	0.866	0.533	0.333	1.14	0.232	1.372
1	1.049	1.315	0.133	1.358	0.864	0.494	1.68	0.375	2.055
1 1/4	1.380	1.660	0.140	2.164	1.495	0.669	2.28	0.649	2.929
1 1/2	1.610	1.900	0.145	2.835	2.036	0.799	2.72	0.882	3.602
2	2.067	2.375	0.154	4.431	3.356	1.075	3.66	1.454	5.114
2 1/2	2.469	2.875	0.203	6.492	4.788	1.704	5.80	2.073	7.873
3	3.068	3.500	0.216	9.621	7.393	2.228	7.58	3.201	10.781
3 1/2	3.548	4.000	0.226	12.568	9.888	2.680	9.11	4.287	13.397
4	4.026	4.500	0.237	15.903	12.730	3.173	10.80	5.516	16.316
5	5.047	5.563	0.258	24.308	20.004	4.304	14.70	8.674	23.374
6	6.065	6.625	0.280	34.474	28.890	5.584	19.00	12.52	31.52
8	7.981	8.625	0.322	58.426	50.030	8.396	28.60	21.68	50.28
10	10.020	10.750	0.365	90.79	78.85	11.90	40.50	34.16	74.66
12	11.938	12.750	0.406	127.67	113.09	15.77	53.60	48.50	102.10
14	13.126	14.000	0.437	153.94	135.33	18.61	63.30	58.64	121.94
16	15.000	16.000	0.500	201.06	176.71	24.35	82.80	76.58	159.38
18	16.876	18.000	0.562	254.47	223.68	30.79	105.00	96.93	201.93
20	18.814	20.000	0.593	314.16	278.01	36.15	123.00	120.46	243.46

Nominal Dia., In.	Circumference, Inches		Sq. Ft. of Surface per Lineal Foot		Contents of Pipe per Lineal Foot		Lineal Feet to Contain		
	Outside	Inside	Outside	Inside	Cu. Ft.	Gal.	1 Cu. Ft.	1 Gal.	1 Lb. of Water
1/8	1.27	0.84	0.106	0.070	0.0004	0.003	2533.775	338.74	35.714
1/4	1.69	1.14	0.141	0.095	0.0007	0.005	1383.789	185.00	22.222
3/8	2.12	1.55	0.177	0.129	0.0013	0.010	754.360	100.85	12.048
1/2	2.65	1.95	0.221	0.167	0.0021	0.016	473.906	63.36	7.576
3/4	3.29	2.58	0.275	0.215	0.0037	0.028	270.034	36.10	4.310
1	4.13	3.29	0.344	0.274	0.0062	0.045	166.618	22.28	2.667
1 1/4	5.21	4.33	0.435	0.361	0.0104	0.077	96.275	12.87	1.541
1 1/2	5.96	5.06	0.497	0.422	0.0141	0.106	70.733	9.46	1.134
2	7.46	6.49	0.622	0.540	0.0233	0.174	42.913	5.74	0.688
2 1/2	9.03	7.75	0.753	0.654	0.0332	0.248	30.077	4.02	0.482
3	10.96	9.63	0.916	0.803	0.0514	0.383	19.479	2.60	0.312
3 1/2	12.56	11.14	1.047	0.928	0.0682	0.513	14.565	1.95	0.233
4	14.13	12.64	1.178	1.052	0.0884	0.660	11.312	1.51	0.181
5	17.47	15.84	1.456	1.319	0.1390	1.040	7.198	0.96	0.115
6	20.81	19.05	1.734	1.585	0.2010	1.500	4.984	0.67	0.080
8	27.09	25.07	2.258	2.090	0.3480	2.600	2.878	0.38	0.046
10	33.77	31.47	2.814	2.622	0.5470	4.100	1.826	0.24	0.029
12	40.05	37.70	3.370	3.140	0.7850	5.870	1.273	0.17	0.021
14	47.12	44.76	3.930	3.722	1.0690	7.030	1.067	0.14	0.017
16	53.41	51.52	4.440	4.310	1.3920	9.180	0.814	0.11	0.013
18	56.55	53.00	4.712	4.420	1.5530	11.120	0.644	0.09	0.010
20	62.83	59.09	5.236	4.920	1.9250	14.400	0.517	0.07	0.008

FIG. 5-67. Dimensional and capacity data for schedule 40 steel pipe.

Diameter, In.			Wall Thickness, In.	Cross-Sectional Area, Sq. In.			Weight per Foot, Lb.		
Nominal	Actual Inside	Actual Outside		Outside	Inside	Metal	Of Pipe	Of Water in Pipe	Of Pipe and Water
1/8	0.215	0.405	0.095	0.129	0.036	0.093	0.314	0.016	0.330
1/4	0.302	0.540	0.119	0.229	0.072	0.157	0.535	0.031	0.566
3/8	0.423	0.675	0.126	0.358	0.141	0.217	0.738	0.061	0.799
1/2	0.546	0.840	0.147	0.554	0.234	0.320	1.087	0.102	1.189
3/4	0.742	1.050	0.154	0.866	0.433	0.433	1.473	0.213	1.686
1	0.957	1.315	0.179	1.358	0.719	0.639	2.171	0.312	2.483
1 1/4	1.278	1.660	0.191	2.164	1.283	0.881	2.996	0.555	3.551
1 1/2	1.500	1.900	0.200	2.835	1.767	1.068	3.631	0.765	4.396
2	1.939	2.375	0.218	4.431	2.954	1.477	5.022	1.280	6.302
2 1/2	2.323	2.875	0.276	6.492	4.238	2.254	7.661	1.830	9.491
3	2.900	3.500	0.300	9.621	6.605	3.016	10.252	2.870	13.122
3 1/2	3.364	4.000	0.318	12.568	8.890	3.678	12.505	3.720	16.225
4	3.826	4.500	0.337	15.903	11.496	4.407	14.983	4.970	19.953
5	4.813	5.563	0.375	24.308	18.196	6.112	20.778	7.940	28.718
6	5.761	6.625	0.432	34.474	26.069	8.405	28.573	11.300	39.873
8	7.625	8.625	0.500	58.426	45.666	12.760	43.388	19.800	63.188
10	9.564	10.750	0.593	90.79	71.87	18.92	64.400	31.130	95.530
12	11.376	12.750	0.687	127.67	101.64	26.03	88.600	44.040	132.640
14	12.500	14.000	0.750	153.94	122.72	31.22	107.000	53.180	160.180
16	14.314	16.000	0.843	201.06	160.92	40.14	137.000	69.730	206.730
18	16.126	18.000	0.937	254.47	204.24	50.23	171.000	88.500	259.500
20	17.938	20.000	1.031	314.16	252.72	61.44	209.000	109.510	318.510

Nominal Dia., In.	Circumference, Inches		Sq. Ft. of Surface per Lineal Foot		Contents of Pipe per Lineal Foot		Lineal Feet to Contain		
	Outside	Inside	Outside	Inside	Cu. Ft.	Gal.	1 Cu. Ft.	1 Gal.	1 Lb. of Water
1/8	1.27	0.675	0.106	0.056	0.00033	0.0019	3070	527	101.01
1/4	1.69	0.943	0.141	0.079	0.00052	0.0037	1920	271	32.26
3/8	2.12	1.328	0.177	0.111	0.00098	0.0073	1370	137	16.39
1/2	2.65	1.715	0.221	0.143	0.00162	0.0122	616	82	9.80
3/4	3.29	2.330	0.275	0.194	0.00300	0.0255	334	39.2	4.69
1	4.13	3.010	0.344	0.251	0.00500	0.0374	200	26.8	3.21
1 1/4	5.21	4.010	0.435	0.334	0.00880	0.0666	114	15.0	1.80
1 1/2	5.96	4.720	0.497	0.393	0.01230	0.0918	81.50	10.90	1.31
2	7.46	6.090	0.622	0.507	0.02060	0.1535	49.80	6.52	0.78
2 1/2	9.03	7.320	0.753	0.610	0.02940	0.220	34.00	4.55	0.55
3	10.96	9.120	0.916	0.760	0.0460	0.344	21.70	2.91	0.35
3 1/2	12.56	10.580	1.047	0.882	0.0617	0.458	16.25	2.18	0.27
4	14.13	12.020	1.178	1.002	0.0800	0.597	12.50	1.675	0.20
5	17.47	15.150	1.456	1.262	0.1260	0.947	7.95	1.055	0.13
6	20.81	18.100	1.734	1.510	0.1820	1.355	5.50	0.738	0.09
8	27.09	24.000	2.258	2.000	0.3180	2.380	3.14	0.420	0.05
10	33.77	30.050	2.814	2.503	0.5560	4.165	1.80	0.241	0.03
12	40.05	35.720	3.370	2.975	0.7060	5.280	1.42	0.189	0.02
14	47.12	39.270	3.930	3.271	0.8520	6.380	1.18	0.157	0.019
16	53.41	44.970	4.440	3.746	1.1170	8.360	0.895	0.119	0.014
18	56.55	50.660	4.712	4.220	1.4180	10.610	0.705	0.094	0.011
20	62.83	56.350	5.236	4.694	1.7550	13.130	0.570	0.076	0.009

FIG. 5-68. Dimensional and capacity data for schedule 80 steel pipe.

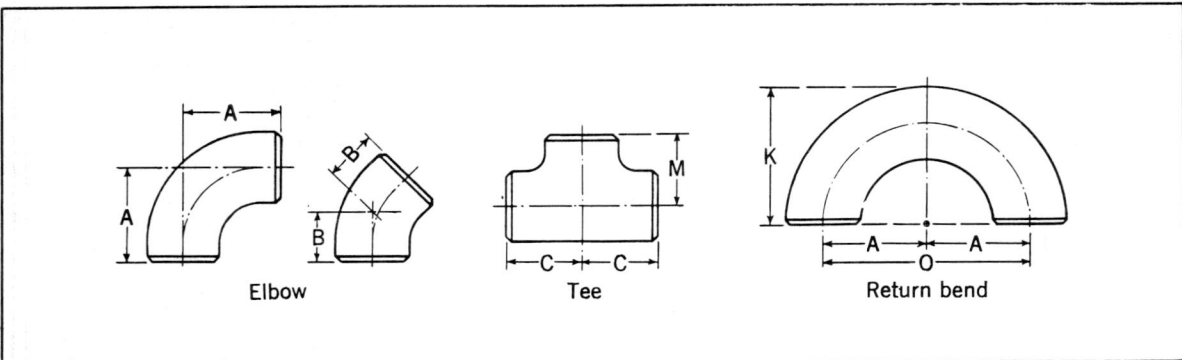

Nominal Pipe Size	Outside Diameter at Bevel	Center-to-End				180° Returns*	
		90-Deg Elbows A	45-Deg Elbows B	Tees, Run C	Tees, Outlet M	O	K
1	1.315	1½	⅞	1½	1½	3	2 3/16
1¼	1.660	1⅞	1	1⅞	1⅞	3¾	2¾
1½	1.900	2¼	1⅛	2¼	2¼	4½	3¼
2	2.375	3	1⅜	2½	2½	6	4 3/16
2½	2.875	3¾	1¾	3	3	7½	5 3/16
3	3.500	4½	2	3⅜	3⅜	9	6¼
3½	4.000	5¼	2¼	3¾	3¾	10½	7¼
4	4.500	6	2½	4⅛	4⅛	12	8¼
5	5.563	7½	3⅛	4⅞	4⅞	15	10 5/16
6	6.625	9	3¾	5⅝	5⅝	18	12 5/16
8	8.625	12	5	7	7	24	16 5/16
10	10.750	15	6¼	8½	8½	30	20⅜
12	12.750	18	7½	10	10	36	24⅜
14	14.000	21	8¾	11	Not standard	42	28
16	16.000	24	10	12		48	32
18	18.000	27	11¼	13½		54	36
20	20.000	30	12½	15		60	40
24	24.000	36	15	17		72	48

All dimensions given in inches.
* Dimension $A = \frac{1}{2}$ dimension O.
† Extracted from ASA B16.9 by permission of the publisher, The American Society of Mechanical Engineers.

FIG. 5-69. Steel butt-welding fittings for elbows, tees, and returns†

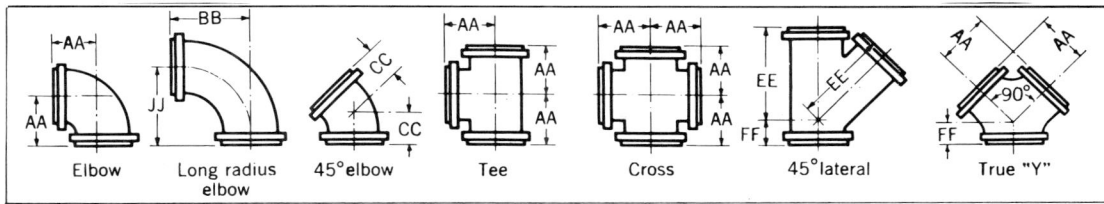

Nom. Pipe Dia.	Inside Dia. of Fitting	Min. Wall Thickness	Flange Dia.	Flange Thickness	Dimensions to Contact Surface of Raised Face				
					AA	BB	CC	EE	FF
150 Pound Fittings									
1	1	1/4	4-1/4	7/16	3-1/2	5	1-3/4	5-3/4	1-3/4
1-1/4	1-1/4	1/4	4-5/8	1/2	3-3/4	5-1/2	2	6-1/4	1-3/4
1-1/2	1-1/2	1/4	5	9/16	4	6	2-1/4	7	2
2	2	1/4	6	5/8	4-1/2	6-1/2	2-1/2	8	2-1/2
2-1/2	2-1/2	1/4	7	11/16	5	7	3	9-1/2	2-1/2
3	3	1/4	7-1/2	3/4	5-1/2	7-3/4	3	10	3
3-1/2	3-1/2	1/4	8-1/2	13/16	6	8-1/2	3-1/2	11-1/2	3
4	4	1/4	9	15/16	6-1/2	9	4	12	3
5	5	9/32	10	15/16	7-1/2	10-1/4	4-1/2	13-1/2	3-1/2
6	6	9/32	11	1	8	11-1/2	5	14-1/2	3-1/2
8	8	5/16	13-1/2	1-1/8	9	14	5-1/2	17-1/2	4-1/2
10	10	11/32	16	1-3/16	11	16-1/2	6-1/2	20-1/2	5
12	12	3/8	19	1-1/4	12	19	7-1/2	24-1/2	5-1/2
14	13-1/4	13/32	21	1-3/8	14	21-1/2	7-1/2	27	6
16	15-1/4	7/16	23-1/2	1-7/16	15	24	8	30	6-1/2
18	17-1/4	15/32	25	1-9/16	16-1/2	26-1/2	8-1/2	32	7
20	19-1/4	1/2	27-1/2	1-11/16	18	29	9-1/2	35	8
24	23-1/4	9/16	32	1-7/8	22	34	11	40-1/2	9
300 Pound Fittings									
1	1	1/4	4-7/8	11/16	4	5	2-1/4	6-1/2	2
1-1/4	1-1/4	1/4	5-1/4	3/4	4-1/4	5-1/2	2-1/2	7-1/4	2-1/4
1-1/2	1-1/2	1/4	6-1/8	13/16	4-1/2	6	2-3/4	8-1/2	2-1/2
2	2	1/4	6-1/2	7/8	5	6-1/2	3	9	2-1/2
2-1/2	2-1/2	1/4	7-1/2	1	5-1/2	7	3-1/2	10-1/2	2-1/2
3	3	9/32	8-1/4	1-1/8	6	7-3/4	3-1/2	11	3
3-1/2	3-1/2	9/32	9	1-3/16	6-1/2	8-1/2	4	12-1/2	3
4	4	5/16	10	1-1/4	7	9	4-1/2	13-1/2	3
5	5	3/8	11	1-3/8	8	10-1/4	5	15	3-1/2
6	6	3/8	12-1/2	1-7/16	8-1/2	11-1/2	5-1/2	17-1/2	4
8	8	7/16	15	1-5/8	10	14	6	20-1/2	5
10	10	1/2	17-1/2	1-7/8	11-1/2	16-1/2	7	24	5-1/2
12	12	9/16	20-1/2	2	13	19	8	27-1/2	6
14	13-1/4	5/8	23	2-1/8	15	21-1/2	8-1/2	31	6-1/2
16	15-1/4	11/16	25-1/2	2-1/4	16-1/2	24	9-1/2	34-1/2	7-1/2
18	17	3/4	28	2-3/8	18	26-1/2	10	37-1/2	8
20	19	13/16	30-1/2	2-1/2	19-1/2	29	10-1/2	40-1/2	8-1/2
24	23	15/16	36	2-3/4	22-1/2	34	12	47-1/2	10

All dimensions given in inches.
† Extracted from ASA B16.5-1957 by permission of the publisher, The American Society of Mechanical Engineers. A raised face of 1/16-inch is included in (a) thickness of flanges, (b) "center-to-contact-surface" dimensions; hence the "center-to-contact-surface" dimensions are the same as the "center-to-flange-edge" dimensions for this type of facing. Where facings other than the 1/16-inch raised face are used, the "center-to-flange-edge" dimensions shall remain unchanged.

FIG. 5-70. 150 and 300 lb steel flanged fittings for other than ring joint facings.†

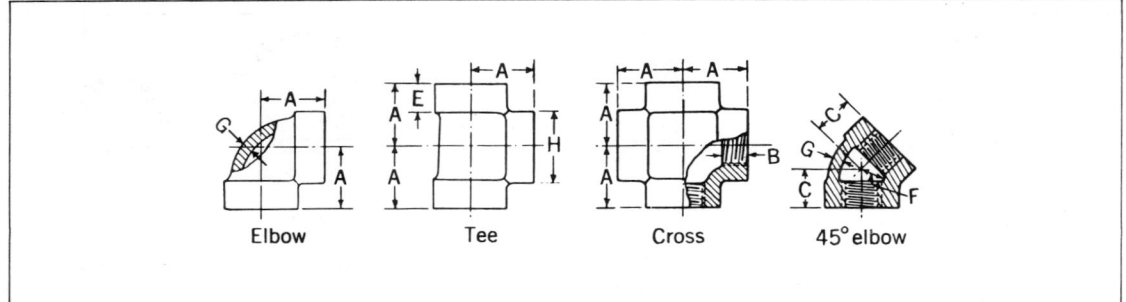

Nom. Pipe Dia.	A	C	B Min.	E Min.	F Min.	F Max.	G	H Min.
Fittings for 125 Pounds per Square Inch								
¼	0.81	0.73	0.32	0.38	0.540	0.584	0.110	0.93
⅜	0.95	0.80	0.36	0.44	0.675	0.719	0.120	1.12
½	1.12	0.88	0.43	0.50	0.840	0.897	0.130	1.34
¾	1.31	0.98	0.50	0.56	1.050	1.107	0.155	1.63
1	1.50	1.12	0.58	0.62	1.315	1.385	0.170	1.95
1¼	1.75	1.29	0.67	0.69	1.660	1.730	0.185	2.39
1½	1.94	1.43	0.70	0.75	1.900	1.970	0.200	2.68
2	2.25	1.68	0.75	0.84	2.375	2.445	0.220	3.28
2½	2.70	1.95	0.92	0.94	2.875	2.975	0.240	3.86
3	3.08	2.17	0.98	1.00	3.500	3.600	0.260	4.62
3½	3.42	2.39	1.03	1.06	4.000	4.100	0.280	5.20
4	3.79	2.61	1.08	1.12	4.500	4.600	0.310	5.79
5	4.50	3.05	1.18	1.18	5.563	5.663	0.380	7.05
6	5.13	3.46	1.28	1.28	6.625	6.725	0.430	8.28
8	6.56	4.28	1.47	1.47	8.625	8.725	0.550	10.63
10	*8.08	5.16	1.68	1.68	10.750	10.850	0.690	13.12
12	*9.50	5.97	1.88	1.88	12.750	12.850	0.800	15.47
Fittings for 250 Pounds per Square Inch								
¼	0.94	0.81	0.43	0.49	0.540	0.584	0.18	1.17
⅜	1.06	0.88	0.47	0.55	0.675	0.719	0.18	1.36
½	1.25	1.00	0.57	0.60	0.840	0.897	0.20	1.59
¾	1.44	1.13	0.64	0.68	1.050	1.107	0.23	1.88
1	1.63	1.31	0.75	0.76	1.315	1.385	0.28	2.24
1¼	1.94	1.50	0.84	0.88	1.660	1.730	0.33	2.73
1½	2.13	1.69	0.87	0.97	1.900	1.970	0.35	3.07
2	2.50	2.00	1.00	1.12	2.375	2.445	0.39	3.74
2½	2.94	2.25	1.17	1.30	2.875	2.975	0.43	4.60
3	3.38	2.50	1.23	1.40	3.500	3.600	0.48	5.36
3½	3.75	2.63	1.28	1.49	4.000	4.100	0.52	5.98
4	4.13	2.81	1.33	1.57	4.500	4.600	0.56	6.61
5	4.88	3.19	1.43	1.74	5.563	5.663	0.66	7.92
6	5.63	3.50	1.53	1.91	6.625	6.725	0.74	9.24
8	7.00	4.31	1.72	2.24	8.625	8.725	0.90	11.73
10	8.63	5.19	1.93	2.58	10.750	10.850	1.08	14.37
12	10.00	6.00	2.13	2.91	12.750	12.850	1.24	16.84

All dimensions given in inches.
† Extracted from ASA B16.4 by permission of the publisher, The American Society of Mechanical Engineers.
* This applies to elbows and tees only.

FIG. 5-71. 125 and 250 lb cast iron screwed fittings.†

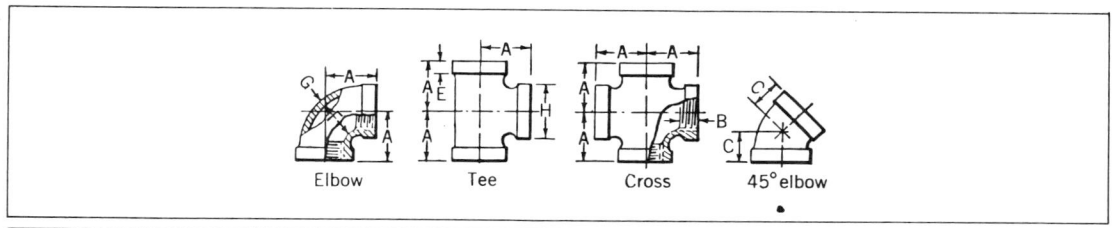

Pipe Dia.	A	C	B Min.	E Min.	F Min.	F Max.	G Min.	H Min.
1/8	0.69	—	0.25	0.200	0.40	0.43	0.090	0.693
1/4	0.81	0.73	0.32	0.215	0.54	0.58	0.095	0.844
3/8	0.95	0.80	0.36	0.230	0.67	0.71	0.100	1.015
1/2	1.12	0.88	0.43	0.249	0.84	0.89	0.105	1.197
3/4	1.31	0.98	0.50	0.273	1.05	1.10	0.120	1.458
1	1.50	1.12	0.58	0.302	1.31	1.38	0.134	1.771
1 1/4	1.75	1.29	0.67	0.341	1.66	1.73	0.145	2.153
1 1/2	1.94	1.43	0.70	0.368	1.90	1.97	0.155	2.427
2	2.25	1.68	0.75	0.422	2.37	2.44	0.173	2.963
2 1/2	2.70	1.95	0.92	0.478	2.87	2.97	0.210	3.589
3	3.08	2.17	0.98	0.548	3.50	3.60	0.231	4.285
3 1/2	3.42	2.39	1.03	0.604	4.00	4.10	0.248	4.843
4	3.79	2.61	1.08	0.661	4.50	4.60	0.265	5.401
5	4.50	3.05	1.18	0.780	5.56	5.66	0.300	6.583
6	5.13	3.46	1.28	0.900	6.62	6.72	0.336	7.767

Straight and Reducing Couplings (Cast)

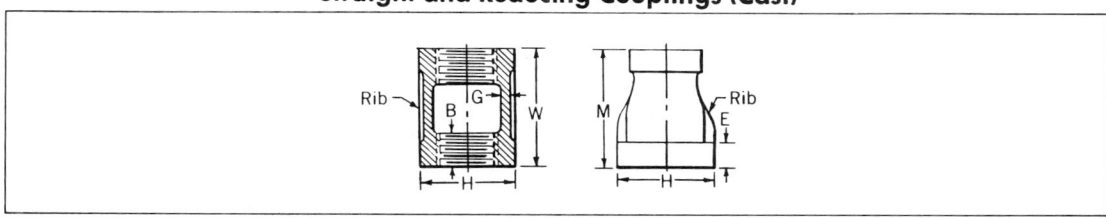

Pipe Dia.	B Min.	E Min.	G Min.	H Min.	Rib Thickness	W	M
1/8	0.25	0.200	0.090	0.693	0.090	0.96	—
1/4	0.32	0.215	0.095	0.844	0.095	1.06	1.00
3/8	0.36	0.230	0.100	1.015	0.100	1.16	1.13
1/2	0.43	0.249	0.105	1.197	0.105	1.34	1.25
3/4	0.50	0.273	0.120	1.458	0.120	1.52	1.44
1	0.58	0.302	0.134	1.771	0.134	1.67	1.69
1 1/4	0.67	0.341	0.145	2.153	0.145	1.93	2.06
1 1/2	0.70	0.368	0.155	2.427	0.155	2.15	2.31
2	0.75	0.422	0.173	2.963	0.173	2.53	2.81
2 1/2	0.92	0.478	0.210	3.589	0.210	2.88	3.25
3	0.98	0.548	0.231	4.285	0.231	3.18	3.69
3 1/2	1.03	0.604	0.248	4.843	0.248	3.43	4.00
4	1.08	0.661	0.265	5.401	0.265	3.69	4.38

All dimensions given in inches.
† Extracted from ASA B16.3 by permission of the publisher, The American Society of Mechanical Engineers.

FIG. 5-72. 150 lb malleable iron screwed fittings.†

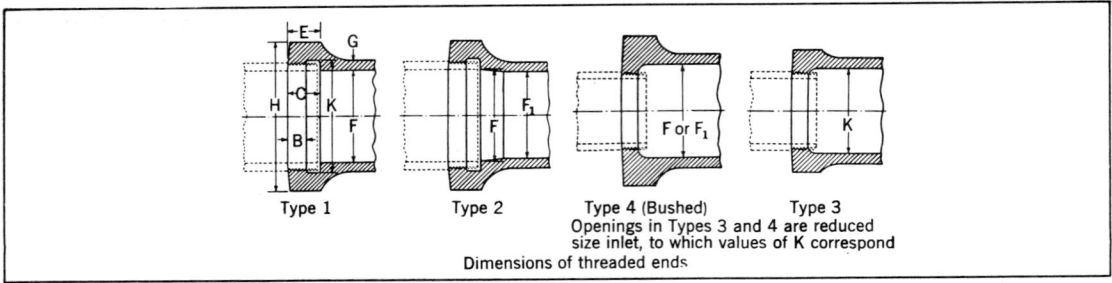

Dimensions of threaded ends

Nominal Pipe Size	Length of Threads B	Total Length of Thread Chamber to Shoulder C	Width of Band E	Inside Diameter of Fitting F	F₁	Metal Thickness G	Outside Diameter of Band H	Recess Diameter Max K	Recess Diameter Min K
1¼	0.420	0.7068	0.71	1.380	1.25	0.185	2.39	1.730	1.660
1½	0.420	0.7235	0.72	1.610	1.50	0.200	2.68	1.970	1.900
2	0.436	0.7565	0.76	2.067	2.00	0.220	3.28	2.445	2.375
2½	0.682	1.1375	1.14	2.469	2.50	0.240	3.86	2.975	2.875
3	0.766	1.2000	1.20	3.068	3.00	0.260	4.62	3.600	3.500
4	0.844	1.3000	1.30	4.026	4.00	0.310	5.79	4.600	4.500
5	0.937	1.4063	1.41	5.047	5.00	0.380	7.05	5.663	5.563
6	0.958	1.5125	1.51	6.065	6.00	0.439	8.28	6.725	6.625
8	1.063	1.7125	1.71	7.981	8.00	0.550	10.63	8.625	8.725

Dimensions of Tees, Crosses, and Y-Branches

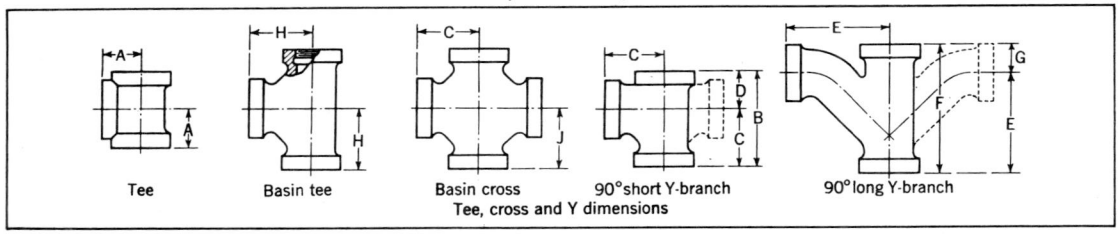

Tee, cross and Y dimensions

Nominal Pipe Size	Center to End of Tee A	90 Deg Short Y-Branch End to End B	90 Deg Short Y-Branch Center to End C	90 Deg Short Y-Branch Center to End D	90 Deg Long Y-Branch Center to End E	90 Deg Long Y-Branch End to End F	90 Deg Long Y-Branch Center to End G	Center to End of Basin Tee H	Center to End of Basin Cross J
1-1/4	1- 3/4	3- 3/4	2- 1/4	1- 1/2	3-5/8	4-3/4	1-1/8	2- 5/16	2- 5/16
1-1/2	1-15/16	4- 1/4	2- 1/2	1- 3/4	4-1/8	5-3/8	1-1/4	2-11/16	2- 5/16
2	2- 1/4	5- 3/16	3- 1/16	2- 1/8	5-1/4	7	1-3/4	3- 1/2	3- 1/2
2-1/2	2-11/16	6- 5/16	3-11/16	2- 5/8	6-1/4	8-1/4	2	4- 1/4	—
3	3- 1/16	7- 1/4	4- 1/4	3	7-1/2	9-7/8	2-3/8	—	—
4	3-13/16	8- 3/4	5- 3/16	3- 9/16	9-7/8	13	3-1/8	—	—
5	4- 1/2	10- 5/16	6- 1/8	4- 3/16	12-1/4	15-3/4	3-1/2	—	—
6	5- 1/8	11-15/16	7- 1/8	4-14/16	14-5/8	18-3/4	4-1/8	—	—

All dimensions are given in inches.
† Extracted from ASA B16.12 by permission of the publisher, The American Society of Mechanical Engineers.

FIG. 5-73. Cast iron screwed drainage fittings.†

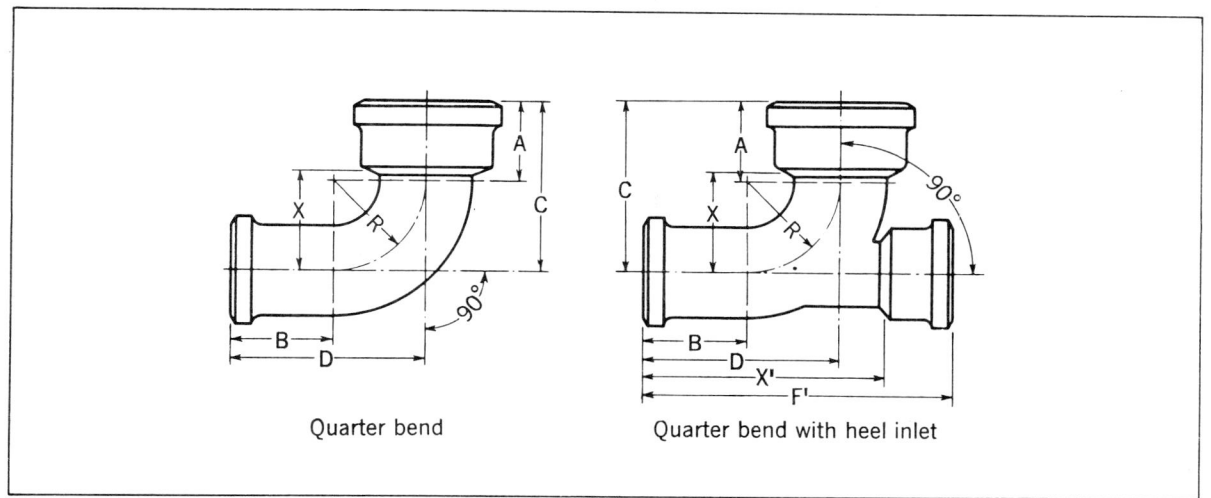

Quarter Bends

Size	A	B	C	D	R	X	Weight, Pounds
2	2¾	3	5¾	6	3	3¼	5
3	3¼	3½	6¾	7	3½	4	10
4	3½	4	7½	8	4	4½	15
5	3½	4	8	8½	4½	5	19
6	3½	4	8½	9	5	5½	24
8	4⅛	5½	10⅛	11½	6	6⅝	51
10	4⅛	5½	11⅛	12½	7	7⅝	78
12	5	7	13	15	8	8¾	111
15	5	7	14½	16½	9½	10¼	169

Quarter Bends with Heel Inlets

Size	A	B	C	D	F'	R	X	X'	Weight, Pounds
3 × 2	3¼	3½	6¾	7	11½	3½	4	9	13
4 × 2	3½	4	7½	8	13	4	4½	10½	18
4 × 3	3½	4	7½	8	13¼	4	4½	10½	19
5 × 2	3½	4	8	8½	14¼	4½	5	11¾	22
5 × 3	3½	4	8	8½	14½	4½	5	11¾	24
5 × 4	3½	4	8	8½	14¾	4½	5	11¾	25
6 × 2	3½	4	8½	9	15	5	5½	12½	27
6 × 3	3½	4	8½	9	15¼	5	5½	12½	29
6 × 4	3½	4	8½	9	15½	5	5½	12½	30

All dimensions are given in inches.
Dimensions D, X and X^1 are laying lengths.
† From Commercial Standard 188.53 of the Department of Commerce.

FIG. 5-74. Cast iron soil-pipe quarter bends.†

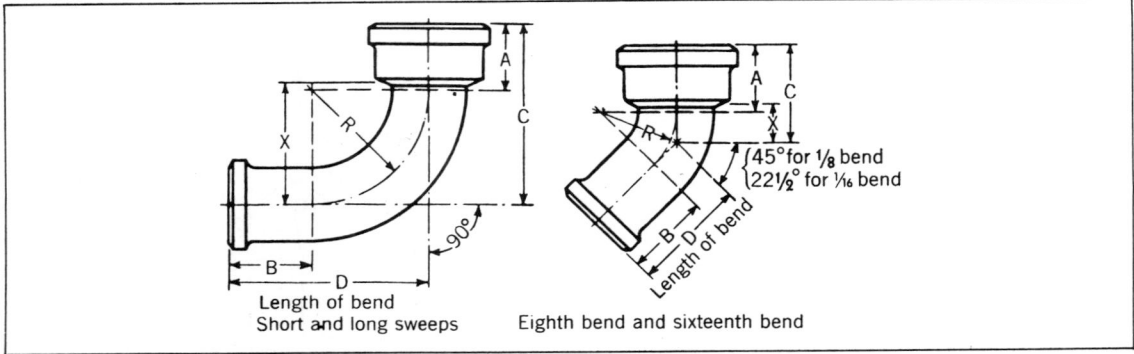

Size	Dimensions						Weight, Pounds
	A	B	C	D	R	X	
Short Sweeps, Quarter Bend							
2	2-3/4	3	7-3/4	8	5	5-1/4	5
3	3-1/4	3-1/2	8-3/4	9	5-1/2	6	9
4	3-1/2	4	9-1/2	10	6	6-1/2	12-1/2
5	3-1/2	4	10	10-1/2	6-1/2	7	16
6	3-1/2	4	10-1/2	11	7	7-1/2	20
8	4-1/8	5-1/2	12-1/8	13-1/2	8	8-5/8	38
10	4-1/8	5-1/2	13-1/8	14-1/2	9	9-5/8	62
12	5	7	15	17	10	10-3/4	89
15	5	7	16-1/2	18-1/2	11-1/2	12-1/4	130
Long Sweeps, Quarter Bend							
2	2-3/4	3	10-3/4	11	8	8-1/4	6-1/2
3	3-1/4	3-1/2	11-3/4	12	8-1/2	9	11
4	3-1/2	4	12-1/2	13	9	9-1/2	15
5	3-1/2	4	13	13-1/2	9-1/2	10	19-1/2
6	3-1/2	4	13-1/2	14	10	10-1/2	24
8	4-1/8	5-1/2	15-1/8	16-1/2	11	11-5/8	45
10	4-1/8	5-1/2	16-1/8	17-1/2	12	12-5/8	72
12	5	7	18	20	13	13-3/4	101
15	5	7	19-1/2	21-1/2	14-1/2	15-1/4	147
Eighth Bends							
2	2-3/4	3	4	4-1/4	3	1-1/2	3-1/4
3	3-1/4	3-1/2	4-11/16	4-15/16	3-1/2	1-15/16	5-1/2
4	3-1/2	4	5-3/16	5-11/16	4	2-3/16	8-1/2
5	3-1/2	4	5-3/8	5-7/8	4-1/2	2-3/8	10-1/2
6	3-1/2	4	5-9/16	6-1/16	5	2-9/16	13
8	4-1/8	5-1/2	6-5/8	8	6	3-1/8	28
10	4-1/8	5-1/2	7	8-3/8	7	3-1/2	44
12	5	7	8-5/16	10-5/16	8	4-1/16	64
15	5	7	8-15/16	10-15/16	9-1/2	4-11/16	92

All dimensions are given in inches.
Dimensions D and X are laying lengths.
† From Commercial Standard 188.53 of the Department of Commerce.

FIG. 5-75. Cast iron soil-pipe sweeps and bends.†

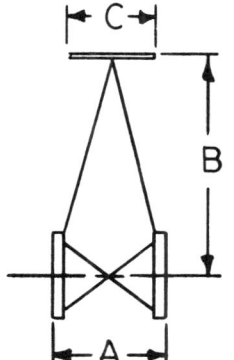

NON-RISING STEM WEDGE-DISC

SIZE	A	B	C
1/4	1 3/4	3 5/8	1 3/4
3/8	1 13/16	3 5/8	1 3/4
1/2	2	4	2 1/16
3/4	2 1/4	4 5/8	2 9/16
1	2 11/16	5 3/8	2 3/4
1 1/4	3	6 5/16	3 1/16
1 1/2	3 1/4	7 1/8	3 5/8
2	3 11/16	8 3/8	4 1/16

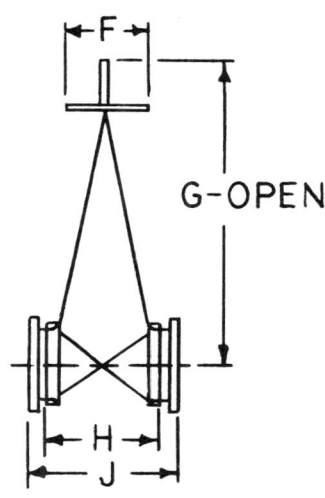

SCREWED OR FLANGED
I.B.B.M. VALVES O.S. & Y.

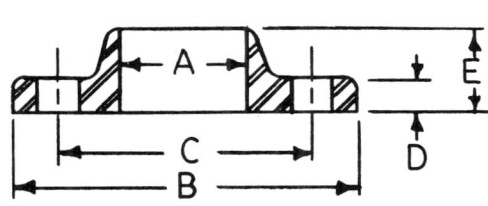

WELDING NECK FLANGES

SIZE	A	B	C	D	E
2 1/2	2.47	7	5 1/2	7/8	2 3/4
3	3.07	7 1/2	6	15/16	2 3/4
3 1/2	3.55	8 1/2	7	15/16	2 13/16
4	4.03	9	7 1/2	15/16	3
5	5.05	10	8 1/2	15/16	3 1/2
6	6.07	11	9 1/2	1	3 1/2
8	7.98	13 1/2	11 3/4	1 1/8	4
10	10.02	16	14 1/4	1 3/16	4
12	12.00	19	17	1 1/4	4 1/2
14	13.25	21	18 3/4	1 3/8	5
16	15.25	23 1/2	21 1/4	1 7/16	5
18	17.25	25	22 3/4	1 9/16	5 1/2
20	19.25	27 1/2	25	1 11/16	5 11/16
24	23.25	32	29 1/2	1 7/8	6

SIZE	F	G	H	J
2 1/2	8	17	6 1/2	7 1/2
3	9	19 1/2	6 3/4	8
3 1/2	9	21 1/4	7 3/8	8 1/2
4	10	23 3/4	7 5/8	9
5	12	28 3/4	8 1/2	10
6	12	32 5/8	8 7/8	10 1/2
8	16	41 7/8	10 1/4	11 1/2
10	20	50 1/4	....	13
12	20	58	....	14
14	20	67 3/4	....	15
16	22	76 1/4	....	16
18	24	83 1/2	....	17
20	24	91 1/4	....	18
24	30	109	....	20

FIG. 5-76. Valve dimension — 125 # W.S.P.

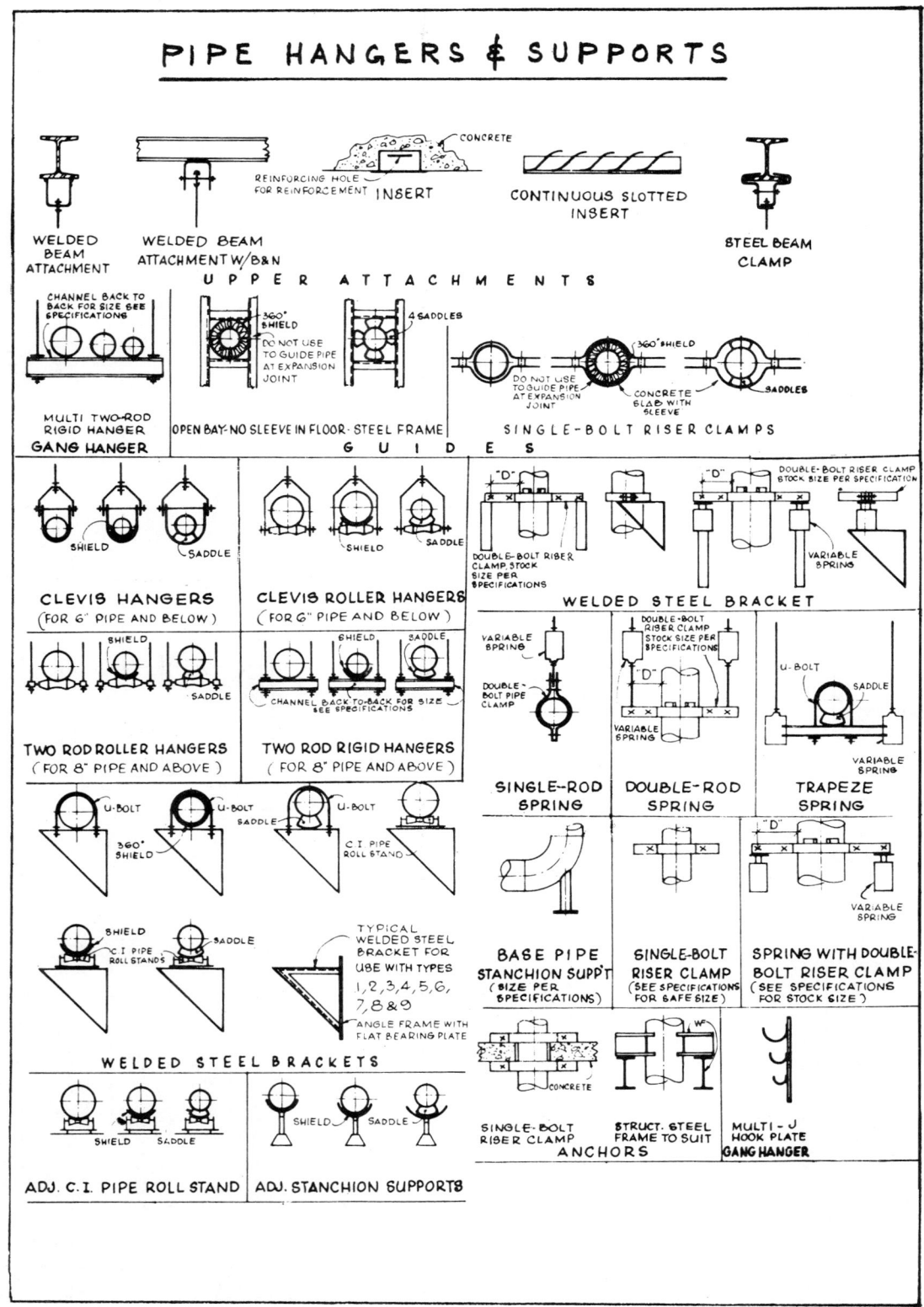

Fig. 5-77. Pipe hangers and supports.

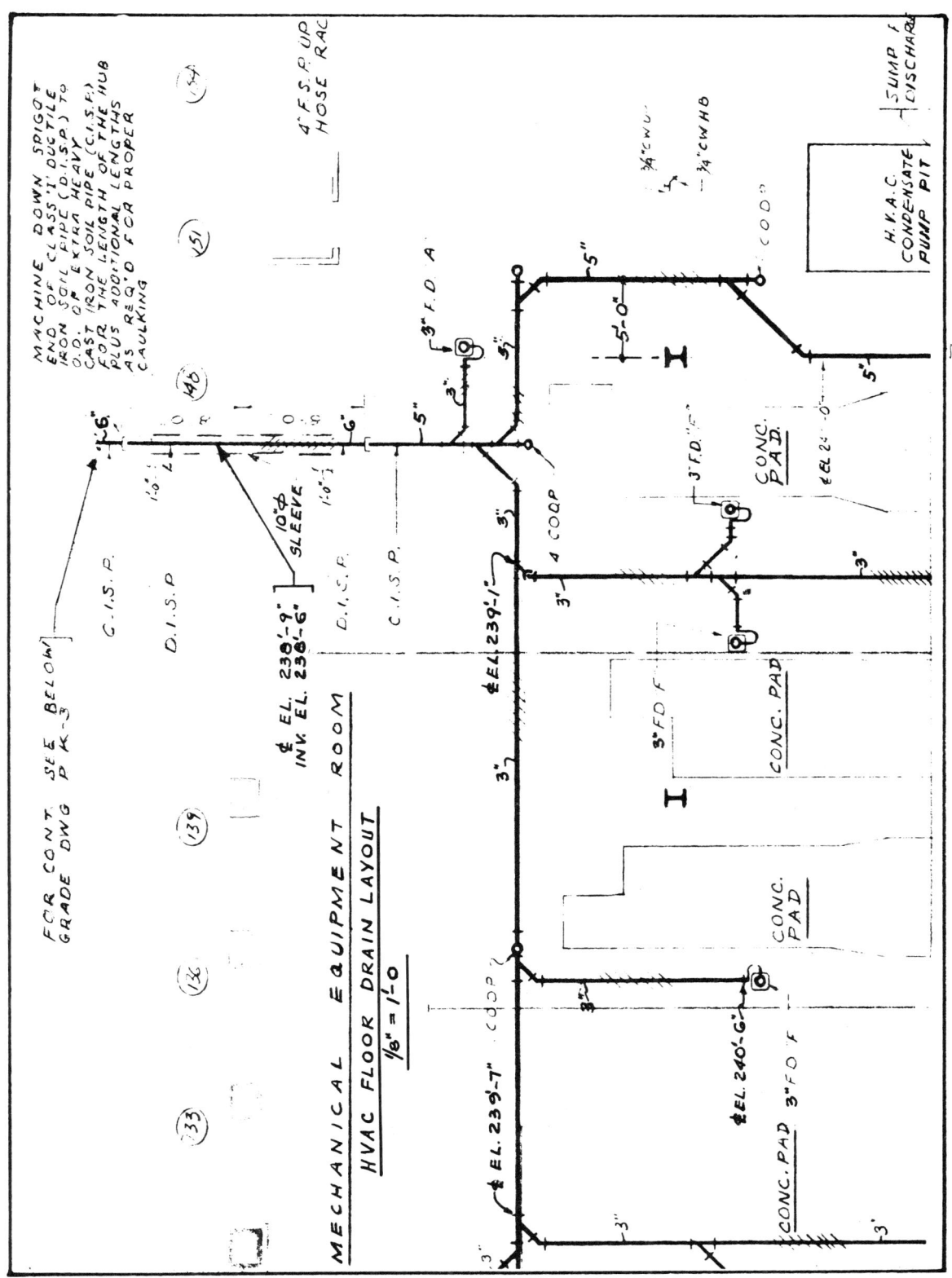

Fig. 5-78. Typical plumbing drawings.

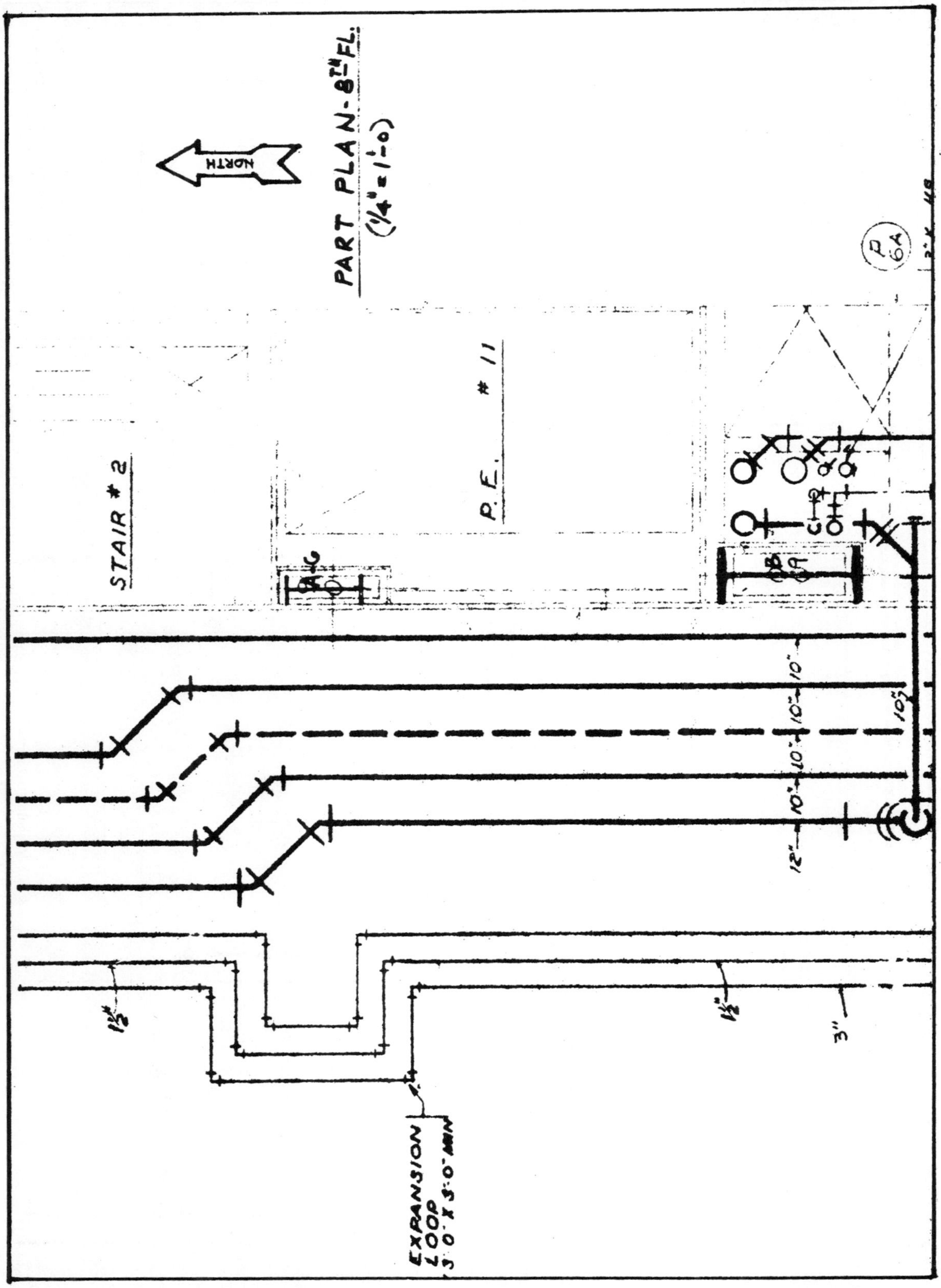

Fig. 5-79. Typical plumbing drawings.

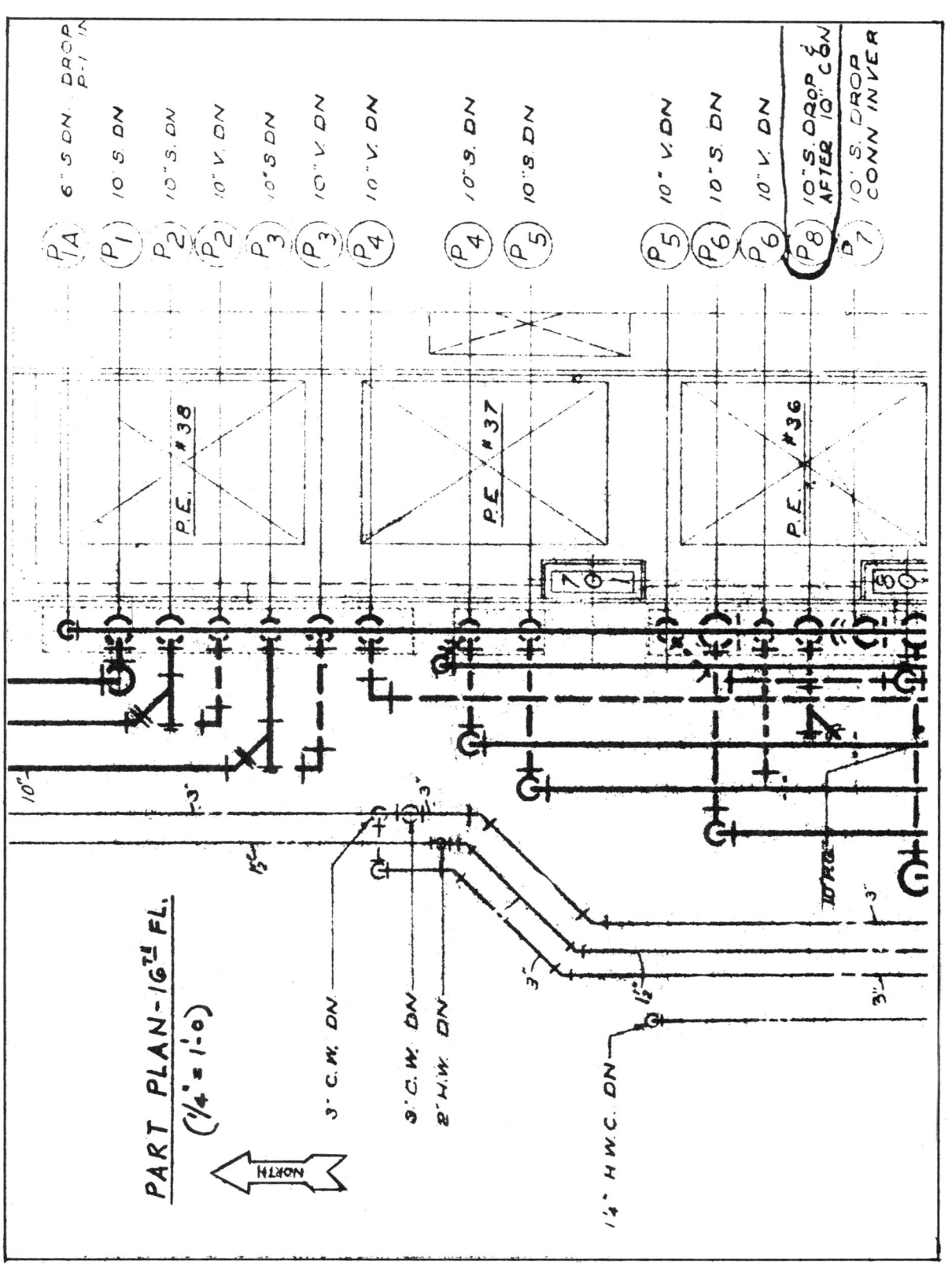

Fig. 5-80. Typical plumbing drawings.

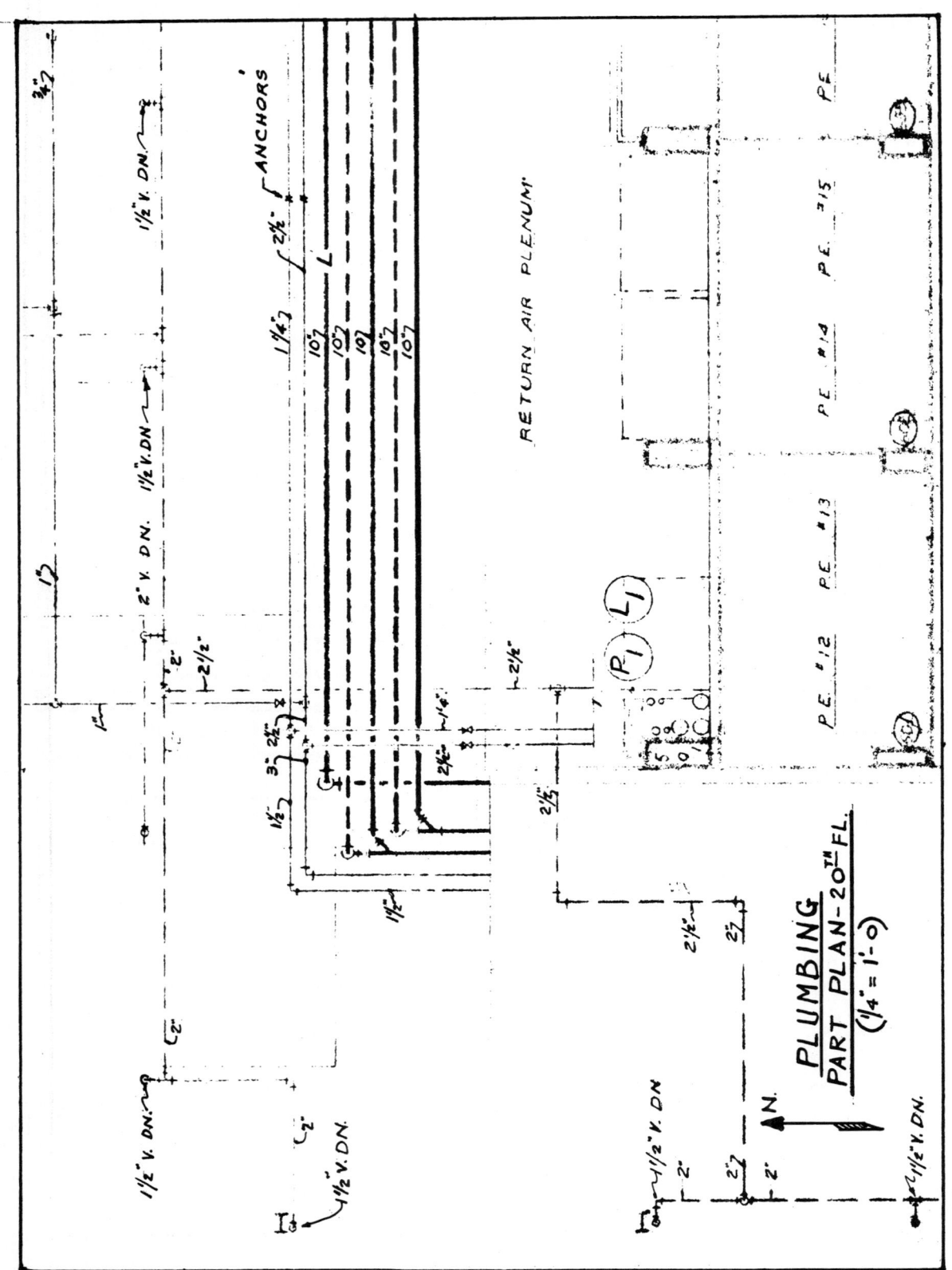

Fig. 5-81. Typical plumbing part plan.

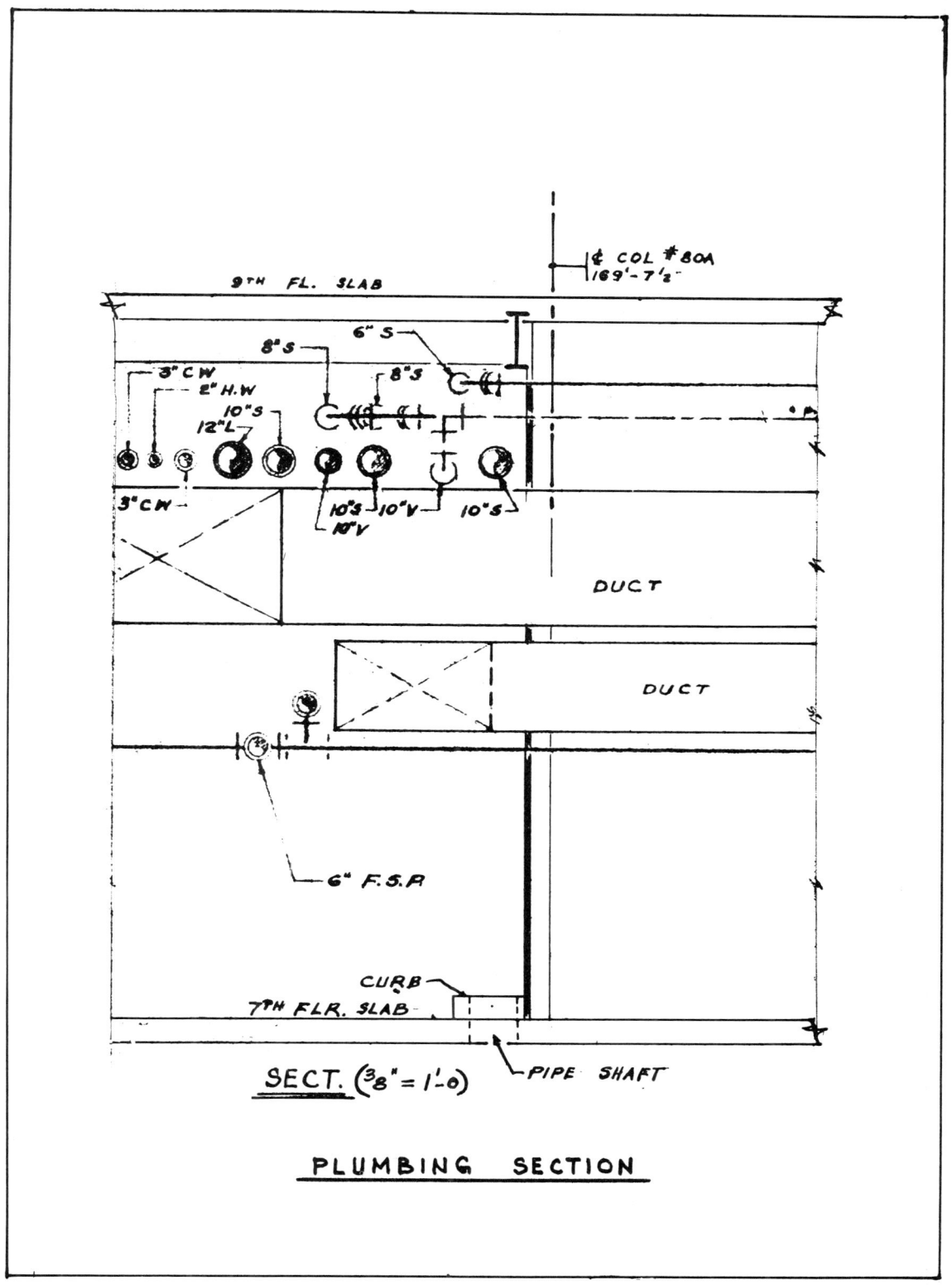

Fig. 5-82. Typical plumbing section drawing.

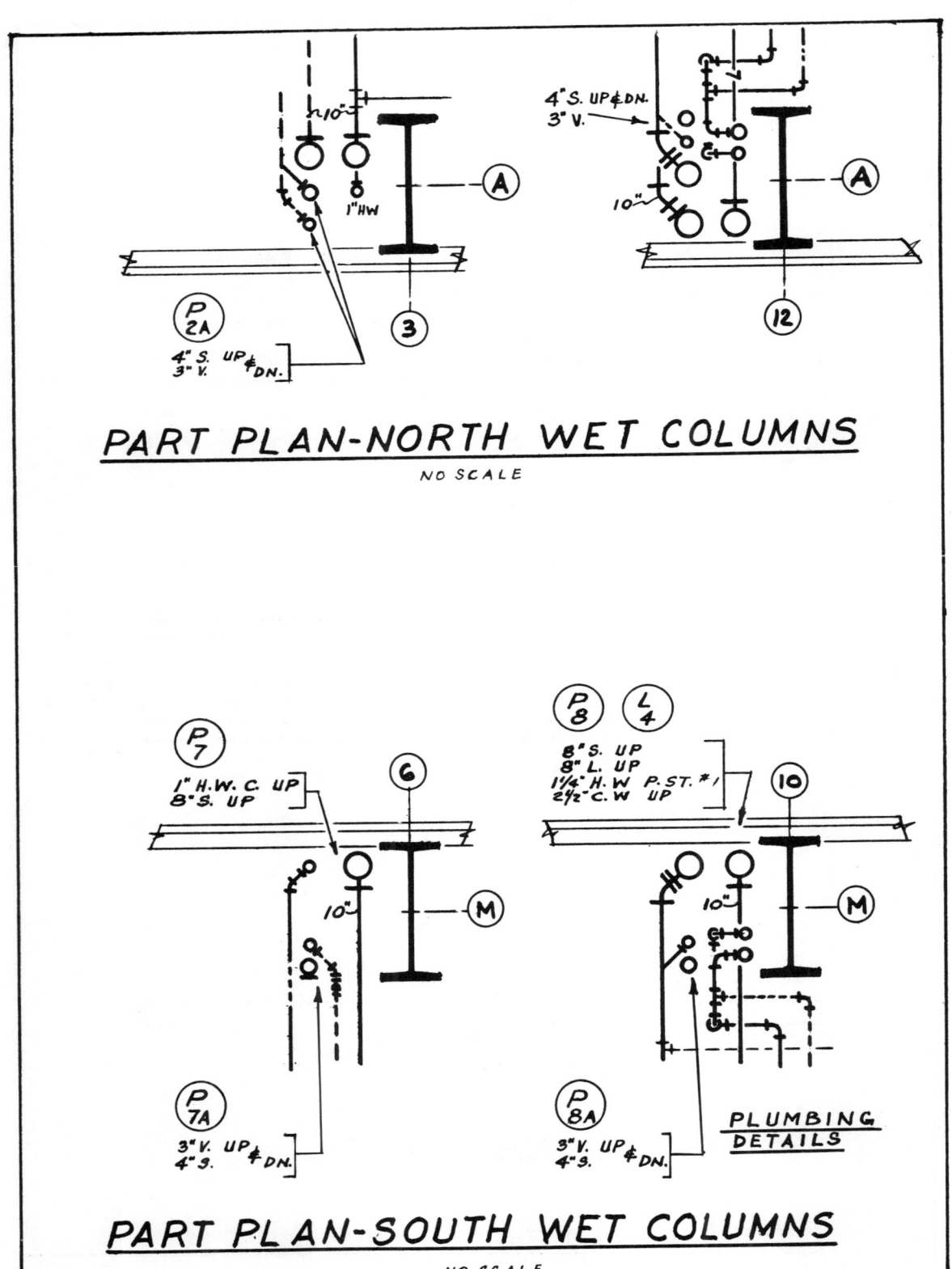

Fig. 5-83. Typical plumbing part plans.

LIGHTING FIXTURE LIST

FIXTURE TYPE DESIGNATION	FIXTURE DESCRIPTION	FIXTURE MANUFACTURERS & CATALOGUE NUMBERS	QUANTITY PER FIXTURE	LAMPS — I = INCANDESCENT / F = FLUORESCENT / M = MERCURY	DESIGNATION & WATTS (LAMP MANUFACTURERS' STANDARD ABBREVIATION)	VOLTS (INPUT TO LAMP AUXILIARY IF USED)	REMARKS
A	PENDANT OR SURFACE MOUNTED 4 FOOT LONG FLUORESCENT FIXTURE WITH INJECTION MOLDED ACRYLIC PRISMATIC WRAPAROUND DIFFUSER	LIGHTOLIER B72P24 WAKEFIELD PHR-214-AA OR EQUAL	2	F	F40WW	120	WHERE FIXTURE IS PENDANT MOUNTED, TOP OF FIXTURE IS TO BE 4" BELOW BOTTOM OF BEAMS UNLESS OTHERWISE NOTED
B	RECESSED INCANDESCENT LOW BRIGHTNESS DOWNLIGHT WITH MINIMUM CEILING FLANGE	GOTHAM #816 OR EQUAL	1	I	150A	120	
C	RECESSED FLUORESCENT 2'-0" X 2'-0" FIXTURE WITH FRAMELESS ACRYLIC PRISMATIC INJECTION MOLDED DIFFUSER	LIGHTOLIER #B61N15 WAKEFIELD THT-4225 B-6250 OR EQUAL	4	F	F20T 12/WW	120	
D	SAME AS TYPE "A" EXCEPT FOR WIDTH AND NUMBER OF LAMPS	LIGHTOLIER #B73S44 WAKEFIELD PHR-414-AA OR EQUAL	4	F	F40WW	120	WHERE FIXTURE IS PENDANT MOUNTED, TOP OF FIXTURE TO BE 1½" BELOW BOTTOM OF BEAMS UNLESS OTHERWISE NOTED.
E	RECESSED FLUORESCENT 1'-0" X 4'-0" FIXTURE WITH FRAMELESS ACRYLIC PRISMATIC INJECTION MOLDED DIFFUSER	LIGHTOLIER P61Q15 SMITHCRAFT TF-12HF 2-40 OR EQUAL	2	F	F40WW	120	
F	SAME AS TYPE "E" EXCEPT 2'-0" X 4'-0" AND NUMBER OF LAMPS	LIGHTOLIER #B61R15 SMITHCRAFT TF-22HF 4-40 OR EQUAL	4	F	F40WW	120	
G	SURFACE-MOUNTED KEYLESS PORCELAIN LAMP RECEPTACLE	P & S #110 OR EQUAL	1	I	100A	120	

Fig. 5-84. Lighting schedule.

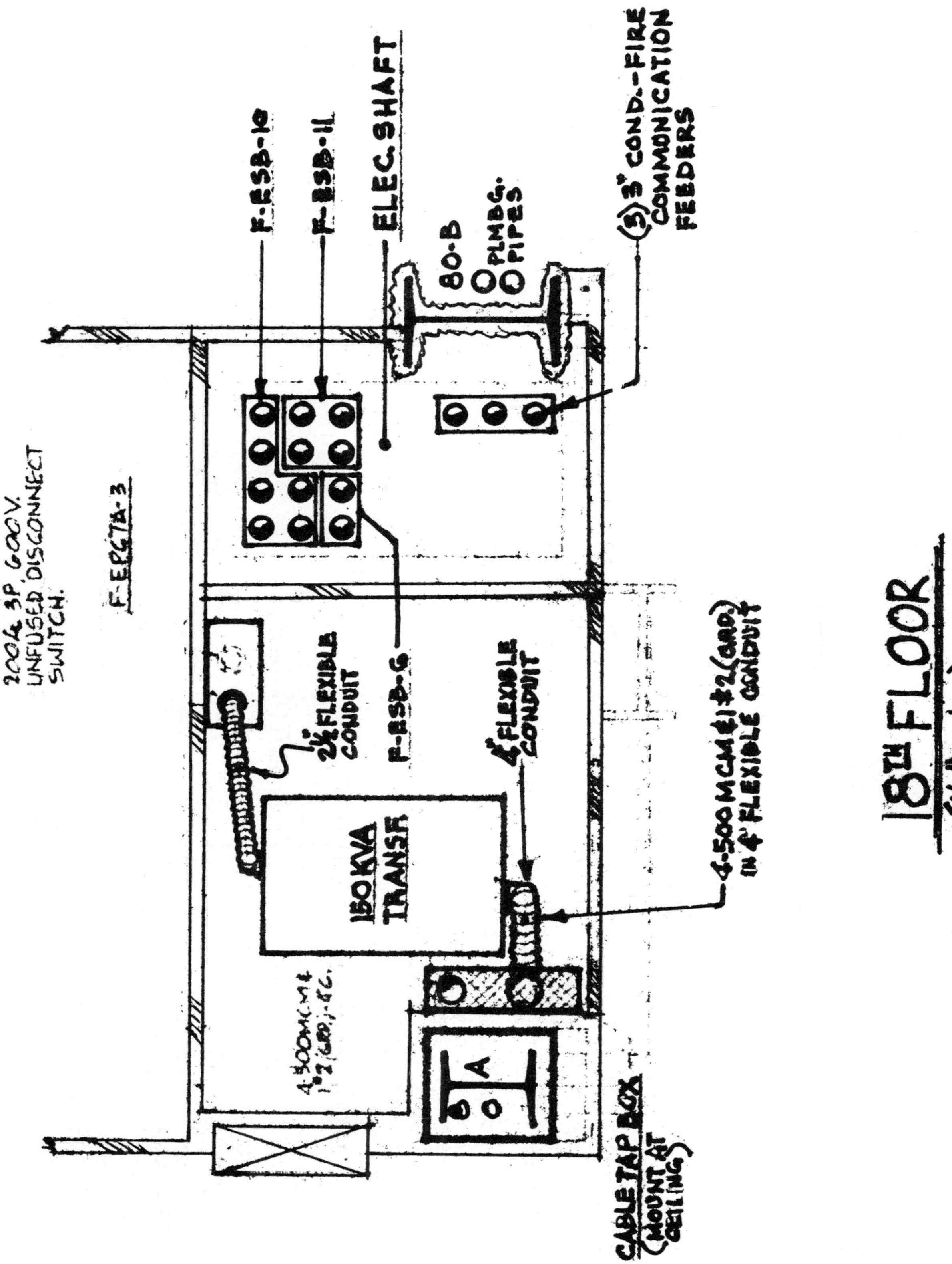

Fig. 5-85. Typical electrical closet.

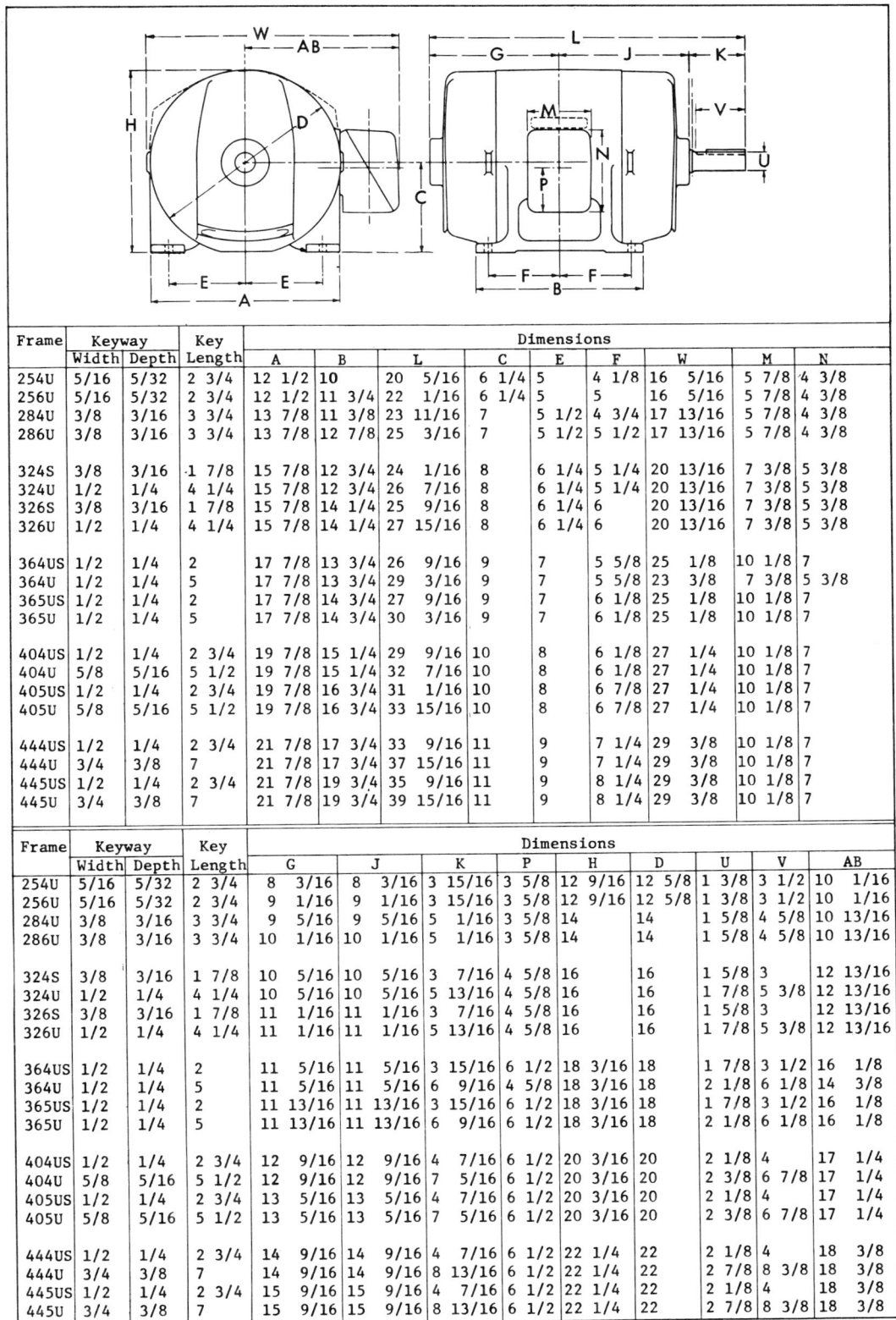

Frame	Keyway Width	Keyway Depth	Key Length	Dimensions A	B	L	C	E	F	W	M	N
254U	5/16	5/32	2 3/4	12 1/2	10	20 5/16	6 1/4	5	4 1/8	16 5/16	5 7/8	4 3/8
256U	5/16	5/32	2 3/4	12 1/2	11 3/4	22 1/16	6 1/4	5	5	16 5/16	5 7/8	4 3/8
284U	3/8	3/16	3 3/4	13 7/8	11 3/8	23 11/16	7	5 1/2	4 3/4	17 13/16	5 7/8	4 3/8
286U	3/8	3/16	3 3/4	13 7/8	12 7/8	25 3/16	7	5 1/2	5 1/2	17 13/16	5 7/8	4 3/8
324S	3/8	3/16	1 7/8	15 7/8	12 3/4	24 1/16	8	6 1/4	5 1/4	20 13/16	7 3/8	5 3/8
324U	1/2	1/4	4 1/4	15 7/8	12 3/4	26 7/16	8	6 1/4	5 1/4	20 13/16	7 3/8	5 3/8
326S	3/8	3/16	1 7/8	15 7/8	14 1/4	25 9/16	8	6 1/4	6	20 13/16	7 3/8	5 3/8
326U	1/2	1/4	4 1/4	15 7/8	14 1/4	27 15/16	8	6 1/4	6	20 13/16	7 3/8	5 3/8
364US	1/2	1/4	2	17 7/8	13 3/4	26 9/16	9	7	5 5/8	25 1/8	10 1/8	7
364U	1/2	1/4	5	17 7/8	13 3/4	29 3/16	9	7	5 5/8	23 3/8	7 3/8	5 3/8
365US	1/2	1/4	2	17 7/8	14 3/4	27 9/16	9	7	6 1/8	25 1/8	10 1/8	7
365U	1/2	1/4	5	17 7/8	14 3/4	30 3/16	9	7	6 1/8	25 1/8	10 1/8	7
404US	1/2	1/4	2 3/4	19 7/8	15 1/4	29 9/16	10	8	6 1/8	27 1/4	10 1/8	7
404U	5/8	5/16	5 1/2	19 7/8	15 1/4	32 7/16	10	8	6 1/8	27 1/4	10 1/8	7
405US	1/2	1/4	2 3/4	19 7/8	16 3/4	31 1/16	10	8	6 7/8	27 1/4	10 1/8	7
405U	5/8	5/16	5 1/2	19 7/8	16 3/4	33 15/16	10	8	6 7/8	27 1/4	10 1/8	7
444US	1/2	1/4	2 3/4	21 7/8	17 3/4	33 9/16	11	9	7 1/4	29 3/8	10 1/8	7
444U	3/4	3/8	7	21 7/8	17 3/4	37 15/16	11	9	7 1/4	29 3/8	10 1/8	7
445US	1/2	1/4	2 3/4	21 7/8	19 3/4	35 9/16	11	9	8 1/4	29 3/8	10 1/8	7
445U	3/4	3/8	7	21 7/8	19 3/4	39 15/16	11	9	8 1/4	29 3/8	10 1/8	7

Frame	Keyway Width	Keyway Depth	Key Length	Dimensions G	J	K	P	H	D	U	V	AB
254U	5/16	5/32	2 3/4	8 3/16	8 3/16	3 15/16	3 5/8	12 9/16	12 5/8	1 3/8	3 1/2	10 1/16
256U	5/16	5/32	2 3/4	9 1/16	9 1/16	3 15/16	3 5/8	12 9/16	12 5/8	1 3/8	3 1/2	10 1/16
284U	3/8	3/16	3 3/4	9 5/16	9 5/16	5 1/16	3 5/8	14	14	1 5/8	4 5/8	10 13/16
286U	3/8	3/16	3 3/4	10 1/16	10 1/16	5 1/16	3 5/8	14	14	1 5/8	4 5/8	10 13/16
324S	3/8	3/16	1 7/8	10 5/16	10 5/16	3 7/16	4 5/8	16	16	1 5/8	3	12 13/16
324U	1/2	1/4	4 1/4	10 5/16	10 5/16	5 13/16	4 5/8	16	16	1 7/8	5 3/8	12 13/16
326S	3/8	3/16	1 7/8	11 1/16	11 1/16	3 7/16	4 5/8	16	16	1 5/8	3	12 13/16
326U	1/2	1/4	4 1/4	11 1/16	11 1/16	5 13/16	4 5/8	16	16	1 7/8	5 3/8	12 13/16
364US	1/2	1/4	2	11 5/16	11 5/16	3 15/16	6 1/2	18 3/16	18	1 7/8	3 1/2	16 1/8
364U	1/2	1/4	5	11 5/16	11 5/16	6 9/16	4 5/8	18 3/16	18	2 1/8	6 1/8	14 3/8
365US	1/2	1/4	2	11 13/16	11 13/16	3 15/16	6 1/2	18 3/16	18	1 7/8	3 1/2	16 1/8
365U	1/2	1/4	5	11 13/16	11 13/16	6 9/16	6 1/2	18 3/16	18	2 1/8	6 1/8	16 1/8
404US	1/2	1/4	2 3/4	12 9/16	12 9/16	4 7/16	6 1/2	20 3/16	20	2 1/8	4	17 1/4
404U	5/8	5/16	5 1/2	12 9/16	12 9/16	7 5/16	6 1/2	20 3/16	20	2 3/8	6 7/8	17 1/4
405US	1/2	1/4	2 3/4	13 5/16	13 5/16	4 7/16	6 1/2	20 3/16	20	2 1/8	4	17 1/4
405U	5/8	5/16	5 1/2	13 5/16	13 5/16	7 5/16	6 1/2	20 3/16	20	2 3/8	6 7/8	17 1/4
444US	1/2	1/4	2 3/4	14 9/16	14 9/16	4 7/16	6 1/2	22 1/4	22	2 1/8	4	18 3/8
444U	3/4	3/8	7	14 9/16	14 9/16	8 13/16	6 1/2	22 1/4	22	2 7/8	8 3/8	18 3/8
445US	1/2	1/4	2 3/4	15 9/16	15 9/16	4 7/16	6 1/2	22 1/4	22	2 1/8	4	18 3/8
445U	3/4	3/8	7	15 9/16	15 9/16	8 13/16	6 1/2	22 1/4	22	2 7/8	8 3/8	18 3/8

FIG. 5-86. Electric motor sizes. (All dimensions are approximate and are given in inches.)

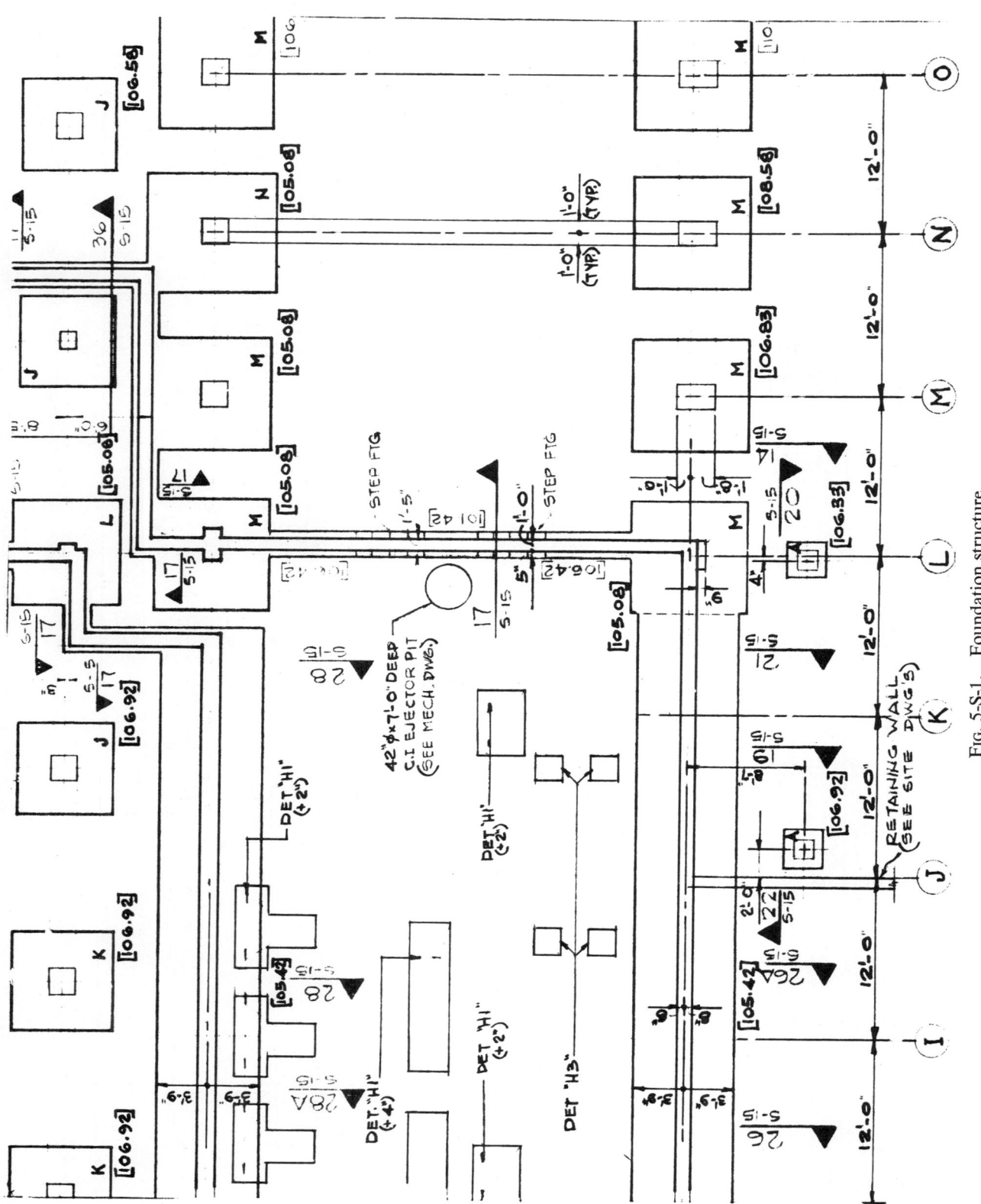

Fig. 5-S-1. Foundation structure.

SECTIONS

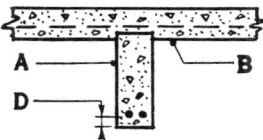 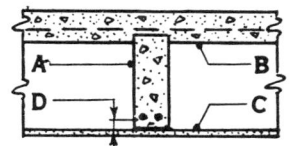 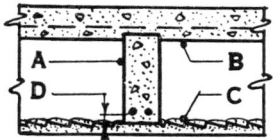

Fig. 5-S-2a. Floors of reinforced concrete slabs on precast concrete joists. A — Precast concrete joists at least 8 inches deep and spaced not more than 30 inches on centers, B — Reinforced concrete slab, C — Ceiling, attached or suspended, and D — Thickness of concrete protection under reinforcing steel in joists.

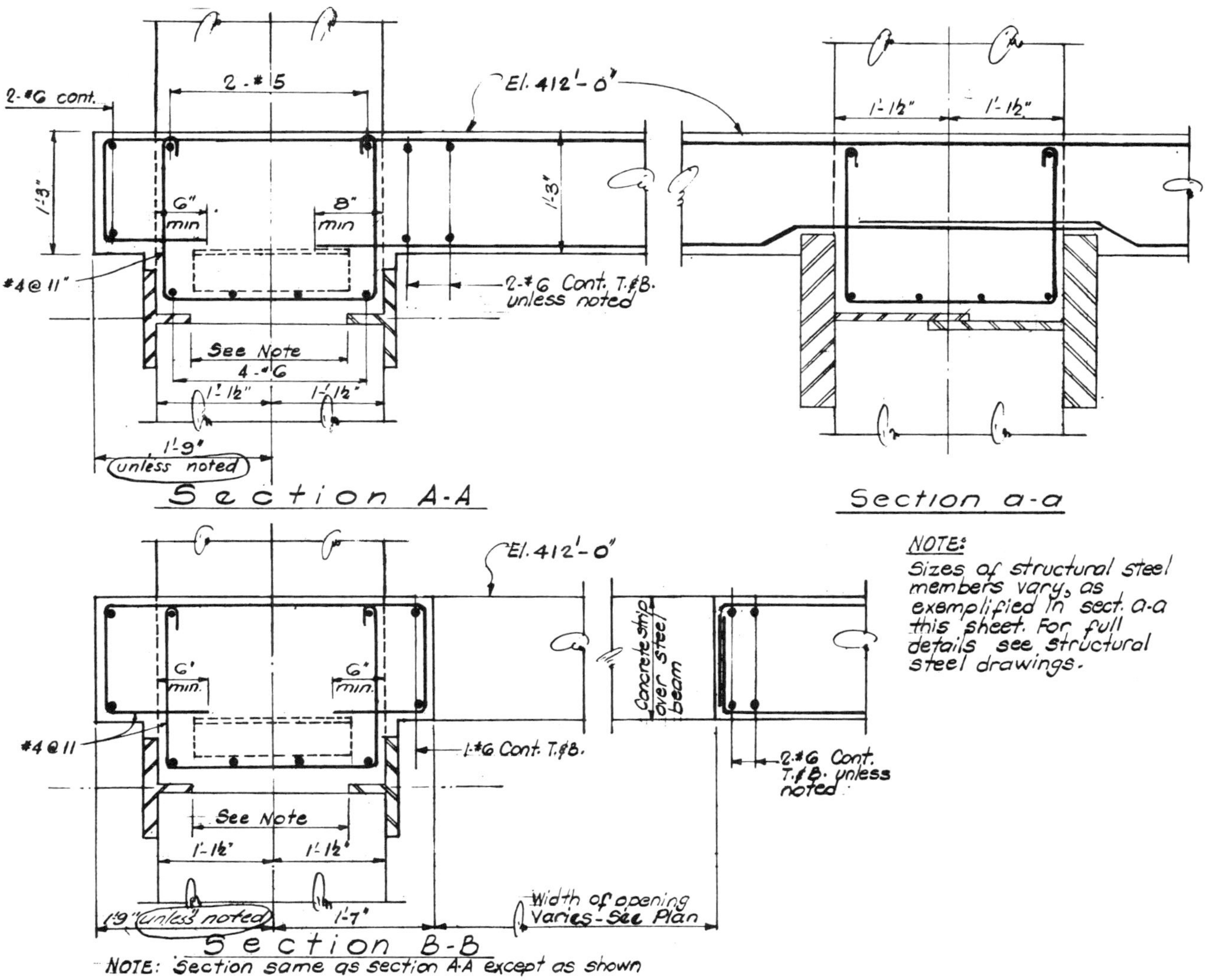

NOTE: Sizes of structural steel members vary, as exemplified in sect. a-a this sheet. For full details see structural steel drawings.

Fig. 5-S-2b. Structural concrete "section" drawing.

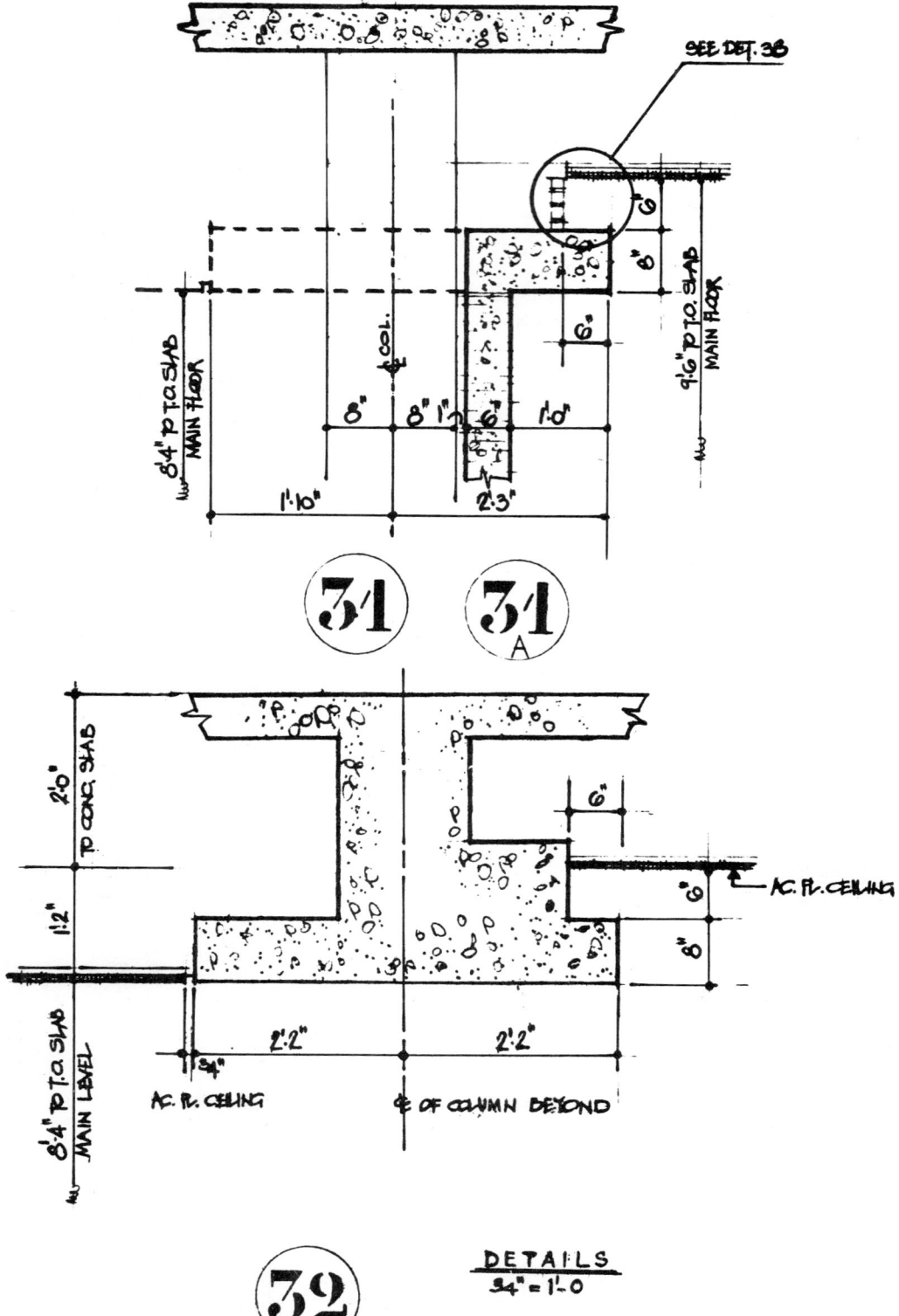

Fig. 5-S-3. Reinforced concrete details.

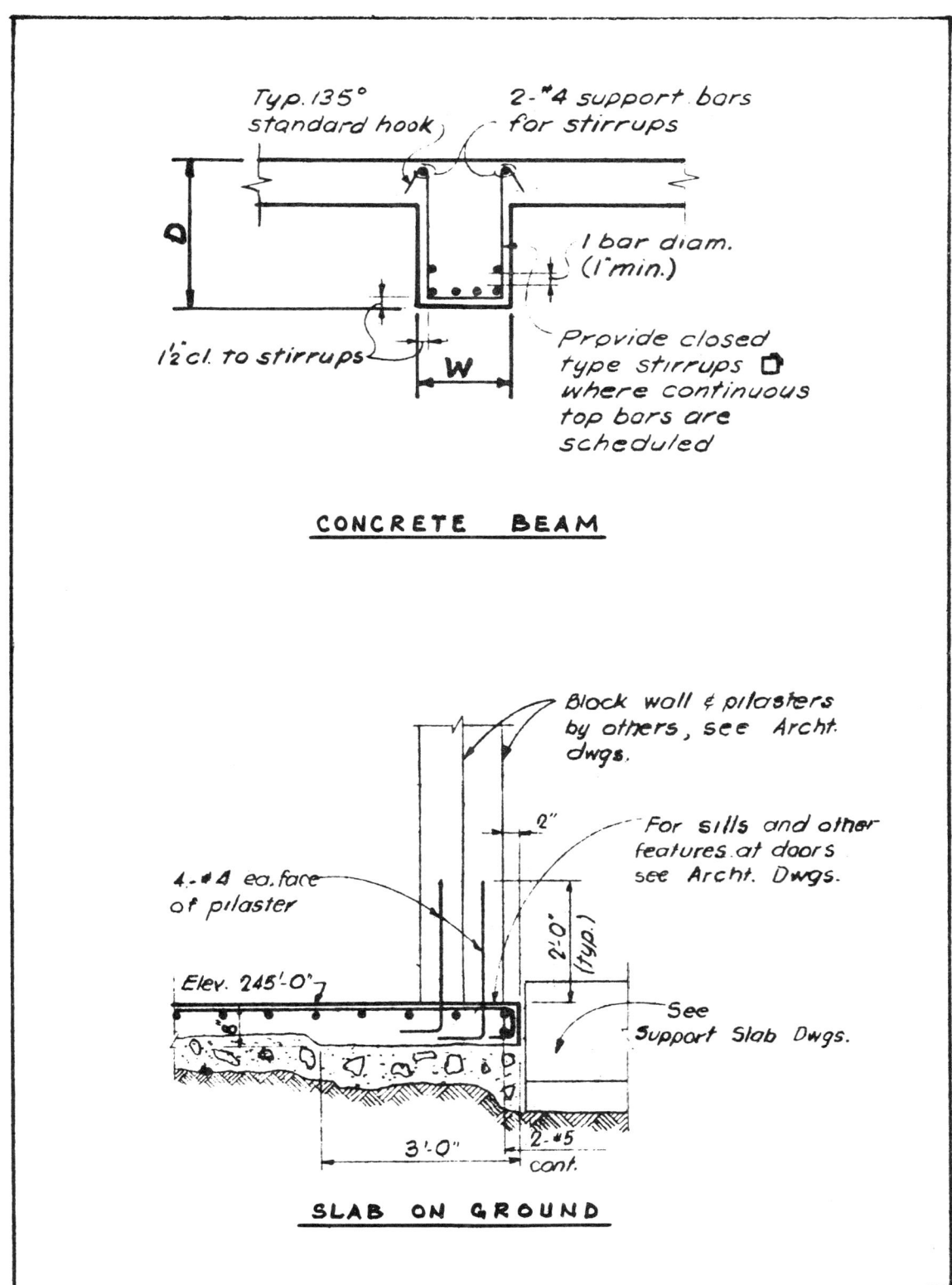

FIG. 5-S-4. Reinforced concrete details.

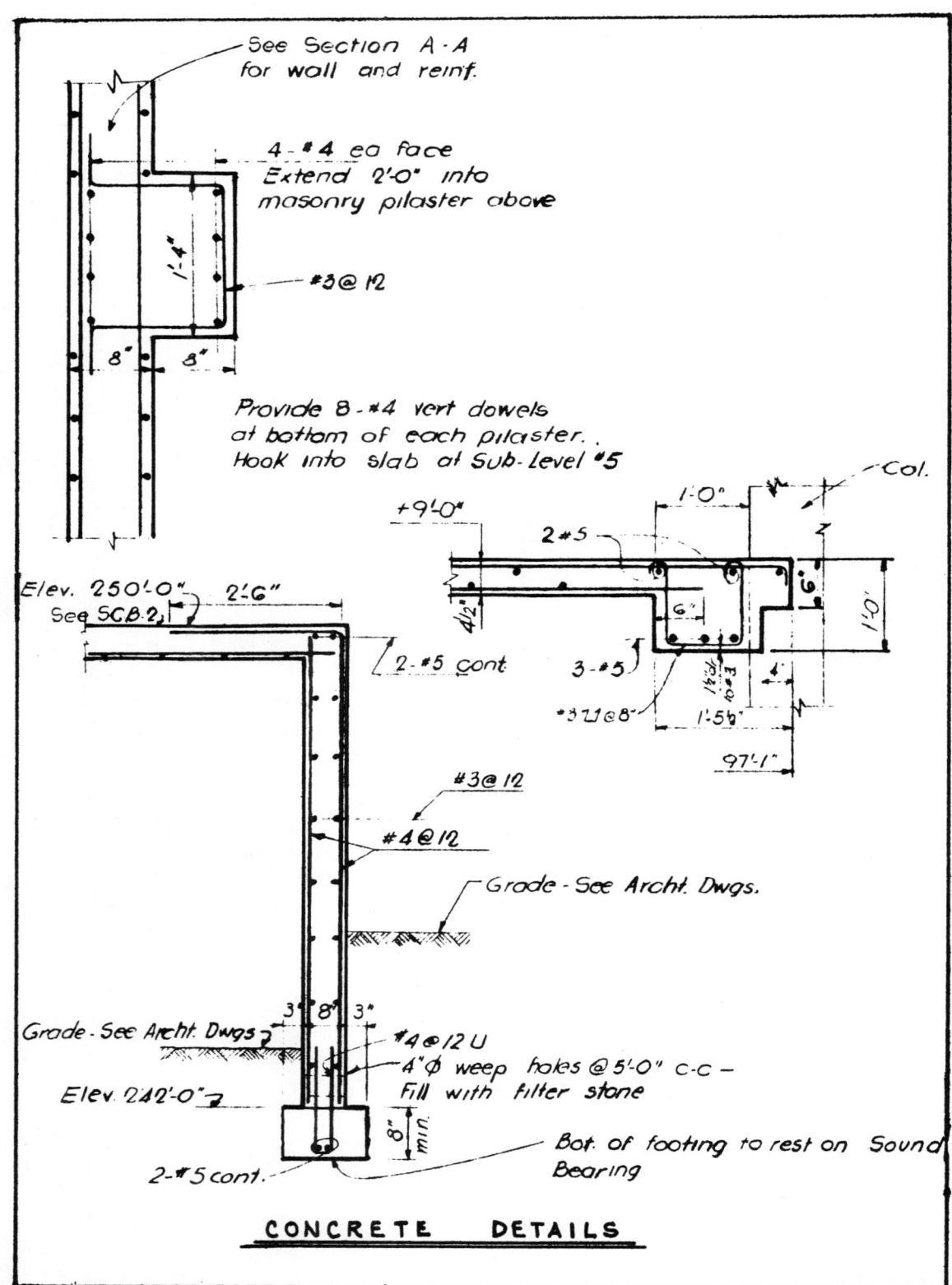

Fig. 5-S-5. Reinforced concrete details.

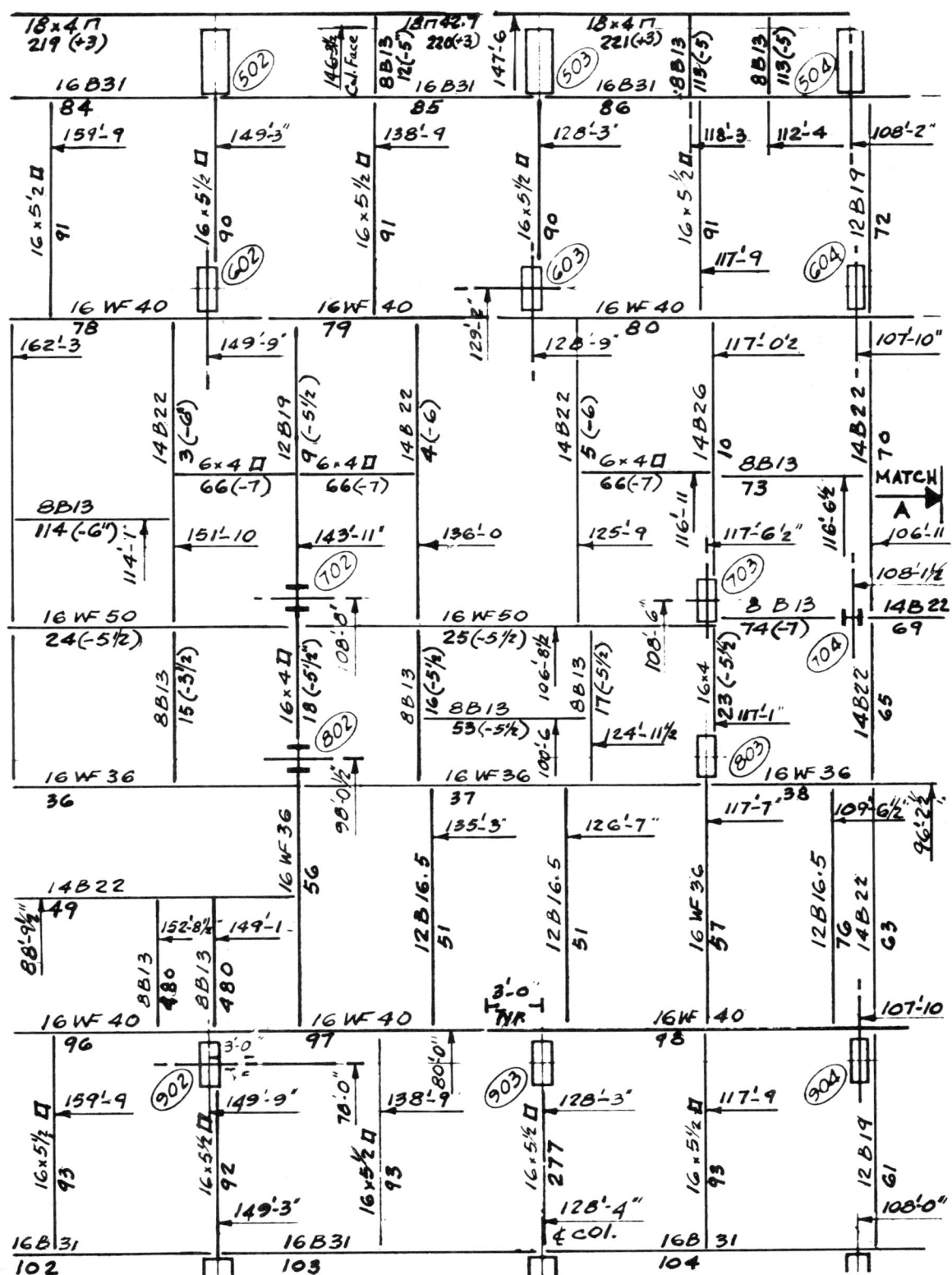

Fig. 5-S-6. Steel framing plans.

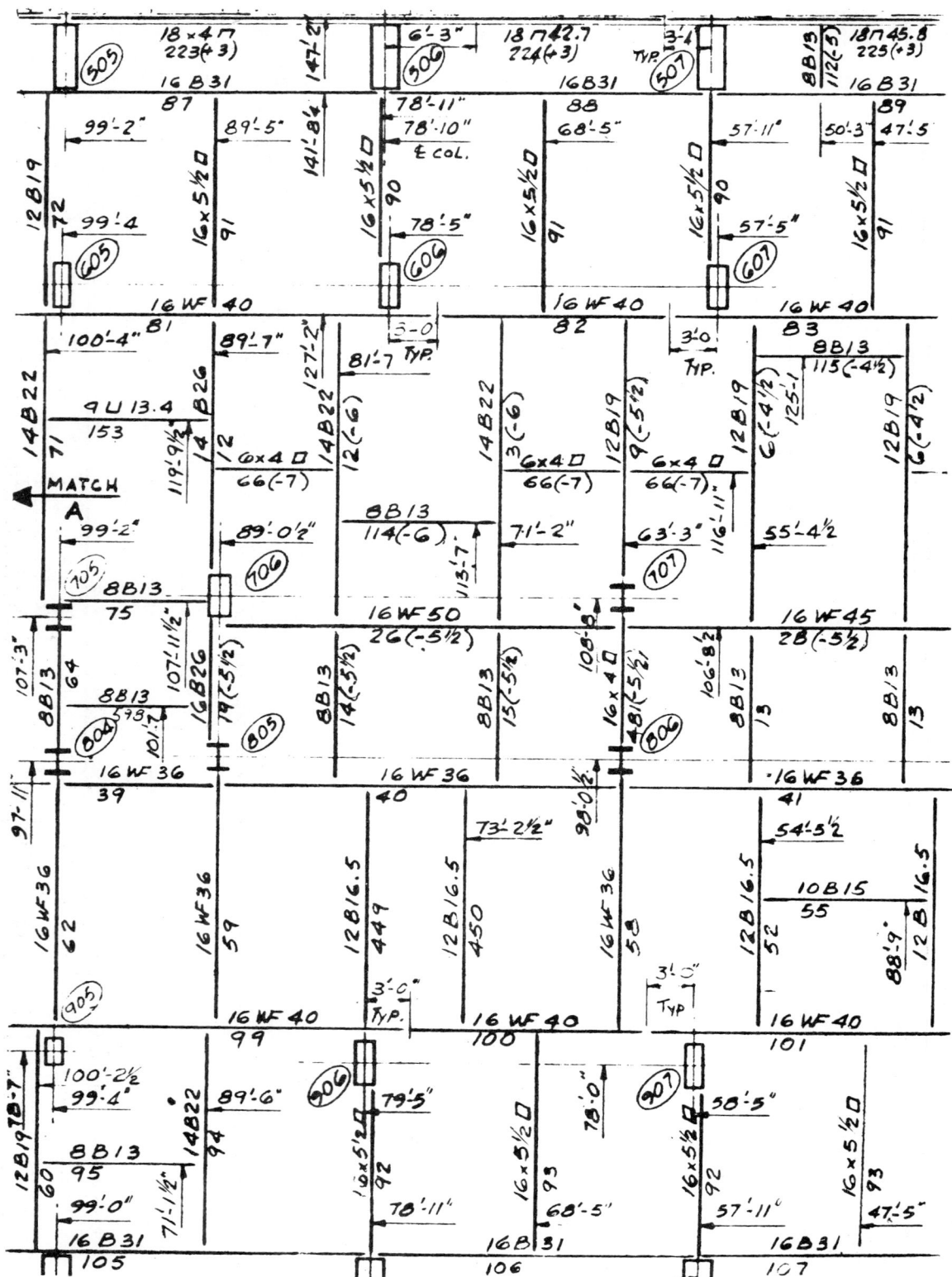

Fig. 5-S-7. Steel framing plans.

Fig. 5-S-8. Dimensions for detailing.

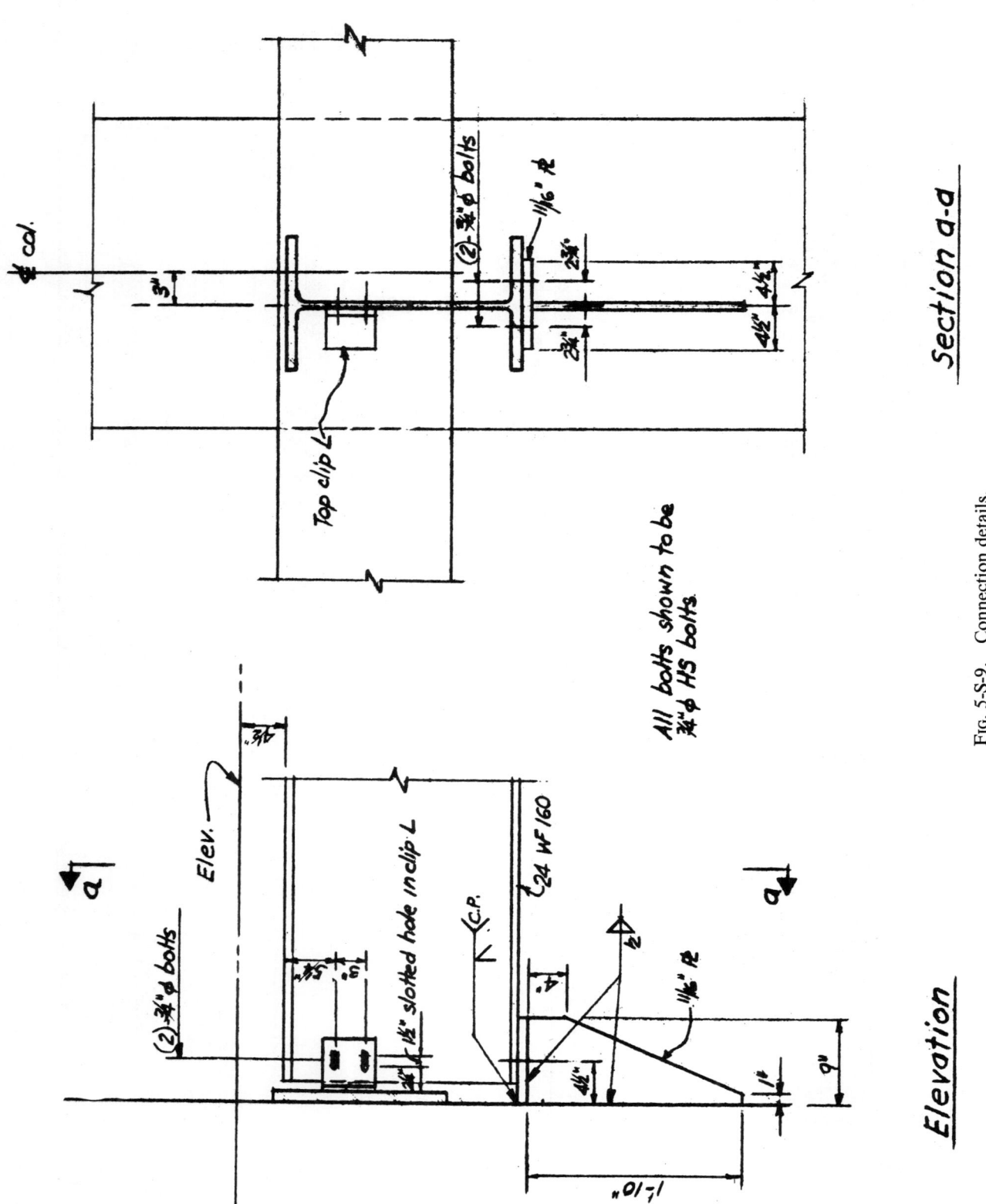

Fig. 5-S-9. Connection details.

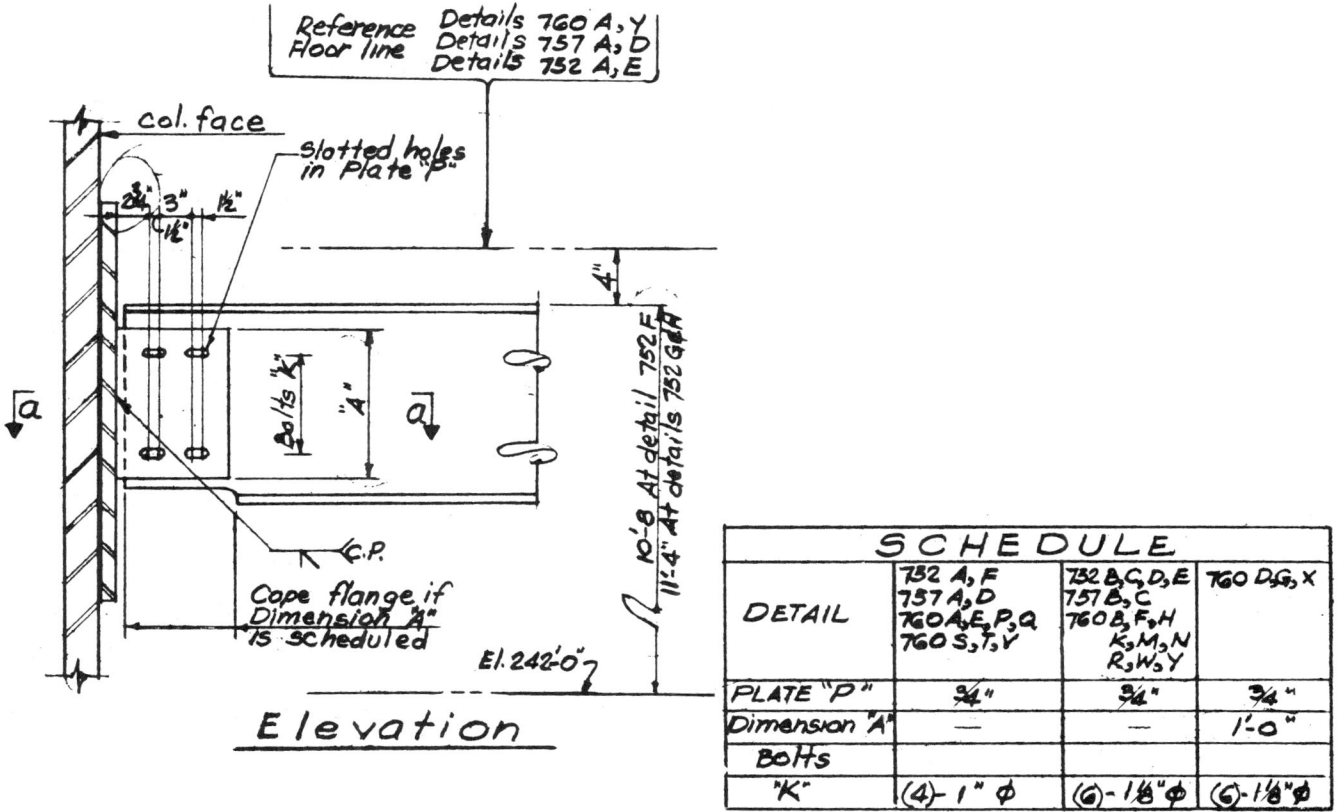

Fig. 5-S-10. Connection details.

(Text continued from page 204.)

Reinforced concrete columns often have column caps which protrude outward from and below the underside of the slab. Ducts at close proximity to columns and close to slab must be checked for clearance in these cases.

Typical reinforced concrete details are shown in Figs. 5-S-2 to 5-S-5. The abundance of information for reinforcing bars is not needed by the sheet metal draftsman.

A steel-framing part plan of 10th and 11th floors of an office building is shown in Figs. 5-S-6 and 5-S-7. Columns, channels, and beams are located by *summation measurements*. These are measurements taken from a common point of 0'-0". Dimensions between members are easily determined by subtraction. Rectangular shapes and I-beams are tied in to centerlines while channels and angles are dimensioned to the heel.

Structural shapes are shown by nominal size and weight per lineal foot. To find the actual dimensions, the tables such as in Fig. 5-S-8 are used. For example, the size of a 16WF40, given in the chart, is 16" x 7".

Connections to columns can be potential obstructions and care must be taken to check special details. Figures 5-S-9 to 5-S-12 show typical connection details.

Structural drawing notes must always be checked to determine the fireproofing concrete thickness, especially where space for ducts is limited. The concrete aggregate varies the rating in hours for thickness used. Spray-on fireproofing, generally one-inch thick, is often used as a substitute for concrete. Fireproofing details of structural steel are shown in Figs. 5-S-13 to 5-S-17.

Figure 5-S-18 shows a steel-shop drawing with

(Text continued on page 247.)

Fig. 5-S-11. Connection details.

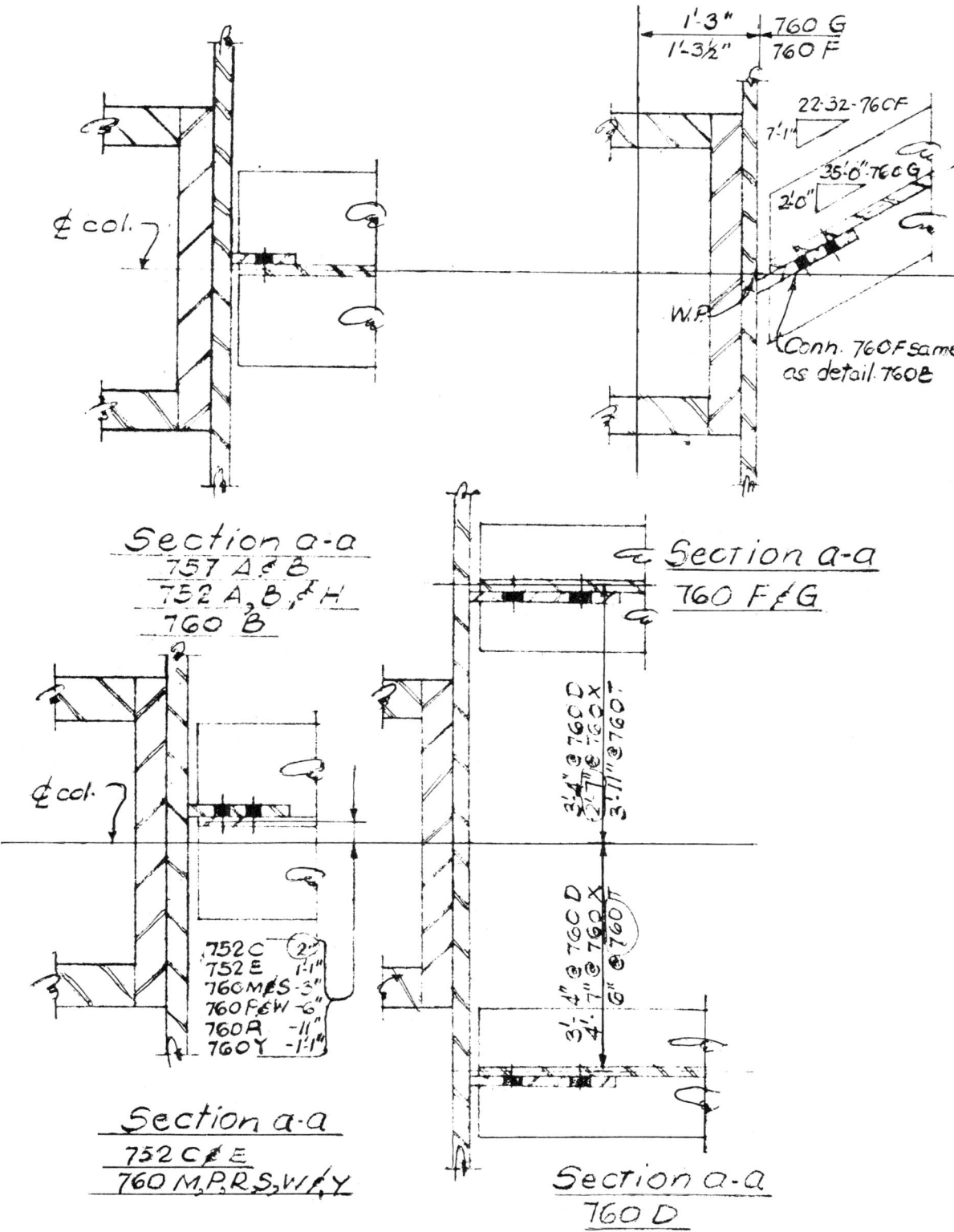

Fig. 5-S-12. Connection details.

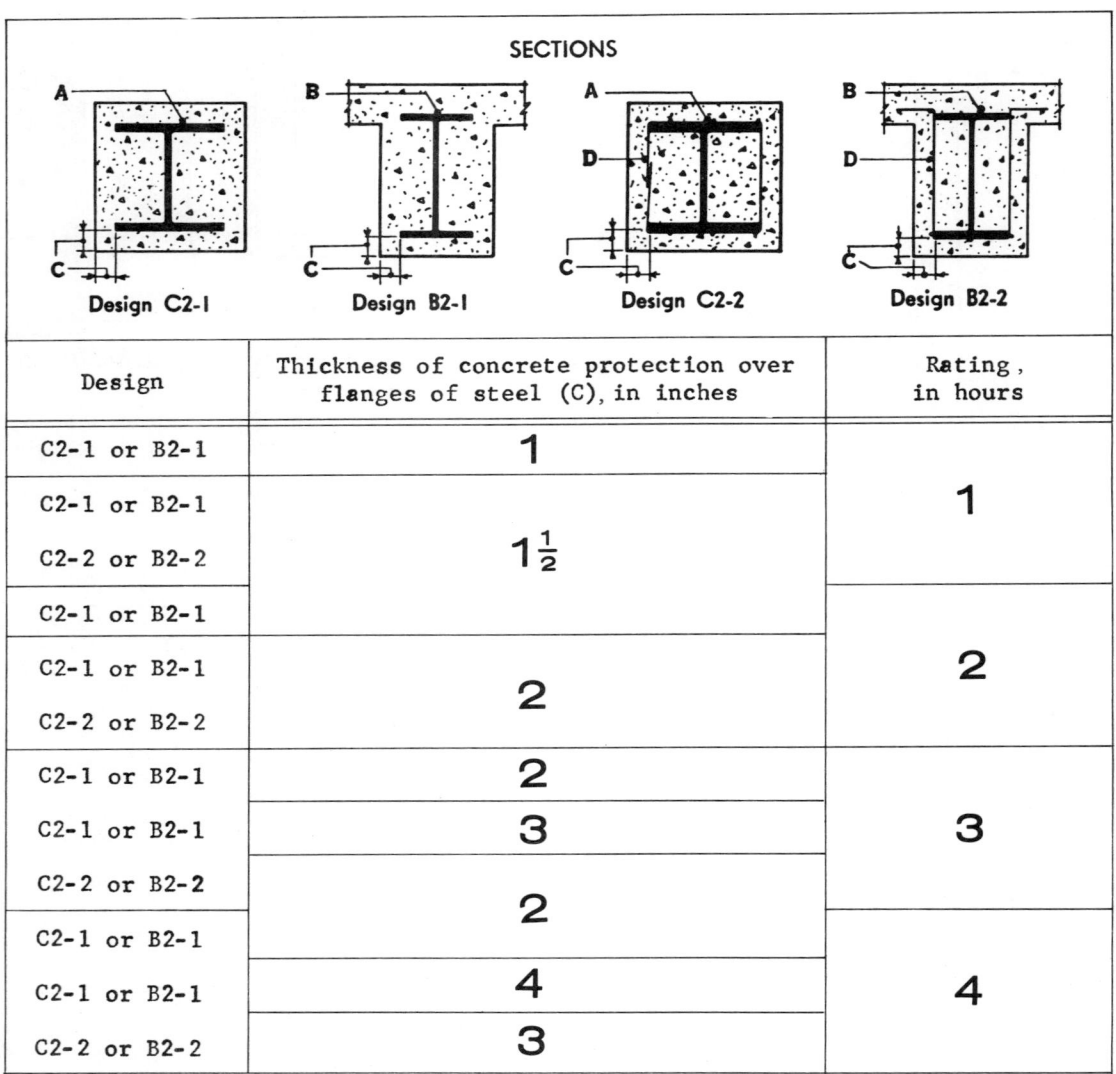

Fig. 5-S-13. Steel columns, beams, and girders with concrete protection. A — Columns of any size or shape, B — Beams or girders of any size or shape with concrete protection cast separately or integrally with floor slab, C — Thickness of concrete protection over flange of structural steel as given in table (thickness does not include plaster), and D — Wire mesh or steel ties.

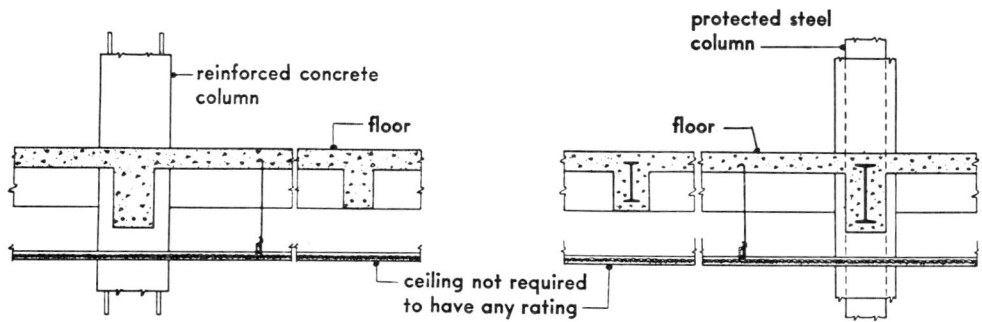

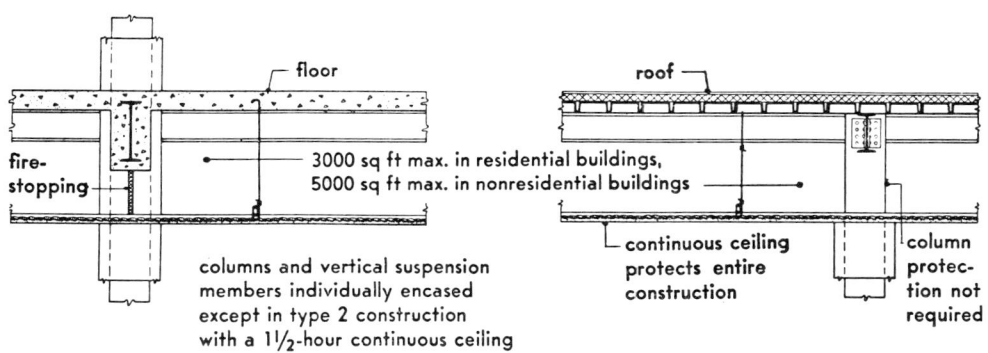

Fig. 5-S-14. Protection of columns, beams, and girders.

SECTIONS

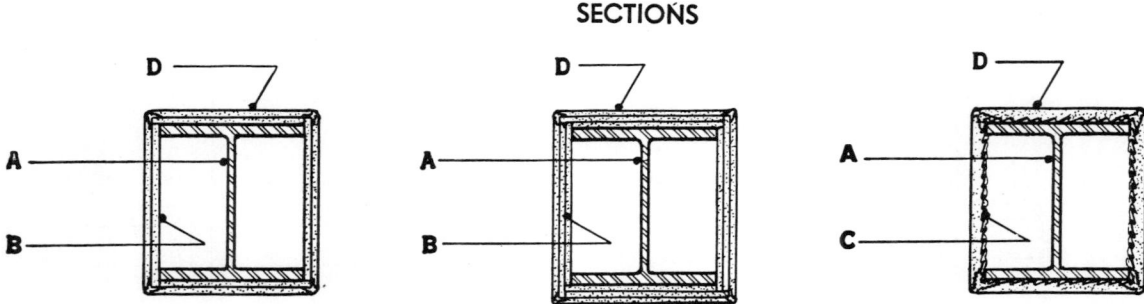

Fig. 5-S-15. Steel columns with plaster protection. A — Steel columns, B — Gypsum lath attached to column with wire ties or wire mesh, C — Metal lath, and D — Plaster.

SECTIONS

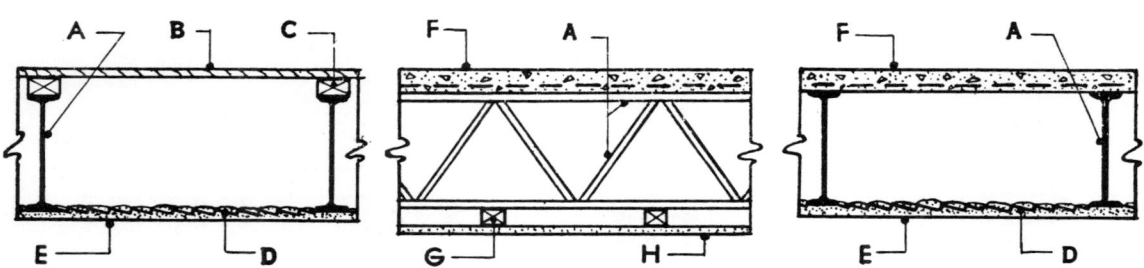

Fig. 5-S-16. Floors of steel-joist construction with protective ceilings. A — Bar joists or light steel beams spaced 16 inches on centers for gypsum lath attached directly to the joists or beams, and 24 inches on centers when metal lath is used, B — T & G wood floor, C — Wood nailer blocks, D — Gypsum lath, metal lath, or precast concrete or gypsum tile attached directly to or suspended below the joists or beams, E — Plaster, F — Floor slab, G — Steel channels or wood furring strips, and H — Gypsum wallboard.

SECTIONS

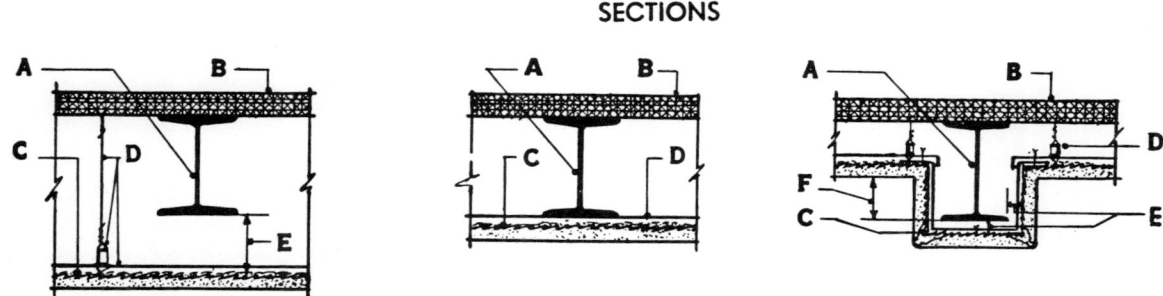

Fig. 5-S-17. Protective ceilings for steel beams, girders, and trusses. A — Structural steel, including beams, girders, and trusses, B — Floor consisting of any combination of noncombustible materials and having a total thickness of at least 2 inches, C — Metal lath, D — Ceiling support consisting of runner and furring channels with ties, E — Space between structural steel and lath, and F — Distance of projection of beam below the ceiling.

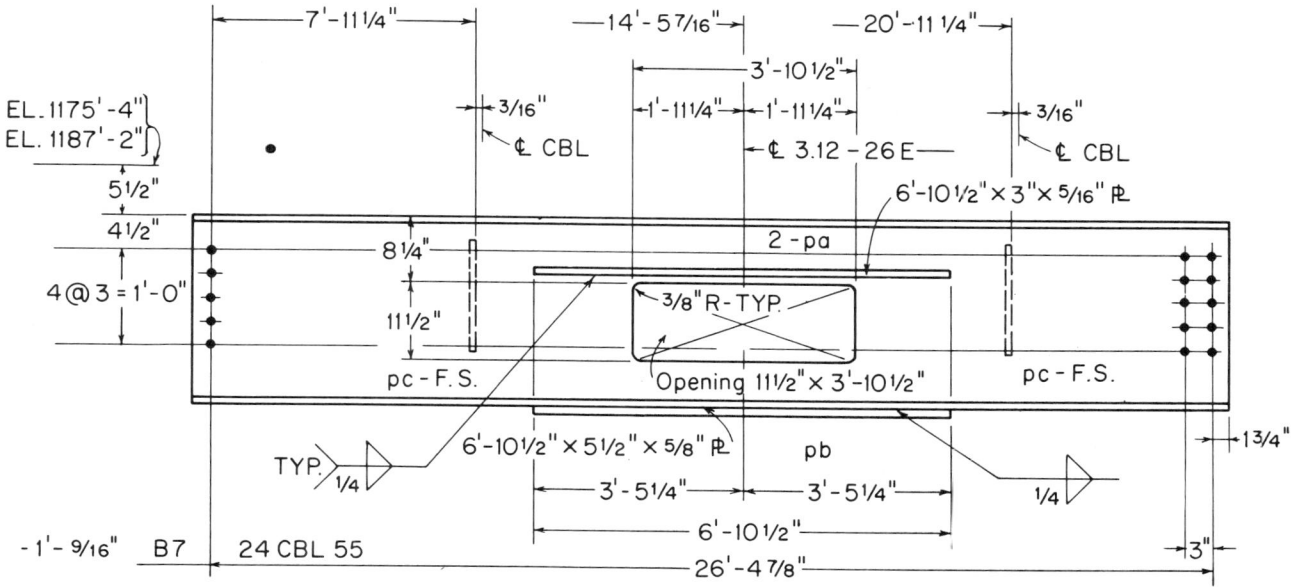

FIG. 5-S-18. BEAM. Penetration.

(Text continued from page 241.)

a cutout for a 32" x 10" duct. When slab and steel elevations vary, the draftsman should check detailed sections. (See Fig. 5-S-19.)

The steel erector's shop drawings are often desirable for checking critical conditions when close working conditions occur.

Joists may be erected by a different contractor than the steel erector. This is important to the sheet metal draftsman, particularly if ducts are passing through the joists and spacing is critical. Joist erector's shop drawings must be checked accordingly, so that bridging and bracing are cleared by ducts.

Metal decking is often used as a base form for concrete slabs and the thickness should be allowed in floor-to-floor calculations for tie-in dimensions from bottom of deck to top of duct runs.

THE ARCHITECTURAL DRAWINGS

A typical schedule of drawings, Fig. 5-A-1 indicates the scope of each type by prefixes as: G — general, A — architectural, S — structural, etc.

The draftsman should have similar drawing schedules readily available so that reference information of specific drawings can be located efficiently.

Project Key Plans (Fig. 5-A-2) can be prepared and checked from Site Plan and Site Details drawings. The building area contained on a shop drawing should be clearly marked off on the Key Plan of each drawing. This clearly identifies the building segment and also provides field crews with a visual aid for materials handling and distribution at the job site.

If the HVAC drawings indicate ducts below grade, the Utility Plans and Details must be checked so that duct layouts can be routed to avoid work of the other trades. Specifically, piping and electrical conduit tie-ins to existing mains are located by the survey engineers. Plumbing inverts and pitch requirements must be considered where close working conditions are encountered. These are generally shown in decimal feet elevations from the survey datum line. (See Figs. 5-A-3 and 5-A-4.)

Floor-plan room layouts and finishes, building elevations and sections, wall sections, details, reflected ceiling plans, types of ceilings, and ceiling heights are shown on the architectural drawings. See Figs. 5-A-5 to 5-A-7, for wall sections. Any architectural feature which affects the alignment or location of an air inlet or outlet should be dimensioned and located on the shop drawing. For

(Text continued on page 250.)

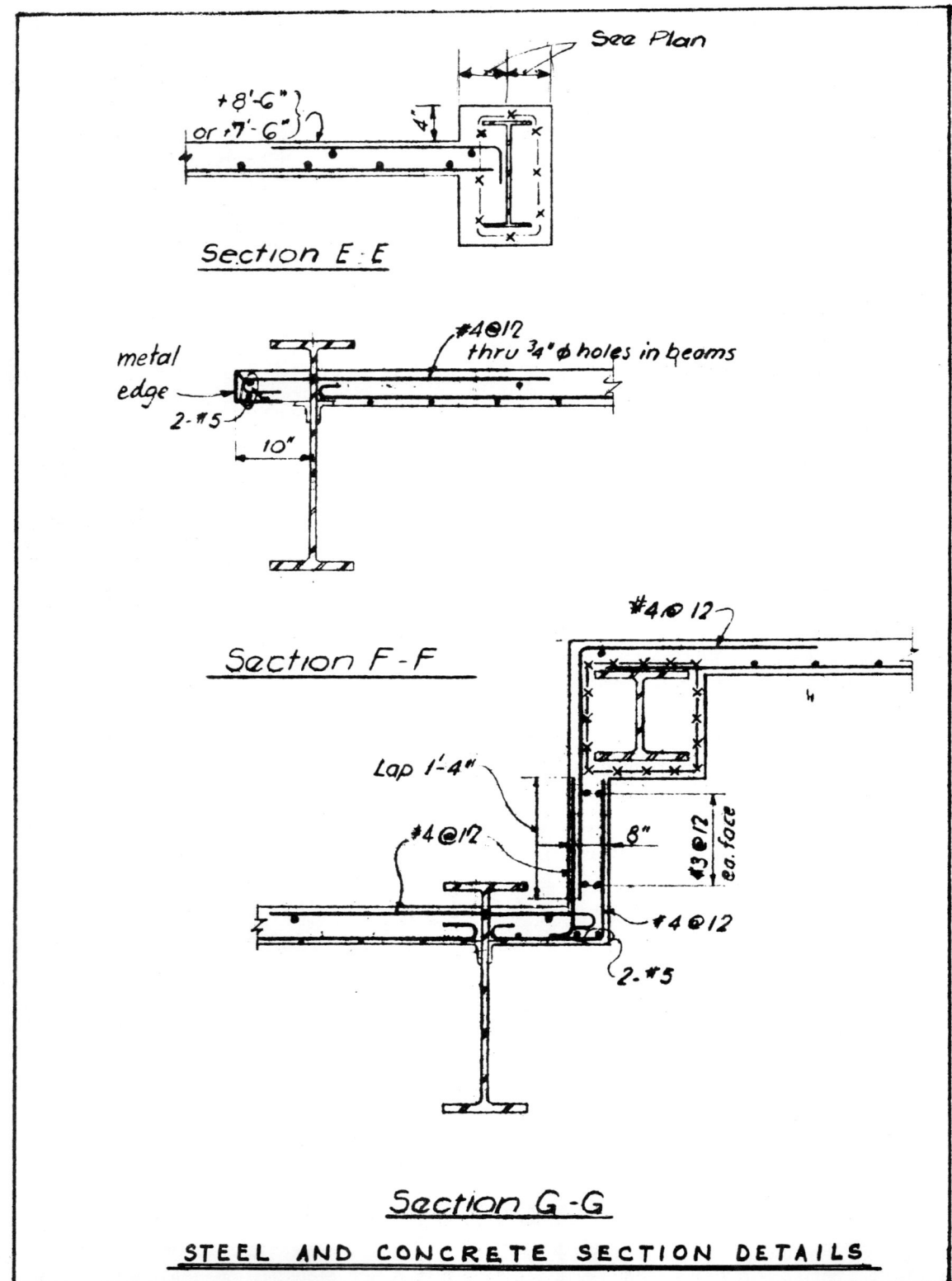

Fig. 5-S-19. Steel and concrete section details.

Type	No.	Drawing Title	Type	No.	Drawing Title
General	G-1	Title Sheet	Structural	S-1	Foundation & Ground Floor Plan
	G-11	Site Plan		S-2	First Floor Framing Plan
	G-12	Site Details		S-3	Second Floor Framing Plan
	G-21	Utility Plan		S-4	Roof Framing Plan
	G-22	Power Signal & Grounds Lighting Plans & Details		S-5	Typical Details, General Notes, & Miscellaneous Details
	G-23	Utility Details		S-6	Column & Beam Schedule
Architectural	A-1	Ground Floor Plan & Finish Schedule	Plumbing	P-1	Ground Floor Plan
	A-2	First Floor Plan & Finish Schedule		P-2	First Floor Plan
	A-3	Second Floor Plan & Finish Schedule		P-3	Second Floor & Roof Plan
	A-4	Roof Plan & Roof Details		P-4	Details
	A-5	Elevations		P-5	Riser Diagrams
	A-6	Sections	Heating, Ventilating, and Air Conditioning	M-1	Ground Floor Plan
	A-7	Wall Sections & Details		M-2	First Floor Plan
	A-8	Windows & Entrance Frames & Details		M-3	Second Floor Plan
	A-9	Stair & Elevator Details		M-4	Mechanical Equipment Room & Sections
	A-10	Stair Tower Details		M-5	Details
	A-11	Toilet & Kitchenette Details		M-6	Schedules
	A-12	Interior Details		M-7	Control Diagrams & Symbols
	A-13	Ground Floor Laboratory Layouts	Electrical	E-1	Ground Floor Plan — Lighting and Symbol List
	A-14	First Floor Laboratory Layouts		E-2	Ground Floor Plan — Power
	A-15	Second Floor Laboratory Layouts		E-3	First Floor Plan — Lighting & Power
	A-16	Laboratory Equipment Elevations		E-4	Second Floor Plan — Lighting & Power
	A-17	Laboratory Equipment Elevations		E-5	Riser Diagrams — Schedules & Details
	A-18	Laboratory Equipment Elevations			
	A-19	Reflected Ceiling Plans			
	A-20	Miscellaneous Details & Lettering			
	A-21	Door Schedule & Details			

FIG. 5-A-1. Typical index of project drawings.

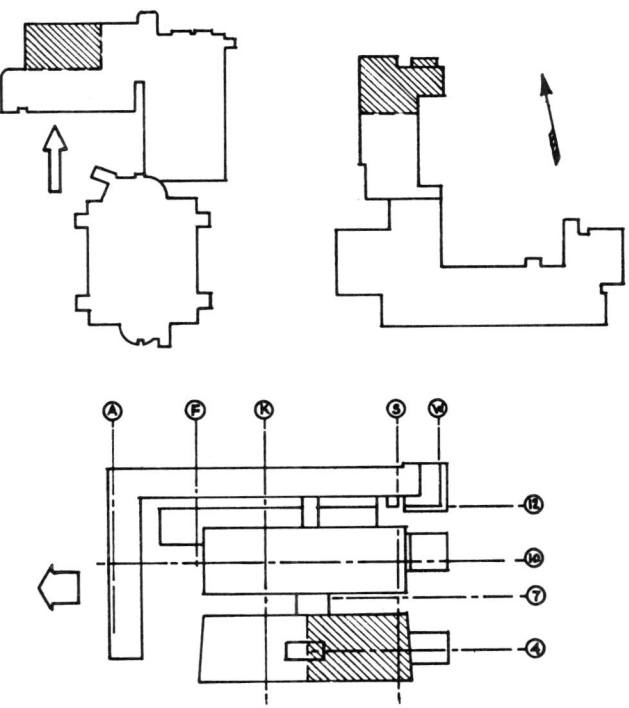

FIG. 5-A-2. Typical project key plans.

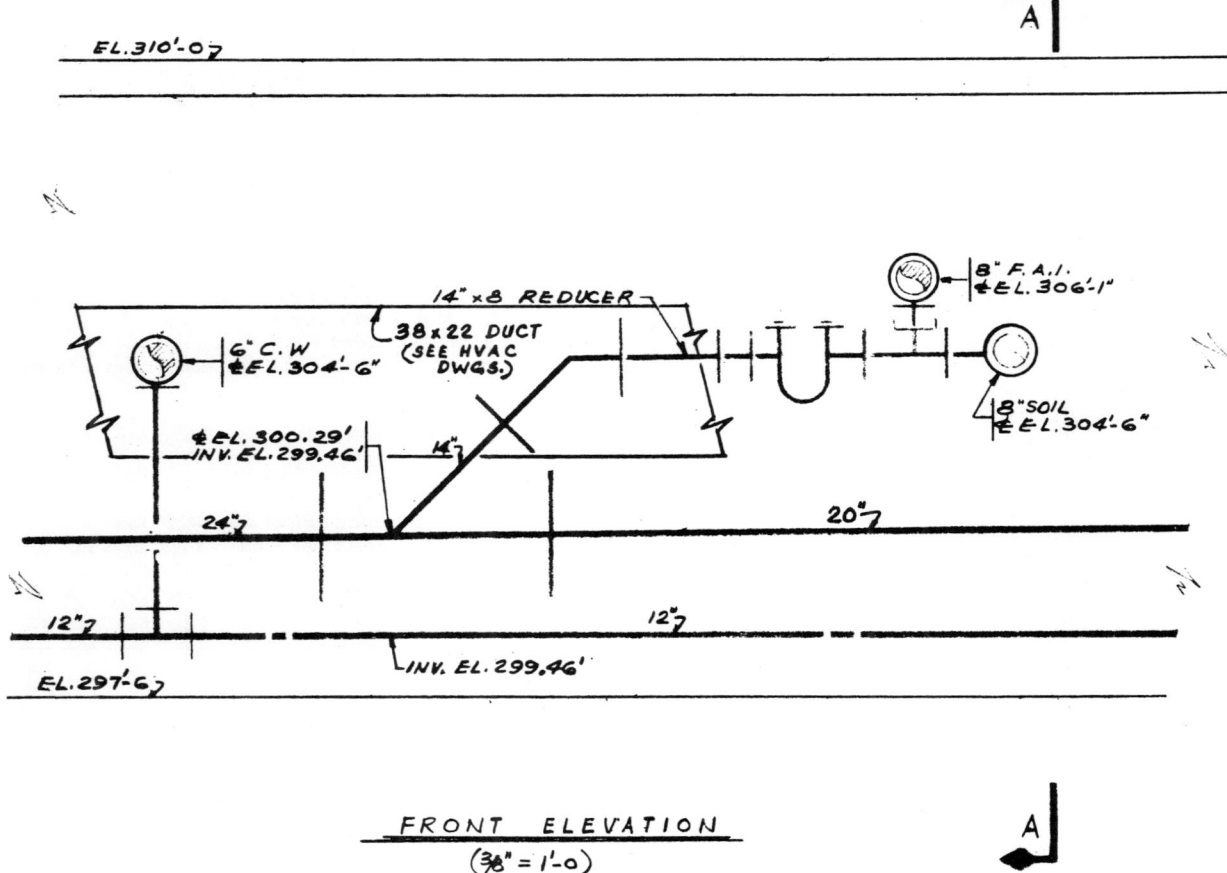

Fig. 5-A-3. Plumbing inverts.

example, a supply register which ends at a room partition should be tied in to a building structural member such as a column centerline or a beam centerline. This must be done because the ducts will be erected before the wall-partition layouts are made. When definite ceiling patterns are not shown on the architectural drawings, the draftsman should request verification by the architect for critical tie-in dimensions of registers, grilles, or diffusers.

Summation dimensions are often shown on architectural drawings. Lightly drawn module lines, spaced ten feet apart vertically and horizontally for the entire height and width of the shop drawing, are convenient for plotting dimensions for this system. Summation dimensions can be easily located from the closest module line.

Concrete pads for casings and fan equipment in the mechanical rooms are often indicated on the architectural floor-plan drawings by definite dimensions. As sizes may vary, the draftsman must verify the approved equipment cuts, and check layouts as required.

It is advisable to prepare pad layout shop drawings which will be compatible to the final, approved equipment. Electrical motor conduit stub-up locations and plumbing drain requirements should be indicated on these drawings.

While duct-system layouts are only shown in general arrangements on the HVAC design draw-

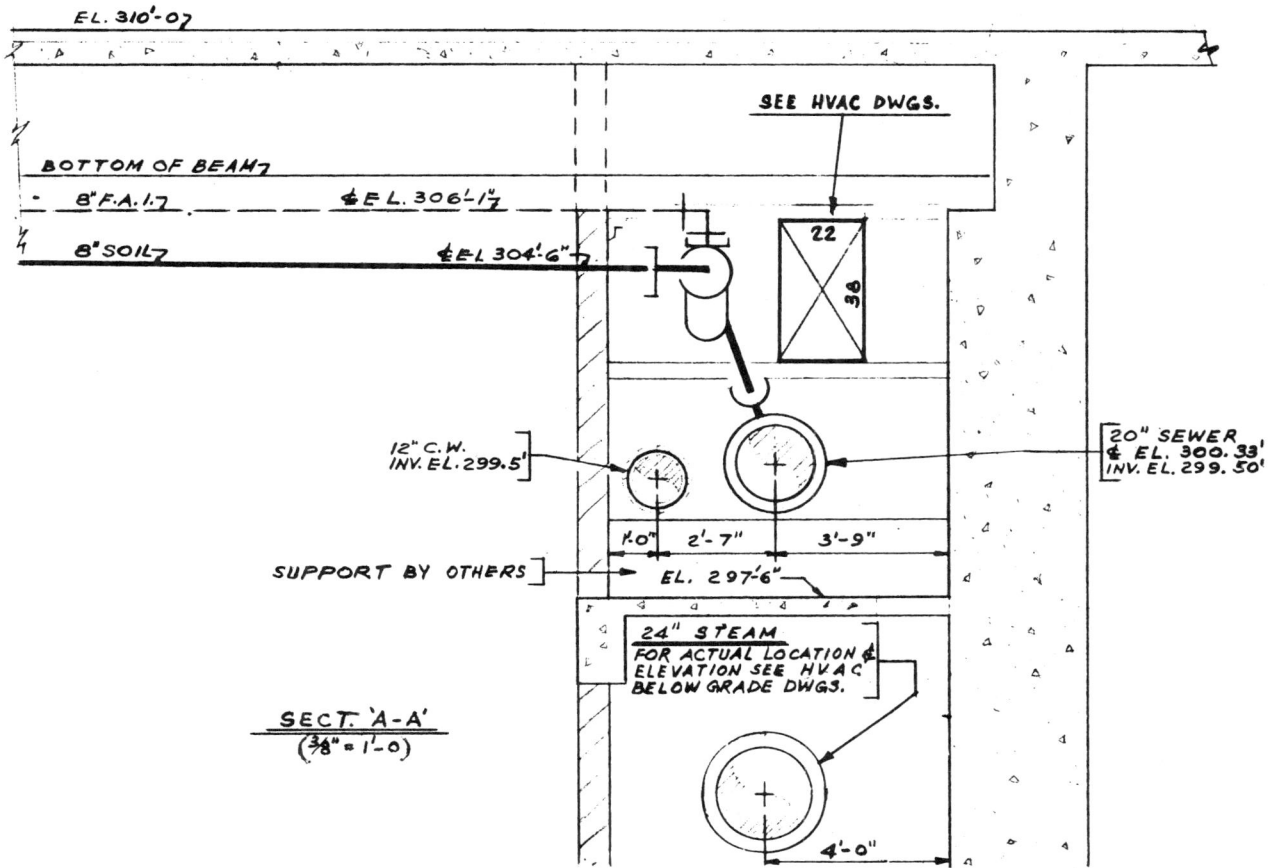

Fig. 5-A-4. Plumbing inverts.

SECTIONS

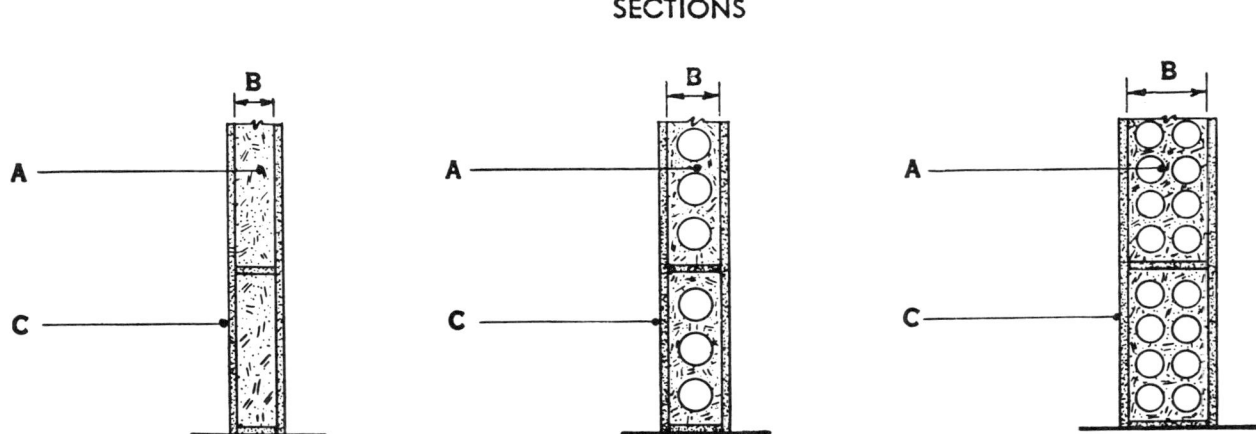

Fig. 5-A-5. Partitions of gypsum tile or block: nonbearing. A — Gypsum block (block laid in gypsum-sand mortar), B — Nominal thickness of wall without plaster, and C — Gypsum-sand, gypsum-perlite, or gypsum-vermiculite plaster at least ½-inch thick and having at least three parts of aggregate to one part of fibered gypsum cement.

SECTIONS

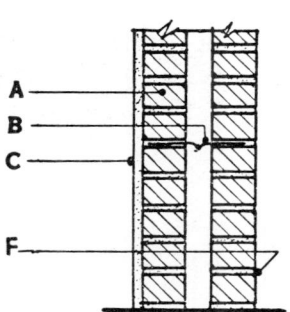 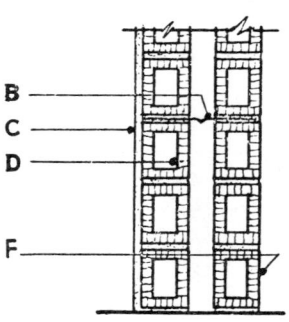 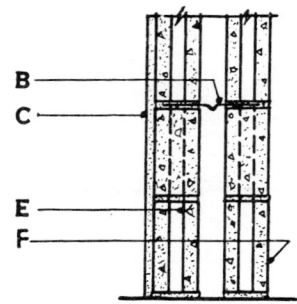

Fig. 5-A-6. Walls of cavity type: bearing. A — Clay brick, B — Corrosion-resistant metal ties spaced to provide one tie each square foot of wall surface, C — Gypsum plaster at least ½-inch thick and not more than three parts of aggregate to one part of fibered gypsum, D — Structural clay load-bearing tile, E — Concrete masonry units of load-bearing grade, and F — Exterior face of wall.

SECTIONS

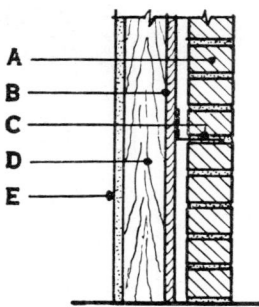 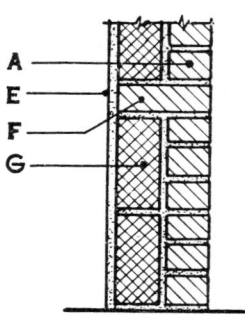 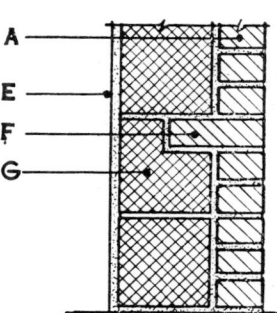

Fig. 5-A-7. Walls of faced or veneered construction: bearing. A — Brick, B — Sheathing, C — Corrosion-resistant metal ties spaced 24 inches on centers; vertically and horizontally, D — Wood or steel studs, E — Plaster of any kind (at least ½-inch thick), or gypsum wallboard (at least ⅜-inch thick), F — Masonry bond, and G — Masonry backing unit.

ings, it is the primary purpose of the sheet metal draftsman to analyze and interpret the structural and architectural drawings, prepare the working shop-drawings, and also coordinate work to be installed by the other trades.

GENERAL NOTES FOR ARCHITECTURAL DRAWINGS

1. All elevations indicated on drawings refer to the U.S. Coast & Geodetic Survey Datum of Mean Sea Level (1929 adjustment).
2. All figures and dimensions on plans are to rough face of walls or block partition (except for centerlines of columns and masonry openings).
3. The term "finished floor" refers to top of finished concrete slab exclusive of resilient tile flooring. Where concrete slab is depressed to receive wood, stone, or ceramic tile the term "finished floor" refers to the level of the finished wood, stone, or ceramic-tile flooring.
4. Floor slabs under water closets and urinals shall not be poured until rough plumbing is in place. The masonry contractor shall consult with the plumbing contractor before erecting partitions behind toilet fixtures.
5. Interior partitions shall not be built to full height until all ducts, pipe, etc., are in place.

6. Interior partitions shall be braced securely to the structure, and be tightly wedged to concrete ribs or ceiling.

7. Fan and mechanical-room partitions shall not be built until all equipment is in place. Sizes and locations of concrete bases to be coordinated with HVAC contractor.

8. Where heating and plumbing pipes are concealed in hung or furred ceilings, these ceilings shall be installed only after contractor has tested the pipes.

9. All brick coursing is based on 3-brick courses = 8 inches.
All block coursing is based on 1-block course = 8 inches.

10. When door openings are shown next to partitions, the minimum distance shall be not less than 6″ unless otherwise noted or detailed.

11. All doors leading to an exit enclosure shall have a fire rating of 3/4 hour, unless otherwise indicated.

12. The dimensions indicated for recessed convector enclosures, fire-hose cabinets, drinking fountains, etc., are approximate. General contractor to coordinate these dimensions with equipment purchased.

13. Consult mechanical, plumbing, and electrical drawings for location of cleanouts and access doors in partitions and hung ceilings, not shown on architectural drawings, for requirements of movable equipment indicated as not in contract, and for sizes and locations of openings in foundation and exterior walls.

14. See architectural drawings for exact location of electrical fixtures, heating and ventilating registers, diffusers, etc.

ARCHITECTURAL ABBREVIATIONS

AC BL	Acoustical Block
ACOUS	Acoustic
AC PLAS	Acoustical Plaster
AC T	Acoustical Tile
ALUM	Aluminum
ASB	Asbestos
AFF	Above Finished Floor
A T	Asphalt Tile
BAL	Balcony
B B	Bulletin Board
BD	Board
BLKG	Blocking
BM	Beam
BR	Brick
BSMT	Basement
CAB	Cabinet
C & H ST	Coat & Hat Strip
CAV	Cavity
CB	Ceiling Break
CBL	Concrete Block
CEM	Cement
CH B	Chalk Board
₵	Centerline
CL	Closet
CLG	Ceiling
COL	Column
COMP	Compressed
CONC	Concrete
CONSTN	Construction
CONT	Continuous
CONTR	Contract
CONV	Convector
CORR	Corridor
CTS	Centers
C W	Cold Water
D.B.	Double Bubbler
DET	Detail
D F	Drinking Fountain
DIA	Diameter
DIM	Dimension
DN	Down
DR	Drain
DR RM	Dressing Room
DUPL	Duplicating
DUP REC	Duplex Receptacle
DWG	Drawing
EDG	Edging
EL	Elevation
ELEC	Electric
ELEV	Elevation

EQ	Equipment	L S J	Long Span Joist
EXP JT	Expansion Joint	LTG	Lighting
EXT	Exterior	M A	Metal Angle
EXTN	Extension	MAX	Maximum
F.A.	Fire Alarm	MECH	Mechanical
F.B.	Flemish Bond	MET	Metal
F.E.	Fire Extinguisher	MEZZ	Mezzanine
F DR	Floor Drain	MIN	Minimum
F.G.	Fire Gong	M L	Metal Lath
F H	Flat Head	M.L.F.	Metal Lath Furring
F H C	Fire Hose Cabinet	M O	Masonry Opening
F H R	Fire Hose Reel	N I C	Not In Contract
FIN	Finished	NO	Number
FL	Floor	NOM	Nominal
FOLD	Folding	N T S	Not To Scale
FP	Fireproofing		
FP S C	Fireproof Self Closing	O C	On Center
FTG	Footing	O D	Outside Diameter
FX	Fixed	OH	Overhead
		OPG	Opening
GA	Gage	OPP HAND	Opposite Hand
GALV	Galvanized	OZ	Ounce
GEN	General		
GRAV	Gravel	P & R	Projected & Recessed
G.T.	Glazed Tile	P	Plaster
		PART	Partition
H	High	℞	Plate
H & V	Heating & Ventilating	PLAS LAM	Plastic Laminate
HC	Hung Ceiling	PLAT	Platform
HCF	Hard Cement Finish	PREP	Preparation
H CONC	Hardened Concrete	PROJ	Projection
H M	Hollow Metal	PT	Paint
H R	Hand Rail	PVC	Polyvinylchloride
HVAC	Heating, Ventilating & Air Conditioning	R	Riser
H W	Hot Water	RAD	Radius
H W PL	Hard White Plaster	REC	Recess
		REG	Register
ID	Inside Diameter	REINF	Reinforced
INSUL	Insulation	REQD	Required
INT	Interior	REMOV	Removable
INTERM	Intermediate	RET	Retaining
		R H	Round Head
JAN	Janitor	RM	Room
JT	Joint		
		S	Saddle
LAB	Laboratory	S.B.	Single Bubbler
L C	Lead Coated	SECT	Section
LECT	Lecture		

DESIGN DRAWINGS AND SPECIFICATIONS

SQ	Square
S.S.	Stainless Steel
ST	Steel
STOR	Storage
STRUCT	Structural
SUSP	Suspended
T	Tread
T & G	Tongue & Groove
TEMP	Tempered
THK	Thick
T O W	Top Of Wall
T.S.	Top Of Slab
TV	Television
TYP	Typical
U.O.N.	Unless Otherwise Noted
V	Volts
VAC	Vacuum
V A T	Vinyl Asbestos Tile
V.C.T.	Vitreous Ceramic Tile
VERM	Vermiculite
VEST	Vestibule
W	Wide
W B	Waste Basket
W D	Window Dimension
WD	Wood
WIN	Window
W.L.	Wire Lath
WP	Waterproofing

SAMPLE SPECIFICATIONS

Ductwork — Low-Pressure

Construct low-velocity ductwork in accordance with the specification table. (Fig. 5-A-8.)

Ducts exposed to weather shall be copper.

Exhaust ducts from shower rooms and dishwasher machine shall be of aluminum and of watertight construction.

Longitudinal seams shall be corner-Pittsburgh, intermediate groove, or inside standing.

Uninsulated exposed ducts shall have no longitudinal seam below the duct horizontal centerline.

Transverse connecting and stiffening angles shall be riveted, tack- or spot-welded to the duct at intervals of 6", maximum.

Flush Seam — Duct Construction

Construct flush seam ductwork of galvanized steel in accordance with the specification table. (Fig. 5-A-9.)

Construct rectangular duct, longitudinal joints with Pittsburgh type seams. Locate all longitudinal joints at the top corners of the duct.

Hang the duct by 1" x $\frac{1}{8}$" flat bars riveted or welded to the sides of the duct. The lowest rivet or weld shall be within 1" of the bottom of the duct. Make fastenings to the duct a minimum of 6" on centers.

Ductwork — Medium-Pressure and Airtight

Construct airtight ductwork of galvanized steel in accordance with the specification table. (Fig. 5-A-10.)

Construct rectangular duct, longitudinal joints with Pittsburgh type seams.

For duct systems 2 to 4 in. wg, the longitudinal seams are to be sealed airtight with an approved sealing compound.

For duct systems 4.1 to 8 in. wg, the longitudinal seams are to be welded airtight.

For duct systems 2 to 4 in. wg, transverse connecting angles shall be riveted, tack or spot welded to the duct at intervals of 6", minimum and sealed airtight with an approved sealing compound.

For duct systems 4.1 to 8 in. wg, transverse connecting angles shall be continuous-welded to the duct.

All transverse connections shall be sealed airtight with an approved sealing compound.

Straight round-duct sections shall be helical grooved, spiral conduit.

Transverse connections shall have bead stiffening and have a 2-inch lap for sizes to 12 inches in diameter; a 4-inch lap for diameter sizes 12 inches and larger.

Connections shall be sealed airtight with an approved sealing compound and tape.

All elbows to have centerline radius equal to 1.5 times duct diameter.

Duct Hangers and Support

Hang round ducts of 24-inch diameter, and smaller, by two duct clamps. Each clamp shall be

METAL GAGES			DUCT DIMENSION, INCHES	TRANSVERSE JOINTS		TRANSVERSE BRACING	HANGERS		
GALV. STEEL GAGE	ALUM. B & S GAGE	COPPER COLD ROLLED		CONSTRUCTION	SPACING, FEET		ROD DIA., INCH	SHELF ANGLE	MAX. SPACING
26	24	16 OZ.	UP THRU 12	1" POCKET LOCK HEMMED "S" SLIP	8	NONE	3/8	1 1/2 X 1 1/2 X 1/8" OR 1 X 1/8" STRAP	8' - 0"
24	22	24 OZ.	13 THRU 18	DRIVE SLIP (VERTICAL ONLY)		NONE			
			19 THRU 30	1" POCKET LOCK HEMMED "S" SLIP 1" BAR SLIP	4 8	1 X 1 X 1/8" ANGLE CENTERED BETWEEN JOINTS			
22	20	32 OZ.	31 THRU 42	1" POCKET LOCK 1" STANDING "S" SLIP 1" REINFORCED BAR SLIP	4 8	NONE 1 X 1 X 1/8" ANGLE CENTERED BETWEEN JOINTS	3/8	1 1/2 X 1 1/2 X 1/8"	6' - 0"
			43 THRU 54	1 1/2" POCKET LOCK 1 1/2" STANDING "S" SLIP 1 1/2" REINFORCED BAR SLIP	4 8	NONE OR CROSS BREAK 1 1/2 X 1 1/2 X 1/8" ANGLE CENTERED BETWEEN JOINTS	1/2	2 X 2 X 1/8" OR EQUIVALENT CHANNEL	6' - 0"
20	18	36 OZ.	55 THRU 60	1 1/2" REINFORCED BAR SLIP 1 1/2" STANDING "S" SLIP	4 8	1 1/2 X 1 1/2 X 1/8" ANGLES CENTERED BETWEEN JOINTS 1 1/2 X 1 1/2 X 1/8" ANGLES ON 2' - 0" CENTERS BETWEEN JOINTS	1/2	2 X 2 X 3/16" OR EQUIVALENT CHANNEL	4' - 0"
			61 THRU 84	1 1/2" ANGLE REINFORCED POCKET LOCK					
18	16	48 OZ.	85 THRU 96	1 1/2" REINFORCED BAR SLIP 1 1/2" COMPANION ANGLES	4 8	1 1/2 X 1 1/2 X 3/16" ANGLES CENTERED BETWEEN JOINTS 1 1/2 X 1 1/2 X 3/16" ANGLES ON 2' - 0" CENTERS BETWEEN JOINTS	1/2	3 X 3 X 1/4"	4' - 0"
			97 THRU 120	2" ANGLE REINFORCED POCKET LOCK 2" COMPANION ANGLES	4 8	2 X 2 X 1/4" ANGLES CENTERED BETWEEN JOINTS 2 X 2 X 1/4" ANGLES ON 2' - 0" CENTERS			

FIG. 5-A-8. Specification table for conventional low-pressure ducts.

U.S. STD. GAGE	DUCT DIMENSION, INCHES	TRANSVERSE JOINTS				INTERNAL BRACING	HANGERS		
		TOP	SIDES	BOTTOM	MAX. SPACING		ROD OR STRAP	INTERNAL SUPPORT BAR	MAX. SPACING
26	UP THRU 18	FLUSH SEAM END SLIP	FLUSH SEAM END SLIP	FLUSH SEAM END SLIP	8' - 0"	NONE	3/8" ROD OR 1 X 1/8" STRAP	NONE	8' - 0"
24	19 THRU 30	BAR STIFFENED	BAR STIFFENED	BAR STIFFENED	4' - 0"	NONE			6' - 0"
22	31 THRU 42	FLUSH SEAM END SLIP 1 1/4 X 3/8" BAR	FLUSH SEAM END SLIP 1 1/4 X 3/8" BAR	FLUSH SEAM END SLIP 1 1/4 X 3/8" BAR	8' - 0"	1 1/4 X 3/8" BAR BETWEEN JOINTS AND AT HANGERS			
20	43 THRU 60	ANGLE REINFORCED STANDING "S" SLIP 1 1/2 X 1 1/2 X 1/4" ANGLE	BAR STIFFENED FULL "S" SLIP 1 1/4 X 3/8" BAR	BAR STIFFENED FULL "S" SLIP 1 1/4 X 3/8" BAR	3' - 0"	1 X 1/8" FLAT GALVANIZED EVERDUR BRAZED TO DUCT ON CENTERS BETWEEN JOINT & AT HANGERS	3/8" ROD	1 1/4 X 3/8" FLAT GALVANIZED BAR AT HANGER	4' - 0"

FIG. 5-A-9. Specification table for flush-seam duct construction.

formed to contact 180 degrees of the duct perimeter and bolted together at the top and bottom of the duct. Suspend the duct clamp from one ⅜-inch-diameter threaded rod.

Hang round duct of 25-inch diameter, and larger, by duct clamps as described above. Clamps shall be bolted at the horizontal sides of the duct and suspended by two ⅜-inch diameter threaded rods.

Support vertical rectangular-ducts at intervals of not more than twelve feet; and at each story by structural angles bolted to the two larger sides of the ducts, or to the entire duct perimeter, when the minimum side-dimension is 24 inches or greater.

Support vertical round-ducts of 24-inch diameter, and less, by duct clamps as previously described. Ends shall be extended and bent apart to provide a stable footing and then fastened to the floor slab or supporting steel.

Support vertical round-ducts of 25-inch diameter, and larger, by a rolled structural angle, bolted or spot welded to the duct on 6-inch centers and fastened to the floor slab or supporting steel.

Turning Vanes in Elbows

Construct turning vanes of the same material as the ducts in which they are installed.

Construct turning vanes for low- and medium-pressure systems of 20-gage galvanized steel, or the equivalent thickness for other duct materials, as shown in the specification tables.

Turning vanes shall be manufactured double-vanes or shop-fabricated turning vanes constructed to the same standards. Submit samples of shop-fabricated units for approval.

Reinforce joints to frames by welding or brazing, for turning vanes in high-pressure systems.

Elbows in kitchen exhaust systems shall not have turning vanes.

Duct Access Doors

Provide doors in ductwork, equipment housings, and connections thereto, for access to all apparatus and accessories, air filters, coils, automatic controls, automatic dampers and damper motors, fire dampers, grease fittings, supply and exhaust fans, bearings and connections, and all other areas and equipment requiring periodic inspection or service.

Unless otherwise noted, access doors in casings shall be 58 inches high by 18 inches wide, and access doors in ducts shall be 16 x 20 inches, minimum, for ducts 18 inches wide, and larger. Smaller ducts shall have doors with clear opening equal to duct width.

Construct and install access doors of the same materials as the ductwork and casings in which they are provided, and to withstand the same test pressures without deformation, vibration or leakage.

Provide doors in insulated casings and ductwork of the double insulated, reinforced panel type with a minimum of 18-gage sheet metal on both sides of a core of 6 lb density mineral-fiber rigid insulation.

U.S. GAGE TYPE MP	U.S. GAGE TYPE AT	DUCT DIMENSION, INCHES	TRANSVERSE JOINTS HORIZONTAL & VERTICAL	TRANSVERSE JOINTS SPACING, FEET	TRANSVERSE BRACING	HANGERS ROD DIA.	HANGERS SHELF ANGLE	HANGERS MAX. SPACING
24	26	UP THRU 12	1" POCKET LOCK	8	NONE	3/8"	1 1/2 X 1 1/2 X 1/8" OR 1 X 1/8" STRAP	8' - 0"
22	24	13 THRU 18						
		19 THRU 30	1" POCKET LOCK	4	NONE			
				8	1 X 1 X 1/8" ANGLE CENTERED BETWEEN JOINTS			
20	22	31 THRU 42	1" POCKET LOCK	4	NONE	3/8"	1 1/2 X 1 1/2 X 1/8"	6' - 0"
				8	1 X 1 X 1/8" ANGLE CENTERED BETWEEN JOINTS			
		43 THRU 54	1 1/2" POCKET LOCK	4	NONE	1/2"	2 X 2 X 1/8" OR EQUIVALENT CHANNEL	6' - 0"
				8	1 1/2 X 1 1/2 X 1/8" ANGLE CENTERED BETWEEN JOINTS			
18	20	55 THRU 60						
18	20	61 THRU 84	1 1/2" ANGLE REIN-FORCED POCKET LOCK	4	1 1/2 X 1 1/2 X 1/8" ANGLES CENTERED BETWEEN JOINTS	1/2"	2 X 2 X 3/16" OR EQUIVALENT CHANNEL	4' - 0"
				8	1 1/2 X 1 1/2 X 1/8" ANGLES CENTERED 2' - 0" BETWEEN JOINTS			
16	18	85 THRU 96	1 1/2" ANGLES RE-INFORCED POCKET LOCK	4	1 1/2 X 1 1/2 X 3/16" ANGLES CENTERED BETWEEN JOINTS			
			1 1/2" COMPANION ANGLES	8	1 1/2 X 1 1/2 X 3/16" ANGLES ON 2' - 0" CENTERS BE-TWEEN JOINTS			
16	18	97 THRU 120	2" ANGLES RE-INFORCED POCKET LOCK	4	2 X 2 X 1/4" ANGLES CENTERED BETWEEN JOINTS	1/2"	3 X 3 X 1/4" OR EQUIVALENT CHANNEL	4' - 0"
			2" COMPANION ANGLES	8	2 X 2 X 1/4" ANGLES ON 2' - 0" CENTERS			

FIG. 5-A-10. Specification table for medium-pressure and airtight ducts.

For access doors in ductwork which are less than 24 inches in height, provide two brass butt-hinges and two cam latches.

For access doors in casings and access doors 24 inches in height and over, provide two heavy bronze butt-hinges and two pairs of lever type latches operable from both sides of the door.

Fit doors closely. Securely attach round soft-rubber gasketing to the doors by cement and countersunk rivets, for a continuous airtight seal.

Where access doors are concealed in hung ceiling, provide indicator buttons in the ceiling immediately below the access door.

Belt Guards

Belt drives exposed to contact by personnel shall be constructed of properly supported, and easily removed $1\frac{1}{4}''$ x $1\frac{1}{4}''$ x $\frac{1}{8}''$ galvanized angle-iron frames and $\frac{3}{4}''$ No. 16 galvanized expanded-metal mesh. Round off and finish all guard edges.

Provide openings for the insertion of a tachometer in all machinery guards covering the ends of motor or equipment shafts. Elongate the motor tachometer openings to allow for adjustment of belt tension.

Double-Wall Apparatus Casings

Double-wall casing panels shall be fabricated of galvanized steel sheets formed to completely enclose $1\frac{1}{2}''$-thick rigid fireproof 6 lb density, mineral-fiber insulation cemented to the sheet metal. Weld panels at 2-inch intervals and seal seams with an approved sealing compound.

Install all portions of casings with galvanized steel cadmium-plated rivets, screws, and bolts.

Panel Joints and Finishes:

1. Exterior Corner Joints: Continuous $1\frac{1}{2}''$ x $1\frac{1}{2}''$ x $\frac{1}{8}''$ galvanized steel angle
2. At Concrete Base: Continuous shop formed 14-gage galvanized steel base channel
3. Butt Joints: 22-gage galvanized formed tees and hemmed finishing strips
4. Fastener Spacing: Rivets, screws and bolts; maximum 6-inch centers
5. Welds: Maximum 2-inch centers.

Caulk joints and openings airtight and watertight. Where pipes are required to penetrate the panels, provide 22-gage, galvanized steel sleeve; pack space between sleeve and pipe with mineral-fiber insulation. On outside of the panel, provide $\frac{1}{4}$-inch rubber gasket and 10-gage gasket ring. Seal all of the openings with an approved sealing compound.

Access doors shall be 58 inches x 18 inches, minimum, and mounted 12 inches above bottom of casing channel.

Coordination of Work

It is necessary that the mechanical and electrical contractors work in the closest of harmony to avoid conflicts between the trades as to locations of ducts, heating, plumbing, and sprinkler piping, pneumatic tube lines, and electrical fixtures and conduits. The various contractors shall coordinate their work to eliminate pyramiding down from the underside of the slab so that space above hung ceilings will be efficiently used. Clearances for recessed ceiling lights shall be determined by the electrical contractor and furnished to each of the mechanical contractors.

Shop drawings of all trades shall be coordinated prior to submission to the architects and engineers. Lead time allowance for shop-drawing approvals should be approximately two calendar weeks.

BALANCING AIR QUANTITIES

General

Air system balancing shall be performed by an organization or by personnel qualified by experience and practice to perform this service. Evidence of qualification shall be submitted with bid proposal.

Balancing shall be scheduled for approval and completion by a minimum of three calendar weeks before building completion.

All air-handling systems, components, calibrated automatic controls, and heating and cooling mediums shall be completely installed and operable prior to air-balancing procedure.

Fan-curve operating data and final air readings shall be submitted in duplicate to the architects and engineers.

Air velocity and pressure shall be measured by a pitot tube and a manometer or U-gage, and an Anemotherm air meter.

Fan and motor speeds shall be measured with a Strobo Tach or direct reading tachometer.

Fan-motor amperage and voltage shall be measured with direct connected or clamp-on instruments.

Items to Check before Balancing

Motor shall be protected against overloading by proper-size devices in electrical control systems.

Belts and drives shall be properly aligned and tensioned.

Fan impellers shall have uniform, parallel clearance at inlet cones.

All drive bearings shall be lubricated.

Volume, fire, and automatic dampers shall be properly set and operable.

High-pressure ducts should be tested for leaks before insulation is installed.

Air filters to be installed as specified.

Air-Balancing Procedure

Supply, return, and exhaust systems for a common space shall be simultaneously operating with 100 percent outside air or full recirculation.

Set fans at rated speeds for design volumes and pressures.

Check manufacturers' fan curves for volume, pressure, and horsepower.

Measure motor amperage and voltage.

Take temperature readings in return, outside, mixed, and supply-air ducts.

Measure total system airflow in supply main by a pitot tube traverse.

Plane of traverse shall be a minimum distance of eight equivalent duct diameters, downstream of a duct fitting.

For rectangular ducts pitot-tube readings shall be taken at centers of equal rectangles at maximum distances of 6 inches, and with a minimum of 16 readings.

For traverse points in round duct, see Fig. 8 in Air Moving and Conditioning Association Standard 210-67.

Total air volume for all systems must be design-volume minimum or up to 5 percent greater than design-volume maximum.

Allowable leakage for low and medium pressure systems is 3 percent, and for high pressure systems, 0 percent.

Adjust fan speed to conform with design air-volume.

If fan speed had to be increased, measure motor amperage and voltage.

In accordance with manufacturers' instructions and data, check induction and terminal units.

In conjunction with automatic temperature-control contractor, adjust all control dampers.

Adjust manual volume dampers and splitters in each sub-main for design air-volumes.

Measure and adjust each air inlet and outlet as often as necessary to obtain desired air quantities.

Test Data

Submit the following test data, arranged by system, to the architect and the engineer:

Fan:

Manufacturer and model number
cfm, design
cfm, actual
rpm
Inlet static pressure
Discharge static pressure
Fan curves showing variation of cfm with static pressure at operating rpm and hp requirements.

Motor:

Manufacturer and model number
Horsepower
Phase
Frequency
NEMA code letter
Rated volts
Actual volts
Rated amperes
Actual amperes
Locked rotor amperes

Starter:

Manufacturer and model number
Heater size
Line voltage

Ampere rating
Control voltage
Frequency

Outdoor Air Intake
Size of intake
Actual free area
Face velocity
Outdoor air temperature
Return air temperature
Mixed air temperature with averaged traverse readings

Air Outlet or Inlet
Manufacturer and model number
Nominal size
Actual free area
Face velocity
cfm, design
cfm, actual
cfm, percent difference

Induction Unit Data:
Manufacturer and model number
Nozzle pressure
cfm, rated
cfm, actual

Outlets, inlets, or terminal units shall be smoke tested for performance of air-velocity pattern, air-temperature pattern, and noise level at request of architect or engineer.

CHAPTER 6

Preparation of the Shop Drawing

The purpose of a sheet metal shop drawing is to impart an exact layout of duct fittings and components to aid in shop fabrication and field erection. The structural, architectural, and mechanical design drawings are used for research and reference in preparation of the shop drawing. When the sheet metal shop drawing is thoroughly coordinated with work of the other trades and approved by the architect and engineer, the end result is an efficient and workmanlike installation. An extensive series of shop drawings is required on many jobs such as schools, hospitals, office buildings, and industrial plants.

To Change Design Drawing to Shop-Drawing Size

While design drawings, especially on large projects, are generally drawn to ⅛-inch scale, it is preferable to use a scale of ⅜-inch or larger for shop drawings, to assure accuracy and clarity. An initial study of the job scope should be made for proper breakdown of the HVAC drawings to compensate for the scale differences. For example, if design drawings are made ⅛-inch equal to 1 foot and shop drawings are made ½-inch to 1 foot, the ratio is 1 to 4, and the HVAC drawings can be divided into parts one-quarter the size of the shop-drawing paper.

However, if space for sections is needed on the shop-drawing paper because of complicated areas, for example, divisions of the HVAC drawing should be reduced accordingly. On the other hand, a small scale, say, ¼ inch, is helpful where large ducts of relatively straight runs are shown in big areas. The HVAC drawing is then divided to that ratio, 1 to 2.

Duplication, or perhaps partial duplication of shop drawings, is often possible, especially in high-rise office building work. Structural and/or architectural backgrounds may be similar, then transparent prints (sepias) can be made to avoid having to redraw plans for each floor. Building areas with opposite-hand duct layouts may also be duplicated by reverse-print transparencies.

The shop drawing phase should follow a schedule which conforms to the construction requirements of the building project and also takes into consideration the coordination of work with other trades.

A drawing may be required for the size and location of duct openings through poured concrete slabs and walls. Conventional ducts should have openings with a minimum of 1-inch clearance, although openings for ducts with matched angle connections should allow for frame clearance. It is therefore, necessary to check duct opening sizes which may be shown on the structural drawings, for adequate clearances. In preparing shop drawings for ducts below grade, a careful check of the electrical and plumbing mechanical drawings must be made so that coordination with these respective trades will allow sufficient lead-time for the general construction schedule.

Joint and fitting lengths are affected by fabrication requirements and types of connections used for conventional low pressure, high pressure, or heavy-gage ducts. An examination of the HVAC drawings and job specifications provides the draftsman with this information.

If acoustical lining is indicated, the thickness and type should be so noted on the shop drawing. Generally, ducts are increased to allow for lining thickness, although the plans and specifications should be checked for verification of the intent of the design engineer.

Drawing Layout and Specifications

Approved HVAC equipment cuts should be available for layout dimensions and reference: Automatic damper schedules, filters, coils, fans, air-conditioning units, vibration eliminators, sound traps, re-heat equipment, terminal units, grilles, registers, diffusers, light or ceiling air-units, and also for boilers, furnaces, emergency diesel generators, cooling towers and evaporative condensers. (See Chapter 5.)

Types of materials and gage specifications should be checked carefully for boiler breechings, ducts below grade or those exposed to weather, and diesel, fume-hood, and kitchen exhaust systems. Watertight duct construction is generally used for shower rooms, dishwasher equipment, or other types of exhaust systems which contain water-saturated air. Horizontal ducts may also be pitched downward to drain connections.

Although fire dampers are usually shown on HVAC drawings, a clause is often included in the specifications for fire dampers to be installed as per local building codes. If a fire damper location is doubtful, a note should be shown prominently on the drawing to request confirmation by the architect and engineer. Each fire damper in a duct must be accessible through an adequately sized access door for inspection and fusible-link maintenance.

As a shop drawing is developed, space conditions may offer possibilities to change a duct size, shown on the HVAC drawing, to an equivalent but more economical size. For example, a 44 x 12 may be changed to 24 x 20, yet have the same capacity. It is also not improbable to re-route a duct run on the drawing for the improvement of the overall job.

Prominent notes to the attention of the architect, engineer, or contractors should be indicated on the drawing. These usually are to confirm a dimension on the shop drawing which is not shown on the structural or architectural drawings, to indicate a specific location which is inadequate for duct clearance, to indicate work to be done by others, and to identify locations where coordination of work by other trades is critical.

GENERAL NOTES FOR DRAFTING PROJECTS NOS. 1 TO 7

1. Use ½" = 1′-0″ scale for floor plan drawings and ¾" = 1′-0″ scale for mechanical equipment rooms (M.E.R.)
2. Details may be 1″ = 1′0″, or 1½" = 1′0″.
3. Reference drawings of ⅛" = 1′0″ and ¼" = 1′-0 scales are multiplied by 4 and 2 respectively, to determine *minimum* drawing-sheet sizes.
4. Anticipate whether space for sections or details should be allowed on the drawing.
5. All drawings shall have a ½-inch border, and a title box at bottom right corner.
6. Use standard abbreviations and symbols. (See Figs. 1-21 to 1-27.)
7. Layout should be planned so that the entire drawing fits properly on the sheet.
8. Begin layout, using and duplicating drawings ('S' Dwgs.) as reference.
9. Rough-in penciled lines lightly (use 3H, or 4H lead).
10. Calculate heights of floor to bottom of slab, and floor to bottom of beams then indicate these on the drawing.
11. Draw partitions, walls, etc., from architectural ('A' Dwgs.).
12. Check reflected ceiling (RC, or ARC Dwgs.)
13. Indicate whether or not spaces have hanging ceilings, also their type and height.
14. If ducts are above a hanging ceiling, calculate available space required for ducts and clearances.
15. Start duct layout from design drawings (HVAC).
16. Check plumbing, and electrical drawings ('P' and 'E') for coordination of work to be installed by those respective contractors.
17. Show duct plus dimensions for Finished Floor to Bottom of Duct (FIN FL to BTM of duct).

18. Show minus dimensions for Bottom of Slab to Top of Duct (BTM of SL to TOD).
19. 'Tie in' ducts by indicating dimensions to centerline of nearest column.
20. Indicate drawing number when duct is continued on another drawing.
21. Tally lineal-duct dimensions and check with tie ins.
22. Locate and show: fire dampers, access doors, splitters, volume dampers, boots, and duct lining.
23. Number all duct pieces.

Drafting Project No. 1

This will be a shop drawing of an "in-between" portion of an air-conditioned supply duct in a new building under construction. The basement mechanical-equipment room shop-drawing 10 (MER-10, imaginary) ended the slip connection west of COL A-3, and Shop DWG 4 (imaginary) ended the large end as shown. (See Fig. 6-1, part 1.)

This supply system is a conventional low-pressure duct constructed of standard gage galvanized iron — as per the ASHRAE Guide and Data Book. Use 1½-inch slip connections. The duct will be covered with 1-inch fiberglas insulation. The basement finished floor elevation = 108'-4". The finished ground floor elevation = 120'-2". The top of steel of ground-floor framing is at elevation 119'-11". Ground-floor slab is 8-inch reinforced concrete. All structural steel shall have a minimum of 2-inch concrete fireproofing. Draw a plan and connect the 20" x 30" to the 30" x 20". Draw section as shown by 'A-A'.

Drafting Project No. 2

This will be in the same building as No. 1, but at the 2nd floor level. Figure 6-1, part 2 indicates a 24" x 18" air-conditioned supply duct from the 2nd floor, M.E.R. DWG 28. Because of a co-ordination problem with piping, the duct was ended west of Col. 3. A 3-inch steam riser is located in the column recess and elbows south at the ceiling, turns east below the 18-inch beam, then to an elbow and up through a pipe sleeve, to the 3rd floor.

TYPICAL COMPUTATION PROCEDURE

Figure 6-1-1 is used for the following examples:

1. *Floor to floor*
   - ELEV of floor above   120'- 2"
   - ELEV of floor   108'- 4"
   - Difference   11'-10" = FIN FL to FIN FL

2. *Floor to bottom of slab*
   - Floor to floor   11'-10"
   - Slab thickness   8"
   - Difference   11'- 2" = FIN FL to BOS

3. *Bottom of beam*
   - Top of steel below FIN FL   3"
   - 21 WF 73 = 8¼ × 21¼   21¼" deep
   - Fireproofing   2"
   - Sum   26¼" = 2'-2¼"
   - Floor to floor   11'-10"
   -    2'- 2¼"
   - Difference   9'- 7¾" = BTM of BM ABV FIN FL
   - Floor to bottom of slab   11'- 2"
   -    9'- 7¾"
   - Difference   1'- 6¼" BTM of slab to BTM of BM

PREPARATION OF THE SHOP DRAWING

4. *Top of duct to bottom of slab*
   For 20 x 30 Duct:
BTM of duct to FIN FL	7'- 8½"
Depth of duct to FIN FL	2'- 6"
Sum	10'- 2½"
FIN FL to BTM of slab	11'- 2"
	10'- 2½"
Difference	0'-11½" = TOD to BOS

   For 30 x 20 Duct:
BTM of duct to FIN FL	7'-11"
Depth of duct	1'- 8"
Sum	9'- 7"
FIN FL to BTM of slab	11'- 2"
	9'- 7"
Difference	1'- 7" = TOD to BOS

5. Requirements for lineal feet of duct

   ```
           14'- 6"
          +4'- 3"
           18'- 9"
              -3"
           18'- 6" Total
   (94" joints = 15'- 8")
           18'- 6"
          -15'- 8"
            2'-10" = Length for transformation fitting
   ```

6. *Fitting notes*

   (A) Differences in elevation (in direction of airflow)

   For bottom:
   ```
       7'-11"
      -7'- 8½"
       0'- 2½" = (BU 2½")
   ```

   For top:
   ```
       1'- 7"
        -11½"
       0'- 7½" = (TD 7½")
   ```

   (B) Offset for side of duct
   ```
       1'-11"
      -1'- 6"
       0'- 5"
   ```
   Difference of duct widths = 30" − 20" = 10"
   Fitting is equal break (EB).

GENERAL AND PROCEDURAL NOTES

1. Notes and construction details are similar to those for DWG 1.
2. Connect the ducts in rooms 2-5 and 2-6.
3. Hold cross-section area of duct in any offsets or transitions.
4. The ceiling diffuser collar is to be made OD to fit into the diffuser neck.

(Text continued on page 270.)

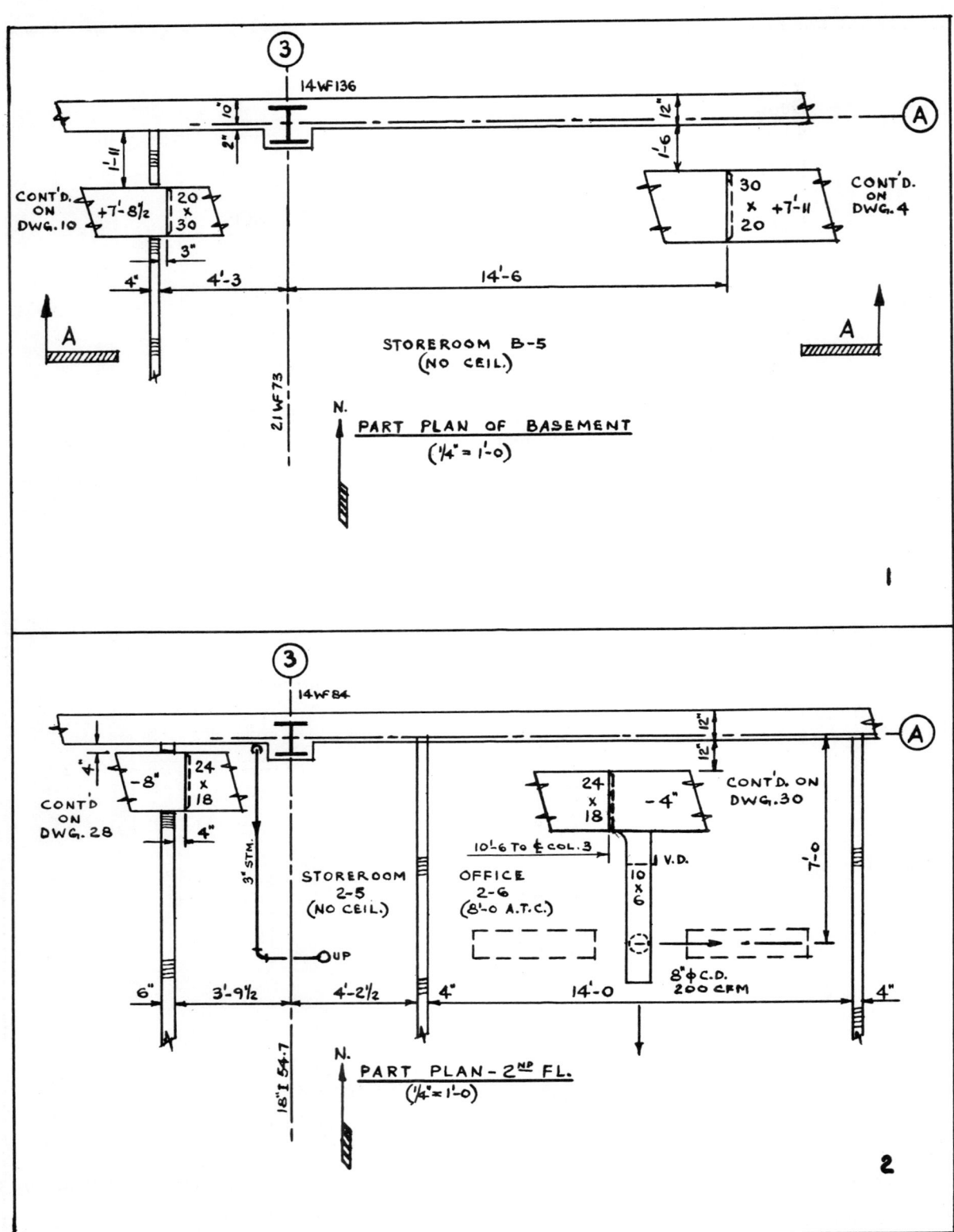

Fig. 6-1. Part plans for drafting Projects 1 and 2.

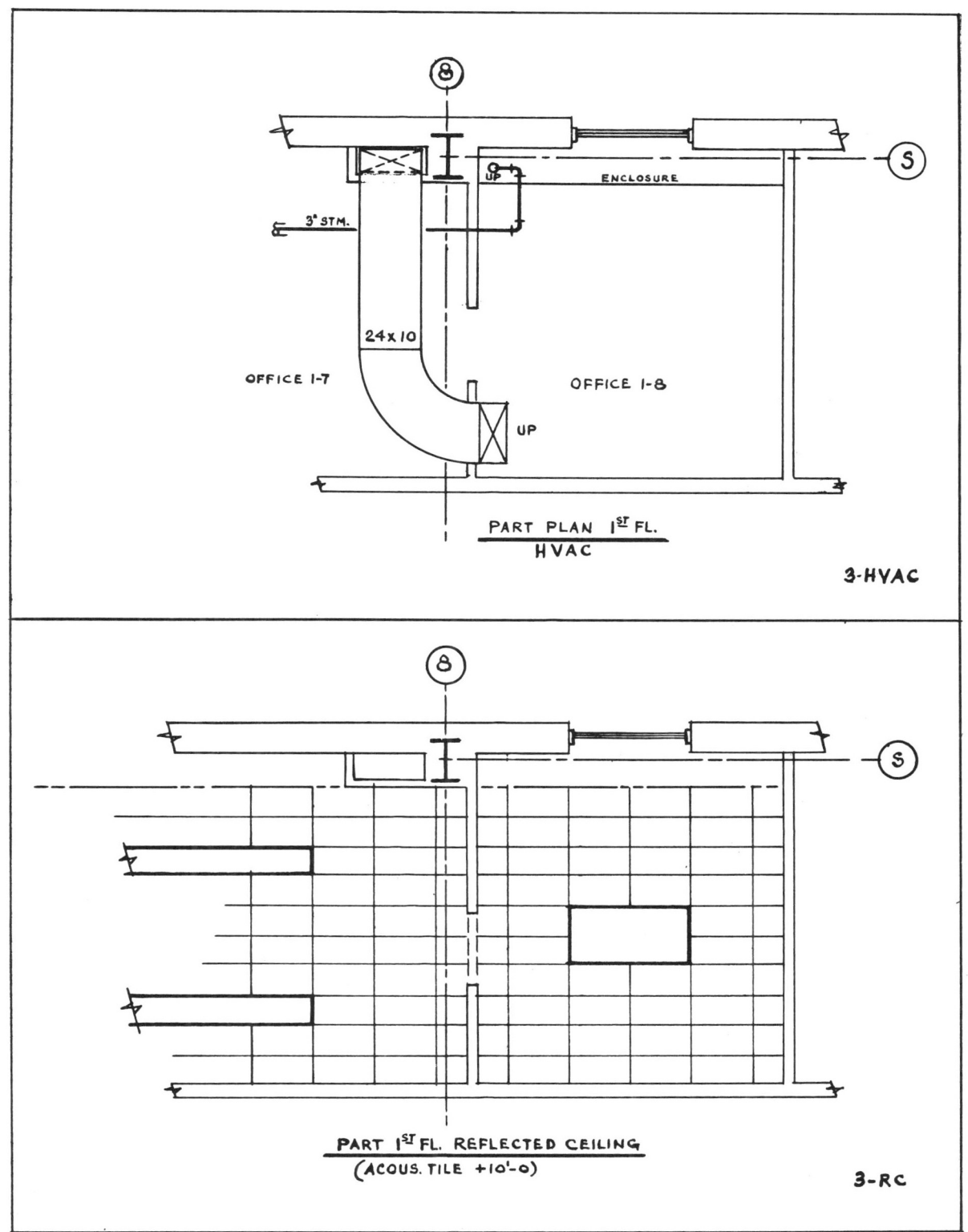

Fig. 6-2. Part plans — HVAC and reflected ceiling — Project 3.

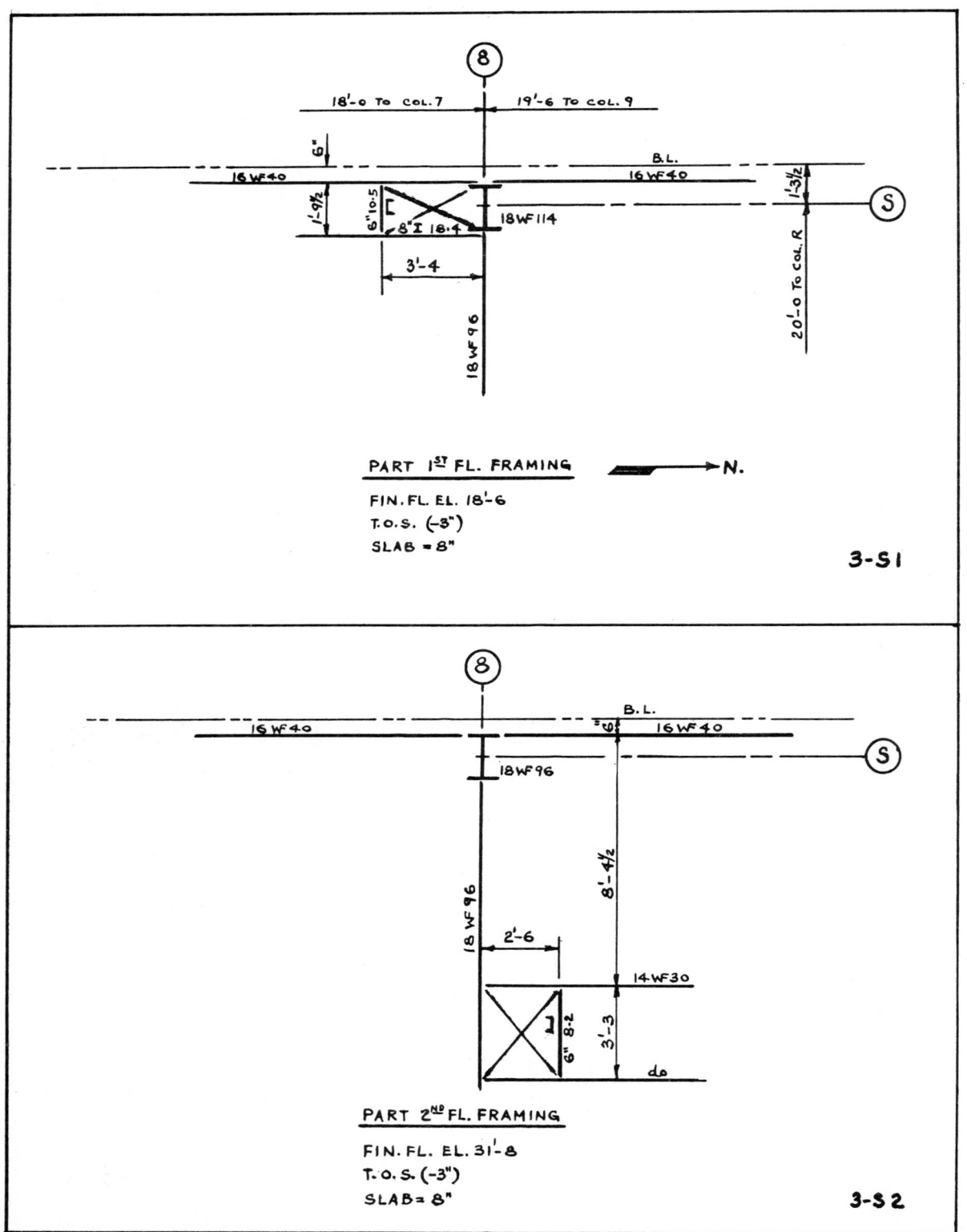

Fig. 6-3. Part plans — Structural — Project 3.

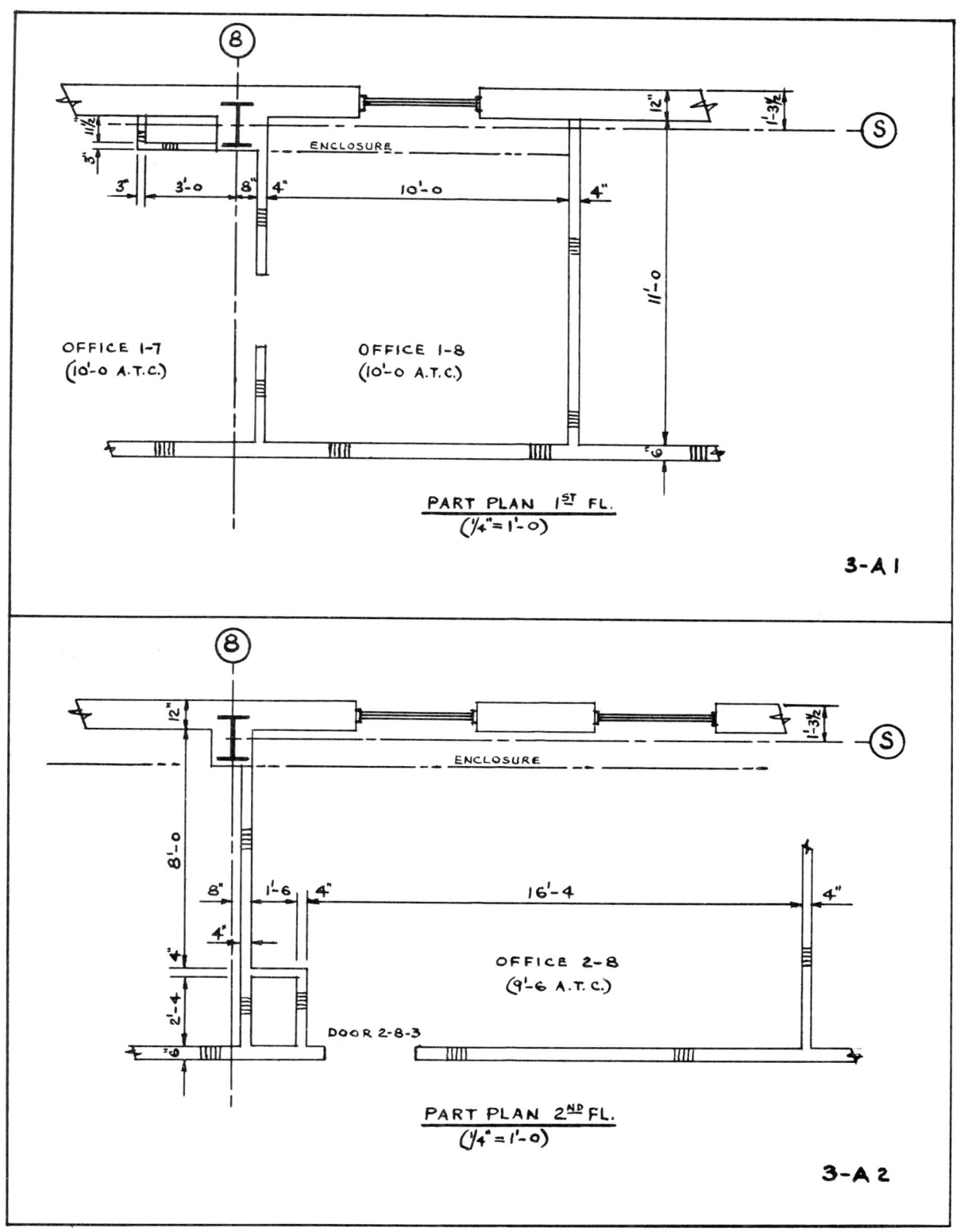

Fig. 6-4. Parts plans — Architectural — Project 3.

(Text continued from page 265.)

5. Calculate boot width x 6 inches high for 10" x 6" branch tap, and indicate on drawing.
6. Draw duct sections as required.
7. Allow 1 inch for insulation on steam pipe.
8. Allow 3½ inches for ceiling construction.
9. Ceiling tiles are 12" x 24" lay-in type.
10. Finished 2nd FL is at EL 143'-2"
11. Finished 3rd FL is at EL 154'-2"
12. TOS for 3rd FL framing is at EL 153'-11"

Drafting Project No. 3

See Figs. 6-2, 6-3, and 6-4 for Reference Plates 3-HVAC, 3-RC, 3-S1, 3-S2, 3-A1, and 3-A2. These plates are shown exactly as portions of an actual set of project prints.

Plate 3-HVAC (Heating, Ventilating, and Air-conditioning drawing)

Basement shop DWG 6 (imaginary) has ended slip-end of 24" x 10" supply duct at EL 18'-10½". End the slip-end of 24" x 10" through the 2nd FL at EL 32'-2½" and indicate that it will be continued on Shop DWG 16 (imaginary). Allow 1½-inch thick insulation for 3 inch steam piping. Coordinate the duct layout with pipe.

Plate 3-RC (part plan of ceiling layout)

Ceiling is acoustical pan type. Allow 3½ inches for ceiling construction. Venetian blind pocket is in line with duct closure at COL S-8.

Plate 3-S1, and 3-S2 (structural)

North orientation arrow is reference for all Project 3 plates; steel shall have a minimum of 2-inch concrete fireproofing. TOS is 3 inches below FIN FL.

Plate 3-A1, and 3-A2 (architectural)

FIN 1st FL is at EL 18'-6". FIN 2nd FL is at 31'-8". No electrical or plumbing coordination is required.

Drafting Project No. 4

See Figs. 6-5 and 6-6 for ⅛" = 1'-0 scale drawings. Orientation of the building is shown on 4-A and 4-S.

Plate 4-HVAC

Supply duct from riser is shown as a positive nest (calculate size required.) Note that the riser decreases as it goes down, which indicates that the fan room is above. (When actually working from a set of design drawings, the draftsman would check the air-riser diagram for his own verification.) The 15-inch DIA diffuser has a 90° baffle to prevent a downdraft at the riser shaft. Diffuser collars shall be OD to fit into the necks of the diffusers.

The return-air register collar shall have a "boot-tap" into the riser. (Calculate boot size.) The kitchen-exhaust riser is 10-ga black iron with matched angle connections. The supply and exhaust ducts are conventional low pressure of standard gage galvanized steel and 1½ inch slip connections.

Although not shown, fire-stops are required at riser shaft penetrations. Therefore, a 30" x 24" fusible-link-register shall be used for return air, and a fire damper and access door shall be installed in the 30" x 8" supply duct.

Show piping for coordination purposes. Draw a plan and two sections at riser shaft — one looking north and one looking east.

Plate 4-RC

Allow 3½ inches for ceiling construction. Note that the office ceiling has equal borders at north and south walls, and equal borders at east and west walls. Indicate ceiling openings required for ceiling diffusers.

Plate 4-A

FIN. 7th FL EL = 97.2083'
FIN. 8th FL EL = 109.4583'

Plate 4-S

All columns are 14 WF 78
Top of steel 7th FL framing EL = 96.875'
Top of steel 8th FL framing EL = 109.125'
5-inch concrete slab top — 3 inches below FIN FL
3-inch concrete topping fill for finish
All steel shall have a minimum of 2-inch con-

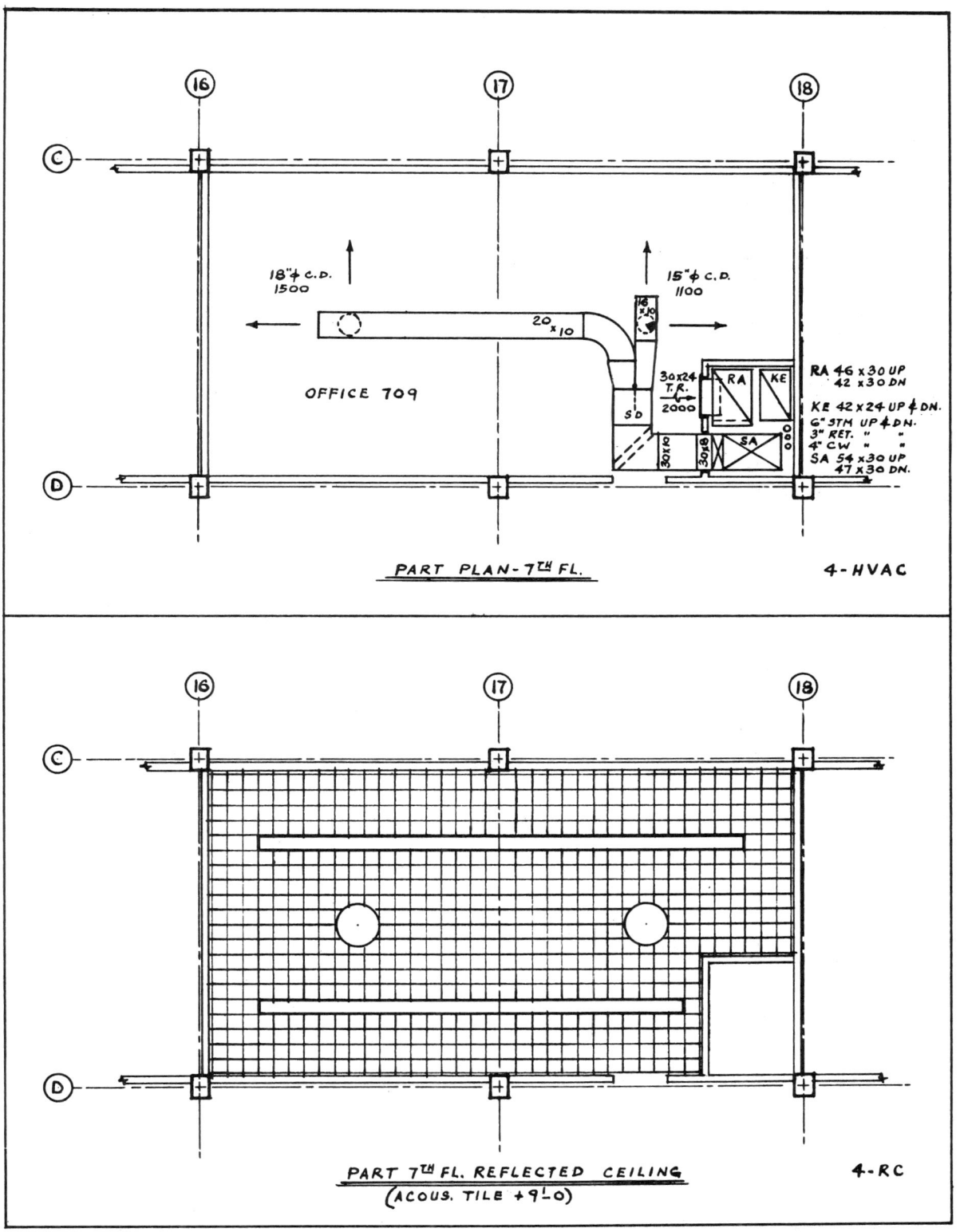

Fig. 6-5. Part plans — HVAC and reflected ceiling — Project 4.

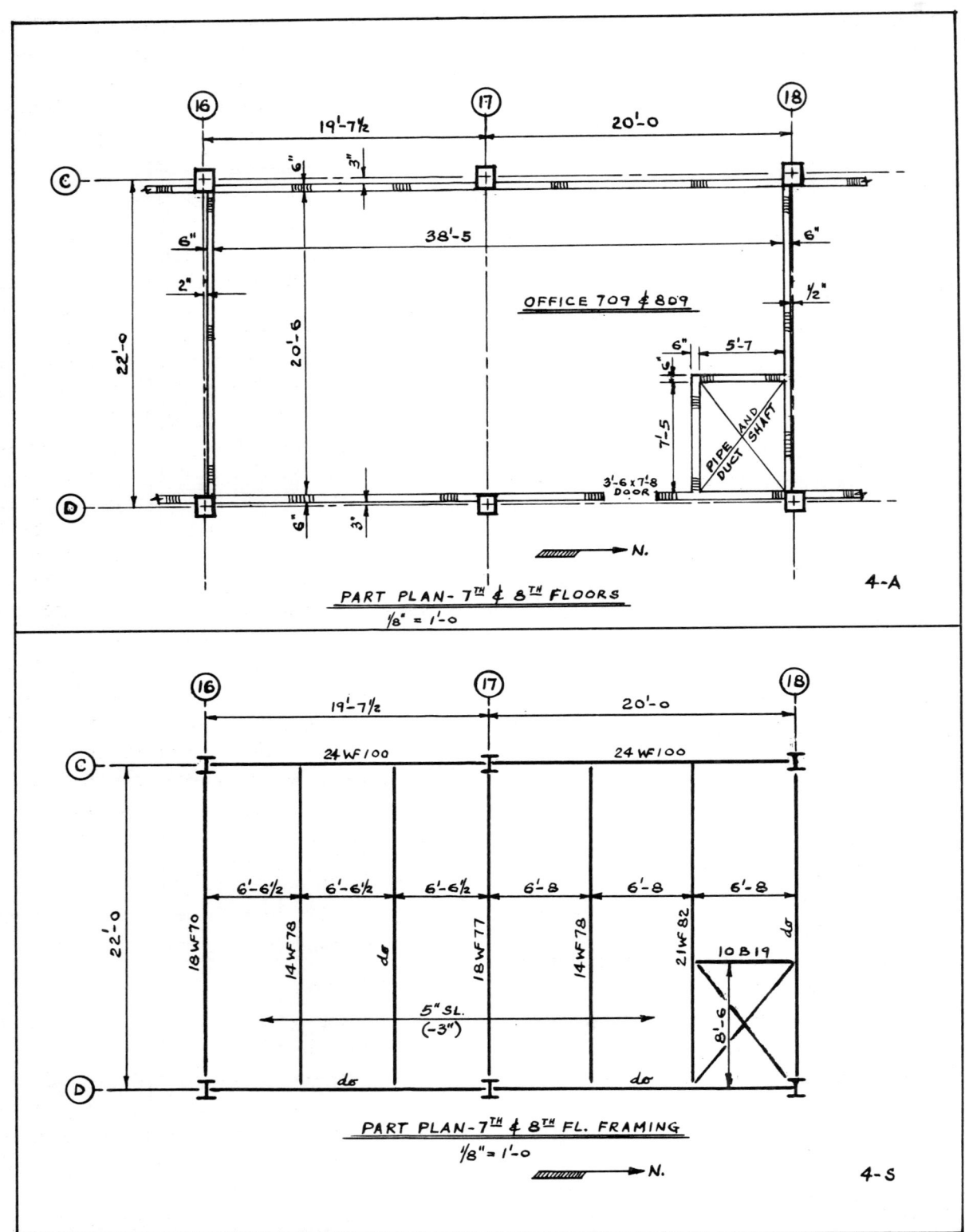

Fig. 6-6. Part plans — Architectural and Structural — Project 4.

crete fireproofing. No electrical or plumbing coordination is required.

Drafting Project No. 5

This project will consist of two drawings. The first will be a floor plan of all ducts for the area and be drawn to a scale of $\frac{1}{2}'' = 1'\text{-}0''$. The second drawing will be for the equipment room plan, details, and for shop fabrication and field erection notes.

The general theme of this project is for an air-heated, air-conditioned, one story in height, and ventilated office section of a modern industrial plant, and having a concrete floor slab on grade.

Plate 5-HVAC (See Fig. 6-7.)

1. The heating and air-conditioning is accomplished by a two-zone overhead duct system; one for the perimeter rooms and one for the interior.
2. Return air shall be through louvers in the lower section of perimeter doors unless otherwise noted.
3. A central return register (60" x 36") is located in the corridor.
4. The fresh-air intake louver shall be of 18-oz. cold-rolled copper with ½-inch square copper mesh behind louver blades.
5. Thermostatically controlled motorized dampers are in the outside- and return-air ducts, to properly proportion the air quantities to the air intake of the AC unit.
6. The air-conditioning unit is a 15-ton nominal capacity water-cooled condenser type and will be mounted on a vibration eliminator mat, directly on the floor.
7. Flexible duct connectors shall be provided for unit intake and discharge.
8. Duct furnaces shall be gas fired and vented through the roof as indicated. Two heating unit sizes will be used: Exterior zone #225, and Interior zone #125.
9. A by-pass duct arrangement will be required so that low-temperature air will not flow through the duct furnaces during the cooling season.
10. Return air inlets: Horizontal 45° face blades and register. Return air outlets: Grille only. Supply air outlets. Double deflection registers: Supply diffusers — Type 'P-A' with dampers and grids.
11. *Prepare a schedule such as:*

Inlet or Outlet Schedule

Room	Type	Size	cfm	Collar No.	Mounting Ht.	Remarks

12. Splitters shall be indicated at each neck "take-off." Minimum length of splitter to be 'N,' and they shall be controlled manually by quadrants mounted on top or bottom of ducts — whichever is more accessible for balancing air-quantities.
13. Two exhaust fans shall be installed on roof curbs. Top of curbs will be at EL 28.625". Toilets 9 and 10 will be exhausted through masonry wall openings provided by the general contractor, but installed by the HVAC contractor. Conference Room 7 will be exhausted as indicated. Both fans are $\frac{1}{12}$ hp, 600 cfm at ¼" S.P.
14. All ducts through non-air-conditioned space shall be insulated with 1" fiberglas.
15. Sheet metal work shall be of galvanized iron, as per ASHRAE standards.
16. Exterior zone thermostat is in room 14. Interior zone thermostat is in room 13.

Plate 5-S (See Fig. 6-8.)

1. All columns are 8 WF 31.
2. All steel will have 2" concrete fireproofing
3. Roof slab is 5"-thick reinforced concrete

Plate 5-A (See Fig. 6-9.)

1. Exterior walls are 8-inch glazed tile, finished on both sides.
2. All interior partitions will be cinder-concrete block, plastered both sides, with the exception of the equipment-room and manufacturing-area sides, which will be unplastered.
3. Entrance doors into Corridor 1 will be all glass; — two double swing doors, each 2'-6".

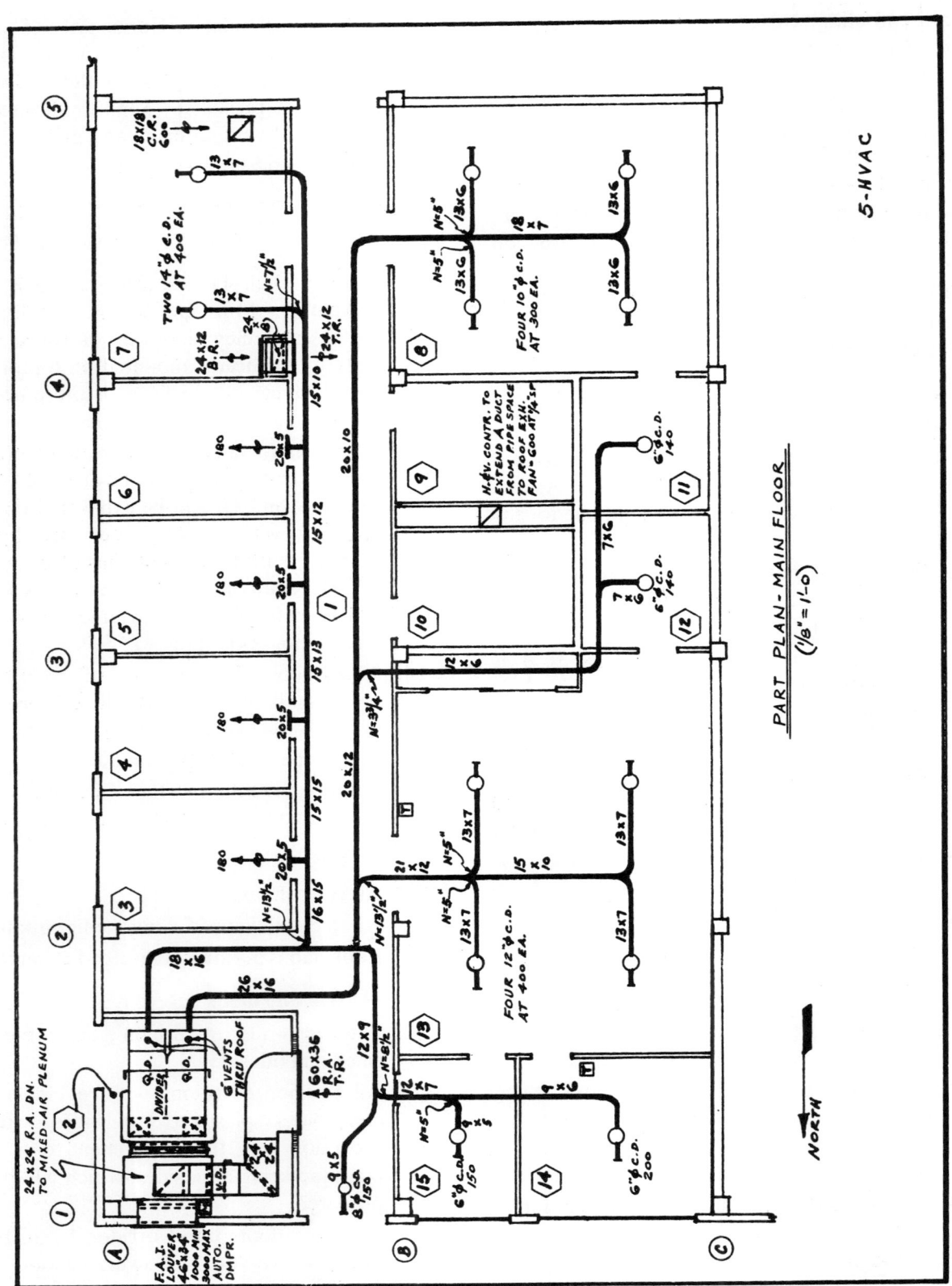

Fig. 6-7. Part plan — HVAC — Project 5.

Ch. 6 PREPARATION OF THE SHOP DRAWING 275

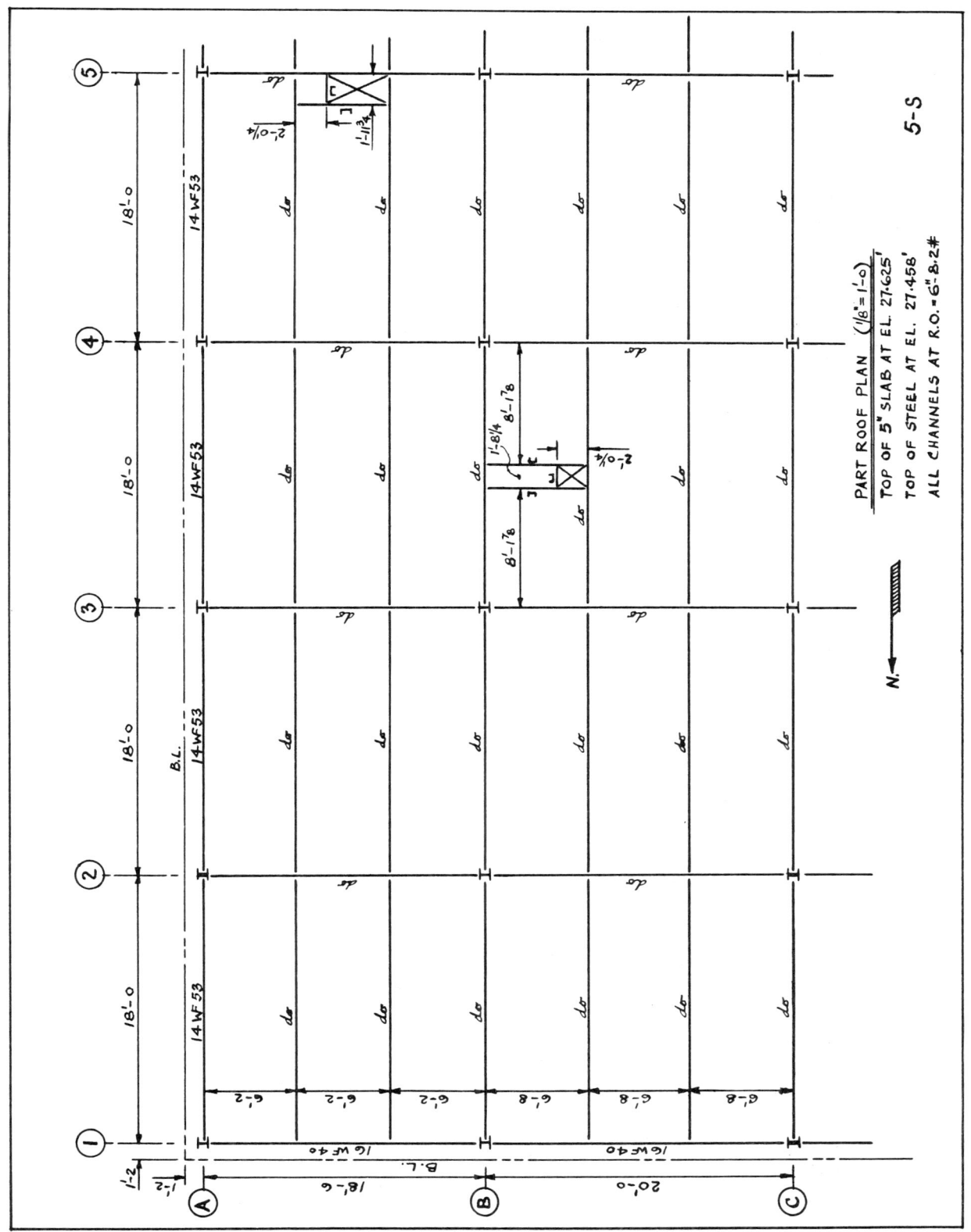

Fig. 6-8. Part plan — Structural — Project 5.

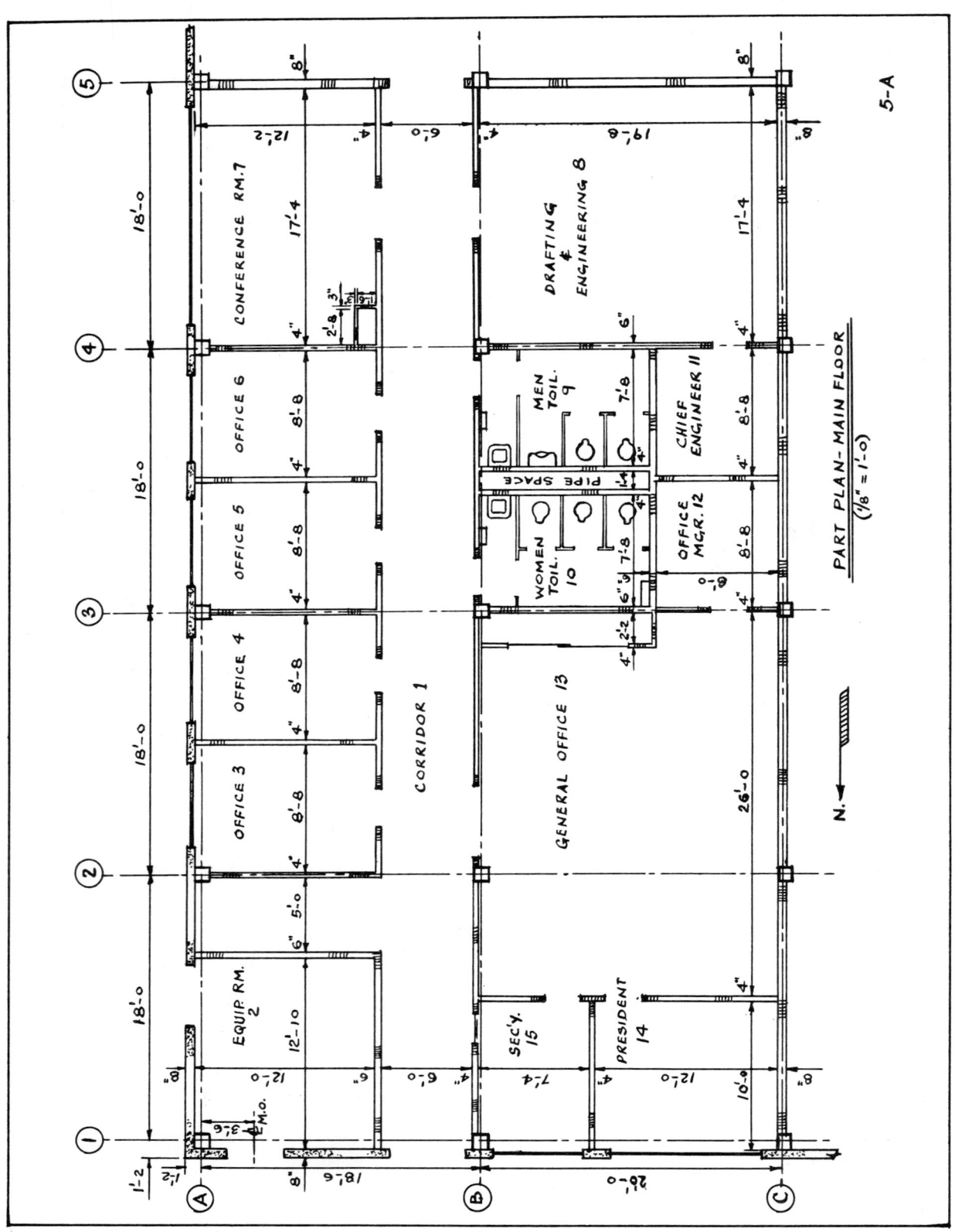

Fig. 6-9. Part plan — Architectural — Project 5.

4. Conference Room 7 door shall be 3'-6" wide, unlouvered, flush metal.
5. Room 8 door shall be 3'-6" wide, louvered, flush metal, inswing, left-hinged.
6. Two double swing 2'-6" doors, louvered, glass-top flush-metal doors will be for Room 13.
7. Equipment Room 2 to have 4'-0" steel panels, outswing, left-hinged.
8. All other doors to be 2'-6" wide, louvered, flush metal, inswing, left-hinged.
9. All doors 6'-8" high except Room 2, which will be 7'-6" high.
10. All windows will be 4'-0" high, fixed, tinted glass. Top of sills at EL 18.7017'.
11. Fresh-air-intake louver opening will be 3'-10½" wide by 2'-10½" high. Bottom of opening shall be at EL 16.125'.
12. All suspended ceilings shall be metal-pan acoustical type of 12-inch-square pan.
13. Allow 3¼ inches for ceiling construction.
14. Finished ceiling heights are as follows:
    Rooms 8 and 13 = 9'-6"
    Rooms 3, 4, 5, 6, 7, 14, 15 = 9'-0"
    Corridor 1 and Rooms 11, 12 = 8'-0"
    No ceiling in Rooms 2, 9, 10 — closet in Room 13
15. FIN FL is at EL 15.125'.
16. Roof will be of 5-inch reinforced concrete with 3-inch rigid insulation, 5-ply, 20-year-bonded roofing and marble chips.
17. Top of roof slab EL is 27.625'.

Plate 5-P (See Fig. 6-10.)

1. A wall-mounted gas meter is 24" x 24" x 9" deep, BTM 4'-6" AFF. In plan view the centerline of the meter is 7'-6" from centerline of COL, line A.
2. The hot-water tank is 24 inches in diameter and 72 inches high; bottom of tank 12 inches AFF.
3. Indicate coordination notes on shop drawing for plumbing contractor to check with his piping layout.
4. Water pipes are insulated with 1" fiberglas.

Plate 5-EL (See Fig. 6-11.)

1. 24-inch x 24-inch lights will project 6-inches into ceilings. Allow 8-inches minimum, from finished ceiling line to bottom of duct.
2. Fixtures will be standard lengths with multiples of 4'-0", except in Room 8 which will have special, long 15'-0" lighting.
3. All fixtures will be 12-inches wide and be fully recessed into the ceiling. Allow 9-inches from finished ceiling line for clearance.
4. Ceiling contractor will lay out the ceiling pattern for coordination of lighting strip as indicated.

Plate 5-C (See Fig. 6-12.)

1. The return-air opening in the back of the air-conditioning unit has a flange, turned in square on (4) sides. The shop drawing shall indicate the type of connection required on the sheet-metal collar.
2. Unit will be mounted on 1-inch-thick vibration eliminator mat.
3. Both views of duct furnaces are elevations.
4. Indicate maximum outside dimensions (including flashing) for roof fan curbs.

Drafting Project No. 6.

Two hot-water heating boilers of a large industrial plant require a smoke breeching as shown on Plate 6-HVAC (Fig. 6-13). The boiler room piping on the ¼" = 1'-0" scale drawing is at high level only. Oil piping for the oil-fired boilers is not shown and is at the north boiler end. Two induced-draft fans discharge the flue gases into a common 28" x 28" breeching, and thence into a low level chimney. The breeching shall be 10-gA black iron, with 2" x 2" x ³⁄₁₆" matched angle connections with welded corners and ⅜-inch bolts on 6-inch centers, minimum. Hangers will be ½" DIA rod suspended from inserts in slab above and

(*Text continued on page 282.*)

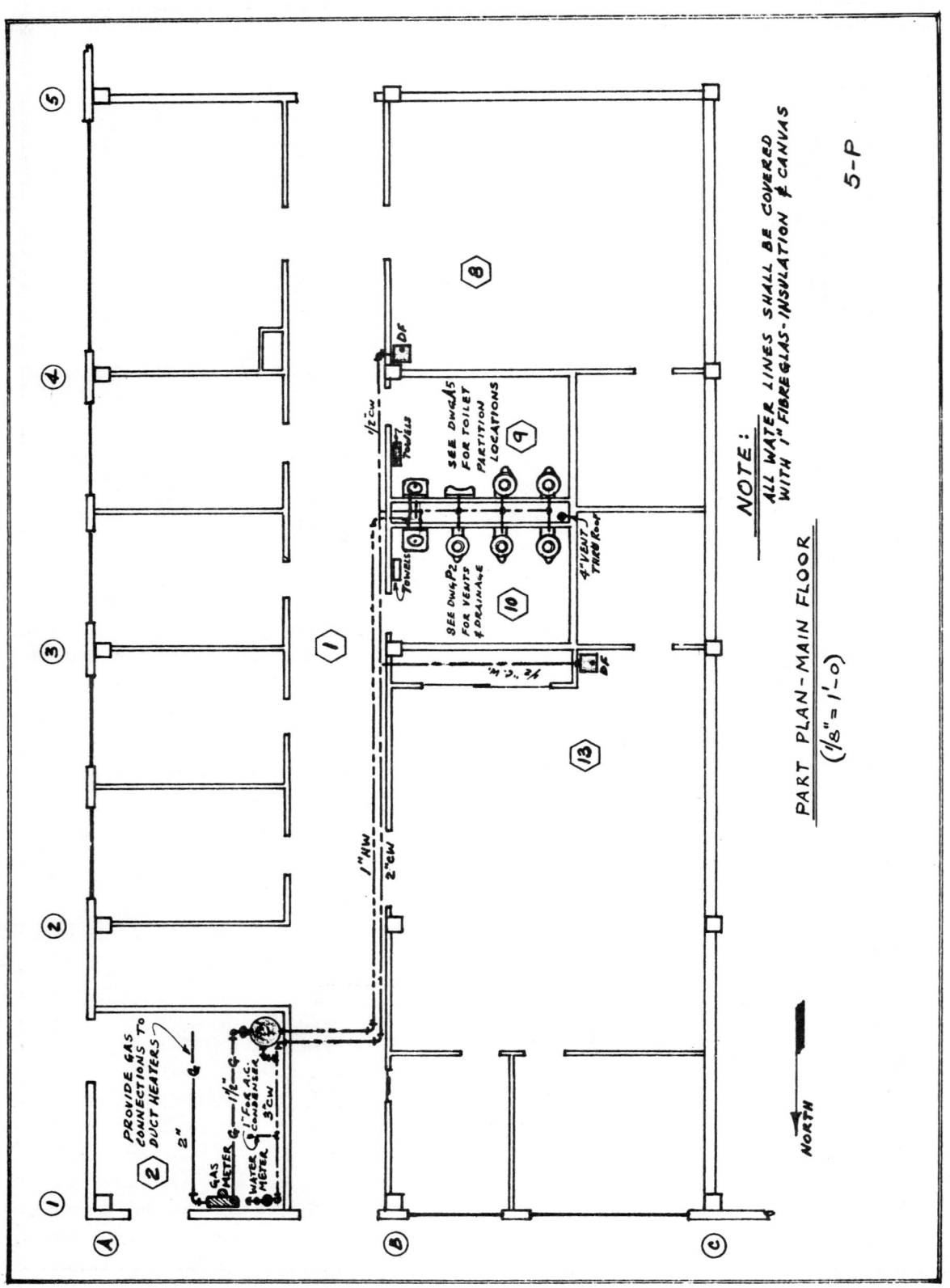

Fig. 6-10. Part plan — Plumbing — Project 5.

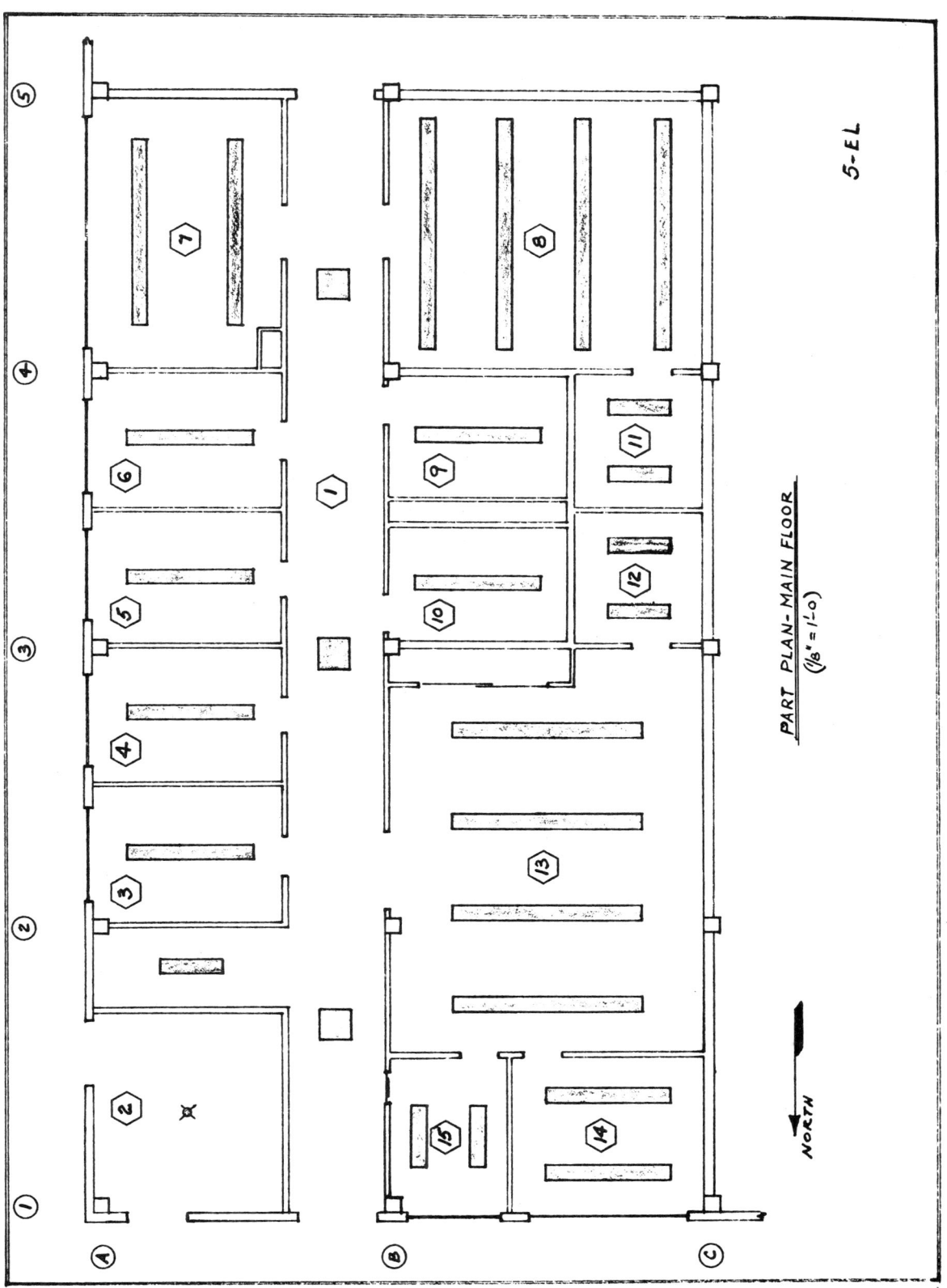

Fig. 6-11. Part plan — Electrical — Project 5.

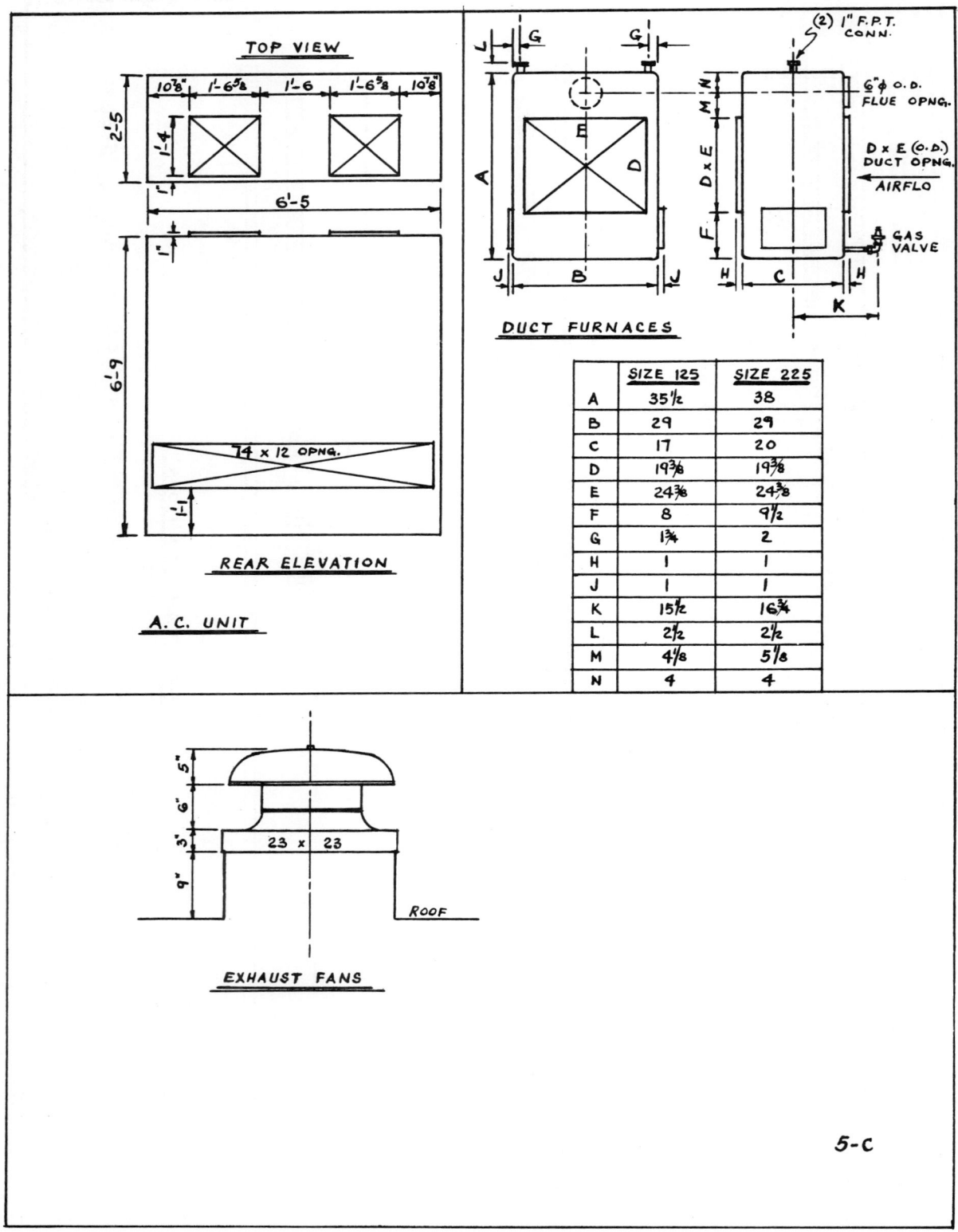

Fig. 6-12. Equipment cuts — Project 5.

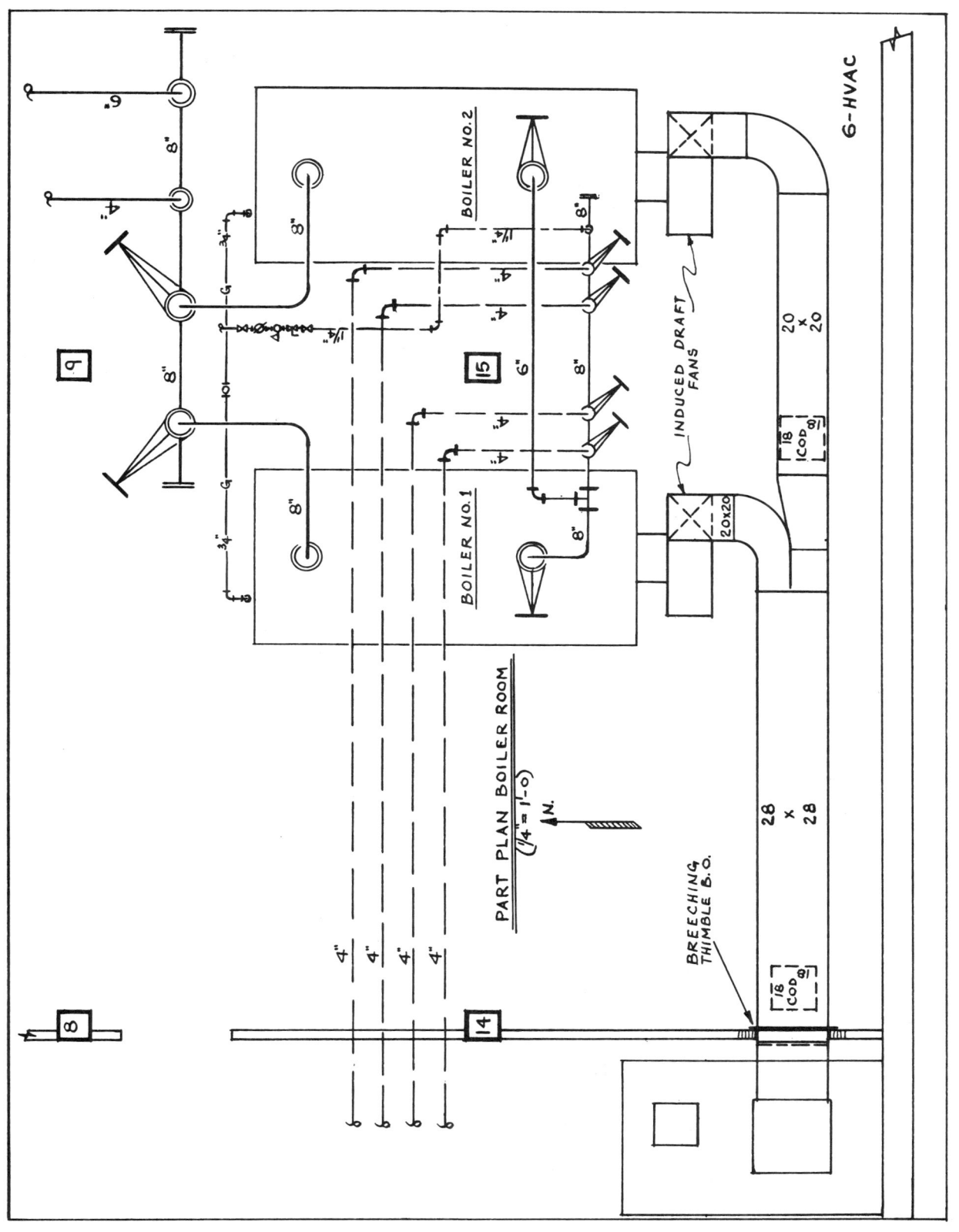

Fig. 6-13. Part plan — HVAC — Project 6.

(Text continued from page 277.)

4-inch channels as a cradle support. Hangers shall be 6'-0" on center. Allow 2½-inches for exterior breeching insulation. Hanger rods must clear insulation by 1-inch.

Clean-out doors shall be hinged and made gas-tight by cam-lock handles on three sides. Frames of doors shall clear insulation and finish flush. No expansion joint is required. A thimble at the chimney will be furnished and installed by the general contractor. A slide fit into the thimble for a minimum of 8-inches will be caulked on four sides with asbestos rope by HVAC contractor.

The induced-draft-fan discharge connections are at EL 12'-6½". Connections are 18" x 20" with 2" x 2" x ½" angles. The 20-inch dimension is east-west. The throat of the fan scroll is 6-inches east of boiler centerline. North side of the fan connections are 6'-0" south of COL 15 centerline. Boilers are equidistant from COL 15, and 13'-0" center to center. Use ACN connection to breeching, 8-inches long, minimum.

Structural Plates 6S-B and 6S-R (See Fig. 6-14) are self explanatory.

Architectural Plate 6A-B (See Fig. 6-15) locates breeching spool as shown, and plumbing Plate 6-P (Fig. 6-15), indicates piping at high level only. Coordination will be required with the plumbing contractor. Show piping on your drawing. Indicate notes as necessary.

Prepare a slab insert layout after the breeching drawing is completed.

Drawing Project No. 7

Five shop drawings will be prepared for this project — two for the auditorium ceiling ducts, one for Fan Room No. 1 and the north portion of the projection room, one for Fan Room No. 2 and the south portion of the projection room with details and sections on each, and one for casing details and equipment pads. This project is for the heating and ventilation of a new, high school auditorium. Supply and return systems are low-pressure galvanized construction. The entire supply duct will be internally lined with 1-inch duct board for acoustic and thermal treatment. Dimensions on drawings are net clear-duct sizes.

Plates 7-HV-A and 7-HV-N (Figs. 6-16 and 6-17.)

Ceiling diffusers are 18-inch DIA with distributing grids, and 31-inch DIA ceiling openings are required. Duct collars are OD — Outside Dimension (See Fig. 6-23). Return-air bottom registers 28" x 12", are horizontal face-bar, key operated opposed-blade registers. 14" x 4" return-air openings at lighting troffers shall have manual volume-dampers. All supply and return branches are positive nests. The sizes have been omitted so that calculations will be done during preparation of the shop drawings. Splitters with top-mounted, locking quadrant controls shall be at each nest connection. The 40" x 12" return-air duct on Plates 7-HV-A and 7-HV-N, penetrates the projection room floor-slab and connects into a 45" x 13" masonry shaft below the auditorium floor slab.

This shaft runs east to west and (40) 10-inch DIA mushroom type, floor inlets at 94 cfm each, are used. This is for information only, as no shop drawing is required for work below the projection room.

On Plate 7-HV-N, air from fan RE-4 tees to a 40" x 60" RA and a 60" x 60" spill-air duct. The louver is by HVAC contractor. Turn in a 1½-inch flange on four sides, for duct collar connection to louver frame. Automatic dampers will be opposed-blade type with motors in the duct. Exhaust fan E-5 is mounted to a 12-inch-square collar which increases to 18" x 12" at the louver. The fan operates at 250 cfm. Coordinate the 2-inch steam pipe with ducts.

Plate 7-HV-S (See Fig. 6-18.)

A minimum-maximum outside-air intake damper is located directly behind the louver; damper to be sized for 25 percent minimum, outside air. The return-air duct and outside-air duct enter the apparatus casing for a mixing affect of the air-stream. This casing is of double-wall construction with 1½-inch-thick fiberglas insulation between an inner wall of 18-ga galvanized, and an outer wall of 20-ga galvanized. Panels shall be made completely airtight at joints with sealing compound and metal T-bars. Access doors are 20" x 60". The casing shall be mounted on a 6-inch-high concrete

(Text continued on page 288.)

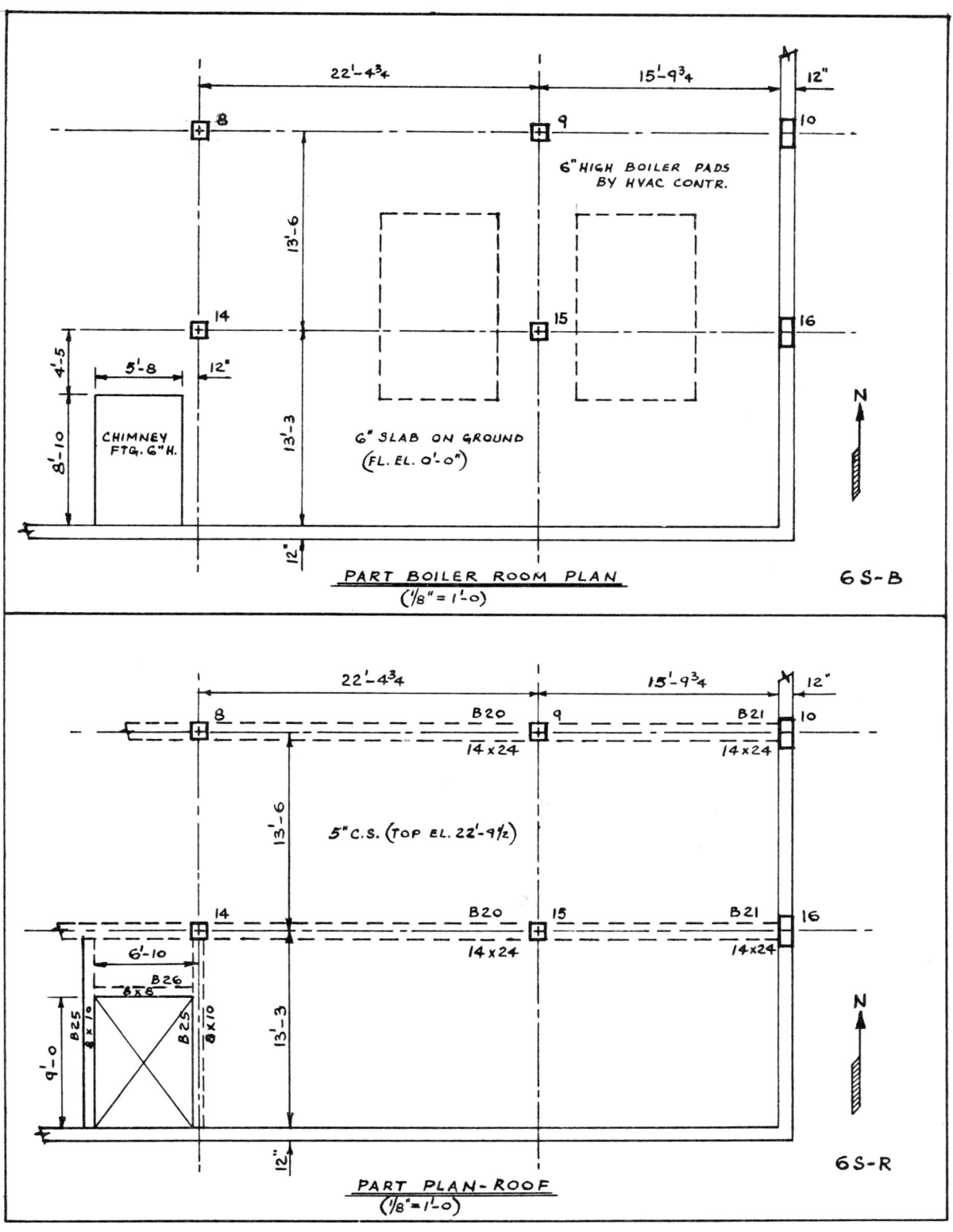

Fig. 6-14. Part plans — Structural — Project 6.

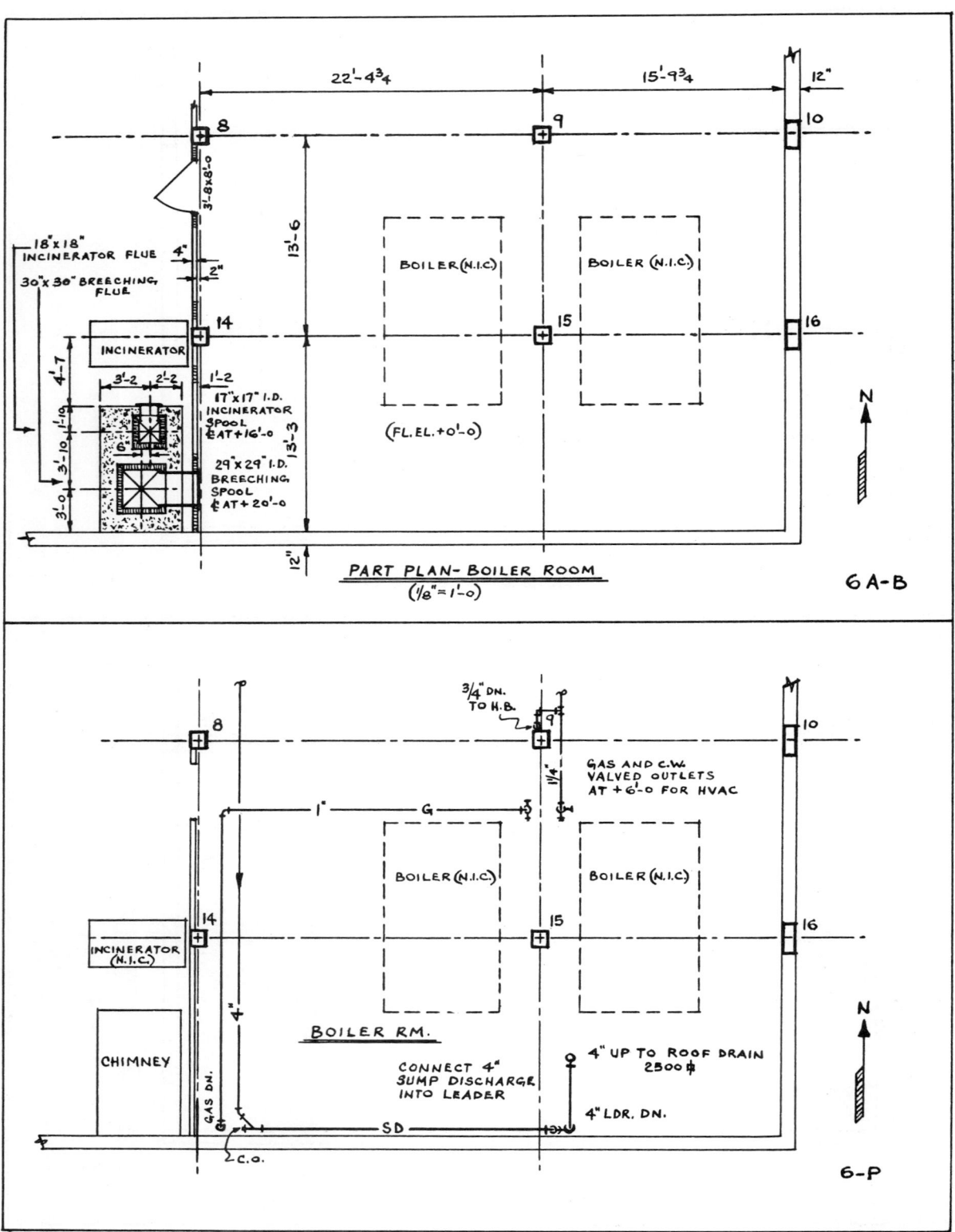

Fig. 6-15. Part plans — Architectural and Plumbing — Project 6.

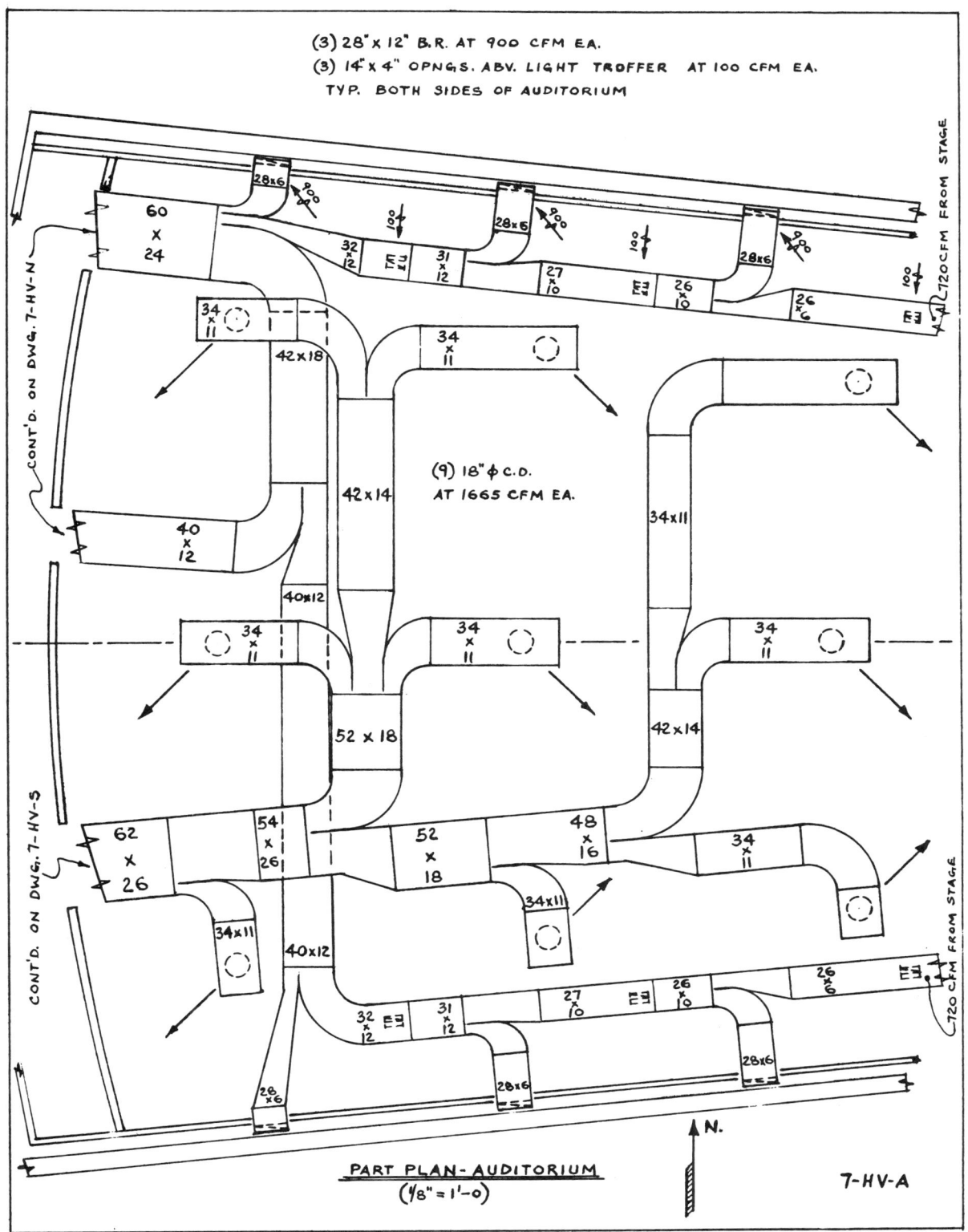

Fig. 6-16. Mechanical — Project 7.

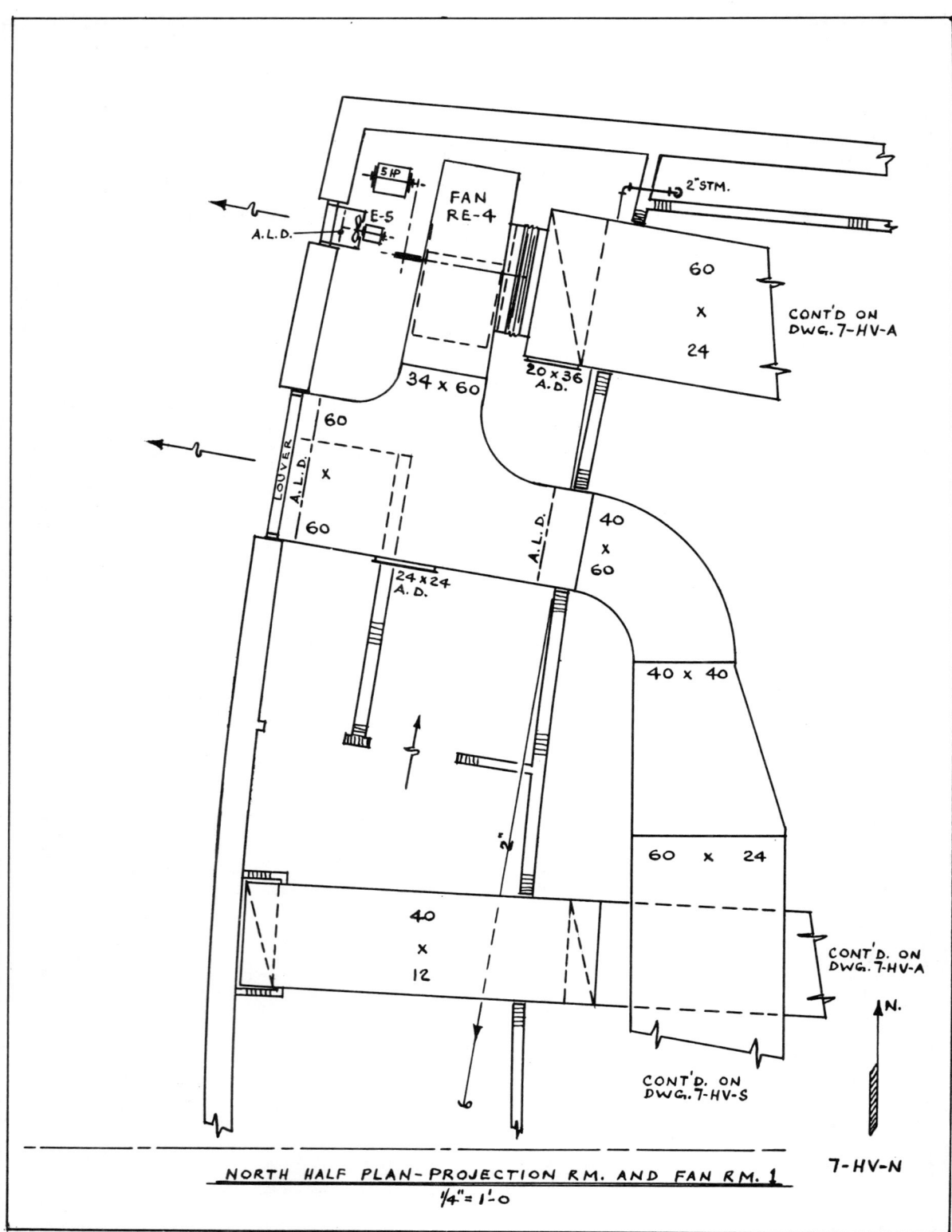

Fig. 6-17. Mechanical — Project 7.

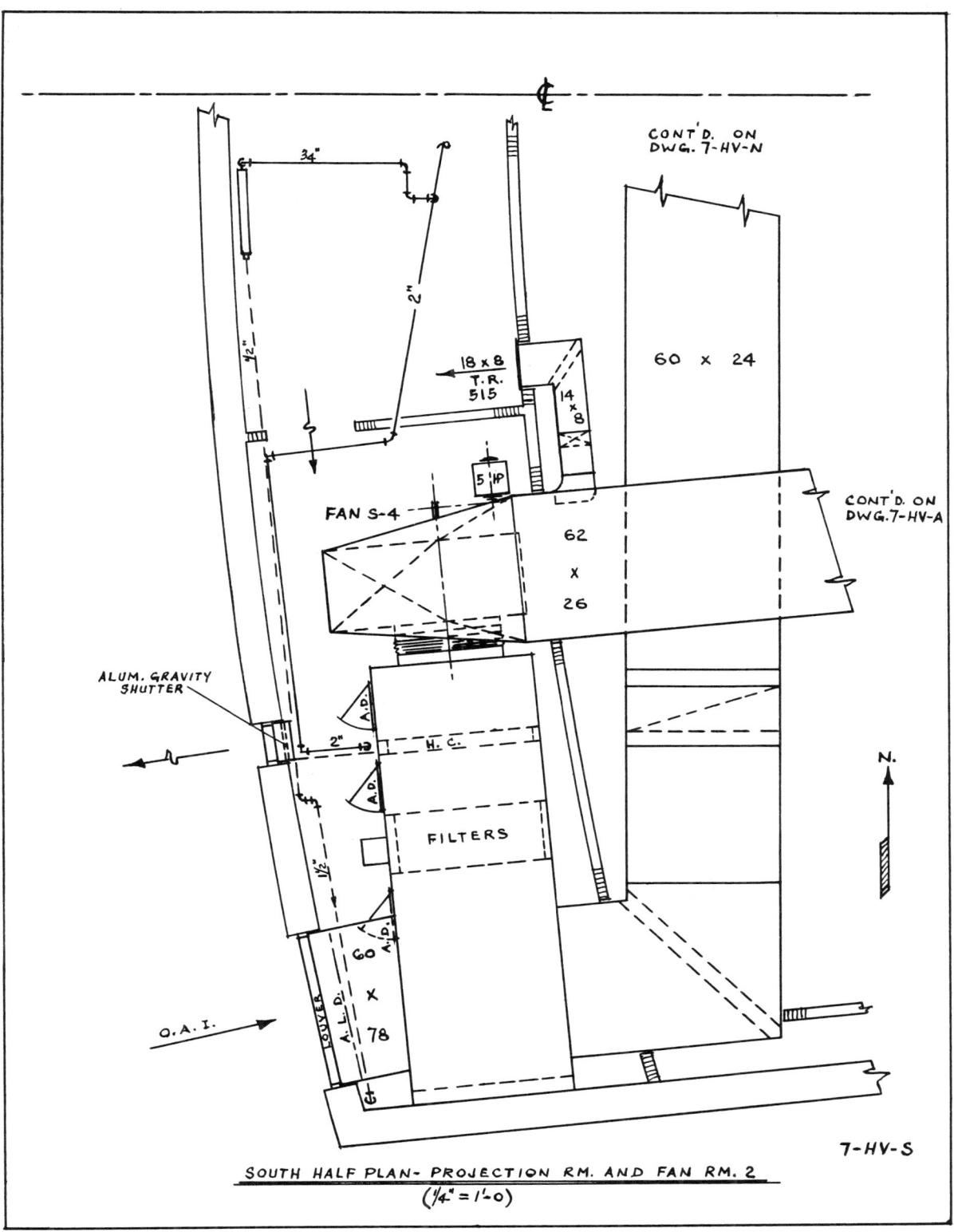

Fig. 6-18. Mechanical — Project 7.

(Text continued from page 282.)

curb for the entire perimeter, and extended to the perimeter wall. An interior panel of similar construction shall be installed, approximately 6-inches from the building wall. The heating coils are of one bank and three coils high, and shall be on a 2″ x 2″ x ¼″ angle stand, 12-inches high, minimum. Safing for the perimeters of the filters and coils will be 22-ga galvanized and shall be installed to make an airtight seal. Supply fan S-4 and return fan RE-4 are to be mounted on spring type vibration eliminators, 5-inches high. A concrete pad 4-inches high will be provided by general contractor for each fan; layout of pads by HVAC. Flexible connections, a minimum of 6-inches long, are to be shown for fan inlets and discharges. A positive internal boot for the projection-room branch tap shall be calculated and dimensions shown on the shop drawing.

Plates 7-A1 and 7-A2 (See Figs. 6-19 and 6-20)

Architectural drawings for the lower portion and upper level of the auditorium, respectively. These plates are self-explanatory.

Plate 7-S (See Fig. 6-21)

The roof framing half plan, with the south half as shown, but of opposite hand. The trusses are 40-inch-long spans mounted on the 16-inch structural walls. Each truss is level in the north to east direction. The roof is 2-inch-thick, pre-cast concrete planks, with 1½-inch rigid insulation, 3 layers of felt, and white marble chips. Elevations indicated are from the top of the concrete planking.

Plate 7-ARC (See Fig. 6-22)

Reflected ceiling part plan. The small circles are 12-inch DIA spotlights, flush mounted, and project 16-inches into the ceiling. Ceiling diffusers are the larger circles. Provide opening sizes for the ceiling contractor; allow 3½ inches for ceiling construction.

Plate 7-HV-EC (Fig. 6-23)

Layout for equipment cuts. The air filter is of the automatic renewable media type with the drive mechanism outside of the casing. Provide cutout as required. The supply and return fans are of the same size, and each has 5hp motors at position 'Y'. Fan S-4 is Counter Clockwise Upblast (CCWUB) and fan RE-4 is Clockwise Top Angular Upblast (CWTAU). For purposes of this project, fan drive centers may be scaled from the plates. Heating coil cut is for (1) coil with front and end views shown. Complete all drawings for this project in a manner for submission to the architect and engineer for approval.

FIELD MEASURING

While job site measurements are only occasionally required to prepare shop drawings for duct systems of new buildings, they are of primary importance for drawings of alterations or additions to existing buildings.

Field measurements of a new job may be needed for verification of dimensions which were not shown on the design drawings, or to field-check smoke breechings, kitchen exhaust, and similar ducts which must be very accurate for an efficient installation. A slip-on type connection should be "strapped" with a tape so the stretchout can be noted for shop information. Duct connections and equipment locations and obstructions to be avoided, should be measured in plan and elevation views and also "tied in" to structural members of the building. Particular notes should be made for conditions of collar connections to louvers.

For a field trip, the draftsman should have a print of the shop drawing to be checked, the HVAC design drawing, a 6-ft or 8-ft folding rule, a 10-ft tape, a 50-ft or 100-ft tape, plumb bob, pad, pencils, and yellow crayon or chalk. At the job, a long, thin piece of wood such as a furring strip, is handy as an auxiliary measuring tool for high ceilings and slabs.

Figure 6-1, part 2 can be used as a typical example of a field check, except that dimensions have to be indicated for plan and elevation locations of the steam pipe and the width and depth of the 18-inch I-beam. These dimensions should also be "minus" (from the bottom of slab) to the centerline of the pipe and to the bottom of the concrete fire-

(Text continued on page 295.)

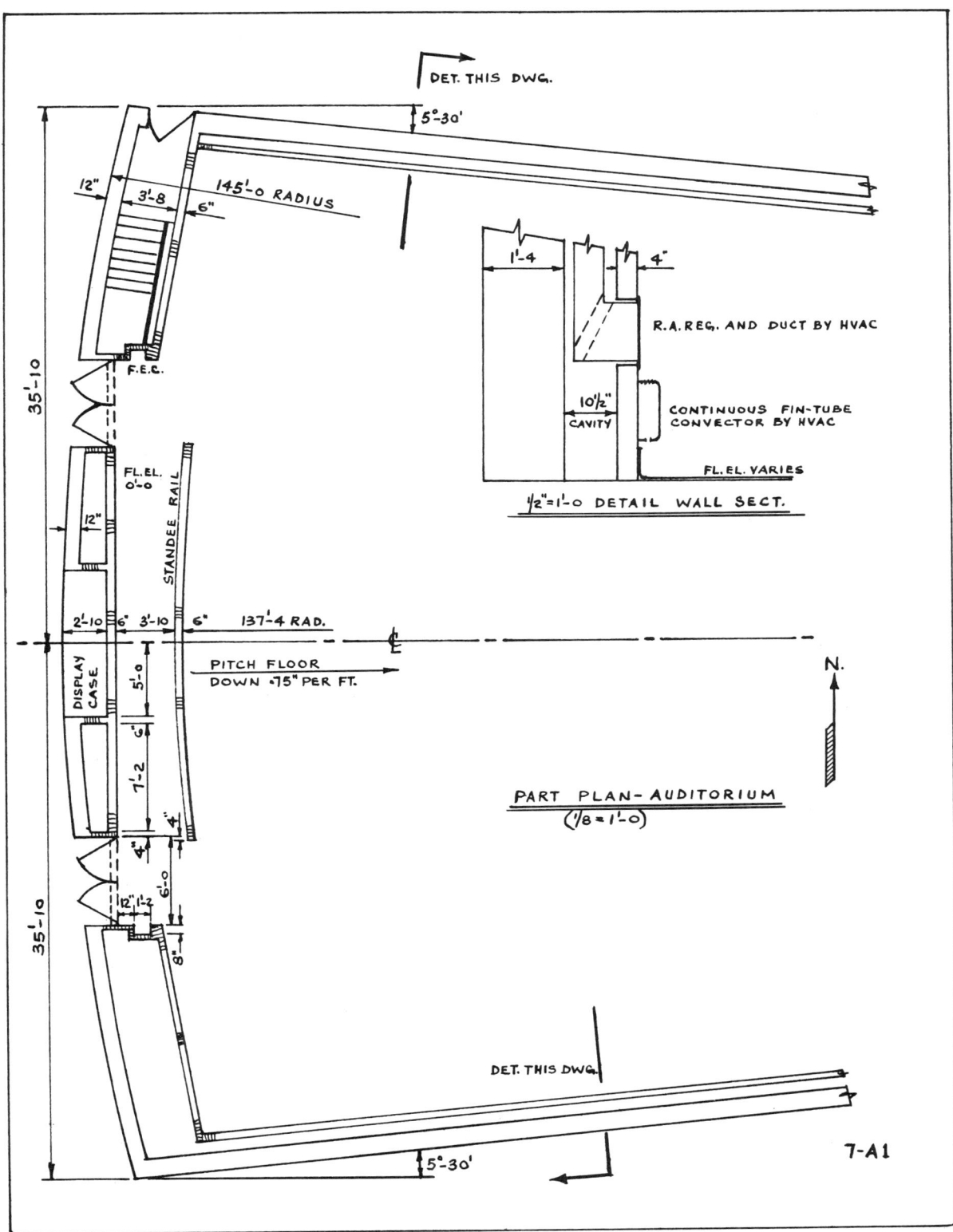

FIG. 6-19. Architectural — Project 7.

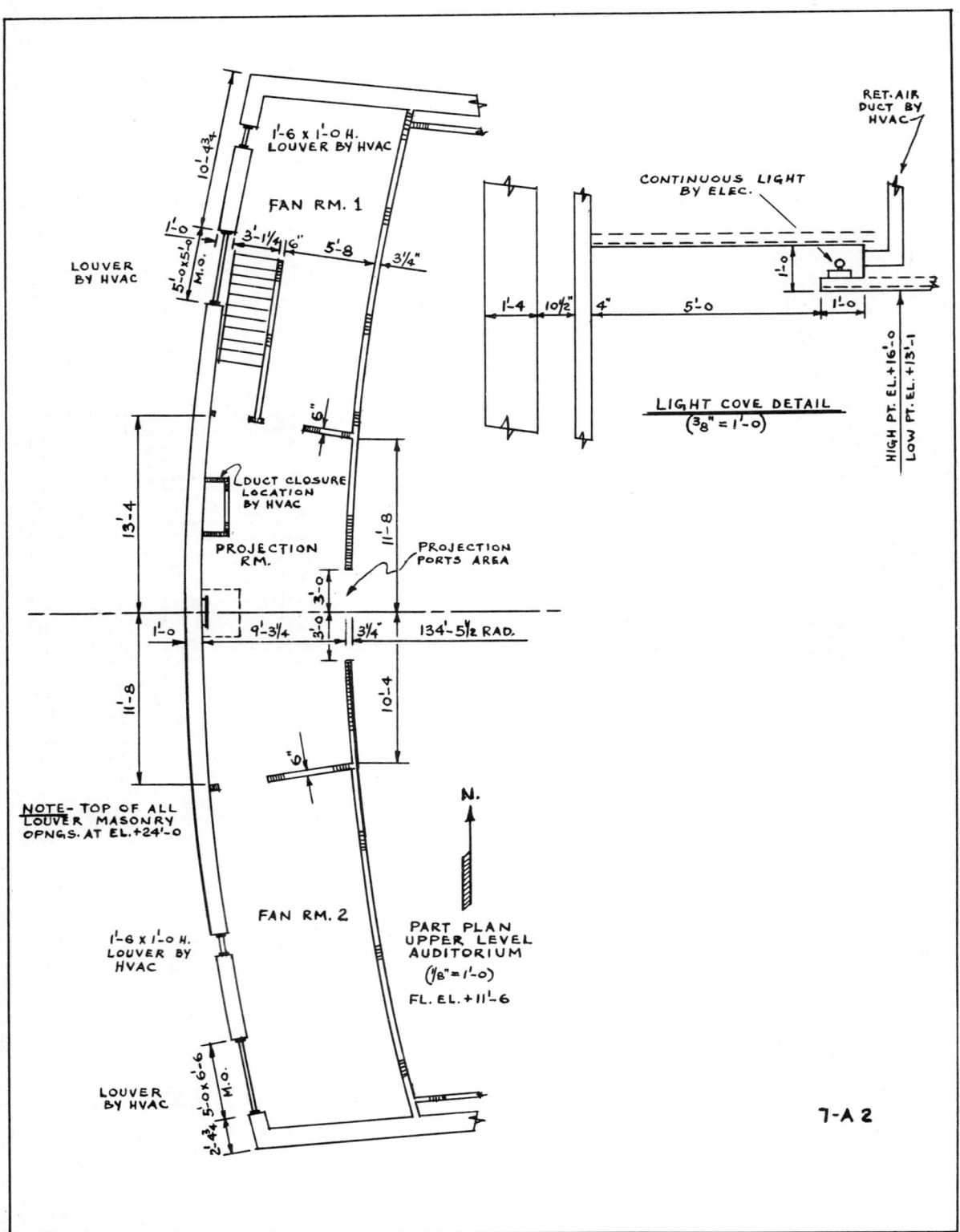

Fig. 6-20. Architectural — Project 7.

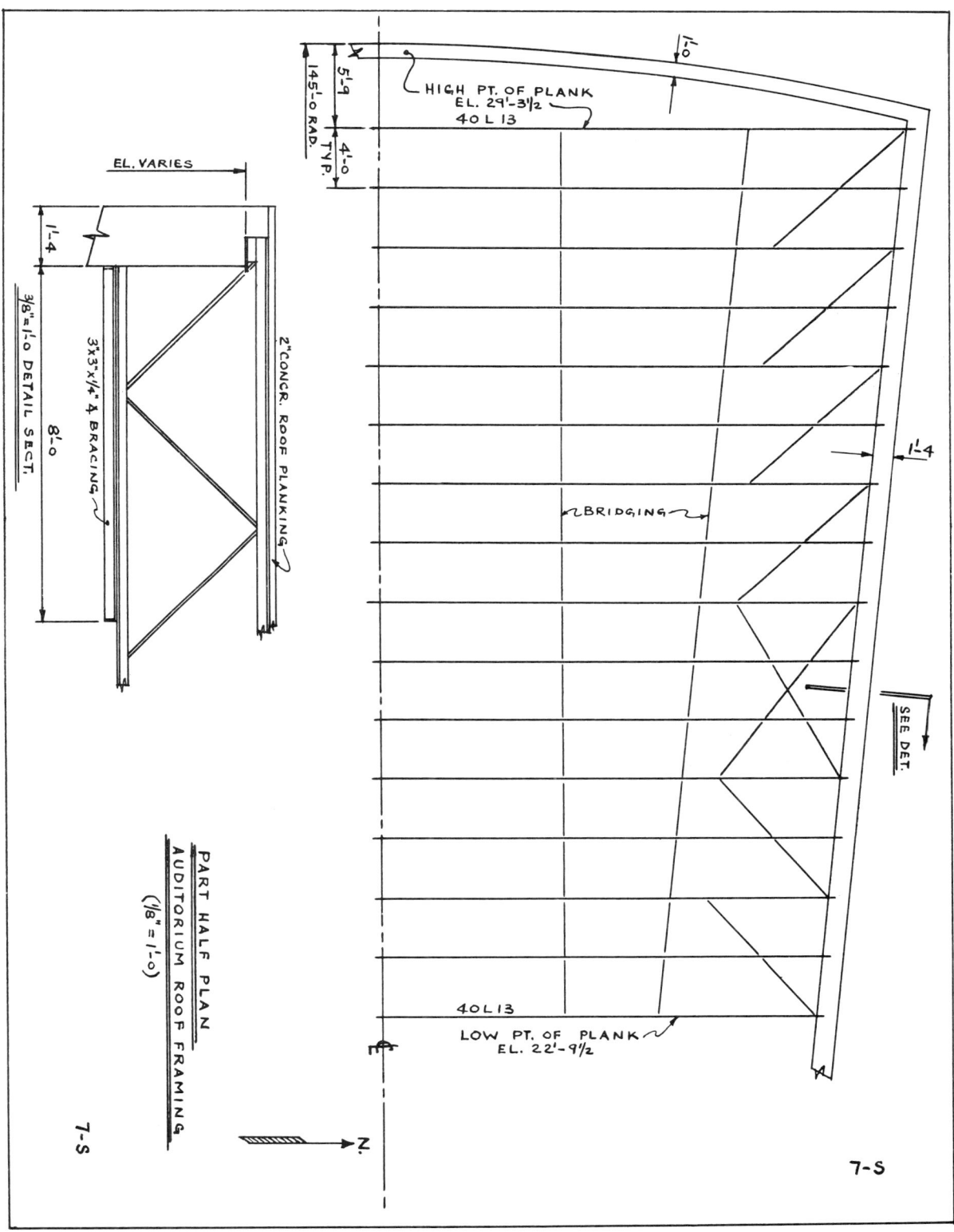

Fig. 6-21. Structural — Project 7.

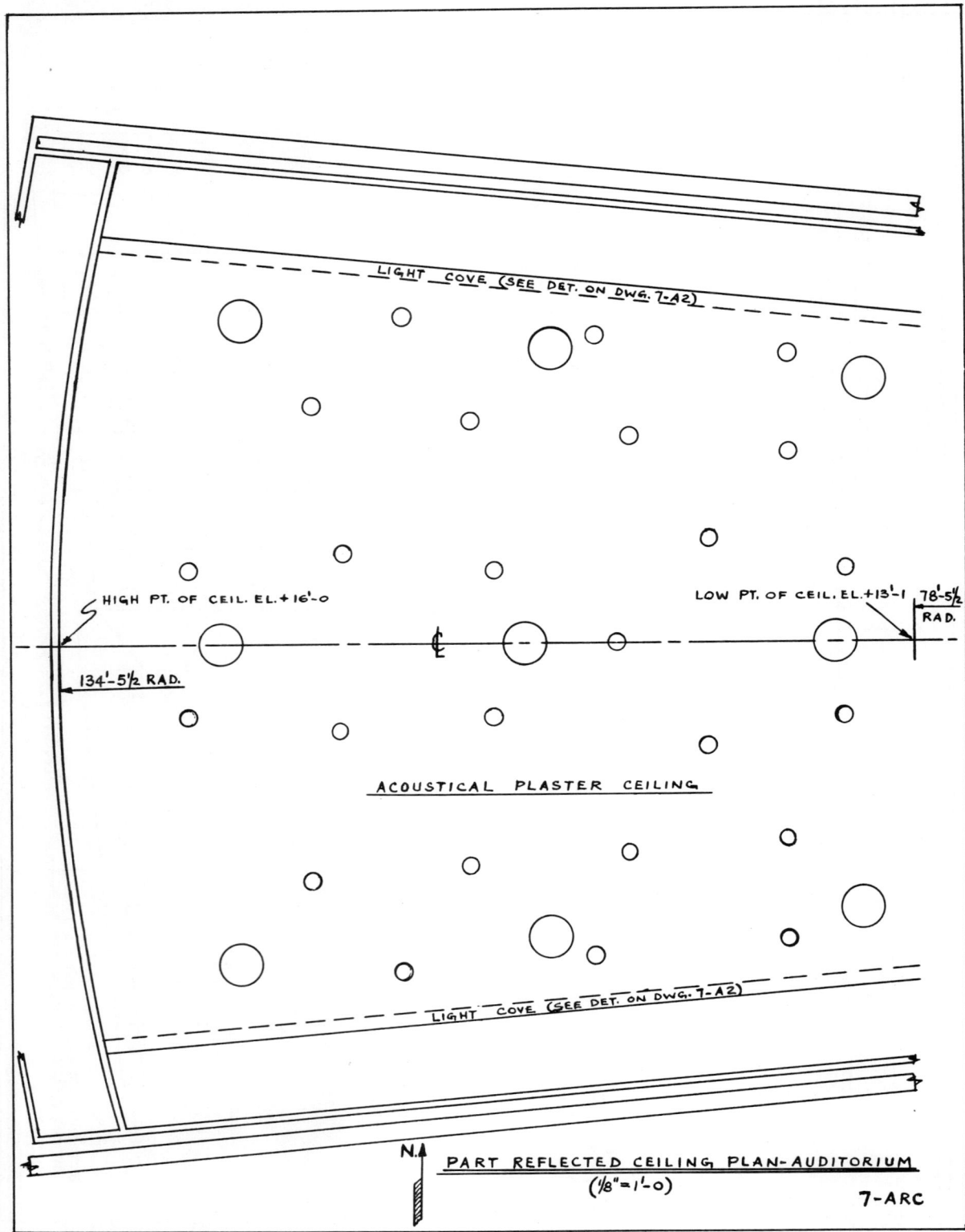

Fig. 6-22. Reflected ceiling — Project 7.

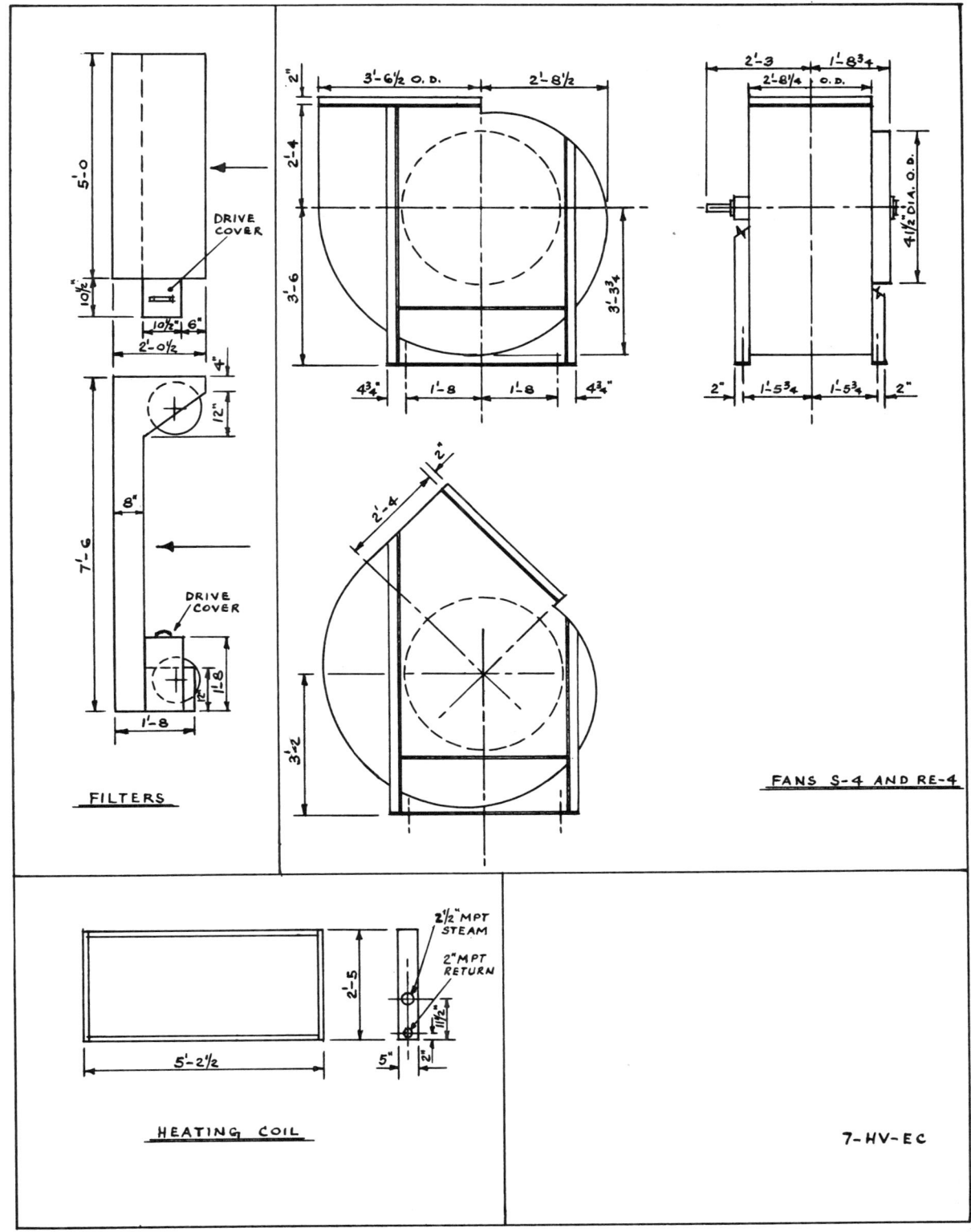

FIG. 6-23. Equipment cuts — Project 7.

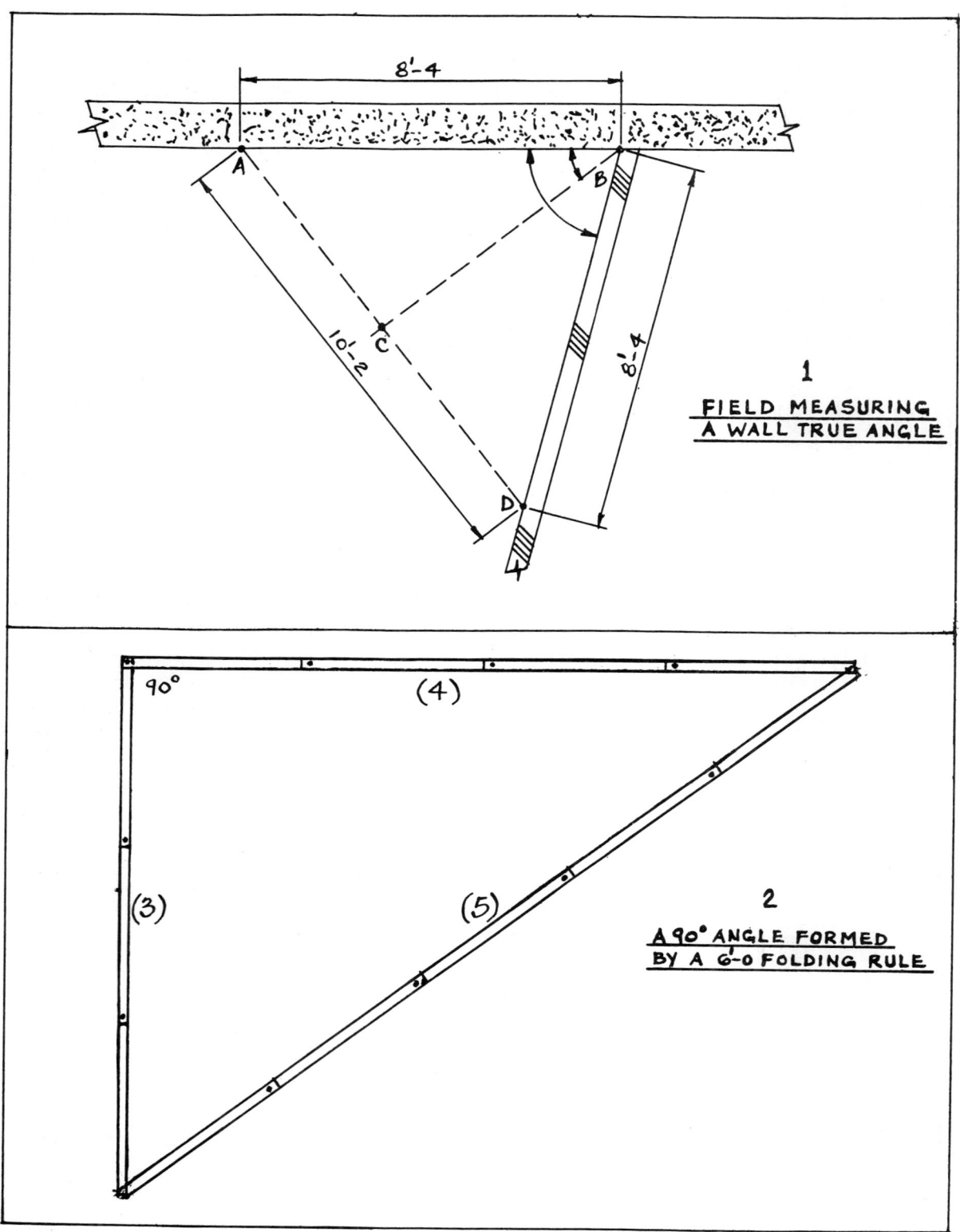

Fig. 6-24. Aids for field measuring.

(Text continued from page 288.)

proofing, respectively. The column projection measurement is of prime importance because the duct is 4-inches from the north wall, and an offset fitting is required.

On alteration or addition jobs, "as-built" drawings may be available from the office of the plant engineer. These drawings indicate actual installations of the various mechanical trades and are excellent for reference. Dashed lines are used to show ducts, piping, and equipment at close proximity to work which will be drawn. Separate symbols or notes should distinguish ducts to be removed and discarded from those which will be relocated.

When the shop drawings are coordinated with other trades, submitted and returned approved from the architect and engineer, they should be field-checked as previously explained. In addition, notes should be made of types of connections and existing duct locations which will be reconnected. Ceiling, piping, or duct hangers in the path of a new duct-run must be field measured, as they are not indicated on "as-built" drawings. Duct openings through existing concrete or masonry walls should be located, indicated, and checked.

Out-of-square walls can be measured mathematically. (See Fig. 6-24, part 1.)

Field Measuring a Wall True-Angle

1. Unknown angle = ABD
2. Mark off 8'-4 along each side of wall as AB, DB
3. Measure AD (in this example = 10'-2)
4. Bisect angle ABD, as BC
5. AC = DC
6. Sin angle ABC = $\frac{AC}{AB} = \frac{5'-1}{8'-4} = \frac{61}{100} = .61$
7. From trig tables sin of .61 = 37°-36'
8. Angle ABD = 75°-12', say 75°.

Practical application:

Find angle of wall where AD = 12'-0

Sin of angle ABC = $\frac{6'-0}{8'-4} = \frac{72}{100} = .72 = 46°-4'$

Angle of wall = 46° × 2 = 92°.

A makeshift square for field measuring can be formed by a conventional 6'-0 folding rule. (See Fig. 6-24, part 2.) Walls can also be checked for "square" by any measurements of 3-4-5 ratios. When no "as-built" drawings are available, the job site must be field-measured before shop drawings are prepared. Careful, sufficient, and accurate measurements will result in well-prepared shop drawings.

Appendix

TERMINOLOGY FOR SHEET METAL DRAFTSMEN

Air Conditioning — The process of treating air in an enclosed space to control, simultaneously, its temperature, humidity, cleanliness, and distribution, and to satisfy the requirements of the space for human comfort conditions or for a particular manufacturing operation

Ambient Air — The surrounding air, encompassing all sides

Anemometer — A rotating-vane instrument for measuring air velocity

Aspect Ratio, Rectangular Duct — For straight duct, the ratio of the $AR = \dfrac{W}{D}$ long side to the short side. For elbows, it is the ratio of the depth to the cheek width $\left(AR = \dfrac{D}{W}\right)$

Attenuation Box — An airtight device which reduces the sound- and pressure-levels of conditioned air for distribution to a space

Baffle — A sheet-metal partition to deflect airflow

Blow — The horizontal distance of an airstream from a supply outlet to a terminal velocity of 75 rpm at 6½ feet ABV (above) the floor

Boot — a ratio-size scoop installed in the airstream at the connection of a clinch-collar-branch to the main

Branch — A duct connection to or from the main run

By-pass — A path of air diverted around a system element

Casing — An airtight sheet-metal enclosure which contains one or more pieces of apparatus of the air-handling-system

Ceiling Diffuser — A ceiling outlet which distributes air in a desired pattern of planes

Central Fan System — A system — located in an equipment room — for heating, ventilating, or air conditioning which distributes air through ducts to various parts of a building

Cheek — The portion of a rectangular-duct elbow in the plane of the turn

Collar — A short section of duct which connects to another duct or piece of equipment

Compound Offset — A duct fitting which changes its path in plan and ELEV (elevation), but remains parallel to the original line of the run

Condensation — Liquid formed from a vapor by the extraction of heat

Condenser — A chamber used to liquefy a gas

Cooling Tower — An apparatus which cools water by evaporation to the outside air, to be used in the refrigeration cycle of an air-conditioning system

Coupling — A short collar which is used as a connection piece for round duct

Curb — A concrete peripheral band, generally 4″ x 4″ or 6″ x 6″, used as a sub-base of an apparatus casing

Cuts — Detailed DWGS (drawings) of related equipment in air-handling systems

Damper — A device of one or more blades used to control the volume of air in a duct system

Datum — A fixed plane from which heights in a building are measured

Developed Length — The true length of a duct along its centerline

Direct Expansion — An air conditioning system which uses a cooling coil (evaporator), wherein the refrigerant is used as the medium of heat transfer

Direct Fired Unit — A device in which air passes over a heat exchanger of metal plates which were heated by the flame itself

Drawband — Flat bar, or metal strips, with bolted ends, which are used to make airtight connections on round objects

Electrolytic CONN (connection) — A duct connection of dissimilar metals separated by a gasket of nonconducting material to prevent premature corrosion

Equivalent Length — The lineal feet of ST (straight) duct of a DIA (diameter) offering the same resistance to flow as a particular fitting or piece of apparatus in an air-handling system

Evaporator — The part of a cooling system in which the refrigerant liquid is vaporized, such as a direct expansion coil, or the shell- and tube-chilled water section

Exhaust Air — Air which is expelled from a building

Filter — A device which removes impurities from the air

Fire Damper — A damper in a duct system normally held open by a fusible link which melts at a preset temperature and thence allows the damper blades to close by gravity.

Fitting — A duct section that changes position or size from one end to the other

Flexible Conduit — A factory made nonrigid round duct generally used for connections to terminal boxes in HP (high pressure) systems

Flexible CONN (connection) — Canvas, asbestos, or neoprene connections to and from vibration-isolated equipment

Free Area — Actual, net clear space for air pressure

Fusible Link — A connection joined by metal of a low melting point for use in fire dampers

Gooseneck — An air inlet or air discharge fitting on the exterior of a BLDG (building). The OPNG (opening) is at an acute angle to the horizontal, so as to prevent the entry of snow or rain.

Grille — A decorative or functional open-type cover for an air outlet or inlet.

Head — A sheet metal closure for the end of a branch or main, usually a recessed pan type on rectangular ducts

Heel — The outside, long portion of an elbow at RT (right angle) to the radius of the turn

Humidifier — A device to place additional moisture into an airstream

Inch of Water — A pressure equivalent in a duct system which would raise a column of 62°F water 1" high, or equal to 5.197 lbs per sq ft

Induction — The movement of ambient air caused by the velocity of the primary air outlet

Inertia Base — A concrete base of approximately three (3) times the weight of an object supported and mounted on floating spring type VE (vibration eliminators)

Infiltration — Air leakage into a building

Insert — A hanging device placed onto a slab deck form prior to the concrete pour

Insulation (Thermal) — A material used to retard the flow of heat

Joggle — Two opposite bends, each less than 90°, in a sheet of metal so that the surfaces are parallel

Joint — A straight section of duct with the same opening on both ends

Knocked Down — Sheet metal parts that will be assembled on the job

Lateral — A round duct fitting with the branch at less than 90° to the line of the run

Linear Diffuser — A long, narrow, supply-air ceiling outlet

Lining — An interior duct covering for thermal and acoustical values

Louver — A wall-type air intake or discharge of horizontal, angular metal slats spaced in a parallel overlapping arrangement to exclude rain and snow

Louver Damper — A volume control of parallel multiple blades

Make-up Air — Outside air brought into the building to replenish the volume discharged by an exhaust system

Mixing Box — An airtight device supplied by a hot-air duct and a cold-air duct. A thermostat actuates proportioning dampers which mix the air in the box to the required temperature and thence distribute it to the space

Moisture Eliminators — An arrangement of a series of vertical angular metal slats formed with edges to divert the saturated air leaving the cooling coils so that water droplets impinge on the slats and drain into a base pan

Multiple Elbow — A rectangular-duct elbow with connections at equally spaced degree angles, the sum of which is the degree of the elbow

Offset — A fitting or fittings which change the location of the duct in plan or elevation with the line of the run parallel to the original

Pitot Tube — An instrument consisting of two tubes, one within the other, for air pressure measurements in duct systems

Plenum — An air space or enclosure containing a pressure other than ambient

Quadrant DMPR (damper) — A manual damper with a device to position the damper from fully closed to fully open

Radius of Diffusion — Distance in feet, from the centerline of a ceiling diffuser which is safely served with air motion

Recirculated Air — Air that has been supplied to a space and thence returned to the equipment for redistribution

Recovery EL (elbow) — A round pipe elbow, that repositions a branch duct from a rotated lateral to a level plane

Reducer — A duct fitting with a cross section area larger on one end than on the other

Register — A decorative or functional open-type cover for an air outlet or inlet with an internal damper for volume adjustments

Return Air — Air that has circulated in a treated space and shall flow back to the air-handling apparatus

Riser — Vertical ducts for a distance of more than one floor

Rotate — The turn of a round duct FTG (fitting) to change the centerline elevation of the run, or branch

Run — A continuous line of duct joints and fittings in a fairly straight path

Safing — Metal filler pieces fitted along the perimeter of filters, coils, and other equipment to prevent passage of by-pass air

Set — See "Offset."

Sleeve — A metal form which provides a clear opening in a concrete pour for a duct or pipe

Solid — A flush-type sheet metal closure for the end of a branch or main duct

Sound Absorber — A device which reduces the sound level of airflow in a duct

Spiral Pipe — Straight sections of round duct formed from continuous strip metal drawn through a machine and lock-seamed into a helix

Split System — A combination system of air ducts and hot water or steam perimeter radiation

Splitter — A hinged sheet of metal to divert air into a branch duct as required

Static Pressure — In a duct, air pressure that exerts itself in all directions

Superstructure — That portion of an arrangement for suspended equipment which supports the hangers

Supply Air — Air that has passed through the apparatus and shall be distributed to the space

Teardrop — A streamlined, airtight cover for a foreign object in the airstream of a duct system

Tee — A branch connection at 90° to the main

Telescoping Collars — Two short duct sections which fit snugly, one within the other, so as to allow adjustment for a finished alignment

Throat — The inside, short portion of an elbow at RT (right angle) to the radius of the turn

Throw — See "Blow"

Total Pressure — The sum of static and velocity pressures in a duct system

Transition — A duct FTG (fitting) which changes cross section of one end to the other

Trap Seal — A U-shape portion of pipe in which water remains to prevent the passage of air or gas

Turning Vanes — A series of small radius blades evenly spaced along the diagonal, and parallel to the turn of a square throat, square heel elbow

Velocity Pressure — In a duct system, the pressure exerted by air motion in the direction of flow

Velometer — A direct reading air velocity instrument primarily used for testing and balancing.

Ventilation — The introduction, distribution, and movement of outside air in a building

Volume — The three-dimensional measurement of an object

Warm Air Heating — An air system which satisfies comfort design requirements for a building or space during the heating season

Wye — A FTG (fitting) at the end of a duct which splits into two branches at angles less than 90° to the run

Zoning — A system that is controlled to maintain separate design conditions in different portions of a building

HEAT AND POWER

British Thermal Unit is $\frac{1}{180}$ of the heat required to raise the temperature of 1 lb. of water from 32F to 212F. It is substantially equal to the quantity of heat required to raise 1 lb. of water from 63F to 64F

One Btu equals $\frac{1}{3415}$ KWH

1 Btu equals .293 Watthours

1 Watthour equals 3.415 Btu

1 Kilowatt (1000 Watts) equals 1.3405 hp (56.93 Btu per Minute)

1 Mech. hp equals .746 Kw. (42.44 Btu per Minute)

1 Boiler hp equals 33,471.9 Btu per Hour

1 Btu will warm 55 cu ft Dry Air one degree F at 70F Temperature, Barometric Pressure 29.921 in. (Weight .07492 pounds per cu ft)

Specific Heat for Dry Air at Temperature Range from 32 to 212F. equals .24

TEMPERATURE

Fahrenheit degrees equal 9/5 Centigrade degrees plus 32

Centigrade degrees equal 5/9 Fahrenheit degrees minus 32

Absolute Temperature, expressed in Fahrenheit degrees, equals Fahrenheit degrees plus 460

Absolute Temperature, expressed in Centigrade degrees, equals Centigrade degrees plus 273

HEATING VALUE OF VARIOUS FUELS

Natural Gas equals 1,000 to 1,100 Btu per cu ft
Manufactured Gas equals 500 to 550 Btu per cu ft
Blast Furnace Gas equals 85 to 100 Btu per cu ft
Fuel Oil (industrial) equals 145,000 to 150,000 Btu per gal
Bituminous Coal equals 12,000 Btu per lb

WEIGHT AND VOLUME

1 gal (U.S.) equals 231 cu in. (.13368 cu ft)
1 cu ft equals 7.4805 gals (1728 cu in.)
1 cu ft Water at 60F equals 62.37 lbs
1 cu ft Water at 212F equals 59.76 lbs
1 gal Water at 60F equals 8.34 lbs
1 gal Water at 212F equals 7.99 lbs
1 lb equals 16 oz (7000 grains)
1 Bushel equals 1.244 cu ft
1 Short Ton equals 2,000 lbs — 1 Long Ton equals 2,240 lbs

CONVERSION FACTORS

To Find	Multiply	By
Btuh	Kilowatt Hours	3415
Btuh	Watts	3.415
Btu/min	Watts	0.05692
Cubic Feet	Cubic Inches	0.00058
Cubic Feet	Gallons	0.1337
Cubic Feet	Pounds of Water	0.01602
Cubic Inches	Gallons	231
Cubic Inches	Pounds of Water	27.68
Decimal/lin. ft	Inches	0.0833
Foot Pounds/min	Horsepower	33,000
Feet/min	Miles/hr.	88
Gallons	Cubic Feet	7.48052
Gallons	Pounds of Water	0.1198
Gallons	Cubic Inches	0.00433
Miles/hr	Feet/min	0.01136
Ounces/sq in.	Inches of Water	0.5781
Pounds	Gallons of Water	8.3453
Pounds/sq in.	Inches of Water	0.03613
Pounds/sq ft	Inches of Water	5.202
Radians	Degrees, angular	0.01745
Square Inches	Square Feet	144
Square Feet	Square Inches	0.00694
Watts	Btu/min.	17.57
Watts	Horsepower	745.7

PRESSURE

1 lb per sq in. equals 144 lbs per sq ft
1 Atmosphere equals 14.7 lbs per sq in.
1 Inch at Water Gauge Pressure has a corresponding Velocity of 4,005 fpm
1 Mile per hour Wind equals 5,280 ft per hr (88 fpm)

Wind Velocity to Pressure*

V	P	V	P	V	P
5	0.1	35	4.9	65	16.9
10	0.4	40	6.4	70	19.6
15	0.9	45	8.1	75	22.5
20	1.6	50	10.0	80	25.6
25	2.5	55	12.1	100	40.0
30	3.6	60	14.4		

* See formula below.

Formula $P = 0.004\ V^2$
where: V = Wind Velocity, mph
P = Pressure, lbs/sq ft

SPECIFIC HEAT

The quantity of heat, in Btu, required to raise the Temperature of 1 lb of a substance 1 deg F

Air (at 70°F)	.24	Lead-Solid	.031
Aluminum	.22	Lead-Fluid	.037
Bake-dough	.60	Nickel	.11
Brass	.089	Paper	.32
Carbon	.204	Steel	.1175
Copper	.094	Tin-Solid	.056
Glass	.20	Tin-Fluid	.064
Iron-gray	.11	Water (at 70°F)	1.0
Iron-cast	.16	Zinc	.093

Index of Tables and Figures

Accessory constructions, duct, (Fig. 2-19) 42
Aids for field measuring, (Fig. 6-24) 294
Air, changes in various spaces, (Table 4-2) 127
  heat removed per pound of dry, (Table 4-1) 125, 126
Air conditioning, and heating unit, (Fig. 5-48) 198
  direct expansion coil unit, (Fig. 5-49) 198
  standard graphical symbols for, (Fig. 5-62) 207
Air filter, high capacity (and frame), (Fig. 5-58) 204
  high capacity (complete assembly), (Fig. 5-60) 205
Air friction in straight galvanized ducts, (Fig. 4-4a) 140, (Fig. 4-4b) 141
Airtight ducts, specification table for medium pressure and, (Fig. 5-A-10) 258
Air velocities for conventional and high-velocity systems, design, (Table 4-8) 136
Anemostat calculator, (Fig. 4-2) 132
Angles, functions of commonly used, (Table 3-8) 106
  to develop true hip, (Fig. 3-10) 122
Apparatus casing section, (Fig. 5-36) 187
Application of line standards and dimensions, (Fig. 1-2) 4
Architectural, part plans (Project 3), (Fig. 6-4) 269
  part plans (Project 4), (Fig. 6-6) 272
  part plans (Project 5), (Fig. 6-9) 276
  part plans (Project 6), (Fig. 6-15) 284
  part plans (Project 7), (Fig. 6-19) 289, (Fig. 6-20) 290
Areas of round ducts, square foot, (Table 4-3) 128
Arrangements; fan drive and inlet box, (Fig. 5-37) 188
  tubular centrifugal fan drive, (Fig. 5-41) 192
Automatic roll-type filter, (Fig. 5-59) 204
Axial fan cuts, (Fig. 5-42) 193, (Fig. 5-43) 194

Base, fan scroll layout and fan, (Fig. 4-6) 147
Beam penetration, (Fig. 5-S-18) 247

Beams, steel columns, and girders; protection of, (Fig. 5-S-14) 245
  protective ceilings for, (Fig. 5-3-17) 246
  with concrete protection, (Fig. 5-S-13) 244
Bearing walls, cavity type, (Fig. 5-A-6) 252
  veneered or faced, (Fig. 5-A-7) 252
Belt or chain drive, motor position, (Fig. 5-40) 191
Bends, cast iron soil-pipe quarter, (Fig. 5-74) 219
  cast iron soil-pipe sweeps and, (Fig. 5-75) 220
Bevels, rectangular duct elbows and, (Fig. 2-9) 28
Block, partitions, (Fig. 5-A-5) 251
Box, inlet, (Fig. 5-37) 188, (Fig. 5-39) 190
Branch take-off, Y-, (Fig. 2-12) 33
  and register collars, (Fig. 2-13) 34
Breeching supports, smoke, (Fig. 5-8) 158
Butt welding pipe fittings, (Fig. 5-69) 214

Calculator, Anemostat, (Fig. 4-2) 132
Capacity, data for schedule 40 pipe, (Fig. 5-67) 212
  high air filter, (Fig. 5-60) 205
Cap, duct-end stiffeners and collar connections, (Fig. 2-2) 19
Casing details, longitudinal, (Fig. 4-5) 144
Casing section, apparatus, (Fig. 5-36) 187
Cast iron piping, (Fig. 5-73) 218, (Fig. 5-74) 219, (Fig. 5-75) 220
Cavity walls, (Fig. 5-A-6) 252
Ceilings, protective, (Figs. 5-S-16, 5-S-17) 246
  reflected (Project 3), (Fig. 6-2) 267
  reflected (Project 4), (Fig. 6-5) 271
  reflected (Project 7), (Fig. 6-22) 292
Central station system M.E.R., (Figs. 5-32 to 5-34) 183–185
Centrifugal fan, direct-drive, (Fig. 5-51) 199
  drive arrangements, (Fig. 5-41) 192
Chain drive motor position, (Fig. 5-40) 191
Changes of air in various spaces, (Table 4-2) 127
Chart, elbow splitter location, (Fig. 4-3) 137

for rolling standard laterals, (Fig. 3-9) 120
  gain, (Table 3-10) 108–111
Cheek ogee set, equal, (Fig. 3-1) 112
  sets computation, (Fig. 3-2) 114, (Fig. 3-3) 115
Chilled water coils, (Figs. 5-52, 5-53) 200
Cleanable type filter and mounting frame, (Fig. 5-57) 203
Closet, typical electrical, (Fig. 5-85) 230
Coils, chilled water, (Figs. 5-52, 5-53) 200
  direct expansion, (Fig. 5-49) 198
  reheat cuts and duct connections, (Fig. 2-20) 43
Collars, rectangular duct, (Fig. 2-14) 35
Columns, protection, (Figs. 5-S-13 to 5-S-15) 244–246
Computing a rolling offset, (Fig. 3-7) 118
Computing dimensions of a right triangle, (Table 3-7) 106
Concrete, reinforced details, (Figs. 5-S-3 to 5-S-5) 234–236
  reinforced slabs, (Fig. 5-S-2) 233
  sections, (Fig. 5-S-19) 248
Connections, duct, (Figs. 2-2 to 2-4) 19–21
  duct coil, (Fig. 2-20) 43
  running lengths of slip joint, (Table 2-1) 23
  to horizontal-roll filter, (Fig. 5-61) 206
Construction, duct accessory, (Fig. 2-19) 42
  examples of geometrical, (Fig. 1-3) 8, (Fig. 1-4) 9
  faced or veneered wall, (Fig. 5-A-7) 252
  for rectangular duct straight joints, (Fig. 2-5) 22
  geometrical drawing sheet layout, (Fig. 1-5) 11
  steel-joist, (Fig. 5-S-16) 246
Conventional and high-velocity design air velocities, (Table 4-8) 136
Conventional low-pressure duct specification table, (Fig. 5-A-8) 256
Conversions from decimals of a foot to inches, (Table 3-1) 49
Cooling coils, (Figs. 5-52, 5-53) 200
Corner, radius to clear a, (Fig. 3-11) 122

INDEX OF TABLES AND FIGURES

Cuts, axial fan, (Fig. 5-42) 193, (Fig. 5-43) 194
  equipment (Project 5), (Fig. 6-12) 280
  equipment (Project 7), (Fig. 6-23) 293
  reheat coil, (Fig. 2-20) 43

Design air velocities for conventional and high-velocity systems, (Table 4-8) 136
Designation of direction of rotation and discharge of fans, (Fig. 5-38) 189
Designation of position of inlet boxes, (Fig. 5-39) 190
Design drawings, small M.E.R. mechanical, (Fig. 5-30) 181, (Fig. 5-31) 182
Detail, concrete, (Figs. 5-S-3 to 5-S-5) 234–236
  connection, (Figs. 5-S-9 to 5-S-12) 240–243
  enlarged six-inch-scale rule, (Fig. 1-1) 2
  fire damper, (Fig. 5-44) 195, (Fig. 5-45), 196
  longitudinal casing, (Fig. 4-5) 144
Develop the true angle of a hip, (Fig. 3-10) 122
Dimensions, and capacity pipe data, (Fig. 5-67) 212, (Fig. 5-68) 213
  application of line standards and, (Fig. 1-2) 4
  computing right triangle, (Table 3-7) 106
  for detailing, (Fig. 5-S-8) 239
  of standard 45° laterals, (Fig. 2-16) 39
  valve, (Fig. 5-76) 221
Direct-drive centrifugal fan, (Fig. 5-51) 199
Direct expansion coil air-conditioning unit, (Fig. 5-49) 198
Direction of fan rotation, (Fig. 5-38) 189
Discharge designation of fans, (Fig. 5-38) 189
Drafting projects (Projects 1-7), (Figs. 6-1 to 6-23) 266–293
Drainage fittings, (Fig. 5-73) 218
Drawing, mechanical design of small M.E.R., (Fig. 5-30) 181, (Fig. 5-31) 182
  sheet layout for rectangular duct fittings, (Fig. 2-6) 24
  typical plumbing, (Figs. 5-78 to 5-83) 223–228
Drive, arrangements for tubular fans, (Fig. 5-41) 192
  fan and inlet box arrangements, (Fig. 5-37) 188
  motor positions, (Fig. 5-40) 191
Dual-duct type unit, (Fig. 5-47) 197
Duct, accessory constructions, (Fig. 2-19) 42
  connection to horizontal-roll filter, (Fig. 5-61) 206
  drafting symbols, (Figs. 5-1 to 5-7) 151–157
  drawing sheet layout for rectangular fittings, (Fig. 2-6) 24
  end stiffeners, cap, and collar connections, (Fig. 2-2) 19
  equal friction in round and rectangular, (Table 4-9) 138, 139
  friction of air in galvanized, (Fig. 4-4a) 140, (Fig. 4-4b) 141
  losses in rectangular, (Table 4-5) 134
  losses in rectangular 90° elbows (Table 4-7) 135
  losses in round, (Table 4-6) 134
  partial plans of round fittings, (Fig. 2-18) 41
  rectangular collars, (Fig. 2-14) 35
  rectangular elbows, (Fig. 2-8) 27
  rectangular elbows and bevels, (Fig. 2-9) 28
  rectangular nested fittings and a Y-branch take off, (Fig. 2-12) 33
  rectangular offsets, (Fig. 2-10) 30
  rectangular offsets and nesting 90° elbows, (Fig. 2-11) 31
  rectangular straight joints, (Fig. 2-7) 25
  rectangular Y-branches and register collars, (Fig. 2-13) 34
  reheat coil connections, (Fig. 2-20) 43
  seams and edges, (Fig. 2-1) 17
  specification table for flush-seam, (Fig. 5-A-9) 257
  specification table for low pressure, (Fig. 5-A-8) 256
  specification table for medium pressure and airtight, (Fig. 5-A-10) 258
  square foot areas of round, (Table 4-3) 128
  straight joint rectangular construction, (Fig. 2-5) 22
  types of connections, (Fig. 2-3) 20
Ductulator, Trane, (Fig. 4-1) 131

Edges, duct seams and, (Fig. 2-1) 17
Elbows, and bevels, (Fig. 2-9) 28
  butt welding piping, (Fig. 5-69) 214
  losses in rectangular, (Table 4-7) 135
  nesting 90-degree, (Fig. 2-11) 31
  rectangular duct, (Fig. 2-8) 27
  round standards, (Fig. 2-17) 40
  splitter location chart, (Fig. 4-3) 137
Electrical closet, typical, (Fig. 5-85) 230
Electrical, motor sizes, (Fig. 5-86) 231
  part plan (Project 5), (Fig. 6-11) 279
Elevation and plan of a rolling lateral, (Fig. 3-8) 119
Eliminators, vibration, (Fig. 5-54) 201, (Fig. 5-55) 202
End stiffeners, (Fig. 2-2) 19
Enlarged detail, six-inch-scale rule, (Fig. 1-1) 2
Equal cheek sets, (Fig. 3-1) 112, (Fig. 3-2) 114
Equal taper set, (Fig. 3-4) 116
Equipment cuts (Project 5), (Fig. 6-12) 280
  (Project 7), (Fig. 6-23) 293
Equivalents, velocities to velocity pressure, (Table 4-4) 129, 130
  duct friction, (Table 4-9), 138, 139
Examples of geometrical construction, (Fig. 1-3) 8, (Fig. 1-4) 9
Expansion coil air-conditioning unit, direct, (Fig. 5-49) 198

Faced wall construction, (Fig. 5-A-7) 252
Fan arrangements and designations, (Figs. 5-37 to 5-41) 188–192
Fan scroll layout and fan base, (Fig. 4-6) 147
Field measuring aids, (Fig. 6-24) 294
Filter, throw-away type, (Figs. 5-56 to 5-61) 203–206
Fire dampers, (Fig. 5-44) 195, (Fig. 5-45) 196
Fittings, butt welded pipe, (Fig. 5-69) 214
  drainage, (Fig. 5-73) 218
  flanged pipe, (Fig. 5-70) 215
  iron screwed pipe, (Fig. 5-71) 216, (Fig. 5-72) 217
  partial plans of round, (Fig. 2-18) 41
  rectangular, (Fig. 2-12) 33
  round standards, (Fig. 2-15) 38
  standard graphical pipe symbols, (Fig. 5-66) 211
Floor plans and section, (Figs. 5-22 to 5-24) 173–175
Floors, concrete, (Fig. 5-S-2) 233
  steel-joist, (Fig. 5-S-16) 246
Flush-seam duct specification table, (Fig. 5-A-9) 257
Foot to inches, conversions from decimals of a, (Table 3-1) 49
Foundation structure, (Fig. 5-S-1) 232
Framing plans, steel, (Fig. 5-S-6) 237, (Fig. 5-S-7) 238
Friction of air in straight galvanized ducts, (Fig. 4-4a) 140, (Fig. 4-4b) 141
Functions, natural trigonometric, (Table 3-5) 90–104
  of commonly used angles, (Table 3-8) 106

Gain chart, (Table 3-10) 108–111
Galvanized duct friction, (Fig. 4-4a) 140, (Fig. 4-4b) 141
Geometrical construction, examples of, (Fig. 1-3) 8, (Fig. 1-4) 9
  sheet layout, (Fig. 1-5) 11
Girders, protection of steel, (Fig. 5-S-13) 244, (Fig. 5-S-14) 245
  protective ceilings for, (Fig. 5-S-17) 246
Graphical symbols, standard, (Figs. 5-62 to 5-66) 207–211
Grilles, registers, and diffusers, (Figs. 5-12 to 5-19) 163–170
Gypsum tile partitions, (Fig. 5-A-5) 251

Heating and air-conditioning unit, (Fig. 5-48) 198

INDEX OF TABLES AND FIGURES

Heating, standard graphical symbols for, (Fig. 5-63) 208
Heating units, residential, (Fig. 5-50) 199
Heat removed per pound of dry air, (Table 4-1) 125, 126
High-capacity air filters, (Fig. 5-58) 204, (Fig. 5-60) 205
High velocity and conventional design air velocities, (Table 4-8) 136
Hip, to develop true angle of, (Fig. 3-10) 122
Horizontal roll-filter duct connection, (Fig. 5-61) 206
HVAC (Project 3), (Fig. 6-2) 267
    (Project 4), (Fig. 6-5) 271
    (Project 5), (Fig. 6-7) 274
    (Project 6), (Fig. 6-13) 281
    (Project 7), (Figs. 6-16 to 6-18) 285–287

Inches, conversions from decimals of a foot to, (Table 3-1) 49
Inlet box locations and arrangements, (Fig. 5-37) 188, (Fig. 5-39) 190
Inverts, plumbing, (Fig. 5-A-3) 250, (Fig. 5-A-4) 251
Iron pipe fittings, (Fig. 5-71 to 5-75) 216–220

Joints, rectangular duct straight, (Fig. 2-7) 25
    running lengths of slip connection, (Table 2-1) 23
    types of construction for rectangular duct straight, (Fig. 2-5) 22
Joists, floor construction with steel-, (Fig. 5-S-16) 246

Lateral, chart for rolling standard, (Fig. 3-9) 120
    plan and elevation of a rolling, (Fig. 3-8) 119
Layout, fan scroll, (Fig. 4-6) 147
    rectangular duct drawing sheet, (Fig. 2-6) 24
Lengths of slip connection joints, running, (Table 2-1) 23
Lighting schedule, (Fig. 5-84) 229
Lineal diffuser and troffer, (Fig. 5-20) 171, (Fig. 5-21) 172
Line standards, application of, (Fig. 1-2) 4
Location chart, elbow splitter, (Fig. 4-3) 137
Logarithms, condensed table of common, (Table 3-4) 84
Logarithm tables, (Table 3-3) 66–83
Longitudinal casing details, (Fig. 4-5) 144
Losses in ducts, (Tables 4-5 to 4-7) 134, 135
Low pressure duct specification, (Fig. 5-A-8) 256

Malleable iron screwed fittings, (Fig. 5-72) 217
Measuring, aids for field, (Fig. 6-24) 294
Mechanical (Project 7), (Figs. 6-16 to 6-18) 285–287
Medium pressure and airtight duct specification table, (Fig. 5-A-10) 258
M.E.R. central station system, (Figs. 5-32 to 5-35) 183–186
Mixing unit, (Figs. 5-9 to 5-11) 160–162
Motor position, belt or chain drive, (Fig. 5-40) 191
Motor sizes, electric, (Fig. 5-86) 231
Mounting frame, cleanable type filter and, (Fig. 5-57) 203
Multi-zone unit, (Fig. 5-46) 197

Natural trigonometric functions, (Table 3-5) 90–104
Nested fittings, rectangular duct, (Fig. 2-12) 33
Nesting 90° elbows, (Fig. 2-11) 31
Nonbearing partitions, (Fig. 5-A-5) 251
Numbers, products of, (Table 3-9) 106

Partial plans of round duct fittings, (Fig. 2-18) 41
Partitions of gypsum tile or block, (Fig. 5-A-5) 251
Penetration, beam, (Fig. 5-S-18) 247
Pipe and piping, dimensional and capacity data for, (Figs. 5-67 to 5-77) 212–222
    standard graphical symbols for, (Figs. 5-64 to 5-66) 209–211
Plan, and elevation of a rolling lateral, (Fig. 3-8) 119
    and sections, (Figs. 5-22 to 5-29) 173–180
    drafting project, (Figs. 6-1 to 6-23) 266–293
    plumbing, (Figs. 5-81 to 5-83) 226–228
Plans of round duct fittings, (Fig. 2-18) 41
Plumbing, drawings and plans, (Figs. 5-78 to 5-83) 223–228
    inverts, (Fig. 5-A-3) 250, (Fig. 5-A-4) 251
    (Project 5), (Fig. 6-10) 278
    (Project 6), (Fig. 6-15) 284
Position of inlet boxes and motor, (Fig. 5-39) 190, (Fig. 5-40) 191
Powers, roots and reciprocals of numbers, (Table 3-2) 55–64
Precast concrete, (Fig. 5-S-2) 233
Pressure, velocity equivalents of, (Table 4-4) 129, 130
Products of numbers, (Table 3-9) 106
Project key plans, (Fig. 5-A-2) 249
Protection of steel, (Figs. 5-S-13 to 5-S-17) 244–246

Radius to clear a corner, (Fig. 3-11) 122
Reciprocals, powers and roots of numbers, (Table 3-2) 55–64
Rectangular duct, (Figs. 2-5 to 2-14) 22–35
    losses, (Table 4-5) 134
    losses in 90° elbows, (Table 4-7) 135
Reflected ceiling (Project 3), (Fig. 6-2) 267
    (Project 7), (Fig. 6-22) 292
Registers, grilles and diffusers, (Figs. 5-12 to 5-19) 163–170
Reheat coil cuts and duct connections, (Fig. 2-20) 43
Reinforced concrete, (Figs. 5-S-2 to 5-S-5) 233–236
Residential heating units, (Fig. 5-50) 199
Right triangle computations, (Table 3-7), 106
Rolling lateral, chart for, (Fig. 3-9) 120
    plan and elevation of a, (Fig. 3-8) 119
Rolling offset, computing a, (Fig. 3-7) 118
Roll-type filter, (Fig. 5-59) 204, (Fig. 5-61) 206
Roots, powers and reciprocals of numbers, (Table 3-2) 55–64
Rotation and discharge of fans, (Fig. 5-38) 189
Round duct fittings, partial plans of, (Fig. 2-18) 41
Round duct friction equivalent to rectangular, (Table 4-9) 138, 139
Round ducts, losses in, (Table 4-6) 134
    square foot areas of, (Table 4-3) 128
Round elbow standards, (Fig. 2-17) 40
Round fitting standards, (Fig. 2-15) 38
Rubber-in-shear type vibration eliminator, (Fig. 5-54) 201
Running lengths of slip-connection joints, (Table 2-1) 23

Scale rule, enlarged detail of, (Fig. 1-1) 2
Schedule 40 and 80 steel pipe, (Fig. 5-67) 212, (Fig. 5-68) 213
Screwed pipe fittings, (Figs. 5-71 to 5-73) 216–218
Scroll layout, fan (Fig. 4-6) 147
Seams and edges, duct, (Fig. 2-1) 17
Sections and plans, (Figs. 5-22 to 5-34) 173–185
Set, computations for, (Figs. 3-2 to 3-7) 114–118
    ogee, (Fig. 3-1) 112
Sheet layout, for geometrical construction drawing, (Fig. 1-5) 11
    for rectangular duct fitting drawing, (Fig. 2-6) 24
Sizes, electric motor, (Fig. 5-86) 231
Slip-connection joints, running lengths of, (Table 2-1) 23
Smoke breeching support, (Fig. 5-8) 158
Soil pipe, (Fig. 5-74) 219, (Fig. 5-75) 220
Specification tables for ducts, (Figs. 5-A-8 to 5-A-10) 256–258
Spring, vibro-hanger, (Fig. 5-55) 202
Standards and dimensions, application of line, (Fig. 1-2) 4
Standards, dimensions of 45° laterals, (Fig. 2-16) 39

INDEX OF TABLES AND FIGURES

Standards (*continued*)
  round elbow, (Fig. 2-17) 40
  round fitting, (Fig. 2-15) 38
Steel, protection of, (Figs. 5-S-13 to 5-S-17) 244–246
  structural plans and details, (Figs. 5-S-6 to 5-S-12) 237–243
Stiffeners, duct end-, (Fig. 2-2) 19
Straight joints, rectangular duct, (Fig. 2-7) 25
Structural plans (Project 3), (Fig. 6-3) 268
  (Project 4), (Fig. 6-6) 272
  (Project 5), (Fig. 6-8) 275
  (Project 6), (Fig. 6-14) 283
  (Project 7), (Fig. 6-21) 291
Structure, foundation, (Fig. 5-S-1) 232
Supports, pipe hangers and, (Fig. 5-77) 222
Symbols, duct drafting, (Figs. 5-1 to 5-7) 151–157

standard graphical, (Figs. 5-62 to 5-66) 207–211
Systems, design air velocities for conventional and high velocity, (Table 4-8) 136

Take-off, Y-branch, (Fig. 2-12) 33
Taper, unequal and equal set calculations, (Figs. 3-4, 3-5) 116
Throw-away type filter, (Fig. 5-56) 203
Title box, typical, (Fig. 1-5) 11
Trane ductulator, (Fig. 4-1) 131
Triangle, computing a right, (Table 3-7) 106
Trigonometric functions, condensed table of, (Table 3-6) 105
  values of natural, (Table 3-5) 90–104
Troffer and lineal diffuser, (Fig. 5-20) 171, (Fig. 5-21) 172
Tubular centrifugal fans, drive arrangements for, (Fig. 5-41) 192

Valve dimensions, (Fig. 5-76) 221
Velocities for conventional and high-velocity systems, design air, (Table 4-8) 136
Velocity equivalents of velocity pressures, (Table 4-4) 129, 130
Veneered construction of walls, (Fig. 5-A-7) 252
Vertical fire damper detail, (Fig. 5-44) 195
Vibration eliminators, (Fig. 5-54) 201, (Fig. 5-55) 202

Walls, cavity-type, (Fig. 5-8-6) 252
  faced or veneered construction of, (Fig. 5-A-7) 252
Water cooling coils, chilled-, (Figs. 5-52, 5-53) 200

Y-branch take-off, (Fig. 2-12) 33
Y-branches and register collars, (Fig. 2-13) 34